LONDON MATHEMATICAL SOCIETY LECTURE NOTE SERIES

Managing Editor: Professor N.J. Hitchin, Mathematical Institute,
University of Oxford, 24–29 St Giles, Oxford OX1 3LB, United Kingdom

The titles below are available from booksellers, or, in case of difficulty, from Cambridge University Press.

46	*p*-adic Analysis: a short course on recent work, N. KOBLITZ
59	Applicable differential geometry, M. CRAMPIN & F.A.E. PIRANI
66	Several complex variables and complex manifolds II, M.J. FIELD
86	Topological topics, I.M. JAMES (ed)
88	FPF ring theory, C. FAITH & S. PAGE
90	Polytopes and symmetry, S.A. ROBERTSON
93	Aspects of topology, I.M. JAMES & E.H. KRONHEIMER (eds)
96	Diophantine equations over function fields, R.C. MASON
97	Varieties of constructive mathematics, D.S. BRIDGES & F. RICHMAN
99	Methods of differential geometry in algebraic topology, M. KAROUBI & C. LERUSTE
100	Stopping time techniques for analysts and probabilists, L. EGGHE
105	A local spectral theory for closed operators, I. ERDELYI & WANG SHENGWANG
107	Compactification of Siegel moduli schemes, C.-L. CHAI
109	Diophantine analysis, J. LOXTON & A. VAN DER POORTEN (eds)
113	Lectures on the asymptotic theory of ideals, D. REES
114	Lectures on Bochner-Riesz means, K.M. DAVIS & Y.-C. CHANG
116	Representations of algebras, P.J. WEBB (ed)
119	Triangulated categories in the representation theory of finite-dimensional algebras, D. HAPPEL
121	Proceedings of *Groups - St Andrews 1985*, E. ROBERTSON & C. CAMPBELL (eds)
128	Descriptive set theory and the structure of sets of uniqueness, A.S. KECHRIS & A. LOUVEAU
130	Model theory and modules, M. PREST
131	Algebraic, extremal & metric combinatorics, M.-M. DEZA, P. FRANKL & I.G. ROSENBERG (eds)
138	Analysis at Urbana, II, E. BERKSON, T. PECK, & J. UHL (eds)
139	Advances in homotopy theory, S. SALAMON, B. STEER & W. SUTHERLAND (eds)
140	Geometric aspects of Banach spaces, E.M. PEINADOR & A. RODES (eds)
141	Surveys in combinatorics 1989, J. SIEMONS (ed)
144	Introduction to uniform spaces, I.M. JAMES
146	Cohen-Macaulay modules over Cohen-Macaulay rings, Y. YOSHINO
148	Helices and vector bundles, A.N. RUDAKOV *et al*
149	Solitons, nonlinear evolution equations and inverse scattering, M. ABLOWITZ & P. CLARKSON
150	Geometry of low-dimensional manifolds 1, S. DONALDSON & C.B. THOMAS (eds)
151	Geometry of low-dimensional manifolds 2, S. DONALDSON & C.B. THOMAS (eds)
152	Oligomorphic permutation groups, P. CAMERON
153	L-functions and arithmetic, J. COATES & M.J. TAYLOR (eds)
155	Classification theories of polarized varieties, TAKAO FUJITA
158	Geometry of Banach spaces, P.F.X. MÜLLER & W. SCHACHERMAYER (eds)
159	Groups St Andrews 1989 volume 1, C.M. CAMPBELL & E.F. ROBERTSON (eds)
160	Groups St Andrews 1989 volume 2, C.M. CAMPBELL & E.F. ROBERTSON (eds)
161	Lectures on block theory, BURKHARD KÜLSHAMMER
163	Topics in varieties of group representations, S.M. VOVSI
164	Quasi-symmetric designs, M.S. SHRIKANDE & S.S. SANE
166	Surveys in combinatorics, 1991, A.D. KEEDWELL (ed)
168	Representations of algebras, H. TACHIKAWA & S. BRENNER (eds)
169	Boolean function complexity, M.S. PATERSON (ed)
170	Manifolds with singularities and the Adams-Novikov spectral sequence, B. BOTVINNIK
171	Squares, A.R. RAJWADE
172	Algebraic varieties, GEORGE R. KEMPF
173	Discrete groups and geometry, W.J. HARVEY & C. MACLACHLAN (eds)
174	Lectures on mechanics, J.E. MARSDEN
175	Adams memorial symposium on algebraic topology 1, N. RAY & G. WALKER (eds)
176	Adams memorial symposium on algebraic topology 2, N. RAY & G. WALKER (eds)
177	Applications of categories in computer science, M. FOURMAN, P. JOHNSTONE & A. PITTS (eds)
178	Lower K- and L-theory, A. RANICKI
179	Complex projective geometry, G. ELLINGSRUD *et al*
180	Lectures on ergodic theory and Pesin theory on compact manifolds, M. POLLICOTT
181	Geometric group theory I, G.A. NIBLO & M.A. ROLLER (eds)
182	Geometric group theory II, G.A. NIBLO & M.A. ROLLER (eds)
183	Shintani zeta functions, A. YUKIE
184	Arithmetical functions, W. SCHWARZ & J. SPILKER
185	Representations of solvable groups, O. MANZ & T.R. WOLF
186	Complexity: knots, colourings and counting, D.J.A. WELSH
187	Surveys in combinatorics, 1993, K. WALKER (ed)
188	Local analysis for the odd order theorem, H. BENDER & G. GLAUBERMAN
189	Locally presentable and accessible categories, J. ADAMEK & J. ROSICKY
190	Polynomial invariants of finite groups, D.J. BENSON
191	Finite geometry and combinatorics, F. DE CLERCK *et al*
192	Symplectic geometry, D. SALAMON (ed)
194	Independent random variables and rearrangement invariant spaces, M. BRAVERMAN

195 Arithmetic of blowup algebras, WOLMER VASCONCELOS
196 Microlocal analysis for differential operators, A. GRIGIS & J. SJÖSTRAND
197 Two-dimensional homotopy and combinatorial group theory, C. HOG-ANGELONI *et al*
198 The algebraic characterization of geometric 4-manifolds, J.A. HILLMAN
199 Invariant potential theory in the unit ball of C^n, MANFRED STOLL
200 The Grothendieck theory of dessins d'enfant, L. SCHNEPS (ed)
201 Singularities, JEAN-PAUL BRASSELET (ed)
202 The technique of pseudodifferential operators, H.O. CORDES
203 Hochschild cohomology of von Neumann algebras, A. SINCLAIR & R. SMITH
204 Combinatorial and geometric group theory, A.J. DUNCAN, N.D. GILBERT & J. HOWIE (eds)
205 Ergodic theory and its connections with harmonic analysis, K. PETERSEN & I. SALAMA (eds)
207 Groups of Lie type and their geometries, W.M. KANTOR & L. DI MARTINO (eds)
208 Vector bundles in algebraic geometry, N.J. HITCHIN, P. NEWSTEAD & W.M. OXBURY (eds)
209 Arithmetic of diagonal hypersurfaces over finite fields, F.Q. GOUVÊA & N. YUI
210 Hilbert C*-modules, E.C. LANCE
211 Groups 93 Galway / St Andrews I, C.M. CAMPBELL *et al* (eds)
212 Groups 93 Galway / St Andrews II, C.M. CAMPBELL *et al* (eds)
214 Generalised Euler-Jacobi inversion formula and asymptotics beyond all orders, V. KOWALENKO *et al*
215 Number theory 1992–93, S. DAVID (ed)
216 Stochastic partial differential equations, A. ETHERIDGE (ed)
217 Quadratic forms with applications to algebraic geometry and topology, A. PFISTER
218 Surveys in combinatorics, 1995, PETER ROWLINSON (ed)
220 Algebraic set theory, A. JOYAL & I. MOERDIJK
221 Harmonic approximation, S.J. GARDINER
222 Advances in linear logic, J.-Y. GIRARD, Y. LAFONT & L. REGNIER (eds)
223 Analytic semigroups and semilinear initial boundary value problems, KAZUAKI TAIRA
224 Computability, enumerability, unsolvability, S.B. COOPER, T.A. SLAMAN & S.S. WAINER (eds)
225 A mathematical introduction to string theory, S. ALBEVERIO, J. JOST, S. PAYCHA, S. SCARLATTI
226 Novikov conjectures, index theorems and rigidity I, S. FERRY, A. RANICKI & J. ROSENBERG (eds)
227 Novikov conjectures, index theorems and rigidity II, S. FERRY, A. RANICKI & J. ROSENBERG (eds)
228 Ergodic theory of Z^d actions, M. POLLICOTT & K. SCHMIDT (eds)
229 Ergodicity for infinite dimensional systems, G. DA PRATO & J. ZABCZYK
230 Prolegomena to a middlebrow arithmetic of curves of genus 2, J.W.S. CASSELS & E.V. FLYNN
231 Semigroup theory and its applications, K.H. HOFMANN & M.W. MISLOVE (eds)
232 The descriptive set theory of Polish group actions, H. BECKER & A.S. KECHRIS
233 Finite fields and applications, S. COHEN & H. NIEDERREITER (eds)
234 Introduction to subfactors, V. JONES & V.S. SUNDER
235 Number theory 1993–94, S. DAVID (ed)
236 The James forest, H. FETTER & B. GAMBOA DE BUEN
237 Sieve methods, exponential sums, and their applications in number theory, G.R.H. GREAVES *et al*
238 Representation theory and algebraic geometry, A. MARTSINKOVSKY & G. TODOROV (eds)
239 Clifford algebras and spinors, P. LOUNESTO
240 Stable groups, FRANK O. WAGNER
241 Surveys in combinatorics, 1997, R.A. BAILEY (ed)
242 Geometric Galois actions I, L. SCHNEPS & P. LOCHAK (eds)
243 Geometric Galois actions II, L. SCHNEPS & P. LOCHAK (eds)
244 Model theory of groups and automorphism groups, D. EVANS (ed)
245 Geometry, combinatorial designs and related structures, J.W.P. HIRSCHFELD *et al*
246 *p*-Automorphisms of finite *p*-groups, E.I. KHUKHRO
247 Analytic number theory, Y. MOTOHASHI (ed)
248 Tame topology and o-minimal structures, LOU VAN DEN DRIES
249 The atlas of finite groups: ten years on, ROBERT CURTIS & ROBERT WILSON (eds)
250 Characters and blocks of finite groups, G. NAVARRO
251 Gröbner bases and applications, B. BUCHBERGER & F. WINKLER (eds)
252 Geometry and cohomology in group theory, P. KROPHOLLER, G. NIBLO, R. STÖHR (eds)
253 The *q*-Schur algebra, S. DONKIN
254 Galois representations in arithmetic algebraic geometry, A.J. SCHOLL & R.L. TAYLOR (eds)
255 Symmetries and integrability of difference equations, P.A. CLARKSON & F.W. NIJHOFF (eds)
256 Aspects of Galois theory, HELMUT VÖLKLEIN *et al*
257 An introduction to noncommutative differential geometry and its physical applications 2ed, J. MADORE
258 Sets and proofs, S.B. COOPER & J. TRUSS (eds)
259 Models and computability, S.B. COOPER & J. TRUSS (eds)
260 Groups St Andrews 1997 in Bath, I, C.M. CAMPBELL *et al*
261 Groups St Andrews 1997 in Bath, II, C.M. CAMPBELL *et al*
263 Singularity theory, BILL BRUCE & DAVID MOND (eds)
264 New trends in algebraic geometry, K. HULEK, F. CATANESE, C. PETERS & M. REID (eds)
265 Elliptic curves in cryptography, I. BLAKE, G. SEROUSSI & N. SMART
267 Surveys in combinatorics, 1999, J.D. LAMB & D.A. PREECE (eds)
268 Spectral asymptotics in the semi-classical limit, M. DIMASSI & J. SJÖSTRAND
269 Ergodic theory and topological dynamics of group actions on homogeneous spaces, BEKKA & MAYER
270 Analysis on Lie Groups, N. T. VAROPOULOS & S. MUSTAPHA
271 Singular perturbations of differential operators, S. ALBERVERIO & P. KURASOV
272 Character theory for the odd order function, T. PETERFALVI
273 Spectral theory and geometry, E. B. DAVIES & Y. SAFAROV

London Mathematical Society Lecture Note Series. 271

Singular Perturbations of Differential Operators
Solvable Schrödinger Type Operators

S. Albeverio
University of Bonn

P. Kurasov
Stockholm University

CAMBRIDGE
UNIVERSITY PRESS

CAMBRIDGE UNIVERSITY PRESS
Cambridge, New York, Melbourne, Madrid, Cape Town, Singapore,
São Paulo, Delhi, Dubai, Tokyo, Mexico City

Cambridge University Press
The Edinburgh Building, Cambridge CB2 8RU, UK

Published in the United States of America by Cambridge University Press, New York

www.cambridge.org
Information on this title: www.cambridge.org/9780521779128

First published 2000

A catalogue record for this publication is available from the British Library

ISBN 978-0-521-77912-8 Paperback

Human thought, flying on the trapezes of the star-filled universe, with mathematics stretched beneath, was like an acrobat working with a net but suddenly noticing that in reality there is no net, and Martin envied those who attained that vertigo and, with a new calculation, overcame their fear.

V. V. Nabokov, Glory.

Человеческая мысль, летающая на трапециях звездной вселенной, с протянутой под ней математикой, похожа была на акробата, работающего с сеткой, но вдруг замечающего, что сетки, в сущности, нет, - и Мартын завидовал тем, кто доходит до этого головокружения и новой выкладкой превозмогает страх.

В. В. Набоков, Подвиг.

A Mielikki e Solvejg.
con tutto il cuore
e tanta gratitudine

Моим родителям
Борису Владимировичу и
Наталии Павловне

Preface

Singular perturbations of Schrödinger type operators are of interest in mathematics, e.g. to study spectral phenomena, and in applications of mathematics in various sciences, e.g. in physics, chemistry, biology, and in technology. They also often lead to models in quantum theory which are solvable in the sense that the spectral characteristics (eigenvalues, eigenfunctions, and scattering matrix) can be computed. Such models then allow us to grasp the essential features of interesting and complicated phenomena and serve as an orientation in handling more realistic situations.

In the last ten years two books have appeared on solvable models in quantum theory built using special singular perturbations of Schrödinger operators. The book by S. Albeverio, F. Gesztesy, R. Høegh-Krohn and H. Holden [39] describes the models in rigorous mathematical terms. It gives a detailed analysis of perturbations of the Laplacian in $\mathbf{R}^d, d = 1, 2, 3$, by potentials with support on a discrete finite or infinite set of point sources (chosen in a deterministic, respectively, stochastic manner). Physically these operators describe the motion of a quantum mechanical particle moving under the action of a potential supported, e.g., by the points of a crystal lattice or a random solid. Such systems and models are also described in physical terms in the book by Yu.N.Demkov and V.N.Ostrovsky [255], which also contains a description of applications in other areas such as in optics and electromagnetism. Let us also remark that a translation of the book [39] in Russian has been published with additional comments and literature [40]. Since the appearance of [39, 40, 255] several important new developments have taken place. It is the main aim of the present book to present some of these new developments in a unified formalism which also puts some of the basic results of the preceding books into a new light. The new developments concern in particular a systematic study of finite rank perturbations of (self-adjoint) operators (in particular differential operators), of generalized (singular) perturbations, of the corresponding scattering theory as well as infinite rank perturbations and multiple particles (many-body) problems in quantum theory. We also present the theory of point interaction Hamilto-nians, as a particular case of a general theory of singular perturbations of differential operators. This theory has received steadily increasing attention over the years also for its many applications in physics (solid state physics, nuclear physics), electromagnetism (antennas), and technology (metallurgy, nanophysics).

We hope this monograph can serve as a basis for orientation in a rapidly developing area of analysis, mathematical physics and their applications.

October 1998

S.Albeverio (Bonn), P.Kurasov (Stockholm)

Acknowledgments

Work on this monograph has ranged over several years and we are grateful to many persons and institutions who helped us to accomplish it.

We enjoyed the collaboration with many mathematicians and physicists concerning topics discussed in this book. First of all we acknowledge our great indebtedness to the late Raphael Høegh-Krohn, whose original inspiration was essential to our whole project. Special thanks also go to Boris Pavlov and Ludwig Faddeev, who supported the project at all stages. We would also like to thank
J. Boman, J.B. Brasche, Z. Brzeźniak, Ph. Clèment,
L. Dąbrowski, G.F. Dell'Antonio, Yu.N. Demkov, V.V. Evstratov,
P. Exner, S.M. Fei, J.E. Fenstad, R. Figari,
V.A. Geyler, F. Gesztesy, F. Haake, S. Hassi,
H. Holden, J. Janas, W. Karwowski, V. Koshmanenko,
Yu.A. Kuperin, S.T. Kuroda, S. Lakaev, T. Lindstrøm,
K.A. Makarov, Yu.B. Melnikov, R.A. Minlos, A. Motovilov,
S.N. Naboko, H. Neidhardt, E. Nelson, L.P.Nizhnik,
P. Šeba, Yu.G. Shondin, H. de Snoo, A. Teta, and K. Watanabe
for stimulating discussions.

We are grateful for support to the following persons and institutions:
G.F.Dell'Antonio, SISSA, Trieste;
I.Prigogine, Solvay Institute and Univ. Libre de Brussel;
Mathematical Dept., Ruhr Univ., Bochum;
Alexander von Humboldt Foundation, Bonn;
Mathematical Dept., Bonn Univ.;
Dept. of Mathematics, Luleå Univ.;
Swedish Institute, Stockholm;
Dept. Mathematics, Stockholm Univ.;
Dept. of Mathematical and Computational Physics, St.Petersburg Univ.;
Banach Center, Warszawa, Poland.

Material included in the monograph has been discussed at several international conferences and symposia held in different scientific centres all over the world. In particular it was presented in a series of lectures given by the second named author at the international workshops 'Spectral analysis of some second differential and difference operators' held at the Banach Center, Warsaw, in August-September 1997 and 1998 (organized by J.Janas and S.Naboko). We would like to thank the Banach Center for their hospitality. Another course of lectures on the material included was given in January 1996 by the same author at the Department of Mathematical and Computational Physics of St.Petersburg University.

Other presentations of material in this monograph have been given at the 'SISSA Workshops on singular Schrödinger operators' in September 1994 and 'Semiclassical Limit of Quantum Mechanics and Non-linear Schrödinger equation' in July 1998 (organized by G.F. Dell'Antonio, R. Figari and A. Teta), the Oberwolfach meeting on 'Spectral Theory' in July 1998 (organized by M. Van den Bergh, J.A. Van Casteren, M.Demuth) and on the Bad Herrenalb Conference 'Evolution equations and their applications in physical and life sciences', in September 1998 (organized by G. Lumer and L. Weis).

Contents

1 Rank one perturbations   **9**

  1.1   Bounded perturbations   9

        1.1.1   Resolvent analysis   9

        1.1.2   Infinite coupling   14

  1.2   Krein's formula   15

        1.2.1   Bounded and singular perturbations   15

        1.2.2   Scale of Hilbert spaces   17

        1.2.3   Form bounded and form unbounded
                perturbations   19

        1.2.4   Rank one perturbations and the extension
                theory for symmetric operators   21

        1.2.5   The extension theory and Krein's formula   23

        1.2.6 Q-function for rank one perturbations   28

  1.3   Singular rank one perturbations   30

        1.3.1   Form bounded rank one singular perturbations   30

        1.3.2   Family of rank one form unbounded
                perturbations   32

        1.3.3   Singular rank one form unbounded perturbations of
                homogeneous operators   35

        1.3.4   Resolvent formulas   38

  1.4   Approximations of singular rank one perturbations   41

        1.4.1   Norm convergence of the approximations   41

        1.4.2   Strong resolvent convergence of the
                approximations   46

  1.5   Differential operators with rank one singular perturbations . .   49

        1.5.1   Point interactions in dimension three   49

        1.5.2   Perturbations of the first derivative operator   58

        1.5.3   Dirac operator with a pseudopotential   60

2 Generalized rank one perturbations   63

2.1 Krein's formula for the generalized resolvents 63

    2.1.1 Generalized self–adjoint extensions and

           generalized resolvents 63

    2.1.2 Generalized rank one perturbations 64

2.2 Model generalized perturbations 69

    2.2.1 Introduction 69

    2.2.2 Model perturbations I: densely defined restricted oper-

           ators 70

    2.2.3 Model perturbations II: non–densely defined

           construction 81

    2.2.4 Generalized resolvents and model generalized

           perturbations 90

2.3 Generalized point interactions 92

    2.3.1 Generalized point interaction in dimension three 92

    2.3.2 Generalized delta interaction in dimension one 99

3 Finite rank perturbations and distribution theory   111

3.1 Finite rank perturbations 111

    3.1.1 Preliminaries 111

    3.1.2 Form bounded finite rank perturbations 116

    3.1.3 Form unbounded finite rank perturbations 120

    3.1.4 Generalized finite rank perturbations 125

3.2 Point interactions for differential operators and distribution

        theory 130

    3.2.1 Point interactions for differential operators as

           finite rank perturbations 130

    3.2.2 Distribution theory for discontinuous test functions . . 134

    3.2.3 Differential operator of order n in one

           dimension 140

    3.2.4 Second order differential operator in one

           dimension 142

4 Scattering theory for finite rank perturbations   159

4.1 Scattering theory for rank one perturbations 159

    4.1.1 Rank one perturbations and operators with a simple

           spectrum 160

    4.1.2 Invariance of the absolutely continuous

           spectrum 162

    4.1.3 Wave operators and scattering operator

           for rank one perturbations 169

4.2   Scattering for self–adjoint extensions 175
    4.2.1   Wave operators for self–adjoint extensions 175
    4.2.2   Scattering operator for self–adjoint extensions 177
    4.2.3   Scattering matrix for self–adjoint extensions 184
4.3   Scattering theory for finite rank perturbations 188
    4.3.1   Scattering theory for finite rank perturbations 188
    4.3.2   Scattering matrix for rank two perturbations 190
    4.3.3   Scattering matrix for generalized perturbations 191

5   **Krein's formula for infinite deficiency indices and**
  two-body problems   **195**
5.1   Infinite rank perturbations 195
    5.1.1   Krein's formula for infinite deficiency indices 195
    5.1.2   Generalized perturbations of infinite rank 198
    5.1.3   Resolvent formula for functionals 202
5.2   Two-body problems 205
    5.2.1   Two-body operator with interaction of rank one 205
    5.2.2   Two-body operator with generalized interaction of rank
          one 219

6   **Few-body problems**   **227**
6.1   Few-body formula I: self–adjoint extensions 227
    6.1.1   Tensor structure of the few-body Hilbert space   227
    6.1.2   Few-body operator with $\mathcal{H}_{-1}$ interaction 232
    6.1.3   Few-body operator with $\mathcal{H}_{-2}$ interaction 238
    6.1.4   Self-adjointness of the few-body operator
          with strongly separable interactions 244
    6.1.5   Few-body operator with delta interaction. 247
6.2   Few-body formula II: generalized self–adjoint extensions . . . 248
    6.2.1   Generalized unperturbed operator 248
    6.2.2   Inner-cluster generalized interaction 250
    6.2.3   Few-body operator with generalized interaction 256
    6.2.4   Self-adjointness of the few-body operator with infinites-
          imally separable generalized interactions 265

7   **Three-body models in one dimension**   **273**
7.1   Schrödinger operators constructed using self–adjoint extensions 273
    7.1.1   Tensor structure of the Hilbert space 273
    7.1.2   Definition of the Schrödinger operator as
          a self–adjoint operator. 274
    7.1.3   Spectrum, eigenfunctions and Bethe Ansatz 277

7.2   Operators with generalized delta interactions 281
      7.2.1 Two-body generalized delta interactions 281
      7.2.2 Three-body Schrödinger operator with two-body gen-
             eralized delta interactions 283
      7.2.3 Symmetry group 285
      7.2.4 Outgoing wave and Sommerfeld–Maluzhinetz
             transformation 293
      7.2.5 Solution of the functional equations 298
      7.2.6 Properties of the outgoing wave 305
      7.2.7 Spectrum and scattering matrix 315
      7.2.8 Calculation of the scattering matrix 318

A Historical remarks.   327
   A.1 Extension theory for symmetric
        operators 327
        A.1.1 Extension theory and Nevanlinna R-functions 327
        A.1.2 Symplectic structure and von Neumann construction . 330
   A.2 Finite rank perturbations 332
        A.2.1 Definition of singular finite rank perturbations 332
        A.2.2 Generalized singular perturbations 334
        A.2.3 Singular perturbations defined by forms 336
   A.3 Point interactions 337
        A.3.1 Definition of the point interactions in $\mathbf{R}^1, \mathbf{R}^2$ and $\mathbf{R}^3$. 337
        A.3.2 Approximations of point interactions 340
        A.3.3 Point interactions and inverse problems 341
   A.4 Few–body problems 341
        A.4.1 Three–body problems in $\mathbf{R}^3$ and $\mathbf{R}^2$ 341
        A.4.2 Few–body problems in $\mathbf{R}^1$ 344
   A.5 Recent developments 346
        A.5.1 Sphere interactions 346
        A.5.2 Interactions on low dimensional manifolds 346
        A.5.3 Relativistic point interactions 347
        A.5.4 Self–adjoint operators on graphs 348
        A.5.5 Acoustic problems 349
        A.5.6 Magnetic field, Aharonov–Bohm effect and
               anyons 349
        A.5.7 Time dependent interactions 350
        A.5.8 Solid state 350
        A.5.9 Further developments 351

Bibliography   353
Index   427

Introduction

In this monograph we study systematically certain classes of perturbations of given self–adjoint operators in Hilbert spaces. As main examples we consider second order differential operators in L^2-spaces perturbed by finite or infinite rank operators, respectively by certain generalized 'interaction terms'. Typical results concern spectral properties and scattering quantities. The operators we discuss include as special cases Hamiltonians with 'point interactions', i.e., interactions involving potentials of the δ, respectively δ'-type supported by a (finite or infinite) set of isolated points or suitable lower dimensional hypersurfaces. Such Hamiltonians occur, e.g., in the description of quantum mechanical systems in solid state physics, atomic and nuclear physics as well as in the description of electromagnetic phenomena, in the modelling of certain related chemical and biological phenomena, and in the study of quantum chaotic systems.

Concerning these 'point interaction models', in the last decade two specific monographs have appeared along with a few proceedings, books and specialized papers. One of the main aims of the present book is to present a natural continuation of the previous work [39], much in the same rigorous mathematical spirit, and covering some of the developments which occurred after the appearance of [39] (and its Russian improved version [40]). Our present book extends the analysis of [39] (and [40]) in two directions. On one hand we look at the operators discussed in [39] as special cases of a general theory of (singular) perturbations of (differential) operators. On the other hand we present results on specific perturbations of operators, described much in the same spirit of [39]. Concrete spectral properties of the operators are involved (from control over eigenvalues and eigenfunctions to scattering theory). Our discussion includes in particular the case of 'generalized point interactions', of n-th order operators in one dimension, and certain many-body (multiparticle) problems with 'delta interactions', all cases not discussed in [39] and [40].

In the books [39, 40, 255] models of quantum Hamiltonians with point interactions are considered, with special emphasis on the one–body problem with interaction concentrated at many fixed centers. The mathematical background is formed by a series of papers by L.Faddeev, R.Minlos and F.Berezin [730, 731, 135] where point interactions are analysed in connection with the extension for symmetric operators in Hilbert spaces. In particular few-body problems with point interactions have been considered by these authors. It was observed that the spectral structure of the one-body problem with single point potential in $\mathbf{R}^3$ is simple (it includes, in particular, only one eigenvalue or one resonance). During the decade 1984–1994 a series of papers was published by St.Petersburg mathematicians (see [785, 786] for a review). In these papers a new sort of solvable model was introduced based on point interactions with internal structure. These models are related to the theory of generalized extensions of symmetric operators considered first by M.Krein. In fact Krein's formula relating the resolvents of arbitrary extensions of this type plays an important role in the investigation of such operators.

We will now describe the content of this monograph, at the same time taking the opportunity to make some complementary remarks. Let us stress that for historical remarks and background the basic reference we would like to give to the reader is [39] (which also indicates the original references).

In Chapter 1 rank one perturbations are studied. Thus we have a Hilbert space $\mathcal{H}$, a self–adjoint operator $\mathcal{A}$ and we study operators of the form $\mathcal{A}_\alpha = \mathcal{A} + \alpha \langle \varphi, \cdot \rangle \varphi$, with $\varphi \in \mathcal{H}$, $\langle \cdot, \cdot \rangle$ being the scalar product in $\mathcal{H}$, and $\alpha \in \mathbf{R}$. We look upon $\mathcal{A}_\alpha$ (and $\mathcal{A}$) as self–adjoint extensions of the symmetric operator $\mathcal{A}^0$ defined as $\mathcal{A}$ on

$$\text{Dom}\,(\mathcal{A}^0) = \{\psi \in \text{Dom}\,(\mathcal{A}) : \langle \varphi, \psi \rangle = 0\}.$$

The resolvent and spectral properties of $\mathcal{A}_\alpha$ are exhibited (using Krein's type of method). The case of "infinite coupling" corresponding to $\alpha = \infty$ is then shown to lead to a self–adjoint relation $\mathcal{A}_\infty$ instead of a self–adjoint operator. The case where φ does not belong to $\mathcal{H}$, but is still a linear bounded functional on $\text{Dom}\,(\mathcal{A})$ is analyzed in Section 1.2. Such finite rank perturbations are called "singular". In particular the cases of δ and δ'-perturbations of $-\frac{d^2}{dx^2}$ in $L_2(\mathbf{R}, dx)$ are considered (recovering results discussed in [39] from another point of view); we also recall that singular rank one perturbations were first studied in a rigorous mathematical sense by F.A.Berezin and L.D.Faddeev [135]. For the study of singular rank one perturbations it is useful to introduce a scale $\mathcal{H}_k(\mathcal{A})$, $k \in \mathbf{Z}$, of Hilbert spaces associated with $\mathcal{A}$. The vectors φ defining such perturbations belong to $\mathcal{H}_{-2}(\mathcal{A})$. We compare perturbations given by $\varphi \in \mathcal{H}_{-1}(\mathcal{A})$ and those given by $\varphi \in \mathcal{H}_{-2}(\mathcal{A})$. Moreover we relate the case $\varphi \in \mathcal{H}_{-2}(\mathcal{A})$ to the Krein–Krasnosel'skiĭ theory of self–adjoint extensions ($\mathcal{A}^0$ having deficiency indices $(1, 1)$). In Section 1.3 we characterize the domain of self–adjoint extensions in the case of singular rank one perturbations, first in the "form bounded case" and then in the "form un-bounded case". We also study the effect of perturbations in the situation where the original operator $\mathcal{A}$ and the element φ giving the perturbation are homogeneous with respect to a certain group of unitary transformations of the Hilbert space $\mathcal{H}$. We close the section by studying the resolvent of the perturbed operators. In Section 1.4 approximations of singular rank one perturbations by bounded rank one perturbations are discussed, first in the norm sense and then in the strong resolvent sense.

In Section 1.5 the results of the preceding sections are applied to the case of rank one perturbations of differential operators. We start out with a discussion of the basic model of a point interaction, i.e. an operator of the form $L_\alpha = -\Delta + \alpha\delta$, where Δ is the Laplacian acting in $L^2(\mathbf{R}^3)$ and $\alpha \in \mathbf{R}$. In their original work in 1961 F.A.Berezin and L.D.Faddeev [135] already pointed out the necessity of "renormalizing" the coupling constant α to get a non–trivial self–adjoint realization of L_α. In fact they described the self–adjoint realizations of L_α via Krein's theory of self–adjoint extensions of $-\Delta|C_0^\infty(\mathbf{R}^3 \setminus \{0\})$. Other descriptions of the family L_α have been given later on in terms of boundary conditions, as operators associated with Dirichlet forms [58, 59] and in terms of non–standard analysis [26, 79, 771] (in the latter description α appears to be given by $\epsilon + \tilde{\alpha}\epsilon^2$, where ϵ is an infinitesimal and $\tilde{\alpha}$ is the "renormalized coupling constant", belonging to $\mathbf{R}$ or equal to a certain positive infinite number). See also the books [39, 29]. In our treatment of L_α in the present book we look upon L_α as a singular rank one perturbation of $-\Delta$, given by the element $\delta \in \mathcal{H}_{-2}(-\Delta) \setminus \mathcal{H}_{-1}(-\Delta)$. By applying the results from the preceding section we exhibit in this way the domain of L_α, recover the expression for its resolvent and discuss the homogeneous character of the perturbation. We also discuss the boundary conditions satisfied at the origin by elements in the domain of L_α. Finally we introduce a space of generalized functions suitable to describe L_α. Moreover we relate the description of L_α with that of singularly perturbed Schrödinger operators on $L^2(0, \infty)$. We also apply the previous approximation results to the case of the operator L_α, exhibiting norm convergent approximations by $-\Delta$ perturbed by regular rank one perturbations. Strong resolvent convergence can be obtained by following the idea of renormalizing the coupling constant α. This has been described in [39]. In the present book we present another approach, based on [135], which uses Fourier transforms of distributions. We close the chapter by analyzing two problems of the theory of perturbations of first order operators. First we study operators of the form $A_\alpha = (1/i)\,(d/dx) + \alpha\delta$ on $L^2(\mathbf{R})$, then one dimensional Dirac operators with a delta potential.

In Chapter 2 we study generalized rank one perturbations of symmetric operators $\mathcal{A}^0$ in a Hilbert space $\mathcal{H}$. These perturbations are defined as extensions **A** of $\mathcal{A}^0$ such that **A** is a self–adjoint operator in an extended Hilbert space **H** $\supset$ $\mathcal{H}$. Thus whereas in Chapter 1 only extensions were studied with **H** = $\mathcal{H}$ (so-called "standard" extensions), in Chapter 2 the case **H** $\supset$ $\mathcal{H}$, **H** $\neq$ $\mathcal{H}$, is studied. Krein's formula for the restriction to $\mathcal{H}$ of the resolvent of **A** ("generalized" resolvent) is derived and used to establish spectral properties. An explicit construction of generalized rank one perturbation is given in Section 2.2. Under a certain assumption this leads in particular to the interesting class of "operators with internal structure". The class of extensions of this type, given by operators with internal structure, is determined in terms of their resolvents, which are calculated explicitly. We first study the case of extensions of **A** = $\mathcal{A} \oplus \mathcal{A}'$ characterized by two vectors, $\varphi \in \mathcal{H}_{-2}(\mathcal{A}) \setminus \mathcal{H}$ and $\varphi' \in \mathcal{H}_{-2}(\mathcal{A}') \setminus \mathcal{H}'$, where $\mathcal{H}' = $ **H** $\ominus$ $\mathcal{H}$. We then study (Section 2.2.3) the case where $\varphi \in \mathcal{H}'$, where von Neumann's extension theory has to be replaced by the Krasnoselskiĭ extension theory. A characterization of generalized resolvents in terms of generalized perturbations with internal structure is also given (in Section 2.2.4). Applications of this theory to perturbations of differential operators are given in Section 2.3. We consider in particular in this setting the perturbations of $-\Delta$ by generalized point interactions described originally by B.Pavlov in terms of internal structure. We exhibit in particular their spectral properties and the resolvent, and determine their spectral quantities (eigenvalues, eigenfunctions, scattering amplitudes). We also study in detail the special case of the Hamiltonian for two one dimensional particles with a generalized delta interaction in one dimension. The self–adjoint extensions can be characterized in terms of "energy dependent" boundary conditions. We discuss in detail the scattering theory for such interactions. The operators studied in this chapter will also be of importance in Chapter 7 in order to construct the Hamiltonian describing a system of three one dimensional quantum particles with a δ-interaction.

In Chapter 3 singular perturbations of finite rank of self–adjoint operators are investigated, extending the results of Chapter 1 (which was devoted to rank one perturbations). Thus we consider operators of the form $\mathcal{A}_T = \mathcal{A} + T$, where $\mathcal{A}$ is a given self–adjoint operator in a Hilbert space $\mathcal{H}$ and T is a finite dimensional symmetric operator acting from $\mathcal{H}_2(\mathcal{A})$ to $\mathcal{H}_{-2}(\mathcal{A})$ (using the scale of Hilbert spaces mentioned above). We first study the case of a form bounded T. We start by exhibiting $\mathcal{A}_T$ as the self–adjoint restriction of the adjoint $\mathcal{A}^{0*}$ of a certain operator $\mathcal{A}^0$ which is the restriction of the operator $\mathcal{A}$ to a certain explicitly presented domain. Then we consider all self–adjoint extensions of $\mathcal{A}^0$. This theory will be latter applied to point interactions. Form unbounded finite rank perturbations are discussed in Section 3.1.3. We exhibit in particular their resolvent. The results are then extended to "generalized finite rank perturbations with internal structure" in the following subsection. Point interactions as finite rank perturbations of an n-order differential operator on the real line are studied in Section 3.2. Self–adjoint extensions are characterized in terms of "Lagrangian planes" of a boundary form. We develop a distribution theory for discontinuous test functions suitable for the description of the domains of the self–adjoint extensions. In Section 3.2.4 we look at the special case of second order differential operators providing a detailed description of all self–adjoint extensions in terms of explicit boundary conditions. Particular attention is dedicated to homogeneous properties of the extensions. All previous known results on interactions "located at a point" are recovered and extended. In particular δ and δ' interactions are described in a unified way.

In Chapter 4 the scattering theory for finite rank perturbations of self–adjoint operators is developed. The chapter starts out by discussing self–adjoint extensions of symmetric operators with finite deficiency indices. It is shown in particular that finite rank $\mathcal{H}_{-2}$-perturbations of self–adjoint operators leave the absolutely continuous spectrum invariant and any two operators which are self–adjoint extensions of a given symmetric operator with deficiency indices $(1,1)$ have unitary equivalent absolutely continuous parts. It is then shown that a generalized perturbation with internal structure of a self–adjoint operator has an absolutely continuous part equivalent to the latter if the extension space has finite dimension (Theorem 4.1.3). This is then extended to the equality of the continuous parts of any two self–adjoint extensions of a symmetric operator with finite deficiency indices (Theorem 4.1.4). The wave operators for rank one perturbations given by elements of $\mathcal{H}_{-1}(A)$ of a given self–adjoint operator A are computed (Theorem 4.1.5). In Section 4.2 the scattering theory (wave operators, scattering operator) for the case of two different self–adjoint extensions of a given symmetric operator with finite deficiency indices is worked out and applied to the cases of singular and generalized finite rank perturbations. Explicit formulae are given, e.g. in Theorem 4.2.1 and in Section 4.2.3. The special cases of finite rank perturbations is handled in Sections 4.3.1 and 4.3.2 whereas that of generalized perturbations is treated in Section 4.3.3.

In Chapter 5 a generalization of Krein's formula relating the resolvent of two self–adjoint extensions of a given symmetric operator to the case where the deficiency indices are infinite is derived. In Section 5.1.1 a description is given of the adjoint operator $\mathcal{A}^{0*}$ of the restriction of a self–adjoint operator $\mathcal{A}$ to a densely defined symmetric closed operator $\mathcal{A}^0$. An extension $\mathcal{A}^\Gamma$ of $\mathcal{A}^0$ determined by a self–adjoint "boundary" operator Γ in the deficiency subspace is studied. Applications to generalized perturbations of infinite rank are given in Section 5.1.2. A class of decomposable boundary operators is introduced and associated self–adjoint operators are studied. In Section 5.2 the results are specialized to the study of two-body quantum mechanical problems with interaction of rank one. The discussion is based on the concepts of Nevanlinna functions and the Stieltjes transform, which are summarized in Section 5.2.1. The resolvent of the operator $\mathcal{A}^\Gamma$ is computed using the fact that the operator $\mathcal{A}^\Gamma$ is a restriction of the adjoint operator $\mathcal{A}^{0*}$. The case of $\mathcal{H}_{-1}$ perturbations is given special attention. In Section 5.2.2 the considerations are extended to the study of two-body quantum mechanical operators with generalized interactions of rank one, with applications to quantum mechanical models with internal structure (Theorems 5.2.3 and 5.2.4).

In Chapter 6 a Krein type formula is derived for the study of few-body quantum mechanical problems. Operators describing few-body problems with "cluster interactions" of rank one are introduced and studied. In Section 6.1.2 the case of $\mathcal{H}_{-1}$-interactions is studied in detail, whereas in Section 6.1.3 the case of $\mathcal{H}_{-2}$-interactions is investigated. The concept of separable, respectively, strongly separable interactions is introduced. The separable case is handled in Section 6.1.3. In the strongly separable case $\mathcal{H}_{-2}$ as well as $\mathcal{H}_{-1}$ perturbations are handled, and the resolvent is computed (Theorem 6.1.5). The concepts of separable/non-separable are discussed in Section 6.1.5 for the important case of three particles moving in $\mathbf{R}^n$ with two–body and three–body interactions. In Section 6.2 the case of several quantum mechanical particles with generalized interactions is discussed. It is proven that the corresponding few–body operator is selfadjoint and bounded from below even if the interactions are infinitesimally separable (not necessarily strongly separable as for usual singular interactions).

In Chapter 7 we discuss models describing the motion of three quantum mechanical particles interacting through multiparticle forces in one dimension. In Section 7.1 we study the system consisting of three one-dimensional particles with two-particle interactions of the "δ-function" type (point interactions). For simplicity we assume all masses to be equal. A first step (Sections 7.1.1 and 7.1.2) consists in defining the one-parameter family of Schrödinger operators (Hamiltonians) for this system. The construction is achieved following the method described in the preceding chapter. In Section 7.1.3 we study the spectrum and eigenfunctions of the constructed operators. The eigenfunctions are exactly described by the Bethe Ansatz iff all coupling constants α_γ describing the two-body interactions are equal. In this case we exhibit the eigenfunctions explicitly. In Section 7.2 we extend our study to the case of three quantum mechanical particles with generalized two-body delta interactions (as discussed in Chapter 2, Section 2.2). In Sections 7.2.1 and 7.2.2 the corresponding Hamiltonians are constructed. To discuss the corresponding eigenfunctions we restrict ourselves first to the case where all particles are bosons, respectively, fermions. The necessary consideration of Bose, respectively, Fermi symmetry are given in Section 7.2.3. The outgoing wave appearing in the study of eigenfunctions for such a system of bosons is discussed in detail in Section 7.2.4, following [635] (which in turn used analytic methods introduced by Sommerfeld and Malyuzhinets in optics). The calculation of the outgoing wave is reduced essentially to the solution of a system of functional equations, discussed in Section 7.2.5, using methods of classical complex analysis of meromorphic functions. Detailed analytic properties of the solution are exhibit in Theorem 7.2.2. To find the outgoing waves themselves a study of the asymptotics in a space parameter is undertaken in Section 7.2.6. The spectrum and the scattering matrix for the Hamiltonians are derived in Section 7.2.7. The final result for the scattering matrix, after detailed calculations, is expressed in terms of elementary functions. The case of non–identical particles is also briefly discussed.

Appendix A contains historical remarks concerning basic references used in the text and developments after the publication of [39, 40], whereas we refer basically to [39, 40] for previous references.

We would like to mention that we see the role of solvable models constructed by methods of extension theory in mathematical and theoretical physics as complementary to the role played by quasiclassical analysis (respectively geometrical optics in electromagnetic theory). The latter is aimed at handling problems where the typical size of perturbation is greater than the typical wavelength of the process. Vice versa, point interactions and generalized point interactions are appropriate approximations in problems where the typical wavelength of the process is greater than the size of the perturbation. The quasiclassical approximation usually gives explicit formulas for solutions of partial differential operators (wave functions in quantum mechanical problems), but the calculation of the corresponding spectral characteristics needs additional effort and assumptions. In solvable models with point interactions we usually have to deal with a self–adjoint operator which approximates the original Hamiltonian. Then all spectral characteristics of this operator are calculated in explicit form and serve as approximations for the corresponding spectral characteristics of the original problem. This opens the way to spectral modeling, that is producing models which possess relevant spectral properties. In particular, we may fit some spectral properties into a two-body Hamiltonian with a generalized point interaction and then calculate the spectral properties of the corresponding few-body model.

Thus the use of solvable models is natural and useful in numerous problems of few-body scattering, nanoelectronics, the theory of neural networks, as well as in acoustics, hydrodynamics and elasticity – each time, when the typical wavelength is greater than the size of the perturbation. We hope

that the methods described in the present book will find further numerous
applications in the study of such problems.

Chapter 1

Rank one perturbations

1.1 Bounded perturbations

1.1.1 Resolvent analysis

Let us start our investigation of finite rank perturbations of self–adjoint operators with the simplest sort of perturbation – a rank one bounded perturbation. Let A be a self–adjoint (perhaps unbounded) operator in the Hilbert space H with domain $\text{Dom}\,(A)$. Let φ be a vector from the Hilbert space, $\varphi \in H$ and α be a real number, $\alpha \in \mathbf{R}$. A symmetric rank one bounded perturbation of A is the operator defined by the following formula

$$A_\alpha = A + \alpha \langle \varphi, \cdot \rangle \varphi, \tag{1.1}$$

where $\langle \cdot, \cdot \rangle$ denotes the scalar product in the Hilbert space H. The rank one operator $\alpha \langle \varphi, \cdot \rangle \varphi$ is a bounded operator in the Hilbert space and the operator sum A_α is well defined. Actually the operator A_α is self–adjoint on the domain $\text{Dom}\,(A)$ of the operator A. The spectral properties of the perturbed operator can be obtained using its resolvent, which can be calculated using Krein's formula connecting the resolvents of two self–adjoint extensions of one symmetric operator with finite deficiency indices (see Section 1.2.5). In fact the operators A_α and A are two self–adjoint extensions of the symmetric operator A^0 being the restriction of the operator A to the set of all functions orthogonal to the vector φ :

$$\text{Dom}\,(A^0) = \{\psi \in \text{Dom}\,(A) : \langle \varphi, \psi \rangle = 0\}.$$

The operator A^0 is a symmetric nondensely defined operator, since $\varphi \in H$. Self–adjoint extensions of such symmetric operators have been studied by M. A. Krasnosel'skiĭ [561, 562]. The resolvent of the operator A_α can be

9

calculated in this case without using the extension theory for symmetric operators.

Theorem 1.1.1 *Let A be a self-adjoint operator acting in the Hilbert space H and let φ be arbitrary vector from the Hilbert space, $\varphi \in \mathcal{H}$. Then the resolvents of the original operator A and its rank one perturbation $A_\alpha = A + \alpha\langle\varphi,\cdot\rangle\varphi$, $\alpha \in \mathbf{R}$, are related as follows for arbitrary $z, \Im z \neq 0$,*

$$\frac{1}{A_\alpha - z} - \frac{1}{A - z} = -\frac{\alpha}{1 + \alpha F(z)}\left\langle\frac{1}{A - \bar{z}}\varphi, \cdot\right\rangle\frac{1}{A - z}\varphi, \qquad (1.2)$$

where

$$F(z) = \left\langle\varphi, \frac{1}{A - z}\varphi\right\rangle. \qquad (1.3)$$

Proof To calculate the resolvent of the self-adjoint operator A_α we have to solve the following equation

$$h = (A_\alpha - z)f,$$

for a given $h \in H$ and $f \in \text{Dom}\,(A_\alpha) = \text{Dom}\,(A)$. We assume that the imaginary part of the spectral parameter z is positive $\Im z > 0$. We apply the operator $A_\alpha - z$ to the latter equality

$$
\begin{aligned}
h &= \left(A + \alpha\langle\varphi,\cdot\rangle\varphi - z\right)f \\
  &= Af - zf + \alpha\langle\varphi, f\rangle\varphi.
\end{aligned}
$$

By applying the resolvent of the original operator we get

$$\frac{1}{A - z}h = f + \alpha\langle\varphi, f\rangle\frac{1}{A - z}\varphi.$$

Projection on the vector φ leads to the following formula for $\langle\varphi, f\rangle$

$$\langle\varphi, f\rangle = \frac{\langle\varphi, \frac{1}{A-z}h\rangle}{1 + \alpha\langle\varphi, \frac{1}{A-z}\varphi\rangle}.$$

It follows that

$$f = \frac{1}{A - z}h - \frac{\alpha}{1 + \alpha\langle\varphi, \frac{1}{A-z}\varphi\rangle}\left\langle\varphi, \frac{1}{A - z}g\right\rangle\frac{1}{A - z}\varphi,$$

which is exactly formula (1.2). The theorem is proven.

$\square$

Formula (1.2) can be used to calculate the resolvent of the operator $A_\alpha = A + \alpha\langle\varphi, \cdot\rangle\varphi$ even in the case where the vector φ is not an element from the Hilbert space, but a linear functional on the domain $\text{Dom}(A)$. The perturbation can then be defined using the quadratic form $\alpha\langle\varphi, \cdot\rangle\varphi$, where the scalar product is understood as the action of linear functionals. Such generalized perturbations will be studied in the following sections of this chapter. Formula (1.2) can even be used to define the perturbed operator A_α in the case where the perturbation $\alpha\langle\varphi, \cdot\rangle$ is not a bounded operator in the Hilbert space. In the latter case the domains of the operators A and A_α are different in the general situation and the extension theory for symmetric operators starts to play an important role during the investigation of such perturbations. (In fact we have not used the extension theory to derive the resolvent formula (1.2).) See the following section where the case of infinite coupling constant α is considered.

Let us study first the spectral properties of the bounded perturbations defined above. These properties are described by the function

$$F_\alpha(z) = \left\langle \varphi, \frac{1}{A_\alpha - z}\varphi \right\rangle. \tag{1.4}$$

The function $F_0(z) \equiv F(z)$ appears in the denominator in formula (1.2). This function is related to Krein's Q-function

$$Q_\alpha(z) = \left\langle \varphi, \frac{1 + A_\alpha z}{A_\alpha - z}\frac{1}{A_\alpha^2 + 1}\varphi \right\rangle,$$

which will be defined in Section 1.2.5

$$F_\alpha(z) = \left\langle \varphi, \frac{1}{A_\alpha^2 + 1}\varphi \right\rangle + Q_\alpha(z).$$

The function F_α is a Nevanlinna function, i.e. a holomorphic function in $\mathbf{C} \setminus \mathbf{R}$ satisfying the following conditions

$$\overline{F(z)} = F(\bar{z}); \tag{1.5}$$

$$\frac{\Im F(z)}{\Im z} \geq 0, \quad z \in \mathbf{C} \setminus \mathbf{R}. \tag{1.6}$$

Such functions are also called Herglotz and R-functions. Every Nevanlinna function R possesses the representation

$$R(z) = a + bz + \int_{\mathbf{R}} \frac{1 + \lambda z}{\lambda - z}\frac{1}{\lambda^2 + 1}d\sigma(\lambda), \tag{1.7}$$

where $a \in \mathbf{R}, b \geq 0$ and the positive measure $d\sigma(\lambda)$ satisfies

$$\int_{\mathbf{R}} \frac{d\sigma(\lambda)}{\lambda^2 + 1} < \infty. \tag{1.8}$$

The operator A_α is self–adjoint and the vector φ is an element from the Hilbert space. Therefore there exists a spectral measure $d\mu_\alpha$ such that the function $F_\alpha(z)$ is given by the integral

$$F_\alpha(z) = \int_{\mathbf{R}} \frac{d\mu_\alpha(\lambda)}{\lambda - z},$$

where the measure μ_α is finite, i.e.

$$\int_{\mathbf{R}} d\mu_\alpha < \infty.$$

The functions $F_\alpha(z)$ belong to the class $\mathcal{R}_0$ of Nevanlinna functions [500]. The class $\mathcal{R}_0$ is the subset of Nevanlinna functions R with the following properties

$$\sup_{y>0} y \Im R(iy) < \infty,$$

$$\lim_{y \to \infty} R(iy) = 0.$$

Every Nevanlinna function from the class $\mathcal{R}_0$ possesses the following representation

$$R(z) = \int_{\mathbf{R}} \frac{d\sigma(\lambda)}{\lambda - z}$$

where the measure $d\sigma$ is finite, $\int_{\mathbf{R}} d\sigma(\lambda) < \infty$.

Formula (1.2) implies that the functions $F_\alpha(z)$ and $F_0(z)$ are related by the following rational transformation

$$F_\alpha(z) = \frac{F_0(z)}{1 + \alpha F_0(z)}. \tag{1.9}$$

The difference of the resolvents of the original and perturbed operators is a rank one operator and its trace can easily be calculated

$$\mathrm{Tr}\left[\frac{1}{A - z} - \frac{1}{A_\alpha - z}\right] = \frac{\alpha}{1 + \alpha F_0(z)} \left\langle \varphi, \frac{1}{(A - z)^2} \varphi \right\rangle.$$

We use the relation

$$\left\langle \varphi, \frac{1}{(A - z)^2} \varphi \right\rangle = \frac{dF_0(z)}{dz}$$

to get the following formula

$$\text{Tr}\left[\frac{1}{A - z} - \frac{1}{A_\alpha - z}\right] = \frac{d}{dz}\ln\left(1 + \alpha F_0(z)\right),\qquad(1.10)$$

The branch of the logarithm can be fixed arbitrarily. Different branches of the ln function lead to the same result after differentiation.

   The family of measures $\mu_\alpha, \alpha \in \mathbf{R}$, is characterized by the following lemma.

Lemma 1.1.1 *Let A be a self–adjoint operator in H, $A_\alpha = A + \alpha\langle\varphi, \cdot\rangle\varphi$ be its bounded rank one perturbation. Let $d\mu_\alpha(E)$ be the corresponding spectral measure. Let $f \in L_1(\mathbf{R})$. Then $f \in L_1(\mathbf{R}, d\mu_\alpha)$ for almost every α and we have*

$$\alpha \mapsto \int_{\mathbf{R}} f(E)d\mu_\alpha(E) \in L_1(\mathbf{R}, d\alpha)$$

and

$$\int_{\mathbf{R}}\left(\int_{\mathbf{R}} f(E)d\mu_\alpha(E)\right)d\alpha = \int_{\mathbf{R}} f(E)dE.\qquad(1.11)$$

Proof We prove the lemma first for functions of the form

$$f_z(E) = \frac{1}{E - z} - \frac{1}{E + i},$$

where $z \in \mathbf{C}\setminus\mathbf{R}$. The integral on right hand side of (1.11) can be calculated by closing the contour in the upper half plane

$$\int_{\mathbf{R}} f_z(E)dE = \begin{cases} 0 & \Im z < 0 \\ 2\pi i & \Im z > 0. \end{cases}$$

On the other hand

$$\begin{aligned} h_z(\alpha) &= \int_{\mathbf{R}} f_z(E)d\mu_\alpha(E) \\ &= F_\alpha(z) - F_\alpha(-i) \\ &= \frac{1}{\alpha + F_0(z)^{-1}} - \frac{1}{\alpha + F_0(-i)^{-1}}. \end{aligned}$$

The function $F_0(z)$ belongs to the Nevanlinna class and the poles of the function $h_z(\alpha)$ are situated in the same half planes as those of the function $f_z(E)$. Integration of the function $h_z(\alpha)$ with respect to α gives the same result as the integration of $f_z(E)$ with respect to E. The result is proven for every function $f_z(E)$. The statement of the lemma follows from the Stone–Weierstrass approximation theorem.

$\square$

The results obtained here for bounded rank one perturbations will be generalized in what follows for arbitrary perturbations having finite and even infinite rank.

1.1.2 Infinite coupling

We have just considered rank one bounded perturbations of self–adjoint operators given by (1.1). Only finite real parameters α have been considered. If the coupling constant α is infinite then formula (1.1) has only a heuristic meaning. We use instead the resolvent formula to define the perturbation in this case. We are going to show that such a perturbation determines a self–adjoint operator relation, not a self–adjoint operator.

The operator corresponding to the formal expression (1.1) when $\alpha = \infty$ is well defined on the domain $\mathrm{Dom}\,(A^0) = \{\psi \in D(A) : \langle \varphi, \psi \rangle = 0\}$. The domain $\mathrm{Dom}\,(A^0)$ is not dense in the Hilbert space. One can define the perturbed operator in this case using the resolvent formula (1.2). This formula gives the following expression for the perturbed resolvent in the case $\alpha = \infty \Rightarrow 1/\alpha = 0$

$$\frac{1}{A_\infty - z} = \frac{1}{A - z} - \frac{1}{1/\alpha + F_0(z)} \left\langle \frac{1}{A - \bar{z}}\varphi, \cdot \right\rangle \frac{1}{A - z}\varphi$$

$$= \frac{1}{A - z} - \frac{1}{F_0(z)} \left\langle \frac{1}{A - \bar{z}}\varphi, \cdot \right\rangle \frac{1}{A - z}\varphi.$$

(1.12)

The latter expression defines a self–adjoint relation A_∞ in the Hilbert space H, not a self–adjoint operator. To prove this we apply the latter operator equality to the vector φ and get the following equation

$$\frac{1}{A_\infty - z}\varphi = \frac{1}{A - z}\varphi - \frac{1}{F_0(z)} \left\langle \frac{1}{A - \bar{z}}\varphi, \varphi \right\rangle \frac{1}{A - z}\varphi$$

$$= \frac{1}{A - z}\varphi - \frac{1}{F_0(z)}F_0(z)\frac{1}{A - z}\varphi$$

$$= 0$$

due to the formula (1.4). It follows that formula (1.12) does not define the resolvent of any self–adjoint operator. It gives the formula for the resolvent of the self–adjoint relation $A^0 \dot{+} (0, \varphi)$. The latter formula implies that $F_\infty(z) \equiv 0$, which coincides with the limit

$$\lim_{\alpha \to \infty} F_\alpha(z) = \lim_{\alpha \to \infty} \frac{F_0(z)}{1 + \alpha F_0(z)} = 0.$$

In what follows we are going to consider only self–adjoint operators and we try to avoid discussing self–adjoint relations. But we have to keep in mind that using Krein's formula one obtains not only operators, but also operator relations. The operator relations in connection with the finite rank perturbations have been recently studied by H. de Snoo and S. Hassi [452, 453, 454, 455, 457, 459].

1.2 Krein's formula

1.2.1 Bounded and singular perturbations

We have mentioned in the previous section that not only can bounded rank one perturbations be defined in the framework of the theory of self–adjoint operators, but formula (1.2) can define a rank one perturbation of the self–adjoint operator A even if φ is not an element from the Hilbert space. We start with two examples. Consider first the formal linear differential operator

$$B_\alpha \psi = -\frac{d^2}{dx^2}\psi + \alpha \langle \delta, \psi \rangle \delta, \tag{1.13}$$

defined on functions on the real line, where the symbol δ denotes Dirac's delta function and the scalar product $\langle \delta, \psi \rangle$ is defined as the action of the linear functional δ on the function ψ. The standard norm on the domain of the operator B_0 is equal to the norm in the Sobolev space $W_2^2(\mathbf{R})$. Therefore the scalar product $\langle \delta, \psi \rangle$ is well defined since the function ψ is continuous at the origin. Suppose that there exists a self–adjoint operator acting in $L_2(\mathbf{R})$ corresponding to the formal differential expression (1.13). Then this operator coincides with the unperturbed operator $B = -d^2/dx^2$ on the set of functions vanishing at the origin, i.e. on the domain

$$\text{Dom}\,(B^0) = \{\psi \in W_2^2(\mathbf{R}) : \psi(0) = 0\}.$$

The restricted operator $B^0 = B|_{\text{Dom}(B^0)}$ is a symmetric operator with the deficiency indices $(1, 1)$. The restricted operator is densely defined and all its self–adjoint extensions can be calculated using the von Neumann theory. The adjoint operator B^{0*} coincides with the second derivative operator and has domain $\mathrm{Dom}\,(B^{0*}) = \{\psi \in W_2^2(\mathbf{R} \setminus \{0\}) : \psi(-0) = \psi(+0)\}$. To define the self–adjoint operator B_α we apply the linear operator (1.13) to an arbitrary function ψ from the domain of the adjoint operator B^{0*}

$$\psi \in \mathrm{Dom}\,(B^{0*}).$$

The expression

$$B_\alpha \psi = \left(-\frac{d^2}{dx^2} + \alpha\langle\delta,\cdot\rangle\delta\right)\psi$$

is well defined in the distributional sense, since every $\psi \in \mathrm{Dom}\,(B^{0*})$ is a continuous square integrable function due to the Sobolev embedding theorem (see [913]). The result in general is equal to the sum of a square integrable function and a distribution with support at the origin. From the condition that the range of the function ψ belongs to the Hilbert space $L_2(\mathbf{R})$ we get

$$-\psi'(+0) + \psi'(-0) + \alpha\psi(0) = 0.$$

If the function ψ satisfies the latter condition, then $B_\alpha\psi \in L_2(\mathbf{R})$. Consider the restriction of the operator B_α to the domain

$$\mathrm{Dom}\,(B_\alpha) \;=\; \{\psi \in W_2^2(\mathbf{R} \setminus \{0\}) : \psi(-0) = \psi(+0) \equiv \psi(0)$$

$$\psi'(+0) - \psi'(-0) = \alpha\psi(0)\}.$$

The restrictions of the operators B_α and B^{0*} to this domain coincide. This operator is a self–adjoint extension of the operator B^0 and this operator can be considered as a natural definition for the operator B_α in the framework of the theory of self–adjoint operators.

   The above example shows that formula (1.1) can define a rank one perturbation of the self–adjoint operator A even if the vector φ does not belong to the corresponding Hilbert space. But the vector φ cannot be arbitrary. To define the perturbed operator we used the restriction of the original operator B to the domain $\mathrm{Dom}\,(B^0)$. Therefore the vector φ should be a linear bounded functional on the domain of the operator A with the graph norm. Only in this case does the restriction of the self–adjoint operator A have nontrivial deficiency indices. But not every vector $\varphi \in \mathrm{Dom}\,(A)^*$ defines a rank one perturbation in a unique way. Consider the following linear differential operator

$$B'_\alpha \psi = -\frac{d^2}{dx^2}\psi + \alpha\langle\delta^{(1)}, \psi\rangle\delta^{(1)}, \tag{1.14}$$

where $\delta^{(1)}$ denotes the first derivative of the delta function. If the self–adjoint operator in $L_2(\mathbf{R})$ corresponding to the formal expression (1.14) exists, then it coincides with the operator $B = -d^2/dx^2$ restricted to the domain of functions from $W_2^2(\mathbf{R} \setminus \{0\})$, satisfying certain boundary conditions at the origin. Consider the formal expression $B'_\alpha \psi$, where $\psi \in W_2^2(\mathbf{R} \setminus \{0\})$. The scalar product $\langle \delta^{(1)}, \psi \rangle$ is well defined only if the function ψ is continuous at the origin and has continuous first derivative at this point. But if these two conditions are satisfied, then $B'_\alpha \psi$ belongs to the Hilbert space $L_2(\mathbf{R})$ if and only if $\psi'(0) = 0$. The second derivative operator defined on the domain of $C^1(\mathbf{R})$ functions from $W_2^2(\mathbf{R} \setminus \{0\})$ satisfying the latter condition is symmetric but not self–adjoint. Therefore the heuristic rank one perturbation (1.14) does not determine any self–adjoint operator, but only a symmetric operator. The corresponding family of self–adjoint extensions of this symmetric operator is described by one real parameter, which is not determined by the heuristic expression (1.14). One needs additional assumptions on the interaction to define the unique self–adjoint operator in this case.

The two examples considered show that the rank one perturbations can be defined by vectors $\varphi \in \mathrm{Dom}\,(A)^*$. We are going to refer to the perturbations defined by vectors φ which do not belong to the Hilbert space, $\varphi \notin H$ **singular**. The perturbations defined by vectors from the Hilbert space will be called **bounded**. The main difference between singular and bounded rank one perturbations is that the domains of any self–adjoint operator and its rank one singular perturbation are in general different, while the domain of the operator does not change under rank one bounded perturbations (see Section 1.1). We are going to consider differential operators with singular interactions. The first mathematically rigorous study of such operators was carried out by F.A.Beresin and L.D.Faddeev [135]. The Laplace operator with delta interaction in $L_2(\mathbf{R}^3)$ was considered. We are going to discuss this operator in more detail in Section 1.5.1. In the rest of this chapter we concentrate our attention on rank one singular perturbations.

1.2.2   Scale of Hilbert spaces

The scale of Hilbert spaces associated with the self–adjoint operator A acting in the Hilbert space H will be defined using the modulus $|A|$ of the operator A, where

$$|A| \equiv (A^* A)^{1/2}.$$

The operator $|A|$ is positive and self–adjoint, its domain coincides with the domain of the operator A. For $s \geq 0$, $\mathcal{H}_s(A)$ is $\mathrm{Dom}\,(|A|^{s/2})$ with norm equal to the graph norm of the operator

$$\| \psi \|_s = \| (|A| + 1)^{s/2} \psi \|_{\mathcal{H}}. \tag{1.15}$$

The space $\mathcal{H}_s$ with norm $\| \cdot \|_s$ is complete. The adjoint spaces formed by the linear bounded functionals will be denoted by $\mathcal{H}_{-s}(A) = \mathcal{H}_s(A)^*$. The norm in the space $\mathcal{H}_{-s}(A)$ is defined by the formula

$$\| \psi \|_{-s} = \left\| \frac{1}{(|A| + 1)^{s/2}} \psi \right\|_{\mathcal{H}}, \tag{1.16}$$

where the operator $1/(|A| + 1)^{s/2}$ is defined in the generalized sense. Let $\psi \in \mathcal{H}_{-s}(A), \eta \in H = \mathcal{H}_0(A)$. Then

$$\left\langle \frac{1}{(|A| + 1)^{s/2}} \psi, \eta \right\rangle = \left\langle \psi, \frac{1}{(|A| + 1)^{s/2}} \eta \right\rangle.$$

It follows that $\left(1/(|A| + 1)^{s/2} \right) \psi \in H$ and the norm of the functional is given by the formula (1.16).

The operator $(|A| + 1)^{t/2}$ defines an isometry from $\mathcal{H}_s(A)$ to $\mathcal{H}_{s-t}(A)$. Each space $\mathcal{H}_{-s}(A)$ is equal to the completion of the Hilbert space H in the norm (1.16).

In what follows we are going to use the brackets $\langle \cdot, \cdot \rangle$ to denote not only the scalar product in the Hilbert space H, but the action of the functionals. Let $\psi \in \mathcal{H}_{-s}(A), \eta \in \mathcal{H}_s(A)$. Then we define

$$\langle \psi, \eta \rangle \equiv \left\langle \frac{1}{(|A| + 1)^{s/2}} \psi, (|A| + 1)^{s/2} \eta \right\rangle, \tag{1.17}$$

where the bracket on the right hand side denotes the scalar product.

The spaces $\mathcal{H}_s(A)$ form the following chain of triplets

$$... \subset \mathcal{H}_2(A) \subset \mathcal{H}_1(A) \subset H = \mathcal{H}_0(A) \subset \mathcal{H}_{-1}(A) \subset \mathcal{H}_{-2}(A) \subset ....$$

The space $\mathcal{H}_2(A)$ coincides with the domain of the operator A and $\mathcal{H}_1(A)$ is the domain of $|A|^{1/2}$. For every two $s, t; s \leq t$, the space $\mathcal{H}_t(A)$ is dense in $\mathcal{H}_s(A)$ in the norm $\| \cdot \|_s$. The norm in the original Hilbert space H will be denoted

$$\| \cdot \|_0 = \| \cdot \|_H. \tag{1.18}$$

The norm in the space $\mathcal{H}_1(A)$ can be calculated as follows

$$\begin{aligned} \| \psi \|_1^2 &= \langle (|A| + 1)^{1/2} \psi, (|A| + 1)^{1/2} \psi \rangle \\ &= \langle \psi, (|A| + 1) \psi \rangle \\ &= \langle (\sqrt{|A|} + i) \psi, (\sqrt{|A|} + i) \psi \rangle. \end{aligned} \tag{1.19}$$

Similarly we have

$$\| \, \psi \, \|_{-1}^2 = \left\langle \frac{1}{\sqrt{|A|}+i}\psi, \frac{1}{\sqrt{|A|}+i}\psi \right\rangle. \tag{1.20}$$

One can also introduce the norms $\| \cdot \|_2^*$ and $\| \cdot \|_{-2}^*$ in the spaces $\mathcal{H}_2(A)$ and $\mathcal{H}_{-2}(A)$, which are equivalent to the standard norms in these spaces

$$
\begin{aligned}
\| \, \psi \, \|_2^{*2} &= \| \, (A-i)\psi \, \|^2 = \langle \psi, (A^2+1)\psi \rangle; \\
\| \, \psi \, \|_{-2}^{*2} &= \| \, \frac{1}{A-i}\psi \, \|^2 = \langle \psi, \frac{1}{A^2+1}\psi \rangle.
\end{aligned}
\tag{1.21}
$$

In fact the spaces $\mathcal{H}_s(A)$ are Hilbert spaces with the scalar product associated with the standard norm

$$\langle \psi, \varphi \rangle_s = \langle \psi, (|A|+1)^s \varphi \rangle.$$

1.2.3 Form bounded and form unbounded perturbations

It has been shown that rank one perturbations of a given operator A can be defined only by the vectors φ which are bounded linear functionals on the domain of the operator A, i.e. by the vectors φ being elements from the space $\mathcal{H}_{-2}(A)$. If the operator A is positive then every rank one perturbation defined by the vectors φ from $\mathcal{H}_{-1}(A)$ can be defined using the form perturbation theory. Let us explain this in more detail. The linear operator

$$\langle \varphi, \cdot \rangle \varphi : \mathcal{H}_2(A) \to \mathcal{H}_{-2}(A)$$

defines naturally the following sesquilinear positive form

$$\mathbf{V}_\varphi[\psi, \eta] = \langle \varphi, \psi \rangle \langle \eta, \varphi \rangle = \langle \varphi, \psi \rangle \overline{\langle \varphi, \eta \rangle}$$

for $\psi, \eta \in \mathcal{H}_2(A)$. The sesquilinear positive form $V[\psi, \eta]$ will be called **form bounded with respect to the operator** A if and only if the domain Dom(V) of the form is contained in the space $\mathcal{H}_1(A)$ and there exist two positive real constants a and b such that for any $\psi \in$ Dom(V) the following estimate holds:

$$V[\psi, \psi] \le a \, \| \, \psi \, \|_1^2 + b \, \| \, \psi \, \|_H^2 . \tag{1.22}$$

If the constant a can be chosen arbitrarily small, the form V is said to be **infinitesimally form bounded with respect to the operator** A. Note

that to define form bounded and infinitesimally form bounded perturbations
we have actually used the quadratic form of the positive operator $|A|$, since
the norms defined by

$$a \parallel \psi \parallel_1^2 + b \parallel \psi \parallel_H^2$$

and

$$a \langle \psi, |A|\psi \rangle + b \langle \psi, \psi \rangle$$

are equivalent.

Lemma 1.2.1 *Let $\varphi \in \mathcal{H}_{-1}(A)$. Then the sesquilinear from $\mathbf{V}_\varphi[\psi, \eta] = \langle \varphi, \psi \rangle \overline{\langle \varphi, \eta \rangle}$ is infinitesimally form bounded with respect to the operator A.*

Proof The Hilbert space H is dense in $\mathcal{H}_{-1}(A)$ and for any $\epsilon > 0$ there
exists $\varphi_0 \in H$ such that $\parallel \varphi - \varphi_0 \parallel_{-1}^2 \leq \epsilon/2$. For any $\psi \in \mathcal{H}_2(A) \subset \mathcal{H}_1(A)$ the
following estimate proves the lemma

$$\mathbf{V}_\varphi[\psi, \psi] \;=\; |(\varphi, \psi)|^2 \leq 2|(\varphi - \varphi_0, \psi)|^2 + 2|(\varphi_0, \psi)|^2$$

$$\leq \; \epsilon \parallel \psi \parallel_1^2 + 2 \parallel \varphi_0 \parallel^2 \parallel \psi \parallel_H^2 .$$

The lemma is proven.

$\square$

The latter lemma is valid for any self–adjoint operator A. If the operator
A is a positive self–adjoint operator then the KLMN theorem [842] implies
that for any real α the formal expression $A + \alpha \langle \varphi, \cdot \rangle \varphi$ defines a certain self–
adjoint operator A_α. One can prove that perturbations defined by vectors
from $\mathcal{H}_{-2}(A) \setminus \mathcal{H}_{-1}(A)$ are not form bounded.

Lemma 1.2.2 *Let $\varphi \in \mathcal{H}_{-2}(A) \setminus \mathcal{H}_{-1}(A)$. Then the sesquilinear from*

$$\mathbf{V}_\varphi[\psi, \eta] = \langle \varphi, \psi \rangle \overline{\langle \varphi, \eta \rangle}$$

is not form bounded with respect to the operator A.

Proof Suppose that the bilinear form $\mathbf{V}_\varphi$ is form bounded i.e. there exist
positive constants a, b such that

$$\mathbf{V}_\varphi[\psi, \psi] \leq a \parallel \psi \parallel_1^2 + b \parallel \psi \parallel_H^2 .$$

The latter estimate is valid for every $\psi \in \mathcal{H}_2(A)$, which is dense in $\mathcal{H}_1(A)$. It
follows that φ can be extended as a linear bounded functional to the whole

of $\mathcal{H}_1(A)$. We get a contradiction which proves the lemma.

□

Thus we have proven that rank one perturbations of positive operators are uniquely defined if $\varphi \in \mathcal{H}_{-1}(A)$. The perturbations defined by vectors $\varphi \in \mathcal{H}_{-2} \setminus \mathcal{H}_{-1}(A)$ cannot be defined using the form perturbation technique.

In what follows we are going to call **form bounded** all rank one perturbations determined by the vectors φ from the space $\mathcal{H}_{-1}(A)$, even if the operator A is not positive. Rank one perturbations defined by the vectors $\varphi \in \mathcal{H}_{-2}(A) \setminus \mathcal{H}_{-1}(A)$ will be called **form unbounded**. Note that Lemma 1.2.1 implies that every form bounded rank one perturbation is in fact infinitesimally form bounded.

1.2.4 Rank one perturbations and the extension theory for symmetric operators

Consider the self-adjoint operator A and its rank one perturbation $A_\alpha = A + \alpha \langle \varphi, \cdot \rangle \varphi$ restricted to the set of functions $\mathrm{Dom}\,(A^0) = \{\psi \in \mathrm{Dom}\,(A) : \langle \varphi, \psi \rangle = 0\}$. Let us denote the restricted operator by A^0. In the case of a bounded perturbation the operator A^0 is not densely defined. If the rank one perturbation is singular, then the operator A^0 is a densely defined symmetric operator with deficiency indices $(1, 1)$.

Lemma 1.2.3 *Let A be a self-adjoint operator acting in the Hilbert space H and let $\varphi \in \mathcal{H}_{-2}(A) \setminus H$. Then the restriction A^0 of the operator A to the domain of functions $\mathrm{Dom}\,(A^0) = \{\psi \in \mathrm{Dom}\,(A) : \langle \varphi, \psi \rangle = 0\}$ is a densely defined symmetric operator with deficiency indices $(1, 1)$.*

Proof We prove first that the restricted operator is densely defined. The operator A is densely defined and thus for every $f \in H$ there exists a sequence $f_n \in \mathrm{Dom}\,(A) = \mathcal{H}_2(A)$ converging to f in the Hilbert space norm:

$$\lim_{n \to \infty} \| f - f_n \|_H = 0.$$

The functional φ is not a bounded functional on the Hilbert space H. It follows that there exists a sequence $\psi_n \in \mathrm{Dom}\,(A) = \mathcal{H}_2(A)$ with the unit norm $\| \psi_n \|_H = 1$ such that the corresponding sequence $\langle \varphi, \psi_n \rangle$ diverges to infinity. This sequence can be chosen in such a way that $\lim_{n \to \infty} \langle \varphi, f_n \rangle / \langle \varphi, \psi_n \rangle = 0$. Then the sequence $f_n - (\langle \varphi, f_n \rangle / \langle \varphi, \psi_n \rangle) \psi_n$ belongs to the domain of the restricted operator

$$\left\langle \varphi, f_n - \frac{\langle \varphi, f_n \rangle}{\langle \varphi, \psi_n \rangle} \psi_n \right\rangle = 0$$

and converges to the element f in the Hilbert space norm

$$\left\| f_n - \frac{(\varphi, f_n)}{(\varphi, \psi_n)} \psi_n - f \right\|_H \leq \| f_n - f \|_H + \left| \frac{(\varphi, f_n)}{(\varphi, \psi_n)} \right| \to_{n \to \infty} 0.$$

Thus the operator A^0 is densely defined.

The deficiency elements of the operator A_0 at the point $\lambda = i$ are equal to $g_i = (A - i)^{-1} \varphi$. The latter equality has to be understood in the generalized sense, i.e. g_λ is the bounded linear functional which acts on every $\psi \in H$ in accordance with the formula

$$\langle g_i, \psi \rangle \;=\; \langle (A - i)^{-1} \varphi, \psi \rangle$$

$$=\; \langle \varphi, (A + i)^{-1} \psi \rangle$$

$$\leq\; \| \varphi \|_{-2} \left(\left\| \frac{|A|}{A + i} \psi \right\|_{\mathcal{H}} + \left\| \frac{1}{A + i} \psi \right\|_{\mathcal{H}} \right)$$

$$\leq\; 2 \| \varphi \|_{-2} \| \psi \|_0 \,.$$

Let $\psi \in \mathrm{Dom}\,(A^0)$. Then the following equalities hold

$$\langle \psi, A^* g_i \rangle \;=\; \langle A\psi, g_i \rangle \qquad =\; \left\langle \frac{A}{A + i} \psi, \varphi \right\rangle$$

$$=\; \left\langle \frac{-i}{A + i} \psi, \varphi \right\rangle \;=\; i \langle \psi, g_i \rangle.$$

It follows that g_i is the deficiency element for the restricted operator and corresponds to the complex number i. The deficiency element is unique (up to multiplication by complex numbers) and this finishes the proof of the lemma.

$\square$

We see that the self-adjoint operator corresponding to the formal expression $A_\alpha = A + \alpha \langle \varphi, \cdot \rangle \varphi$, $\varphi \in \mathcal{H}_{-2}(A) \setminus H$, is one of the self-adjoint extensions of the symmetric operator A^0 having deficiency indices $(1, 1)$. Let us discuss now the extension theory for such operators.

In what follows we are going to normalize the vector $\varphi \in \mathcal{H}_{-2}(A)$ using the norm $\| \cdot \|_{-2}^*$ defined by (1.21)

$$\left\| \frac{1}{A - i} \varphi \right\| = \big\| \varphi \big\|_{-2}^* = 1. \tag{1.23}$$

Then the deficiency elements $g_{\pm i} = (1/(A \mp i))\, \varphi$ have unit norms in the Hilbert space H.

1.2.5   The extension theory and Krein's formula

In this section we are going to study the self–adjoint extensions of symmetric operators with unit deficiency indices. Let A^0 be a certain densely defined symmetric operator acting in the Hilbert space H. Without loss of generality we suppose that the operator A^0 is closed. Suppose also that the deficiency indices are equal to $(1,1)$ and let g_i and g_{-i} be two normalized deficiency elements corresponding to $\lambda = \pm i$

$$A^{0*} g_{\pm i} = \pm i g_{\pm i};$$

$$\| \, g_{\pm i} \, \|_H = 1.$$

Then the domain of the adjoint operator is equal to the following linear sum $\mathrm{Dom}\,(A^{0*}) = \mathrm{Dom}\,(A^0) \dotplus \mathcal{L}\{g_i, g_{-i}\}$, where $\dotplus$ denotes the direct sum. Every element ψ from the domain of the adjoint operator possesses the following representation

$$\psi = \hat{\psi} + a_+(\psi)g_i + a_-(\psi)g_{-i}, \qquad (1.24)$$

where $\hat{\psi} \in \mathrm{Dom}\,(A^0), a_{\pm}(\psi) \in \mathbf{C}$. The adjoint operator A^{0*} acts as follows on every $\psi \in \mathrm{Dom}\,(A^{0*})$

$$A^{0*}(\hat{\psi} + a_+(\psi)g_i + a_-(\psi)g_{-i}) = A\hat{\psi} + ia_+(\psi)g_i - ia_-(\psi)g_{-i}. \qquad (1.25)$$

All self–adjoint extensions of the operator A^0 can be parametrized by one unimodular parameter v; $|v| = 1$ using the von Neumann theory. Every self–adjoint extension $A(v)$ coincides with the restriction of the operator A^{0*} to the domain $\mathrm{Dom}(A(v)) = \{\psi \in \mathrm{Dom}\,(A^{0*}) : -va_-(\psi) = a_+(\psi)\}$.

Let us denote by A the self–adjoint extension corresponding to $v = 1$. Then the deficiency elements g_i and g_{-i} are related as follows

$$g_{-i} = \frac{A - i}{A + i} g_i. \qquad (1.26)$$

We are also going to use the following representation for the functions $\psi \in \mathrm{Dom}\,(A^{0*})$

$$\begin{aligned} \psi &= \tilde{\psi} + \frac{b(\psi)}{2}(g_i + g_{-i}) \\ &= \tilde{\psi} + b(\psi)\frac{A}{A + i}g_i, \end{aligned} \qquad (1.27)$$

where $\tilde{\psi} \in \mathrm{Dom}\,(A)$. This representation is related to the representation

(1.24) via the formulas

$$
\begin{cases}
\tilde{\psi} & = \hat{\psi} + \dfrac{a_+(\psi) - a_-(\psi)}{2}(g_i - g_{-i}) \\[2mm]
& = \hat{\psi} + (a_+(\psi) - a_-(\psi))\dfrac{i}{A+i}g_i; \\[4mm]
b(\psi) & = a_+(\psi) + a_-(\psi),
\end{cases}
$$

where we have used (1.26). Using representation (1.27) the action of the adjoint operator A^{0*} is given by

$$
\begin{aligned}
A^{0*}\psi & = A^{0*}\left(\tilde{\psi} + b(\psi)\frac{A}{A+i}g_i\right) \\[3mm]
& = A\tilde{\psi} - b(\psi)\frac{1}{A+i}g_i.
\end{aligned}
\tag{1.28}
$$

The latter formula follows directly from (1.25). Let us calculate the boundary form $\langle A^{0*}\psi, \eta\rangle - \langle \psi, A^{0*}\eta\rangle$ of the adjoint operator for two functions $\psi, \eta \in$ Dom(A^{0*})

$$\langle A^{0*}\psi, \eta\rangle - \langle \psi, A^{0*}\eta\rangle$$

$$
\begin{aligned}
= & \left\langle A\tilde{\psi} - b(\psi)\frac{1}{A+i}g_i, \tilde{\eta} + b(\eta)\frac{A}{A+i}g_i\right\rangle \\[3mm]
& - \left\langle \tilde{\psi} + b(\psi)\frac{A}{A+i}g_i, A\tilde{\eta} - b(\eta)\frac{1}{A+i}g_i\right\rangle \\[5mm]
= & \ \langle A\tilde{\psi}, \tilde{\eta}\rangle - \bar{b}(\psi)\left\langle \frac{1}{A+i}g_i, \tilde{\eta}\right\rangle \\[3mm]
& + b(\eta)\left\langle A\tilde{\psi}, \frac{A}{A+i}g_i\right\rangle - \bar{b}(\psi)b(\eta)\left\langle \frac{1}{A+i}g_i, \frac{A}{A+i}g_i\right\rangle \\[3mm]
& - \langle \tilde{\psi}, A\tilde{\eta}\rangle - \bar{b}(\psi)\left\langle \frac{A}{A+i}g_i, A\tilde{\eta}\right\rangle \\[3mm]
& + b(\eta)\left\langle \tilde{\psi}, \frac{1}{A+i}g_i\right\rangle + \bar{b}(\psi)b(\eta)\left\langle \frac{A}{A+i}g_i, \frac{1}{A+i}g_i\right\rangle \\[5mm]
= & \ \overline{\langle (A-i)g_i, \tilde{\psi}\rangle}b(\eta) - \overline{b(\psi)}\langle (A-i)g_i, \tilde{\eta}\rangle.
\end{aligned}
$$

The deficiency element g_i belongs to the Hilbert space. Therefore the vector $(A - i)g_i$ is a bounded linear functional on the domain of the operator A, i.e. belongs to $\mathcal{H}_{-2}(A)$ and the scalar products $\langle (A - i)g_i, \tilde{\psi} \rangle$ and $\langle (A - i)g_i, \tilde{\eta} \rangle$ appearing in the latter formula are well defined. Another way to parametrize the self-adjoint extensions is to use a real parameter γ instead of the unitary parameter v appeared in the von Neumann formulas. Let us denote by A^γ the restriction of the operator A^{0*} to the domain

$$\text{Dom}\,(A^\gamma) = \{\psi \in \text{Dom}\,(A^{0*}) : \langle (A - i)g_i, \tilde{\psi} \rangle = \gamma b(\psi) \}. \qquad (1.29)$$

We are even going to consider infinite values of the parameter $\gamma : \gamma \in \mathbf{R} \cup \{\infty\}$. For $\gamma = \infty$ we put formally

$$\text{Dom}\,(A^\infty) = \{\psi \in \text{Dom}\,(A^{0*}) : b(\psi) = 0 \} \equiv \text{Dom}\,(A),$$

i.e. the operator A^∞ coincides with the operator A. The operators $A(v)$ and A^γ coincide if the parameters v and γ are related as follows

$$v = \frac{\gamma + i}{\gamma - i} \Leftrightarrow \gamma = -i\frac{1 + v}{1 - v}. \qquad (1.30)$$

When γ runs over all real numbers including infinity, the parameter v runs over all unimodular complex numbers. Therefore every self-adjoint extension of A^0 is described by a certain parameter $\gamma \in \mathbf{R} \cup \{\infty\}$. In what follows we are going to use both descriptions of self-adjoint extensions.

The resolvents of two self-adjoint extensions of one symmetric operator are related by Krein's formula [755, 569]:

Theorem 1.2.1 *Let A and B be two self-adjoint extensions of a certain symmetric densely defined operator A^0 with unit deficiency indices. Then there exists a real number $\gamma \in \mathbf{R} \cup \{\infty\}$ such that the resolvents of the operators A and B are related as follows*

$$\frac{1}{B - z} = \frac{1}{A - z} + \frac{1}{\gamma - \langle g_i, \frac{1 + Az}{A - z} g_i \rangle} \left\langle \frac{A - i}{A - \bar{z}} g_i, \cdot \right\rangle \frac{A - i}{A - z} g_i, \quad \Im z \neq 0, \quad (1.31)$$

where g_i is the normalized deficiency element for A^0 at the point $\lambda = i$. If $A = B$, then $\gamma = \infty$ and the resolvents of the self-adjoint operators coincide:

$$\frac{1}{B - z} = \frac{1}{A - z}, \quad \Im z \neq 0. \qquad (1.32)$$

Proof The operator A is a self–adjoint extension of A^0. Therefore the function $((A - i)/(A + i)) g_i$ is a deficiency element for A^0 at the point $\lambda = -i$ and we can choose

$$g_{-i} = \frac{A - i}{A + i} g_i.$$

Let us describe the self–adjoint extensions of A^0 using (1.29). Then the operator A coincides with the operator A^∞. The operator B is also a self–adjoint extension of A^0, therefore there exists a certain real parameter $\gamma \in \mathbf{R} \cup \{\infty\}$ such that $B = A^\gamma$.

If $\gamma = \infty$, then the operators A and B coincide and formula (1.32) obviously holds.

To prove the theorem we have to calculate the resolvent of the operator A^γ, i.e. we have to solve the following equation

$$h = (A^\gamma - z)f, \tag{1.33}$$

for a given $h \in H$. The function f belongs to the domain of the operator A^γ, $f \in \mathrm{Dom}\,(A^\gamma)$. Let $\gamma \neq 0, \infty$. Then every function f from the domain of the operator A^γ possesses the representation

$$f = \tilde{f} + \frac{1}{\gamma}\langle(A - i)g_i, \tilde{f}\rangle \frac{A}{A + i} g_i.$$

Equality (1.33) and formula (1.28) imply that

$$h = A\tilde{f} - z\tilde{f} - \frac{1}{\gamma}\langle(A - i)g_i, \tilde{f}\rangle \frac{1 + Az}{A + i} g_i.$$

Applying the resolvent of the original operator A to the latter equation and then projecting to the element $(A - i)g_i \in \mathcal{H}_{-2}(A)$ we get

$$\langle(A - i)g_i, \tilde{f}\rangle = \frac{\langle \frac{A-i}{A-\bar{z}} g_i, h\rangle}{1 - \frac{1}{\gamma}\langle g_i, \frac{1+Az}{A-z} g_i\rangle}. \tag{1.34}$$

It follows that the function f is given by

$$
\begin{aligned}
f &= \frac{1}{A-z}h + \frac{1}{\gamma}\langle(A-i)g_i, \tilde{f}\rangle\frac{1+Az}{A-z}\frac{1}{A+i}g_i \\
&\quad + \frac{1}{\gamma}\langle(A-i)g_i, \tilde{f}\rangle\frac{A}{A+i}g_i \\[2mm]
&= \frac{1}{A-z}h + \frac{1}{\gamma}\langle(A-i)g_i, \tilde{f}\rangle\frac{A-i}{A-z}g_i \\[2mm]
&= \frac{1}{A-z}h + \frac{\langle\frac{A-i}{A-z}g_i, h\rangle}{\gamma - \langle g_i, \frac{1+Az}{A-z}g_i\rangle}\frac{A-i}{A-z}g_i.
\end{aligned}
$$

The latter formula implies (1.31).

In the exceptional case $\gamma = 0$ the resolvent formula is obviously satisfied. The theorem is proven.

$\square$

In fact we have proven that the resolvent of any self-adjoint extension A^γ of A^0 is given by

$$
\frac{1}{A^\gamma - z} = \frac{1}{A-z} + \frac{1}{\gamma - \langle g_i, \frac{1+Az}{A-z}g_i\rangle}\left\langle\frac{A-i}{A-\bar{z}}g_i, \cdot\right\rangle\frac{A-i}{A-z}g_i, \quad \Im z \neq 0. \quad (1.35)
$$

The function

$$
Q(z) = \left\langle g_i, \frac{1+Az}{A-z}g_i\right\rangle \quad (1.36)
$$

is called **Krein's Q-function**. It is a Nevanlinna function, since the deficiency element g_i has finite norm. Usually the Q-function is defined by the following relation

$$
\frac{Q(\lambda) - \overline{Q}(z)}{\lambda - \bar{z}} = \langle g_\lambda, g_z\rangle \equiv \left\langle\frac{A-i}{A-\lambda}g_i, \frac{A-i}{A-z}g_i\right\rangle. \quad (1.37)
$$

Obviously the latter relation defines the Q-function uniquely up to real constants. In general two self-adjoint extensions of one symmetric operator with deficiency indices $(1,1)$ define a one-parameter family of Q-functions which differ by real constants. In what follows we are going to use definition (1.36) for the Krein's Q-function, i.e. we are going to distinguish the Q-functions

which differ by real constants. We hope that this convention will not cause
any problem for the reader.

Comparing formula (1.36) and representation (1.7) for an arbitrary Nevanlinna function we see that the Q-functions corresponding to self–adjoint extensions of densely defined symmetric operators do not have a linear term in
the asymptotics, i.e. the constants b appeared in (1.7) are always equal to
zero.

A similar analysis can be carried out in the case where the operator A^0 is
not densely defined [562, 561]. The resolvents of the self–adjoint extensions
are related by similar formulas. The difference is that Krein's formula for the
resolvent describes in this case not only the resolvents of all extensions that
are self–adjoint operators but also the resolvents of the self–adjoint relations
which are extensions of the symmetric operator.

1.2.6 Q-function for rank one perturbations

Let us continue discussion of the operator A_α formally given by $A_\alpha = A +
\alpha \langle \varphi, \cdot \rangle \varphi$ already started in Section 1.2.4. Lemma 1.2.3 implies that the de-
ficiency element for the operator A^0 at the point $\lambda = i$ is given by $g_i =
(1/(A - i)) \varphi$. Therefore the Q-function corresponding to the operators A^0
and A is given by

$$Q(z) = \left\langle g_i, \frac{1 + Az}{A - z} g_i \right\rangle = \left\langle \varphi, \frac{1 + Az}{A - z} \frac{1}{A^2 + 1} \varphi \right\rangle. \tag{1.38}$$

If $\varphi \in \mathcal{H}_{-2}(A)$, then the scalar product appeared in the latter formula is well
defined.

Let φ be an element from $\mathcal{H}_{-1}(A)$. One can write the following formula
for the corresponding Q-function

$$Q(z) = \left\langle \varphi, \frac{1}{A - z} \varphi \right\rangle - \left\langle \varphi, \frac{A}{A^2 + 1} \varphi \right\rangle. \tag{1.39}$$

It follows that $Q(z)$ possesses the integral representation

$$Q(z) = a + \int_{-\infty}^{+\infty} \frac{d\tau(\lambda)}{\lambda - z}, \quad \Im z \neq 0, \tag{1.40}$$

where $a \in \mathbf{R}$ and

$$\int_{-\infty}^{+\infty} \frac{d\tau(\lambda)}{1 + |\lambda|} < \infty.$$

Therefore the function $Q(z)$ belongs to the class $\mathcal{R}_1$ of Nevanlinna functions
[569]. The class $\mathcal{R}_1$ is the subset of Nevanlinna functions R with the following

property

$$\int_1^{+\infty} \frac{\Im R(iy)}{y} dy < \infty. \tag{1.41}$$

Obviously the class $\mathcal{R}_1$ contains the class $\mathcal{R}_0$ of Nevanlinna functions.

Let us define another scale of Hilbert spaces associated with the operator A and the vector $\varphi \in \mathcal{H}_{-2}(A) \setminus H$

$$\mathcal{H}_2(A) = \mathrm{Dom}\,(A) \subset H_\varphi(A) \subset H \subset H_\varphi(A)^* \subset \mathrm{Dom}\,(A)^* = \mathcal{H}_{-2}(A). \tag{1.42}$$

Here $H_\varphi(A)$ denotes the domain of the adjoint operator A^{0*}

$$H_\varphi(A) = \mathrm{Dom}\,(A^{0*}).$$

In the spaces $\mathcal{H}_2(A)$ and $\mathcal{H}_{-2}(A)$ we are going to use the norms $\| \cdot \|_2^*$ and $\| \cdot \|_{-2}^*$ defined by (1.21).

Consider the resolvent $1/(A-i)$ of the operator A acting in the generalized sense. Let $\varphi \in \mathrm{Dom}\,(A)^*$. Then $(1/(A-i))\,\varphi$ is the linear functional which acts on every $\psi \in H$ in accordance with the formula

$$\left| \left\langle \frac{1}{A-i}\varphi, \psi \right\rangle \right| = \left| \left\langle \varphi, \frac{1}{A+i}\psi \right\rangle \right| \leq \| \varphi \|_2^* \left\| \frac{1}{A+i}\psi \right\|_2.$$

It follows that $(1/(A-i))\,\varphi$ is a bounded functional on H and thus is an element from the Hilbert space H.

The norm in the space $H_\varphi(A)$ will be defined using representation (1.27) as follows

$$\| \psi \|_{H_\varphi(A)} = \left\| \tilde{\psi} + b(\psi)\frac{A}{A^2+1}\varphi \right\|_{H_\varphi(A)}$$

$$= \| \tilde{\psi} \|_2^* + |b(\psi)|, \tag{1.43}$$

since $H_\varphi(A)$ is a one dimensional extension of the space $\mathrm{Dom}\,(A)$. The space $H_\varphi(A)^*$ is adjoint to $H_\varphi(A)$. Let $\psi \in \mathrm{Dom}\,(A)$. Then

$$\| \psi \|_{B_\varphi(A)} = \| \psi \|_2^*.$$

The inclusion (1.42) are now obvious.

The second scale of spaces is constructed using the functional φ, while the standard scale of Hilbert spaces is determined by the operator A only. This determines the main difference between the two scales of Hilbert spaces.

1.3   Singular rank one perturbations

1.3.1   Form bounded rank one singular perturbations

Let us consider first form bounded rank one perturbations. We have seen that the operator

$$A_\alpha = A + \alpha\langle\varphi, \cdot\rangle\varphi, \quad \varphi \in \mathcal{H}_{-1}(A) \tag{1.44}$$

can be defined using the form perturbation technique if the operator A is positive. Another way to define the self–adjoint operator corresponding to the latter formal expression is to consider the linear operator defined by this expression. Consider the operator A defined in the generalized sense. Then formula (1.44) determines a linear operator on $\mathcal{H}_1(A)$ with the range in the space $\mathcal{H}_{-1}(A)$. The corresponding operator acting in the Hilbert space is defined by the restriction of the linear operator A_α to the following domain

$$\mathrm{Dom}\,(A_\alpha) = \{\psi \in \mathcal{H}_1(A) \subset H : A_\alpha\psi \in H\}. \tag{1.45}$$

The operator A_α restricted in this way is self–adjoint and will be considered as the unique self–adjoint operator corresponding to the heuristic expression (1.44).

Theorem 1.3.1 *Let $\varphi \in \mathcal{H}_{-1}(A) \setminus H$. Then the domain of the self–adjoint operator $A_\alpha = A + \alpha\langle\varphi, \cdot\rangle\varphi$ coincides with the following set*

$$\mathrm{Dom}\,(A_\alpha) = \left\{\psi \in H_\varphi(A) : \langle\varphi, \tilde\psi\rangle = -\left(\frac{1}{\alpha} + \langle\varphi, \frac{A}{A^2+1}\varphi\rangle\right)b(\psi)\right\}. \tag{1.46}$$

A_α is a self–adjoint extension of A^0. For $\alpha = 0$ we have $A_0 = A$.

Proof The linear operator A_α maps the vector space $\mathcal{H}_1(A)$ to the space $\mathcal{H}_{-1}(A)$. Let ψ be an element from $\mathcal{H}_1(A)$. Let us study the question: Under what conditions is the distribution $A_\alpha\psi$ an element from the Hilbert space H? Consider an arbitrary vector η from the domain $\mathrm{Dom}\,(A^0) \subset H$. Then $\langle\eta, A_\alpha\psi\rangle$ is a bounded linear functional on η only if $\psi \in \mathrm{Dom}\,(A^{0*})$, since the following equalities hold

$$
\begin{aligned}
\langle\eta, A_\alpha\psi\rangle &= \langle\eta, A + \alpha\langle\varphi, \psi\rangle\varphi\rangle \\
&= \langle\eta, A\psi\rangle + \alpha\langle\varphi, \psi\rangle\langle\eta, \varphi\rangle \\
&= \langle A\eta, \psi\rangle.
\end{aligned}
$$

We have taken into account that $\langle \eta, \varphi \rangle = 0$ (as an element from $\mathrm{Dom}\,(A^0)$) and the operator A is defined in the generalized sense on the vectors from $\mathcal{H}_1(A)$.

Let $\psi \in \mathrm{Dom}\,(A^{0*}) = H_\varphi(A)$. Then the representation (1.27) is valid and the linear operator acts as follows

$$
\begin{aligned}
A_\alpha \psi &= (A + \alpha \langle \varphi, \cdot \rangle \varphi)\left(\tilde{\psi} + b(\psi)\frac{A}{A^2+1}\varphi\right) \\[2mm]
&= A\tilde{\psi} + \alpha\langle \varphi, \tilde{\psi}\rangle \varphi + b(\psi)\frac{A^2}{A^2+1}\varphi \\[2mm]
&\quad + \alpha b(\psi)\left\langle \varphi, \frac{A}{A^2+1}\varphi\right\rangle \varphi \\[2mm]
&= \left\{ A\tilde{\psi} - b(\psi)\frac{1}{A^2+1}\varphi \right\} \\[2mm]
&\quad + \left[\alpha\langle \varphi, \tilde{\psi}\rangle b(\psi) + \alpha b(\psi)\langle \varphi, \frac{A}{A^2+1}\varphi\rangle \right]\varphi.
\end{aligned}
\tag{1.47}
$$

The expression in the braces $\{\ \}$ belongs to the original Hilbert space H. Therefore the vector element $A_\alpha \psi$ belongs to H if and only if the expression in the square brackets $[\ ]$ is equal to zero, i.e. if the following equality holds

$$
\langle \varphi, \tilde{\psi}\rangle = -\left(\frac{1}{\alpha} + \left\langle \varphi, \frac{A}{A^2+1}\varphi\right\rangle\right)b(\psi).
\tag{1.48}
$$

The parameter

$$
\gamma = -\frac{1}{\alpha} - \left\langle \varphi, \frac{A}{A^2+1}\varphi\right\rangle
\tag{1.49}
$$

is real and the adjoint operator A^{0*} restricted to the domain of functions from $H_\varphi(A)$ satisfying the boundary condition (1.48) is self–adjoint. The restrictions of the operators A_α and A^{0*} to this domain are identical since the expression in the square brackets $[\ ]$ in formula (1.47) vanishes for the elements satisfying the boundary conditions (1.48). Thus we have proven that the self–adjoint operator defined by the formal expression (1.44) is a self–adjoint extension of the operator A^0 described by the parameter γ given by (1.49).

If $\alpha = 0$ then the parameter $\gamma = \infty$ and corresponding operator coincides with the original operator A. The theorem is thus proven.

$\square$

The latter theorem describes the rank one perturbations of the operator A

using the real parameter γ. The unitary parameter v describing the A_α is given by

$$v = \frac{1 + \alpha\langle\varphi, \frac{1}{A+i}\varphi\rangle}{1 + \alpha\langle\varphi, \frac{1}{A-i}\varphi\rangle}. \tag{1.50}$$

Considering different $\alpha \in \mathbf{R} \cup \{\infty\}$ all self–adjoint extensions of the symmetric operator A^0 can be obtained. The formula (1.48) establishes a one–to–one correspondence between the parameters α and γ, $\alpha, \gamma \in \mathbf{R} \cup \{\infty\}$. The parameter α describes all self–adjoint extensions of the symmetric operator A^0 in an additive manner, while the real parameter γ appeared in Krein's formula and the unitary parameter v from the von Neumann formula are not additive.

We have proven once more that the self–adjoint operator corresponding to a singular rank one perturbation is a self–adjoint extension of the symmetric operator A^0, which is a restriction of the original operator A. In the case of form bounded perturbations the self–adjoint operator is uniquely defined even for operators that are not semibounded. Form unbounded perturbations will be studied in the following section.

1.3.2 Family of rank one form unbounded perturbations

Consider a form unbounded rank one perturbation defined by the same formal expression

$$A_\alpha = A + \alpha\langle\varphi, \cdot\rangle\varphi, \quad \varphi \in \mathcal{H}_{-2}(A) \setminus \mathcal{H}_{-1}(A). \tag{1.51}$$

We have shown that any self–adjoint operator corresponding to this formal expression is an extension of the symmetric operator A^0. Considering rank one form bounded perturbations we have determined the unique self–adjoint extension of the operator A^0 which coincides with the linear operator A_α defined in the generalized sense. In the case under consideration the linear operator A_α is not in general defined on the space $H_\varphi(A) = \text{Dom}(A^{0*})$. The reason is that the linear functional φ is not defined on this domain. It is defined on the domain $\text{Dom}(A) = \mathcal{H}_{-2}(A)$. Thus to define the linear operator on the domain $\text{Dom}(A^{0*})$ one has to extend the functional φ. The extension has to be chosen in such a way that the corresponding sesquilinear form is real. The following lemma describes all possible real extensions.

Lemma 1.3.1 *Every extension of the functional φ to the domain $H_\varphi(A) = \text{Dom}(A^{0*})$ is defined by one parameter $c \in \mathbf{C}$. Let*

$$\psi = \tilde\psi + b(\psi)\frac{A}{A^2+1} \in H_\varphi(A).$$

Then the extended functional φ_c acts as follows

$$\langle \varphi_c, \psi \rangle = \langle \varphi, \tilde{\psi} \rangle + \bar{c}b(\psi). \tag{1.52}$$

This extension defines a real quadratic form $Q[\psi, \psi] = \langle \psi, (A/(A^2 + 1)) \psi \rangle$ with domain $\mathrm{Dom}\,(Q) = H \dotplus \mathcal{L}\{\varphi\}$ if and only if the parameter c is real.

Proof The linear functional φ_c defined by formula (1.52) is bounded and defined on any element ψ from the domain of the adjoint operator. The norm in the space $H_\varphi(A)$ has been defined by (1.43). The quadratic form corresponding to this extension is real if the parameter c is real.

Consider now an arbitrary bounded linear extension $\hat{\varphi}$ of the functional φ to the domain of the adjoint operator. Let $\psi = \tilde{\psi} + b(\psi)(A/(A^2 + 1))\varphi$ be an element from the domain $\mathrm{Dom}\,(A^{0*})$ of the adjoint operator. Since the functional $\hat{\varphi}$ is a linear extension of φ, the following equality holds

$$\langle \hat{\varphi}, \psi \rangle = \langle \varphi, \tilde{\psi} \rangle + b(\psi) \left\langle \hat{\varphi}, \frac{A}{A^2 + 1} \varphi \right\rangle.$$

Thus every bounded linear extension of the functional φ is defined by one parameter

$$\bar{c} = \left\langle \hat{\varphi}, \frac{A}{A^2 + 1} \varphi \right\rangle.$$

Consider an arbitrary element $\psi = \tilde{\psi} + q(\psi)\varphi \in \mathrm{Dom}\,(Q)$, where $\tilde{\psi} \in H, q(\psi) \in \mathbf{C}$. Then the quadratic form can be calculated as follows

$$\begin{aligned}
Q[\psi, \psi] &= \left\langle \tilde{\psi} + q(\psi)\hat{\varphi}, \frac{A}{A^2 + 1}(\tilde{\psi} + q(\psi)\hat{\varphi}) \right\rangle \\
&= \left\langle \tilde{\psi}, \frac{A}{A^2 + 1}\tilde{\psi} \right\rangle + \bar{q}(\psi) \left\langle \varphi, \frac{A}{A^2 + 1}\tilde{\psi} \right\rangle \\
&\quad + q(\psi) \left\langle \tilde{\psi}, \frac{A}{A^2 + 1}\varphi \right\rangle + |q(\psi)|^2 \left\langle \hat{\varphi}, \frac{A}{A^2 + 1}\varphi \right\rangle \\
&= \Re \left(\left\langle \tilde{\psi}, \frac{A}{A^2 + 1}\tilde{\psi} \right\rangle + 2q(\psi) \left\langle \varphi, \frac{A}{A^2 + 1}\tilde{\psi} \right\rangle \right) + |q(\psi)|^2 \bar{c}.
\end{aligned}$$

The latter formula shows that the quadratic form is real if and only if the parameter c is real. The lemma is proven.

$\square$

The following definition will be used below.

Definition 1.3.1 *Let $\varphi \in \mathcal{H}_{-2}(A) \setminus \mathcal{H}_{-1}(A)$. Then the functional φ_c is the linear bounded extension of the functional φ to the domain $H_\varphi(A)$ defined by the condition*

$$\left\langle \varphi_c, \frac{A}{A^2+1}\varphi \right\rangle = c, \tag{1.53}$$

where $c \in \mathbf{R}$.

The following theorem describes the domain of the self–adjoint operator corresponding to the formal expression (1.51) and extension (1.53).

Theorem 1.3.2 *Let φ_c be a linear bounded extension of the functional $\varphi \in \mathcal{H}_{-2}(A) \setminus \mathcal{H}_{-1}(A)$. Then the domain of the self–adjoint operator*

$$A_\alpha = A + \alpha\langle \varphi_c, \cdot\rangle\varphi,$$

being a rank one form unbounded perturbation of A, coincides with the following set

$$\mathrm{Dom}\,(A_\alpha) = \left\{ \psi = \tilde{\psi} + b(\psi)\frac{A}{A^2+1}\varphi \in \mathrm{Dom}\,(A^{0*}) : \right.$$
$$\left. \langle \varphi, \tilde{\psi}\rangle = -\left(\frac{1}{\alpha}+c\right)b(\psi) \right\}.$$

A_α is a self–adjoint extension of A^0 if $c \in \mathbf{R}$. For $\alpha = 0$ we have $A_0 = A$.

Proof The proof is similar to that of theorem 1.3.1. The linear operator A_α acts as follows on the domain $\mathrm{Dom}\,(A^{0*}) \ni \psi$

$$A_\alpha\psi = (A + \alpha\langle\varphi_c, \cdot\rangle\varphi)\left(\tilde{\psi} + b(\psi)\frac{A}{A^2+1}\varphi\right)$$

$$= A\tilde{\psi} + \alpha\langle\varphi, \tilde{\psi}\rangle\varphi + b(\psi)\frac{A^2}{A^2+1}\varphi$$

$$+ \alpha b(\psi)\left\langle\varphi_c, \frac{A}{A^2+1}\varphi\right\rangle$$

$$= \left\{A\tilde{\psi} - b(\psi)\frac{1}{A^2+1}\varphi\right\}$$

$$+ \left[\alpha\langle\varphi, \tilde{\psi}\rangle + b(\psi) + \alpha c b(\psi)\right]\varphi.$$

The range of the linear operator A_α does not belong to the Hilbert space. The domain of the self–adjoint operator A_α is equal to the following set

$$\mathrm{Dom}\,(A_\alpha) = \{\psi \in H_\varphi(A) : A_\alpha\psi \in H\}.$$

The element $A_\alpha \psi$ belongs to H if and only if the following condition is satisfied

$$\langle \varphi, \tilde{\psi} \rangle = -(\frac{1}{\alpha} + c)b(\psi).$$

The parameter

$$\gamma = -(\frac{1}{\alpha} + c) \tag{1.54}$$

is real. The operator 4^{0*} restricted to the domain of functions satisfying the latter condition is self-adjoint and coincides with the operator A_α restricted to the same domain. Thus the theorem is proven.

□

The latter theorem describes rank one perturbations using the real parameter γ appearing in the boundary condition. The unitary parameter v describing the same self-adjoint extension of the operator A^0 is given by

$$v = \frac{1 + \alpha(c + i)}{1 + \alpha(c - i)}.$$

Considering different $\alpha \in \mathbf{R} \cup \{\infty\}$ we get all self-adjoint extensions of the operator A^0. In general the extension depends on the parameter c which describes the extension of the functional φ.

We have considered only extensions of the functional φ determined by the real parameters c. One can see that unreal values of this parameter lead to the boundary conditions defining non–self–adjoint operators (the corresponding parameter γ is not real). Considering different extensions of the functional (different values of the constant $c \in \mathbf{R}$) and one particular $\alpha \neq 0$ we also get all except one self-adjoint extensions of the operator A^0. The exceptional extension A^∞ coincides with the original operator A (see Section 1.2.5).

1.3.3 Singular rank one form unbounded perturbations of homogeneous operators

This section is devoted to the investigation of form unbounded rank one perturbations in the case where the original operator and the element φ are homogeneous with respect to a certain group of unitary transformations of the Hilbert space H. The extension of the functional φ in general can be uniquely defined using the homogeneity properties of the operator and its perturbation.

Lemma 1.3.2 *Let the self–adjoint operator A and the vector $\varphi \in \mathrm{Dom}\,(A)^*$ be homogeneous with respect to a certain unitary group $G(t)$, i.e. there exist real constants $\beta, \theta \in \mathbf{R}$ such that*

$$G(t)A = t^{\beta} A G(t); \tag{1.55}$$

$$\langle G(t)\varphi, \psi \rangle = \langle \varphi, G(1/t)\psi \rangle = t^{\theta}\langle \varphi, \psi \rangle \tag{1.56}$$

for every $\psi \in \mathrm{Dom}\,(A)$. Then φ can be extended as a homogeneous linear bounded functional to the domain $H_{\varphi}(A)$ if and only if

$$f(t) = i\frac{1-t^{\beta}}{1-t^{-\beta-2\theta}}\left\langle \varphi, \frac{1}{(A-i)(A-t^{\beta}i)}\varphi \right\rangle \tag{1.57}$$

does not depend on $t \neq 1$.

Proof Consider an arbitrary linear bounded extension φ_c of the functional φ which is defined by one parameter (see Lemma 1.3.1)

$$c = \left\langle \varphi_c, \frac{A}{A^2+1}\varphi \right\rangle.$$

Suppose that this extension is homogeneous and thus satisfies equation (1.56). Then the function $f(t)$ can be calculated

$$
\begin{aligned}
f(t) &= i\frac{1-t^{\beta}}{1-t^{-\beta-2\theta}}\left\langle \varphi, \frac{1}{(A-i)(A-t^{\beta}i)}\varphi \right\rangle \\[2mm]
&= \frac{1}{1-t^{-\beta-2\theta}}\left\langle \varphi_c, \left(\frac{1}{A-i} - \frac{1}{A-t^{\beta}i}\right)\varphi \right\rangle \\[2mm]
&= \frac{1}{1-t^{-\beta-2\theta}}\left\{ \left\langle \varphi_c, \frac{1}{A-i}\varphi \right\rangle - \left\langle \varphi_c, \frac{1}{A-t^{\beta}i}\varphi \right\rangle \right\} \\[2mm]
&= \frac{1}{1-t^{-\beta-2\theta}}\left\{ \left\langle \varphi_c, \frac{1}{A-i}\varphi \right\rangle - t^{-\theta}\left\langle \varphi_c, \frac{1}{A-t^{\beta}i}G(t)\varphi \right\rangle \right\} \\[2mm]
&= \frac{1}{1-t^{-\beta-2\theta}}\left\{ \left\langle \varphi_c, \frac{1}{A-i}\varphi \right\rangle - t^{-\beta-\theta}\left\langle \varphi_c, G(t)\frac{1}{A-i}\varphi \right\rangle \right\} \\[2mm]
&= \left\langle \varphi_c, \frac{1}{A-i}\varphi \right\rangle \\[2mm]
&= c + i\left\langle \varphi, \frac{1}{A^2+1}\varphi \right\rangle.
\end{aligned}
$$

It follows that for any homogeneous extension φ_c the function $f(t)$ is equal to a certain constant determined by the extension. The imaginary part of $f(t)$ is always equal to 1 if the parameter c is real.

Suppose conversely that the function $f(t)$ is equal to a given constant. Let us define the extension of the functional φ by the following condition

$$\left\langle \varphi_c, \frac{A}{A^2+1}\varphi \right\rangle = c = f(t) - i. \tag{1.58}$$

The imaginary part of $f(t)$ is always equal to 1:

$$\begin{aligned}
\Im f(t) &= \frac{1-t^\beta}{1-t^{-\beta-2\theta}} \left\langle \varphi, \frac{A^2-t^\beta}{(A^2+1)(A^2+t^{2\beta})}\varphi \right\rangle \\
&= \frac{1}{1-t^{-\beta-2\theta}} \left(\left\langle \varphi, \frac{1}{A^2+1}\varphi \right\rangle - t^\beta \left\langle \varphi, \frac{1}{A^2+t^{2\beta}}\varphi \right\rangle \right) \\
&= \frac{1}{1-t^{-\beta-2\theta}} (1 - t^\beta t^{-2\beta-2\theta}) \left\langle \varphi, \frac{1}{A^2+1}\varphi \right\rangle \\
&= \left\langle \varphi, \frac{1}{A^2+1}\varphi \right\rangle.
\end{aligned}$$

Therefore the constant c determined by (1.58) is always real. It is necessary to show that the extension of the functional is homogeneous in this case. In fact it is enough to prove this property only for the elements $(1/(A-i))\varphi$ and $(1/(A+i))\varphi$. We have:

$$\left\langle G(1/t)\varphi_c, \frac{1}{A-i}\varphi \right\rangle$$

$$= \left\langle \varphi_c, G(t)\frac{1}{A-i}\varphi \right\rangle$$

$$= t^{\theta+\beta} \left\langle \varphi_c, \frac{1}{A-t^\beta i}\varphi \right\rangle$$

$$= t^{\theta+\beta} \left(\left\langle \varphi_c, \frac{1}{A-i}\varphi \right\rangle + (t^\beta i - i) \left\langle \varphi, \frac{1}{(A-i)(A-t^\beta i)}\varphi \right\rangle \right)$$

$$= t^{\theta+\beta} \left(\left\langle \varphi_c, \frac{1}{A-i}\varphi \right\rangle - (1 - t^{-\beta-2\theta}) \left\langle \varphi_c, \frac{1}{A-i}\varphi \right\rangle \right)$$

$$= t^{-\theta} \left\langle \varphi_c, \frac{1}{A-i}\varphi \right\rangle.$$

Similarly one can prove that

$$\langle G(1/t)\varphi_c, \frac{1}{A+i}\varphi\rangle = t^{-\theta}\langle \varphi_c, \frac{1}{A+i}\varphi\rangle,$$

and the lemma is proven.

$\square$

It has been shown during the proof of the latter theorem that every homogeneous extension of the functional φ is defined by the real parameter c. It follows that every homogeneous extension necessarily defines a real extension of the quadratic form $Q[\psi, \psi] = \langle \psi, (A/(A^2+1))\,\psi\rangle$.

If the unitary group G consists of only two elements

$$G = \{G(1), G(-1)\}$$

then the homogeneous extension can always be constructed and it is unique. This condition is true for example for the first derivative operator and Dirac operators in one dimension with the delta potential. The group of the unitary transformations coincides with the symmetry group with respect to the origin. These operators are studied at the end of this chapter.

Lemma 1.3.2 implies that if the original operator A and the vector $\varphi \in \mathcal{H}_{-2}(A) \setminus \mathcal{H}_{-1}(A)$ are homogeneous and if the corresponding function $f(t)$ is constant, then there exists a unique self–adjoint operator corresponding to the formal rank one perturbation and possessing the same symmetry properties. Therefore in this case the unique self–adjoint operator can be determined even if the rank one perturbation is not form bounded, but we have to use extra assumptions to determine this operator. The function $f(t)$ is not always constant. For example consider the following operator with rank one singular perturbation

$$-\Delta + \alpha\langle\delta, \cdot\rangle\delta,$$

where Δ is the Laplace operator in $L_2(\mathbf{R}^n)$ and δ is the delta function with the support at the origin. If $n = 1, 3$ then $f(t)$ is equal to a constant and the homogeneous extension can be constructed (see Section 1.5). If $n = 2$ then the function $f(t)$ is not constant and no homogeneous extension exists.

1.3.4   Resolvent formulas

The resolvent of the perturbed operator can be calculated explicitly using the general Krein formula (1.31) and taking into account (1.54)

$$\frac{1}{A_\alpha - z} = \frac{1}{A - z} - \frac{1}{1/\alpha + c + \langle \varphi, \frac{1+Az}{A-z}\frac{1}{A^2+1}\varphi\rangle} \left\langle \frac{1}{A-\bar z}\varphi, \cdot \right\rangle \frac{1}{A-z}\varphi. \quad (1.59)$$

The parameter c which appears in the latter formula can be chosen arbitrary for $\mathcal{H}_{-2}(A) \setminus \mathcal{H}_{-1}(A)$ perturbations. For $\mathcal{H}_{-1}(A)$ perturbations instead c is determined according to

$$c = \left\langle \varphi, \frac{A}{A^2 + 1} \varphi \right\rangle.$$

Let us introduce the following Nevanlinna function

$$F(z) = c + \left\langle \varphi, \frac{1 + Az}{A - z} \frac{1}{A^2 + 1} \varphi \right\rangle$$

$$= c + Q(z), \tag{1.60}$$

where $Q(z)$ is the Q-function associated with the operator A and the vector $\varphi \in \mathcal{H}_{-2}(A)$. Using this notation formula (1.59) just coincides with the formula for the resolvent of rank one bounded perturbation (1.2).

If $\varphi \in \mathcal{H}_{-1}(A)$ then the function $F(z)$ is given by

$$F(z) = c + Q(z) = \left\langle \varphi, \frac{A}{A^2 + 1} \varphi \right\rangle + \left\langle \varphi, \frac{1 + Az}{A - z} \frac{1}{A^2 + 1} \varphi \right\rangle = \left\langle \varphi, \frac{1}{A - z} \varphi \right\rangle$$

which is again formula (1.3). For $\varphi \in \mathcal{H}_{-2}(A) \setminus \mathcal{H}_{-1}(A)$ the function $F(z)$ can be calculated using the extension φ_c of the functional φ as follows

$$F(z) = \left\langle \varphi_c, \frac{1}{A - z} \varphi \right\rangle.$$

Let us introduce the function

$$F_\alpha(z) = \left\langle \varphi_c, \frac{1}{A_\alpha - z} \varphi_c \right\rangle$$

describing the rank one perturbations of the operator A_α.

All five critical formulas for the rank one perturbation [904] can be written in the same form for bounded, form bounded and form unbounded perturbations

$$F_\alpha(z) = \frac{F(z)}{1 + \alpha F(z)}; \tag{1.61}$$

$$\frac{1}{A_\alpha - z} \varphi = \frac{1}{1 - \alpha F(z)} \frac{1}{A - z} \varphi; \tag{1.62}$$

$$\frac{1}{A_\alpha - z} = \frac{1}{A - z} - \frac{\alpha}{1 + \alpha F(z)} \left(\frac{1}{A - \bar{z}} \varphi, \cdot \right) \frac{1}{A - z} \varphi; \tag{1.63}$$

$$\mathrm{Tr} \left[\frac{1}{A - z} - \frac{1}{A_\alpha - z} \right] = \frac{d}{dz} \ln(1 + \alpha F(z)). \tag{1.64}$$

$$\int_{\mathbf{R}} [d\mu_\alpha(E)] d\alpha = dE, \tag{1.65}$$

where μ_α is the spectral measure corresponding to the operator A_α. The formulas can be proven following the main lines of Section 1.1.1. We note that the result does not depend on the parameter c, which can be chosen arbitrary for form unbounded perturbations.

We are now going to prove that if the operators A and B are two self–adjoint extensions of one symmetric densely defined operator A^0 having deficiency indices $(1,1)$, then the operator B is a rank one singular perturbation of A.

Theorem 1.3.3 *Let A^0 be a densely defined symmetric operator with the deficiency indices $(1,1)$. Let A and B be two self–adjoint extensions of the operator A^0. Then the operator B is a rank one singular perturbation of the operator A.*

Proof Theorem 1.2.1 implies that the resolvents of the operators A and B are related by formula (1.31), where g_i is the deficiency element for A^0 at the point i. Consider the vector φ given by

$$\varphi = (A - i)g_i.$$

Since g_i belongs to the Hilbert space, the vector φ is an element of the Hilbert space $\mathcal{H}_{-2}(A)$. The closure of the operator A^0 then coincides with the restriction of the operator A to the domain of functions orthogonal to the vector φ. Obviously the vector φ does not belong to the Hilbert space, since the operator A^0 is densely defined.

To prove the theorem it is enough to show that there exists a real constant α such that

$$B = A + \alpha \langle \varphi, \cdot \rangle \varphi.$$

If the parameter γ appearing in Krein's formula is infinite, then the operators A and B coincide. It follows that the operator B is a rank one perturbation of the operator A with the coupling constant α equal to zero.

If the parameter γ is finite, $\gamma \neq \infty$, then we have to distinguish between form bounded and form unbounded perturbations. Suppose that $\varphi \in \mathcal{H}_{-1}(A)$. It follows from Theorem 1.3.1 that the operator A_α coincides with the operator B if

$$\alpha = \frac{-1}{\gamma + \langle \varphi, \frac{A}{A^2+1}\varphi \rangle}.$$

Suppose now that $\varphi \in \mathcal{H}_{-2}(A) \setminus \mathcal{H}_{-1}(A)$. To define the corresponding rank one perturbations we fix the real parameter $c = \langle \varphi_c, (A/(A^2+1))\varphi \rangle$.

Then the operator $A_\alpha = A + \alpha \langle \varphi_c, \cdot \rangle \varphi$ coincides with B if the coupling constant is chosen equal to

$$\alpha = \frac{-1}{\gamma + c}.$$

The theorem is proven.

□

The same result holds in the case when the symmetric operator A^0 is not densely defined. The vector φ belongs to the original Hilbert space in this case and all except one self–adjoint extension of the restricted operator A^0 are defined on the same domain. The exceptional extension is not an operator but an operator relation. It corresponds to the infinite value of the coupling constant α (see Section 1.1.2).

The latter theorem implies that the self–adjoint extensions of any symmetric operator with unit deficiency indices can be considered as rank one perturbations of a self–adjoint operator and can therefore be parametrized by the additive parameter α instead of the nonadditive parameters γ and v appearing in the boundary conditions.

1.4 Approximations of singular rank one perturbations

1.4.1 Norm convergence of the approximations

We are going to discuss how to approximate singular rank one perturbations by bounded ones. More precisely we consider operators

$$A_\alpha = A + \alpha \langle \varphi, \cdot \rangle \varphi \tag{1.66}$$

given by a singular perturbation, i.e. $\varphi \in \mathcal{H}_{-2}(A) \setminus H$. Let $\varphi_n \in H$ be a sequence of functions from the Hilbert space. Consider the sequence of operators with bounded rank one perturbations

$$A_\alpha^n = A + \alpha \langle \varphi_n, \cdot \rangle \varphi_n. \tag{1.67}$$

The self–adjoint operators A_α and A_α^n have in general different domains, since $\mathrm{Dom}\,(A_\alpha^n) = \mathrm{Dom}\,(A)$. But one can consider linear operators defined by (1.66) and (1.67) in the generalized sense. Two different types of convergence will be studied.

Considering A_α and A_α^n only as self-adjoint operators in the Hilbert space we can study the corresponding resolvent operators, which are bounded operators and therefore have common domain H. We say that the operators A_α^n converge to A_α **in the strong resolvent sense** if and only if

$$\lim_{n \to \infty} \left\| \frac{1}{A_\alpha^n - z} - \frac{1}{A_\alpha - z} \right\| = 0 \qquad (1.68)$$

for some $z, \Im z \neq 0$.

Considering the linear operators defined by formal expressions (1.66) and (1.67) in the generalized sense suppose that these operators can be defined as bounded linear operators on a certain normed space D with the range in perhaps a different normed space D'. We say that the operators A_α^n converge to the operator A_α **in the sense of linear operators** if and only if

$$\text{D} \supset \text{Dom}\,(A_\alpha),$$
$$\text{D} \supset \text{Dom}\,(A) = \text{Dom}\,(A_\alpha^n), \qquad (1.69)$$

and the following limit holds

$$\| A_\alpha^n - A_\alpha \|_{B(\text{D} \to \text{D}')} \to_{n \to \infty} 0, \qquad (1.70)$$

where $\| \cdot \|_{B(\text{D} \to \text{D}')}$ denotes the norm of the linear operator acting on D with the range in D'. Note that the operators A_α^n, A_α defined in the original Hilbert space are not necessarily bounded, but these operators could be bounded as operators mapping D to D'.

Consider first the approximations in the sense of linear operators. We start by investigating the question of how to approximate arbitrary $\varphi \in \mathcal{H}_{-2}(A) \setminus H$ by vectors from the Hilbert space.

Lemma 1.4.1 *Let f be an element from $H \setminus \mathcal{H}_2(A)$ and φ be an element from $\mathcal{H}_{-2}(A)$; then for any c there exists a sequence φ_n of elements from H converging to φ in the $\mathcal{H}_{-2}(A)$ norm such that $\langle f, \varphi_n \rangle$ converges to c.*

Proof The original Hilbert space $H = \mathcal{H}_0(A)$ is dense in $\mathcal{H}_{-2}(A)$. It follows that there exists a sequence $\tilde{\varphi}_n$ of elements from H converging in the $\mathcal{H}_{-2}$ norm to φ. If the sequence $\langle f, \tilde{\varphi}_n \rangle = a_n$ converges to c, then the lemma is proven. If it does not then let us consider a sequence $\psi_n \in \mathcal{H}_0(A)$ with unit $\mathcal{H}_{-2}$ norm $\| \psi_n \|_{\mathcal{H}_{-2}} = 1$ such that $|\langle f, \psi_n \rangle|$ diverges to ∞. Such a sequence exists because $f \notin \mathcal{H}_2(A)$. We can then choose a subsequence such that $c - a_n / \langle f, \psi_n \rangle \to 0$. We keep the same notation for the chosen subsequence. Consider then the sequence

$$\varphi_n = \tilde{\varphi}_n + \frac{c - a_n}{\langle f, \psi_n \rangle} \psi_n.$$

The following estimates are valid

$$\| \varphi_n - \varphi \|_{\mathcal{H}_{-2}} \leq \| \tilde{\varphi}_n - \varphi \|_{\mathcal{H}_{-2}} + \left| \frac{c - a_n}{\langle f, \psi_n \rangle} \right|.$$

It follows that φ_n converge to φ in the $\mathcal{H}_{-2}$ norm. At the same time the sequence $\langle f, \tilde{\varphi}_n \rangle = a_n + c - a_n = c$ obviously converges to c, hence the lemma is proven.

$\square$

The convergence in $\mathcal{H}_{-2}(A)$ was crucial for the proof of the lemma. For example the following lemma is valid.

Lemma 1.4.2 *Let f be an element from $\mathcal{H}_1(A)$ and φ be an element from $\mathcal{H}_{-1}(A)$. Then for every sequence $\varphi_n \in \mathcal{H}_0(A)$ converging to φ in the norm of $\mathcal{H}_{-1}(A)$ the sequence $\langle f, \varphi_n \rangle$ converges to $\langle f, \varphi \rangle$.*

Proof The statement of the lemma follows from the fact that strong convergence of bounded functionals implies weak convergence.

$\square$

The operators A_α and A_α^n are defined on the common domain $H_\varphi(A) = \mathrm{Dom}\,(A^{0*})$:

$$\mathrm{Dom}\,(A^{0*}) \supset \mathrm{Dom}\,(A_\alpha),$$

$$\mathrm{Dom}\,(A^{0*}) \supset \mathrm{Dom}\,(A) = \mathrm{Dom}\,(A_\alpha^n).$$

The range of the linear operators A_α^n, A_α belongs to the space $\mathcal{H}_{-2}(A)$ with the standard norm.

Theorem 1.4.1 *Let the sequence $\varphi_n \in H$ converge to φ in $\mathcal{H}_{-2}(A)$ and $\langle \varphi_n, (A/(A^2 + 1))\, \varphi \rangle$ converge to c. Then the sequence of linear operators $A_\alpha^n = A + \alpha \langle \varphi_n, \cdot \rangle \varphi_n$ defined on the domain $H_\varphi(A)$ converges in the operator norm to the operator A_α.*

Proof Consider an arbitrary element $g = \tilde{g} + b(g)\, (A/(A^2 + 1))\, \varphi \in H_\varphi(A)$.

Then the following estimates are valid

$$\| (A_\alpha^n - A_\alpha)g \|_{-2}$$

$$= |\alpha| \, \| \, \langle \varphi_n, g \rangle \varphi_n - \langle \varphi_c, g \rangle \varphi \, \|_{-2}$$

$$= |\alpha| \Big\| \langle \varphi_n, \tilde{g} \rangle \varphi_n + b(g) \Big\langle \varphi_n, \frac{A}{A^2+1} \varphi \Big\rangle \varphi_n$$

$$- \langle \varphi, \tilde{g} \rangle \varphi - b(g) \Big\langle \varphi_c, \frac{A}{A^2+1} \varphi \Big\rangle \varphi \Big\|_{-2}$$

$$\leq |\alpha| \{ |\langle \varphi_n, \tilde{g} \rangle - \langle \varphi, \tilde{g} \rangle| \, \| \, \varphi_n \, \|_{-2} + |\langle \varphi, \tilde{g} \rangle| \, \| \, \varphi_n - \varphi \, \|_{-2}$$

$$+ |b(g)| \, \Big| \Big\langle \varphi_n, \frac{A}{A^2+1} \varphi \Big\rangle - c \Big| \, \| \, \varphi_n \, \|_{-2} + |b(g)| \, |c| \, \| \, \varphi_n - \varphi \, \|_{-2} \}$$

$$\leq |\alpha| \{ \| \, \varphi_n \, \|_{-2} \| \, \varphi_n - \varphi \, \|_{-2} \| \, \tilde{g} \, \|_2 + \| \, \varphi_n - \varphi \, \|_{-2} \| \, \varphi \, \|_{-2} \| \, \tilde{g} \, \|_2$$

$$+ \Big| \Big\langle \varphi_n, \frac{A}{A^2+1} \varphi \Big\rangle - c \Big| \, \| \, \varphi_n \, \|_{-2} \, |b(g)| + |c| \, \| \, \varphi_n - \varphi \, \|_{-2} \, |b(g)| \}$$

$$\leq |\alpha| \{ (\| \, \varphi_n \, \|_{-2} + \| \, \varphi \, \|_{-2} + |c|) \, \| \, \varphi_n - \varphi \, \|_{-2}$$

$$+ \| \, \varphi_n \, \|_{-2} \, \Big| \Big\langle \varphi_n, \frac{A}{A^2+1} \varphi \Big\rangle - 2c \Big| \} \, \| \, g \, \|_{H_\varphi(A)} \, .$$

The sequence φ_n converges to φ in the $\mathcal{H}_{-2}(A)$ norm, the sequence $\| \varphi_n \|_{-2}$ is bounded and the sequence $\langle \varphi_n, (A/(A^2+1)) \varphi \rangle$ converges to c. It follows that the linear operators converge in the operator norm.

$\square$

Theorem 1.4.2 *Let $\varphi \in \mathcal{H}_{-2}(A) \setminus H$. Then there exists a sequence $\varphi_n \in H$ converging to φ in the $\mathcal{H}_{-2}(A)$ norm such that the sequence of linear operators $A_\alpha^n = A + \alpha \langle \varphi_n, \cdot \rangle \varphi_n$ defined on the domain $H_\varphi(A)$ converges in the operator norm to the operator $A_\alpha = A + \alpha \langle \varphi_c, \cdot \rangle \varphi$.*

Proof The element $(A/(A^2+1)) \varphi$ belongs to the Hilbert space but does not belong to the domain of the operator. It follows from Lemma 1.4.1 that there exists a sequence φ_n converging to φ in the $\mathcal{H}_{-2}(A)$ norm and such that $\langle \varphi_n, (A/(A^2 + 1)) \varphi \rangle$ converge to c. It follows from Theorem 1.4.1 that the operators A_α^n converge to A_α in the operator norm.

□

The approximating sequence φ_n can be constructed using the spectral representation of the original operator A. If the element $\varphi \in \mathcal{H}_{-2}(A)$ then there exists a certain measure $d\mu(\lambda)$ such that

$$\left\langle \frac{1}{A-i}\varphi, \frac{1+zA}{A-z}\frac{1}{A-i}\varphi \right\rangle = \int_{-\infty}^{+\infty} \frac{1+z\lambda}{\lambda-z}\frac{1}{\lambda^2+1}d\mu(\lambda)$$

and $\int_{-\infty}^{+\infty} \frac{d\mu(\lambda)}{\lambda^2+1} < \infty$. Consider the spaces $\mathcal{H}_{-1,-2}(A)$ and $\mathcal{H}_{-2,-1}(A)$ formed by the elements from $\mathcal{H}_{-2}(A)$ satisfying the following additional conditions

$$\int_{-\infty}^{0} \frac{|\lambda|d\mu(\lambda)}{\lambda^2+1} < \infty \quad \text{and} \quad \int_{0}^{\infty} \frac{|\lambda|d\mu(\lambda)}{\lambda^2+1} < \infty$$

respectively. The following lemma can be proven.

Lemma 1.4.3 *Let $\varphi \in \mathcal{H}_{-2}(A) \setminus (\mathcal{H}_{-1,-2}(A) \cup \mathcal{H}_{-2,-1}(A))$ then there exist two sequences $c_n, d_n \to \infty$ such that*

$$\lim_{n\to\infty} \int_{-c_n}^{d_n} \frac{\lambda}{\lambda^2+1}d\mu(\lambda) = c.$$

Proof Convergence of the integral $\int_\infty^{+\infty} (1/(\lambda^2+1)) d\mu(\lambda)$ implies that the two sequences

$$(n+1)\int_{n}^{n+1} \frac{1}{\lambda^2+1}d\mu(\lambda),$$

and

$$(n+1)\int_{-n-1}^{-n} \frac{\lambda}{\lambda^2+1}d\mu(\lambda),$$

$n = 1, 2, ...$, have zero limits when $n \to \infty$. The sums of both sequences are diverging, since

$$\int_{n}^{n+1} \frac{\lambda}{\lambda^2+1}d\mu(\lambda) \le (n+1)\int_{n}^{n+1} \frac{1}{\lambda^2+1}d\mu(\lambda),$$

$$\int_{-n}^{-n-1} \frac{|\lambda|}{\lambda^2+1}d\mu(\lambda) \le (n+1)\int_{-n}^{-n-1} \frac{1}{\lambda^2+1}d\mu(\lambda).$$

The sequences have different signs. It follows that the sequence

$$\int_{-n}^{m} \frac{\lambda}{\lambda^2+1}d\mu(\lambda)$$

can converge to any real number when $n, m \to \infty$. The lemma is proven.

□

Consider the approximating sequence of the elements from the Hilbert space $\varphi_n = E_{(-c_n, d_n)}(A)\varphi$, where $E(A)$ denotes the spectral projector for the operator A. The following limit holds

$$\lim_{n \to \infty} \left\langle \varphi_n, \frac{A}{A^2 + 1} \varphi_n \right\rangle = c.$$

The sequence φ_n will be used in the following section to construct the approximations of rank one perturbations in the strong resolvent sense.

1.4.2 Strong resolvent convergence of the approximations

In this section we study the strong resolvent convergence of the operators. We have shown in fact that the difference of the resolvents of the original and perturbed operators has rank one. We prove first that every rank one $\mathcal{H}_{-1}$ perturbation can be approximated in the strong resolvent sense.

Theorem 1.4.3 *Let A be a self–adjoint operator in the Hilbert space $\mathcal{H}$ and φ be an element from $\mathcal{H}_{-1}(A)$. Let the sequence $\varphi_n \in \mathcal{H}$ converge to φ in the norm $\mathcal{H}_{-1}(A)$. Then the sequence of operators $A_\alpha^n = A + \alpha \langle \varphi_n, \cdot \rangle \varphi_n$ converges to the operator $A_\alpha = A + \alpha \langle \varphi, \cdot \rangle \varphi$ in the strong resolvent sense for all z, $\Im z \neq 0$.*

Proof Since the $\{1/(A_\alpha^n - z)\}$ are uniformly bounded it is enough to prove the weak convergence of the resolvents. Consider two arbitrary vectors ψ_1, ψ_2 from the Hilbert space. The convergence in the space $\mathcal{H}_{-1}(A)$ implies

$$\lim_{n \to \infty} \left\langle \frac{1}{A - \bar{z}}(\varphi - \varphi_n), \psi_1 \right\rangle = 0; \tag{1.71}$$

$$\lim_{n \to \infty} \left\langle \psi_2, \frac{1}{A - z}(\varphi_n - \varphi) \right\rangle = 0. \tag{1.72}$$

Moreover the quadratic form of the resolvent converges at the point $z = i$ and similarly at all other points in the resolvent set of A and we have

$$\left| \left\langle \varphi_n, \frac{1}{A - z} \varphi_n \right\rangle - \left\langle \varphi, \frac{1}{A - z} \varphi \right\rangle \right|$$

$$\leq 2 \, \| \varphi_n - \varphi \|_{-1} (\| \varphi_n \|_{-1} + \| \varphi \|_{-1}) \to 0.$$

We have for the difference of the resolvents

$$\left\langle \psi_2, \frac{1}{A_\alpha^n - z} \psi_1 \right\rangle - \left\langle \psi_2, \frac{1}{A_\alpha - z} \psi_1 \right\rangle$$

$$= \frac{\alpha}{1 + \alpha \langle \varphi, \frac{1}{A-z} \varphi \rangle} \left\langle \frac{1}{A - \bar{z}} \varphi, \psi_1 \right\rangle \left\langle \psi_2, \frac{1}{A - z} \varphi \right\rangle$$

$$- \frac{\alpha}{1 + \alpha \langle \varphi_n, \frac{1}{A-z} \varphi_n \rangle} \left\langle \frac{1}{A - \bar{z}} \varphi_n, \psi_1 \right\rangle \left\langle \psi_2, \frac{1}{A - z} \varphi_n \right\rangle.$$

The weak resolvent convergence follows from the formulas (1.71),(1.72) and the convergence of the quadratic form of the resolvent. The denominator in the first quotient does not vanish because $\Im z \neq 0$.

□

Let us study rank one $\mathcal{H}_{-2}$ perturbations.

Theorem 1.4.4 *Let A be a self–adjoint operator and φ be a functional from $\mathcal{H}_{-2}(A), \| (1/(A - i)) \varphi \| = 1$. Let φ_n be any sequence from the Hilbert space converging to φ in $\mathcal{H}_{-2}(A)$ and let $\lim_{n\to\infty} \langle \varphi_n, (A/(A^2 + 1)) \varphi_n \rangle = c$. Then the sequence of self–adjoint operators*

$$A_\alpha^n = A + \alpha \langle \varphi_n, \cdot \rangle \varphi_n$$

converges to A_α in the strong resolvent sense. If

$$\lim_{n\to\infty} \left| \left\langle \varphi_n, \frac{A}{A^2 + 1} \varphi_n \right\rangle \right| = \infty,$$

the operators A_α^n converge to the original operator in the strong resolvent sense.

Proof The first part of the theorem can be proven using the fact that the convergence in $\mathcal{H}_{-2}$ implies weak convergence of the resolvents and formulas (1.71),(1.72) hold for every $\psi_1, \psi_2 \in \mathcal{H}$. Calculations similar to thoes carried out during the proof of Theorem 1.4.3 lead to the result which has to be proven. One has to take into account only that

$$\lim_{n\to\infty} \left\langle \varphi_n, \frac{1}{A-z}\varphi_n \right\rangle$$

$$= \lim_{n\to\infty} \left\langle \varphi_n, \frac{1+Az}{A-z}\frac{1}{A^2+1}\varphi_n \right\rangle + \lim_{n\to\infty} \left\langle \varphi_n, \frac{A}{A^2+1}\varphi_n \right\rangle$$

$$= \left\langle \varphi, \frac{1+Az}{A-z}\frac{1}{A^2+1}\varphi \right\rangle + c$$

$$= c + Q(z)$$

$$= F(z),$$

where $F(z)$ appeared in (1.60). Consider now the case where

$$\lim_{n\to\infty} |(\varphi_n, \frac{1}{A-i}\varphi_n)| = \infty.$$

The difference of the resolvents of the original operator and its rank one perturbation is the rank one operator

$$\frac{1}{A_\alpha^n - z} - \frac{1}{A-z} = -\frac{\alpha}{1+\alpha\langle \varphi_n, \frac{1}{A-z}\varphi_n \rangle} \left\langle \frac{1}{A-\bar{z}}\varphi_n, \cdot \right\rangle \frac{1}{A-z}\varphi_n.$$

The first term on the right hand side of the last equality converges to zero. It follows that the difference of the resolvents converges weakly to zero since $(1/(A-z))\,\varphi_n$ and $(1/(A-\bar{z}))\,\varphi_n$ converge to $(1/(A-z))\,\varphi$ and $(1/(A-\bar{z}))\,\varphi$ respectively. Hence the theorem is proven.

□

Let $\varphi \in \mathcal{H}_{-2}(A) \setminus (\mathcal{H}_{-1,-2}(A) \cup \mathcal{H}_{-2,-1}(A))$. Then it is possible to construct a sequence of operators converging to A_α in the strong resolvent sense, according to the following

Theorem 1.4.5 *Let $\varphi \in \mathcal{H}_{-2}(A) \setminus (\mathcal{H}_{-1,-2}(A) \cup \mathcal{H}_{-2,-1}(A))$. Then there exist two sequences $c_n, d_n \to \infty$ such that $\varphi_n = E_{(-c_n,d_n)}(A)\varphi$ determines the sequence of self-adjoint operators $A_\alpha^n = A + \alpha\langle \varphi_n, \cdot \rangle \varphi_n$ involving bounded perturbations of A converging to the perturbed operator $A_\alpha = A + \alpha\langle \varphi, \cdot \rangle\varphi$ in the strong resolvent sense.*

Proof The statement follows easily from Lemma 1.4.3 and Theorem 1.4.4.

□

The latter theorem shows how to construct the approximating sequence φ_n leading to the approximations of the operator A_α in the strong resolvent sense.

If $\varphi \in \mathcal{H}_{-1}(A)$ then the sequence $\varphi_n \in H$ converging to φ in the $\mathcal{H}_{-1}$ norm defines a sequence of self–adjoint operators converging to the perturbed operator A_α in the strong resolvent sense. If $\varphi \in \mathcal{H}_{-2}(A) \backslash (\mathcal{H}_{-1,-2}(A) \cup \mathcal{H}_{-2,-1}(A))$ then there exists a sequence φ_n converging to φ in the $\mathcal{H}_{-2}$ norm such that the sequence of the corresponding perturbed operators converges to A_α in the strong resolvent sense. If $\varphi \in \mathcal{H}_{-1,-2}(A) \backslash \mathcal{H}_{-1}(A)$ or $\varphi \in \mathcal{H}_{-2,-1}(A) \backslash \mathcal{H}_{-1}(A)$ then every sequence φ_n converging to φ in the $\mathcal{H}_{-2}$ norm defines a sequence of self–adjoint operators converging to the original operator in the strong resolvent sense. It follows that not every form unbounded rank one perturbation can be approximated in the strong resolvent sense by operators with bounded perturbations. For example if the original operator A is semibounded, then the subspace $\mathcal{H}_{-2}(A) \backslash (\mathcal{H}_{-1,-2}(A) \cup \mathcal{H}_{-2,-1}(A))$ is trivial and no form unbounded perturbation of such an operator can be approximated in the strong resolvent sense without the renormalization of the coupling constant. See Section 1.5.1, where such an approximation with the renormalized coupling constant is constructed for the Laplace operator with the delta interaction in $\mathbf{R}^3$.

Approximations in the sense of linear operators can be constructed for every rank one perturbation. If the perturbation is form bounded then every sequence φ_n converging to φ in the $\mathcal{H}_{-1}$ norm determines a sequence of operators converging to the perturbed operator in the norm of linear operators. If $\varphi \in \mathcal{H}_{-2}(A) \backslash \mathcal{H}_{-1}(A)$ then one can prove only the existence of the approximating sequence.

1.5 Differential operators with rank one singular perturbations

1.5.1 Point interactions in dimension three

We consider now the Schrödinger operator in dimension three defined by the heuristic expression:

$$L_\alpha = -\Delta + \alpha \delta, \qquad (1.73)$$

where Δ is the Laplace operator, α is a real coupling constant and δ is a Dirac delta function in dimension three. This operator was studied for the first time from the mathematical point of view by F.A.Beresin and L.D.Faddeev [135]. The operator L_α to be defined in $L_2(\mathbf{R}^3)$ can be considered as a singular rank one perturbation of the Laplace operator because $\delta\varphi = \varphi(0)\delta = \langle\varphi,\delta\rangle\delta$ and the generalized function δ is an element from $\mathcal{H}_{-2}(-\Delta) \setminus \mathcal{H}_{-1}(-\Delta)$ in three dimensional space. Consider the group $S(t), t > 0$, of scaling transformations of $L_2(\mathbf{R}^3)$ defined as follows: for every function $\psi \in D$ and distribution f

$$(S(t)\psi)(x) = t^{3/2}\psi(tx);$$

$$\langle S(t)f,\psi\rangle = \langle f, S(1/t)\psi\rangle.$$

The Laplace operator and the delta function are homogeneous with respect to the group $S(t)$:

$$S(t)\Delta = t^2 \Delta S(c);$$

$$S(t)\delta = t^{-3/2}\delta.$$

The perturbed operator coincides with one of the self–adjoint extensions of the symmetric Laplace operator $-\Delta_0$ defined on functions from $W_2^2(\mathbf{R}^3)$ vanishing at the origin. The domain Dom $(-\Delta_0^*)$ of the adjoint operator $-\Delta_0^*$ coincides with the space $W_2^2(\mathbf{R}^3 \setminus \{0\})$. The distribution δ possesses a unique extension to the set $W_2^2(\mathbf{R}^3 \setminus \{0\})$. To calculate the parameter c defining the extension of the functional $\varphi = \delta$ one has to take into account that the vector δ does not fulfil the normalization condition $\| (1/(A - i))\,\varphi \| = 1$. Formula (1.58) should be modified as follows

$$\left\langle \varphi_c, \frac{A}{A^2 + 1}\varphi \right\rangle = c = f(t) - i\left\|\frac{1}{A - i}\varphi\right\|^2.$$

Therefore the parameter c is equal to

$$\begin{aligned}
c &= i\frac{1 - t^2}{1 - t}\left\langle \frac{1}{-\Delta + i}\delta, \frac{1}{-\Delta - t^2 i}\delta \right\rangle - i\left\|\frac{1}{-\Delta - i}\delta\right\| \\[2mm]
&= i(1 + t)\left\langle \frac{e^{i\sqrt{-i}|x|}}{4\pi|x|}, \frac{e^{it\sqrt{i}|x|}}{4\pi|x|} \right\rangle - i\left\langle \frac{e^{i\sqrt{i}|x|}}{4\pi|x|}, \frac{e^{i\sqrt{i}|x|}}{4\pi|x|} \right\rangle \\[2mm]
&= -\frac{1}{4\pi\sqrt{2}}.
\end{aligned}$$

Any function ψ belongs to the domain of the adjoint operator $\psi \in$ Dom $(-\Delta_0^*)$ if and only if

$$\psi(x) = \tilde{\psi}(x) + \frac{b(\psi)}{2}\left(\frac{e^{(-1/\sqrt{2}+i/\sqrt{2})|x|}}{4\pi|x|} + \frac{e^{(-1/\sqrt{2}-i/\sqrt{2})|x|}}{4\pi|x|} \right),$$

where $\tilde{\psi} \in \text{Dom}\,(-\Delta) = W_2^2(\mathbf{R}^3), b(\psi) \in \mathbf{C}$. Using the homogeneous extension of the delta functional we define the parameter γ which describes the
self–adjoint extension using (1.54)

$$\gamma = -\frac{1}{\alpha} + \frac{1}{4\pi\sqrt{2}}.$$

Therefore the self–adjoint operator corresponding to the formal expression
(1.73) is the restriction of the adjoint operator to the domain of functions
satisfying the boundary condition

$$\langle \delta, \tilde{\psi} \rangle = \left(-\frac{1}{\alpha} + \frac{1}{4\pi\sqrt{2}} \right) b(\psi). \tag{1.74}$$

Let us consider this extension of the linear functional δ in more detail to
underline the main ideas of the calculations. Every function $\psi \in W_2^2(\mathbf{R}^3\backslash\{0\})$
is continuous outside the origin and has the following asymptotics there

$$\psi(x) =_{x\to 0} \frac{\psi_-}{4\pi|x|} + \psi_0 + o(1), \tag{1.75}$$

where the boundary values ψ_-, ψ_0 are equal to

$$\begin{cases} \psi_- = a(\psi), \\ \psi_0 = -a(\psi)/4\pi\sqrt{2} + \tilde{\psi}(0). \end{cases}$$

The linear operator (1.73) is not defined on all such functions. The distribution $\varphi = \delta$ should be extended to the set of all functions having the
asymptotics (1.75). We denote by E the set of all $C^\infty(\mathbf{R}^3 \setminus \{0\})$ functions
with compact support having the asymptotic behaviour (1.75) at the origin.
Convergence in this space is defined using an arbitrary $C_0^\infty(\mathbf{R}^3)$ function χ
equal to one in some neighbourhood of the origin.

Definition 1.5.1 *A sequence* $\{\psi_n\}$ *of functions from E is said to converge
to a function* $\psi \in E$ *if and only if:*

1. $\lim_{n\to\infty} \psi_{n-} = \psi_-$

2. There exists a bounded domain outside which all the functions ψ_n vanish;

3. The sequence $\{\tilde{\psi}_n^{(k)}\}$ *of the regularized derivatives of order k :*

$$\tilde{\psi}_n^{(k)}(x) = (\psi_n(x) - \frac{\chi(x)\psi_{n-}}{4\pi|x|})^{(k)}$$

converges uniformly to

$$\tilde{\psi}^{(k)}(x) = (\psi(x) - \frac{\chi(x)\psi_-}{4\pi|x|})^{(k)}.$$

This definition does not depend on the choice of the function χ. The derivative of any function from E is defined pointwise everywhere outside the origin. We denote by E' the set of all bounded linear forms on E. The set E contains the standard set of test functions $D = C_0^\infty(\mathbf{R}^3)$.

The following lemma follows easily from Lemma 1.3.2.

Lemma 1.5.1 *Let the distribution $\tilde{\delta} \in E'$*

1. be equal to δ on the test functions from D;

2. be a homogeneous distribution;

then this distribution for every function $\psi \in E$ is equal to

$$\tilde{\delta}(\psi) = \psi_0. \tag{1.76}$$

This means that the distribution $\tilde{\delta}$ "does not feel" the singularity of the test function at the origin.

We are going to use the same notation δ for the delta distribution in D' and E' in what follows. This is justified because of the uniqueness of this extension under our assumptions.

Definition 1.5.2 *The delta distribution δ in E' with support at the origin is the linear functional on E defined by the formula (1.76).*

Following Section 1.3 we define the linear operator L_α on the whole Sobolev space $W_2^2(\mathbf{R}^3 \setminus \{0\}) = \mathrm{Dom}\,(-\Delta_0^*)$ using the closure. The corresponding self–adjoint operator, also denoted by L_α, is defined on the following domain

$$\mathrm{Dom}\,(L_\alpha) = \{\psi \in W_2^2(\mathbf{R}^3 \setminus \{0\}) : L_\alpha \psi \in L^2(\mathbf{R}^3)\}.$$

The latter inclusion has to be understood in the distributional sense with D as the set of test functions. It follows that every function ψ from the domain $\mathrm{Dom}(L_\alpha)$ should satisfy the following boundary condition

$$\psi_- + \alpha\psi_0 = 0. \tag{1.77}$$

The latter condition implies (1.74).

Relations with the Schrödinger operator on the half axis

Consider the subsets $E_r \subset E; D_r \subset D$ consisting of functions ψ in E, respectively D, which depend only on the absolute value of the coordinate, i.e. such that $\psi(x) = \psi(|x|)$. The corresponding distribution spaces will be denoted by E'_r, respectively D'_r. The space of square integrable functions on $\mathbf{R}^3$ depending on $|x|$ will be denoted by $L^2_r(\mathbf{R}^3)$. The transformation $T : \psi(|x|) \to \sqrt{4\pi}r\psi(r)$ acting from $L^2_r(\mathbf{R}^3)$ to $L^2(\mathbf{R}_+)$ preserves the L^2 norm. This transformation transforms the set of test functions D_r into the set $\mathbf{D}$ of $C^\infty(\mathbf{R}_+)$ functions with compact support and equal to zero at the origin. The set of test functions E_r is transformed into the set $\mathbf{E}$ of $C^\infty(\mathbf{R}_+)$ functions with compact support having the following asymptotics at the origin:

$$\psi(r) = \frac{\psi_-}{\sqrt{4\pi}} + r\sqrt{4\pi}\psi_0 + o(r).$$

The transformed linear operator $\mathbf{L}_\alpha = T L_\alpha T^{-1}$ is defined by the following formula:

$$
\begin{aligned}
\langle \varphi, \mathbf{L}_\alpha \psi \rangle_{L^2(\mathbf{R}_+)} &= \left\langle T^{-1}\varphi, L_\alpha T^{-1}\psi \right\rangle_{L^2(\mathbf{R}^3)} \\
&= \left\langle \varphi, -\frac{d^2}{dx^2}\psi \right\rangle_{L^2(\mathbf{R}_+)} + \alpha \bar\varphi_0 \psi_0 \\
&= \left\langle \varphi, \left(-\frac{d^2}{dx^2} + \frac{\alpha}{4\pi}\delta^{(1)}\langle \delta^{(1)}, \cdot \rangle \right)\psi \right\rangle_{L^2(\mathbf{R}_+)}
\end{aligned}
$$

Here $\delta^{(1)}$ denotes the derivative of the Dirac delta function, i.e. the functional defined on the functions from $\mathbf{E}$ as follows: $\langle \psi, \delta^{(1)} \rangle = -\psi'(0)$.

Thus the three dimensional delta potential is quite similar to the pseudopotential on the half axis equal to the projector P in $L^2(\mathbf{R}_+)$ into the derivative of the delta function, i.e. $(Pf) = \langle \delta^{(1)}, f \rangle \delta^{(1)}$. We remark that the element $\delta^{(1)}$ belongs to $\mathcal{H}_{-2}(-d^2/dx^2)$, where the operator $-d^2/dx^2$ is defined on the functions from $W_2^2(\mathbf{R}_+)$ which satisfy the Dirichlet boundary condition at the origin. (See Section 3.2 where point interactions of the second derivative operator in $L_2(\mathbf{R})$ are studied in more detail.)

Approximations of the delta potential

It follows from the previous consideration (see Section 1.4) that it is possible to construct an approximation of the operator L_α by rank one perturbations from $\mathcal{H}_0(L) = L_2(\mathbf{R}^3)$. In the case of the Laplace operator in dimension three such an approximation can be constructed explicitly. The sequence of approximations can be chosen from the set of infinitely differentiable functions with compact support. We discuss first the approximation of the operator L_α. Let ω be a $C_0^\infty(\mathbf{R}_+)$ real function with compact support and vanishing at the origin, normalized such that $\int_0^\infty \omega(x)dx = 1$. An approximation of the delta function can be constructed with the help of scaling. We use the following definition $\omega_\epsilon(x) = (1/\epsilon)\,\omega(x/\epsilon), x \in \mathbf{R}_+$. The following calculations show that the sequence $v_\epsilon(x) = d\omega_\epsilon(x)/dx$ converges when $\epsilon \to 0$ to the $\delta^{(1)}$ distribution for any function $\psi \in \mathbf{E}$:

$$\langle v_\epsilon, \psi \rangle = \int_0^\infty v_\epsilon(x)\psi(x)dx$$

$$= \int_0^\infty \omega_\epsilon'(x)\psi(x)dx$$

$$= -\int_0^\infty \omega_\epsilon(x)\psi'(x)dx + \omega_\epsilon\psi|_0^\infty.$$

The integral in the latter formula converges as $\epsilon \searrow 0$ to the value of the function ψ' at the origin. The nonintegral terms are equal to zero because the function ω has, by assumption, zero limits at the origin and at infinity. By closure this result can be extended to all $\psi \in \mathbf{E}$. It follows that for any function $\psi \in \mathbf{E}$ and any test function $\varphi \in \mathbf{D}$ the following limit holds

$$\lim_{\epsilon \to 0} \left\langle \varphi, (-\frac{d^2}{dx^2} + \frac{\alpha}{4\pi}v_\epsilon\langle v_\epsilon, \cdot \rangle)\psi \right\rangle = \langle \varphi, L_\alpha\psi \rangle.$$

Thus the sequence of operators

$$L_{\alpha,\epsilon} = -\frac{d^2}{dx^2} + \frac{\alpha}{4\pi}v_\epsilon\langle v_\epsilon, \cdot \rangle$$

converges to the operator L_α pointwise in the weak operator topology. An approximation of the operator L_α can be constructed using the same functional sequence v_ϵ. We choose a special (but "standard") delta functional sequence equal to

$$V_\epsilon(x) = \frac{-1}{4\pi}\frac{v_\epsilon(|x|)}{|x|}.$$

V_ϵ has compact support and it is easily verified that

$$\int_{\mathbf{R}^3} V_1(x)d^3x = -\int_0^\infty rv_1(r)dr = -\int_0^\infty r\omega'(r)dr$$
$$= -r\omega(r)|_0^\infty + \int_0^\infty \omega(r)dr = 1.$$

Moreover V_ϵ has the usual scaling properties:

$$\tilde{V}_\epsilon(x) = \frac{-1}{4\pi}\frac{v_\epsilon(|x|)}{|x|} = \frac{-1}{4\pi r}\frac{\partial}{\partial r}\omega_\epsilon(r)|_{r=|x|}$$
$$= \frac{-1}{4\pi r}\frac{\partial}{\partial r}\frac{1}{\epsilon}\omega\left(\frac{r}{\epsilon}\right)|_{r=|x|} = \frac{1}{\epsilon^3}V_1\left(\frac{x}{\epsilon}\right).$$

Finally, for any test function ψ continuous in a neighbourhood of the origin the following limit holds

$$\lim_{\epsilon\to 0} V_\epsilon(\psi) = \lim_{\epsilon\to 0}\int_{\mathbf{R}^3} V_\epsilon(x)\psi(x)d^3x = \psi(0).$$

Lemma 1.5.2 *Let ψ be any test function from E. Then the following limit holds*

$$\lim_{\epsilon\to 0}\langle V_\epsilon,\psi\rangle = \psi_0.$$

Proof Every function $\psi \in E$ possesses the following representation

$$\psi(x) = \left(\frac{\psi_-}{4\pi|x|} + \psi_0\right)\chi(x) + \tilde{\psi}(x),$$

where χ has compact support and is equal to one in a neighbourhood of the origin and satisfies

$$\tilde{\psi}(x) = o(1), \quad x \to 0.$$

Then the following limits exist:

$$\lim_{\epsilon\to 0}\int_{\mathbf{R}^3} d^3x V_\epsilon(x)\tilde{\psi}(x) = 0;$$

$$\lim_{\epsilon\to 0}\int_{\mathbf{R}^3} d^3x V_\epsilon(x)\chi(x)\psi_0 = \psi_0;$$

$$\lim_{\epsilon\to 0}\int_{\mathbf{R}^3} d^3x V_\epsilon(x)\frac{\chi(x)}{4\pi|x|}\psi_- = 0.$$

The last limit follows from the orthogonality of the functions $V_\epsilon(x)$ and $1/|x|$ in $L^2(\mathbf{R}^3)$. The lemma is proven.

□

Consider now the sequence of linear operators defined in the generalized sense

$$L_{\alpha,\epsilon} = -\Delta + \alpha V_\epsilon(x)\langle V_\epsilon(x),\cdot\rangle.$$

This sequence of linear operators $L_{\alpha,\epsilon}$ converges as $\epsilon \searrow 0$ to the operator L_α in the weak operator topology. We prove now that the sequence of linear operators $L_{\alpha,\epsilon}$ converges to the operator L_α in the operator norm. All these operators are defined on the domain $\mathrm{Dom}\,(\Delta_0^*)$ and their ranges belong to $\mathcal{H}_{-2}(-\Delta)$. The norms are defined by equations (1.43) and (1.16) respectively.

Lemma 1.5.3 *Let ω be an infinitely differentiable function with compact support on the positive half axis and assume $\omega(0) = 0$ and $\int_0^\infty \omega(r)dr = 1$. Then*

$$V_\epsilon(x) = \left(\frac{-1}{4\pi r} \frac{\partial}{\partial r} \frac{1}{\epsilon} \omega \left(\frac{r}{\epsilon} \right) \right) \Big|_{r=|x|}, \quad x \in \mathbf{R}^3$$

converges to δ in $\mathcal{H}_{-2}(-\Delta)$ when $\epsilon \searrow 0$.

Proof We have to prove that

$$\lim_{\epsilon \to 0} \left\| \frac{1}{-\Delta + 1} (\delta - V_\epsilon) \right\|_{L_2(\mathbf{R}^3)} = 0, \tag{1.78}$$

since the operator $-\Delta$ is positive. The Fourier transform $\hat{V}_\epsilon$ of the function V_ϵ can be calculated at any $p \in \mathbf{R}^3$:

$$\begin{aligned}
\hat{V}_\epsilon(p) &= \int_0^\infty dr r^2 \int_0^\pi d\theta \sin\theta e^{irp\cos\theta} \frac{-2\pi}{4\pi r} \frac{\partial}{\partial r} \frac{1}{\epsilon} \omega \left(\frac{r}{\epsilon} \right) \\
&= \int_0^\infty \cos rp \frac{1}{\epsilon} \omega \left(\frac{r}{\epsilon} \right) dr.
\end{aligned}$$

The function

$$\hat{V}_\epsilon(p) - 1 = \int_0^\infty (\cos rp - 1) \frac{1}{\epsilon} \omega(\frac{r}{\epsilon}) dr$$

is uniformly bounded and tends to zero uniformly on every compact domain $D \subset \mathbf{R}^3$. It follows that, with $g_\epsilon(p) \equiv (1/(p^2 + 1)) (\hat{V}_\epsilon(p) - 1)$,

$$\| g_\epsilon \|_{L_2(\mathbf{R}^3)} \to_{\epsilon \to 0} 0$$

and the limit (1.78) holds.

$\square$

Theorem 1.5.1 *The sequence of linear operators $L_{\alpha,\epsilon}$ converges in the operator norm to the linear operator L_α on $W_2^2(\mathbf{R}^3 \setminus \{0\})$.*

Proof This follows easily from Lemma 1.5.3 and Theorem 1.4.1.

Approximations with the renormalized coupling constant

The operator $-\Delta$ is positive and the functional δ belongs to $\mathcal{H}_{-2}(-\Delta) \setminus \mathcal{H}_{-1}(-\Delta)$. Therefore a point interaction in dimension three can be approximated in the strong resolvent sense or even norm resolvent sense only using a suitable renormalization of the coupling constant. Two approaches have been developed. In the first approach the operator with the point interaction is approximated by the sequence of operators

$$L_{\alpha,\epsilon} = -\Delta + \alpha(\epsilon)V_\epsilon(x).$$

The functions $V_\epsilon(x)$ are obtained from a certain function $V_1(x)$ by unitary scaling. In this approach the interaction term $\alpha\langle\delta,\cdot\rangle\delta = \alpha\delta$ is considered as a singular potential, not as a rank one operator. To get the norm resolvent convergence the coupling constant α should be chosen with a suitable dependence on the scaling parameter ϵ. This approach is described in detail in the book by S.Albeverio, F.Gesztesy, R.Hoegh-Krohn, and H.Holden [39].

We are going to describe here in more detail the second approach where the approximating sequence of operators is constructed using the spectral representation of the Laplace operator. This approach was first developed by F.A.Beresin and L.D.Faddeev [135]. Consider the sequence of functions $u_n(x)$ converging to the $\delta(x)$. The sequence can be constructed using the Fourier transformation which is just the spectral representation for the Laplace operator

$$\hat{u}_n(p) = \left\{ \begin{array}{ll} \frac{1}{(2\pi)^{3/2}}, & p^2 < n^2, \\ 0, & p^2 > n^2, \end{array} \right.$$

where $\hat{u}_n$ denotes the Fourier transform of the function u_n.

Obviously if $\psi \in C_0^\infty(\mathbf{R}^3)$ then

$$\langle\psi, u_n\rangle = \int_{\mathbf{R}^3} \psi(x)u_n(x)d^3x \rightarrow \psi(0) = \langle\psi, \delta\rangle.$$

Consider the operator

$$L^n\psi = -\Delta\psi + \alpha_n u_n(x) \int_{\mathbf{R}^3} \overline{u_n(y)}\psi(y)d^3y.$$

Let us calculate the resolvent of the operators in terms of the Fourier transform for some $z, \Im z \neq 0$

$$(L^n - z)\psi = f$$

$$\Rightarrow (p^2 - z)\hat{\psi}(p) + \alpha_n \hat{u}_n(p) \int \hat{u}_n(q)\hat{\psi}(q)d^3q = \hat{f}(p).$$

It follows that

$$\hat{\psi}(p) = \frac{\hat{f}(p)}{p^2 - z} - \alpha_n \frac{\hat{u}_n(p)}{p^2 - z} \int \hat{u}_n(q)\hat{\psi}(q)dq. \tag{1.79}$$

We multiply the latter equality by $\hat{u}_n(p)$ and integrate with respect to p to get the following equation

$$\int_{\mathbf{R}^3} \hat{u}_n(p)\hat{\psi}(p)d^3p = \frac{\int_{\mathbf{R}^3} \frac{\hat{f}(p)}{p^2-z}\hat{u}_n(p)d^3p}{1 + \alpha_n \int_{\mathbf{R}^3} \frac{(\hat{u}_n(p))^2}{p^2-z}d^3p}.$$

Finally we get the following formula for the resolvent

$$\hat{\psi}(p) = \frac{\hat{f}(p)}{p^2 - z} - \frac{\alpha_n}{1 + \alpha_n \int_{\mathbf{R}^3} \frac{\hat{u}_n(q)^2}{q^2 - z} d^3 q} \left(\int_{\mathbf{R}^3} \frac{\hat{f}(q)}{q^2 - z} \hat{u}_n(q) d^3 q \right) \frac{\hat{u}_n(p)}{p^2 - z}.$$

The resolvents of the operators L^n have a nontrivial limit if and only if the fractions

$$\frac{\alpha_n}{1 + \alpha_n \int_{\mathbf{R}^3} \frac{\hat{u}_n(q)^2}{q^2 - z} d^3 q}$$

converge to a nontrivial limit. The asymptotic of the integral can be computed explicitly using the spherical coordinates

$$\int_{\mathbf{R}^3} \frac{\hat{u}_n(q)^2}{q^2 - z} d^3 q = \frac{4\pi}{8\pi^3} \int_0^n \frac{r^2}{r^2 - z} dr = \frac{1}{2\pi^2} n + o(n), \quad n \to \infty.$$

One can choose for example

$$\frac{\alpha_n}{1 + \alpha_n n / 2\pi^2} = \alpha \Rightarrow \alpha_n = \frac{\alpha}{1 - \alpha n / 2\pi^2}.$$

For this choice of the coupling constant the sequence of the self–adjoint operators L^n converges to the operator L_α in the strong resolvent sense. We note that the sequence of coupling constants α_n is infinitesimal (in the sense that $\alpha_n \to 0$ as $n \to 0$). If the coupling constant does not depend on n then Theorem 1.4.4 implies that the resolvents of the operators L^n converge to the resolvent of the original operator.

1.5.2 Perturbations of the first derivative operator

Consider rank one perturbations defined by the formal expression

$$A_\alpha = \frac{1}{i} \frac{d}{dx} + \alpha \delta = \frac{1}{i} \frac{d}{dx} + \alpha \langle \delta, \cdot \rangle \delta. \tag{1.80}$$

The operator A_α can be considered as a rank one perturbation of the self–adjoint non–semibounded operator $A = 1/id/dx$ with domain $\mathrm{Dom}\,(A) = W_2^1(\mathbf{R})$. The δ measure defines a bounded linear functional on $W_2^1(\mathbf{R})$ due to the embedding theorem. But the element $(1/(A-i))\,\delta = ie^{-x}\Theta(x)$ does not belong to the domain of the operator A. ($\Theta(x)$ denotes here the Heaviside step function.) The restriction A^0 of the operator A to the domain of functions $\mathrm{Dom}\,(A^0) = \{\psi \in W_2^1(\mathbf{R}) : \psi(0) = 0\}$ has deficiency indices $(1,1)$. The deficiency elements are given by

$$\begin{aligned} g_i &= \frac{1}{A-i}\delta &= ie^{-x}\Theta(x), \\ g_{-i} &= \frac{1}{A+i}\delta &= -ie^{x}\Theta(-x). \end{aligned}$$

Every function ψ from the domain of the adjoint operator possesses the standard representation

$$\psi(x) = \tilde{\psi} + \frac{b(\psi)}{2} i \operatorname{sign} x \; e^{-|x|},$$

where $\tilde{\psi} \in W_2^1(\mathbf{R})$. Consider the group of the central symmetries of the real line:

$$G(1) = I, G(-1) = J,$$
$$G(-1)^2 = G(1),$$

where I and J are the identity and inversion operators respectively defined by the following formulas in the generalized sense

$$(If)(x) = f(x);$$

$$(Jf)(x) = f(-x).$$

The original operator and the functional δ are homogeneous with respect to this group

$$AG(t) = tG(t)A,$$
$$G(t)\delta = \delta.$$

The parameters β and θ for this problem are equal to 1 and 0 respectively. The group consists of only two elements and the extension of the functional δ can be defined using the parameter $f(-1)$. The parameter c defining the extension of the functional δ is given by

$$c = f(-1) - i \left\| \frac{1}{A-i}\delta \right\|^2 = i \left\langle \delta, \frac{1}{(A-i)(A+i)}\delta \right\rangle - i \left\langle \frac{1}{A-i}\delta, \frac{1}{A-i}\delta \right\rangle = 0.$$

The latter equality implies that the extension of the delta function, which is an even distribution, vanishes on every odd test function.

It follows from Theorem 1.3.2 that the self–adjoint operator A_α corresponding to the formal expression (1.80) is defined on the domain of functions satisfying the following conditions

$$\tilde{\psi}(0) = -\frac{1}{\alpha}b(\psi).$$

Thus the operator A_α is the self–adjoint operator $1/id/dx$ defined on the following domain

$$\operatorname{Dom}(A_\alpha) = \left\{ \psi \in W_2^1(\mathbf{R} \setminus \{0\}) : \psi(-0) = \frac{1+i\frac{\alpha}{2}}{1-i\frac{\alpha}{2}}\psi(+0) \right\}.$$

The spectral analysis of the operator A_α can be easily carried out.

1.5.3   Dirac operator with a pseudopotential

A similar analysis to that made in the previous section can be carried out for the one dimensional Dirac operator with the delta potential

$$H_\alpha = \begin{pmatrix} m & -i\frac{d}{dx} \\ -i\frac{d}{dx} & -m \end{pmatrix} + V\vec{\delta} \qquad (1.81)$$

$$V = V^* = \begin{pmatrix} v_{11} & v_{12} \\ v_{21} & v_{22} \end{pmatrix},$$

where $v_{11}, v_{22} \in \mathbf{R}, v_{12} = \overline{v_{21}} \in \mathbf{C}$. This family of Dirac operators with pseudopotentials is described by four real parameters. The original operator

$$H = \begin{pmatrix} m & -i\frac{d}{dx} \\ -i\frac{d}{dx} & -m \end{pmatrix}$$

is defined on the two component functions $f = (f_1, f_2) \in L_2(\mathbf{R}) \oplus L_2(\mathbf{R})$ from the domain $\mathrm{Dom}\,(H) = W_2^1(\mathbf{R}) \oplus W_2^1(\mathbf{R})$. Two delta functions δ_1, δ_2 defined as follows $\langle \delta_i, f \rangle = f_i(0), i = 1, 2$, are bounded linear functionals on the domain of the original operator. The delta function $\vec{\delta}$ is the linear map

$$\vec{\delta} : W_2^1(\mathbf{R}) \oplus W_2^1(\mathbf{R}) \to \mathbf{C}^2,$$

$$\langle \vec{\delta}, f \rangle = \begin{pmatrix} f_1(0) \\ f_2(0) \end{pmatrix}.$$

The product of the delta function and an arbitrary continuous function f is equal to

$$\langle f\vec{\delta}, \psi \rangle = \left\langle \vec{\delta}, \begin{pmatrix} f_1\psi_1 \\ f_2\psi_2 \end{pmatrix} \right\rangle$$

$$= \begin{pmatrix} f_1(0)\psi(0) \\ f_2(0)\psi(0) \end{pmatrix}$$

$$= \begin{pmatrix} f_1(0) & 0 \\ 0 & f_2(0) \end{pmatrix} \langle \vec{\delta}, \varphi \rangle,$$

where ψ is an arbitrary test function from $C_0^\infty(\mathbf{R}) \oplus C_0^\infty(\mathbf{R})$. The heuristic expression (1.81) can be written as

$$H_\alpha = \begin{pmatrix} m & -i\frac{d}{dx} \\ -i\frac{d}{dx} & -m \end{pmatrix} + V \,\mathrm{diag}\{\langle \delta_i, \cdot \rangle\}\vec{\delta}. \qquad (1.82)$$

This operator can be considered as a rank two perturbation of the self–adjoint non–semibounded original operator H. In accordance with this approach we restrict the original operator H to the domain of functions $\mathrm{Dom}\,(H^0) = \{\psi \in \mathrm{Dom}\,(H) : \langle \vec{\delta}, \psi \rangle = 0\}$. The restricted operator H^0 has deficiency indices $(2,2)$. The adjoint operator H^{0*} is defined on the domain $W_2^1(\mathbf{R} \setminus \{0\}) \oplus W_2^1(\mathbf{R} \setminus \{0\})$. To determine the perturbed operator, bounded linear functionals δ_i have to be extended to a set of functions which are discontinuous at the origin and continuous outside the origin. The delta functions are homogeneous with respect to the group of central symmetries of the real line and the extension is unique:

$$\langle \delta_i, f \rangle = \frac{f_i(+0) + f_i(-0)}{2}, \ i = 1, 2.$$

This extension allows one to define the perturbed linear operator on the domain $W_2^1(\mathbf{R} \setminus \{0\}) \oplus W_2^1(\mathbf{R} \setminus \{0\})$ since the boundary values at the origin of the functions from this domain are well defined. The domain of the perturbed self–adjoint operator coincides with the set of all function $\psi \in L_2(\mathbf{R}) \oplus L_2(\mathbf{R})$, such that

$$\begin{pmatrix} m & -i\frac{d}{dx} \\ -i\frac{d}{dx} & -m \end{pmatrix} \psi + V\vec{\delta}(x)\psi \in L_2(\mathbf{R}) \oplus L_2(\mathbf{R}).$$

Let us calculate the distribution

$$f = \begin{pmatrix} m & -i\frac{d}{dx} \\ -i\frac{d}{dx} & -m \end{pmatrix} \psi + V\vec{\delta}(x)\psi$$

for any function ψ from the domain of the adjoint operator H_0^*. Every such distribution can be presented in the following form

$$\begin{aligned} f &= \tilde{f} - i \begin{pmatrix} 0 & 1 \\ 1 & 0 \end{pmatrix} \mathrm{diag}\{\psi_1(+0) - \psi_1(-0), \psi_2(+0) - \psi_2(-0)\}\vec{\delta} \\ &+ \frac{1}{2}V \, \mathrm{diag}\{\psi_1(+0) + \psi_1(-0), \psi_2(+0) + \psi_2(-0)\}\vec{\delta}, \end{aligned}$$

where $\tilde{f} \in L_2(\mathbf{R}) \oplus L_2(\mathbf{R})$. The vector f belongs to the Hilbert space if and only if the coefficient in front of the delta function $\vec{\delta}$ is equal to zero. We get the following boundary conditions for the function ψ at the origin:

$$\left(\frac{1}{2}V - i \begin{pmatrix} 0 & 1 \\ 1 & 0 \end{pmatrix} \right) \psi(+0) = -\left(\frac{1}{2}V + i \begin{pmatrix} 0 & 1 \\ 1 & 0 \end{pmatrix} \right) \psi(-0).$$

These boundary conditions can be written in the form:

$$\psi(+0) = \Lambda\psi(-0); \tag{1.83}$$

$$\Lambda = -\left(\frac{1}{2}V - i \begin{pmatrix} 0 & 1 \\ 1 & 0 \end{pmatrix} \right) \left(\frac{1}{2}V + i \begin{pmatrix} 0 & 1 \\ 1 & 0 \end{pmatrix} \right).$$

One can show that

$$\Lambda \begin{pmatrix} 0 & 1 \\ 1 & 0 \end{pmatrix} \Lambda^* = \begin{pmatrix} 0 & 1 \\ 1 & 0 \end{pmatrix},$$

and it follows that the operator H^{0*} restricted to the domain of functions satisfying the boundary conditions (1.83) is self–adjoint ([125, 126]).

Chapter 2

Generalized rank one perturbations

2.1 Krein's formula for the generalized resolvents

2.1.1 Generalized self–adjoint extensions and generalized resolvents

Consider an arbitrary symmetric operator $\mathcal{A}^0$ acting in a certain Hilbert space $\mathcal{H}$. Let the deficiency indices of the operator $\mathcal{A}^0$ be equal. Then the self–adjoint extensions of $\mathcal{A}^0$ acting in the same Hilbert space can be described using von Neumann theory. The case of extensions of the operators having unit deficiency indices was studied in the previous chapter. Our goal in this chapter is to investigate self–adjoint extensions in extended Hilbert spaces. Such extensions are needed to obtain operators with a richer analytical structure of the spectrum. Let us introduce the following definitions.

Definition 2.1.1 *Let $\mathcal{A}^0$ be a symmetric operator acting in the Hilbert space $\mathcal{H}$. An operator* **A** *is called* **a generalized self–adjoint extension** *of the operator $\mathcal{A}^0$ if there exists a Hilbert space* **H** $\supset \mathcal{H}$ *such that the operator* **A** *is a self–adjoint operator in this Hilbert space and the operator $\mathcal{A}^0$ is its symmetric restriction. All extensions of the operator $\mathcal{A}^0$ inside the Hilbert space $\mathcal{H}$ will be called* **standard extensions**.

Obviously the set of generalized extensions includes the set of standard extensions. Only standard self–adjoint extensions have been considered in the previous chapter. Let us denote by $P_{\mathcal{H}}$ the projector in the space **H** onto the space $\mathcal{H}$.

Definition 2.1.2 *Let* **A** *be a self–adjoint operator acting in the extended Hilbert space* **H** $\supset \mathcal{H}$. *Then the restriction of the resolvent of the operator* **A** *to the space* $\mathcal{H}$

$$\mathbf{R}(z) = P_{\mathcal{H}} \frac{1}{\mathbf{A} - z}|_{\mathcal{H}}$$

is called **the generalized resolvent**. *Every generalized resolvent* $\mathbf{R}(z)$ *which is equal to the resolvent of a self–adjoint operator acting in* $\mathcal{H}$ *is called* **the orthogonal generalized resolvent**.

All generalized resolvents corresponding to the standard extensions of a symmetric operator are orthogonal.

Let the operator **A** be equal to the orthogonal sum of two self–adjoint operators $\mathcal{A}$ and $\mathcal{A}^{\perp}$ acting in the Hilbert spaces $\mathcal{H}$ and $\mathcal{H}^{\perp} = \mathbf{H} \ominus \mathcal{H}$. Then the generalized resolvent $\mathbf{R}(z)$ determined by the operator **A** is equal to the resolvent of the self–adjoint operator $\mathcal{A}$

$$\mathbf{R}(z) = \frac{1}{\mathcal{A} - z}$$

and therefore it is orthogonal. This resolvent does not depend on the operator $\mathcal{A}^{\perp}$ and the extension space $\mathcal{H}^{\perp}$. Therefore we are going to use the following

Definition 2.1.3 *A generalized extension* **A** *is* **equivalent** *to a standard extension* $\mathcal{A}$ *if and only if the operator* **A** *is equal to the orthogonal sum of the self–adjoint operators* $\mathcal{A}$ *and* $\mathcal{A}^{\perp}$ *acting in* $\mathcal{H}$ *and* $\mathcal{H}^{\perp} = \mathbf{H} \ominus \mathcal{H}$:

$$\mathbf{A} = \mathcal{A} \oplus \mathcal{A}^{\perp}. \tag{2.1}$$

The orthogonal generalized resolvents (and only these, in general, among the generalized resolvents) satisfy the resolvent identity

$$\mathbf{R}(\lambda) - \mathbf{R}(\mu) = (\lambda - \mu)\mathbf{R}(\lambda)\mathbf{R}(\mu).$$

The generalized self–adjoint extensions of arbitrary symmetric operators and the corresponding generalized resolvents will be studied in the following chapters. We are going to restrict our consideration in the current chapter to the case where the symmetric operator has unit deficiency indices.

2.1.2   Generalized rank one perturbations

Every symmetric operator with unit deficiency indices can be obtained by restricting a certain self–adjoint operator. We are going to consider in this section only densely defined symmetric operators. Consider an arbitrary self-adjoint operator $\mathcal{A}$ with domain Dom $(\mathcal{A})$ acting in the Hilbert space $\mathcal{H}$. Let the vector φ belong to $\mathcal{H}_{-2}(A) \setminus \mathcal{H}$ and consider the symmetric restriction $\mathcal{A}^0$ of the operator $\mathcal{A}$ to the domain Dom $(\mathcal{A}^0) = \{\psi \in \text{Dom}(\mathcal{A}) : \langle \psi, \varphi \rangle = 0\}$. $\mathcal{A}^0$ is then densely defined but not essentially self-adjoint in $\mathcal{H}$. We have proven that the resolvent of every self-adjoint extension $\mathcal{A}^\gamma$ of the operator $\mathcal{A}^0$ is given by Krein's formula

$$\frac{1}{\mathcal{A}^\gamma - z} = \frac{1}{\mathcal{A} - z} + \frac{1}{\gamma - Q(z)} \left\langle \frac{1}{\mathcal{A} - \bar{z}} \varphi, \cdot \right\rangle \frac{1}{\mathcal{A} - z} \varphi, \ z \in \text{Res}(\mathcal{A}),$$

(Res$(\mathcal{A})$ being the resolvent set of $\mathcal{A}$). This formula describes all standard self-adjoint extensions of the operator $\mathcal{A}^0$ and corresponding orthogonal resolvents. The Q-function $Q(z)$ given by (1.36) and the real parameter γ describe the analytical properties of the standard rank one perturbations. In Chapter 1 it was shown that the function $Q(z)$ is a Nevanlinna function without linear term in the representation (1.7). A more general analytical structure can be obtained using the generalized extensions of symmetric operators acting in extended Hilbert spaces. We are going to derive in this section Krein's formula for the generalized resolvents of a symmetric operator with deficiency indices $(1,1)$. It will be proven that the analytical structure of the problem is described in this case by the sum of two Nevanlinna functions $Q^+(z) + Q(z)$, where $Q^+(z)$ is a Nevanlinna function determined by the generalized extension. The function $Q^+(z)$ coincides with the real constant $-\gamma$ for the standard self-adjoint extensions studied in the previous chapter.

We are going to study the family of generalized resolvents of generalized self-adjoint extensions $\mathbf{A}$ of the symmetric operator $\mathcal{A}^0$ described above. Let $f \in \mathcal{R}(\mathcal{A}^0 - zI)$, $z \in \text{Res}(\mathcal{A})$. Then the values of the resolvent of the original operator and of the generalized resolvent are equal: $\mathbf{R}(z)f = (1/(\mathcal{A} - z))f$. Let $h \in \mathcal{R}(\mathcal{A}^0 - \bar{z}I)$. Then the following calculations

$$\left\langle \left(\frac{1}{\mathcal{A} - z} - \mathbf{R}(z) \right) f, h \right\rangle = \left\langle f, \left(\frac{1}{\mathcal{A} - \bar{z}I} - \mathbf{R}(\bar{z}) \right) h \right\rangle = \langle f, 0 \rangle = 0$$

show that

$$\frac{1}{\mathcal{A} - zI} - \mathbf{R}(z) \in \mathcal{L} \left\{ \frac{1}{\mathcal{A} - zI} \varphi \right\},$$

where $\mathcal{L}$ denotes the closed linear hull. It follows that there exists a certain function $p(z)$ such that

$$\mathbf{R}(z) = \frac{1}{\mathcal{A} - z} - p(z) \left\langle \frac{1}{\mathcal{A} - \bar{z}} \varphi, \cdot \right\rangle \frac{1}{\mathcal{A} - z}. \tag{2.2}$$

The function $p(z)$ describes the analytical structure of the restricted resolvent. To investigate the properties of this function we are going to consider the Cayley transforms of the operators $\mathcal{A}$ and $\mathbf{A}$, $z \in \mathrm{Res}(\mathcal{A})$. respectively $z \in \mathrm{Res}(\mathbf{A})$,

$$U(z) \;=\; (\mathcal{A} - z)(\mathcal{A} - \bar{z})^{-1};$$

$$\mathbf{U}(z) \;=\; (\mathbf{A} - z)(\mathbf{A} - \bar{z})^{-1}.$$

The Cayley transforms $U(\bar{z}), \mathbf{U}(\bar{z})$ restricted to the domain $\mathcal{R}(\mathcal{A}^0 - zI)$ coincide. The ranges $U(\bar{z})\,(1/(\mathcal{A} - \bar{z}))\,\varphi$ and $P_{\mathcal{H}}\mathbf{U}(\bar{z})\,(1/(\mathcal{A} - \bar{z}))\,\varphi$ can be different but both belong to the deficiency subspace $\mathcal{L}\{(1/(\mathcal{A} - \bar{z}))\,\varphi\}$. This means that there exists a certain function $q(z)$ such that the following equality holds

$$P_{\mathcal{H}}\mathbf{U}(\bar{z})f = U(\bar{z})f - q(z)\left\langle \frac{1}{\mathcal{A} - \bar{z}}\varphi, f \right\rangle \frac{1}{\mathcal{A} - z}\varphi \qquad (2.3)$$

for every $f \in \mathcal{H}$. Let us study the relations between the functions $p(z)$ and $q(z)$. Consider an arbitrary element $f \in \mathrm{Dom}\,(\mathcal{A}^0)$ and the corresponding $g = (\mathcal{A} - \bar{z})f$. The condition $\langle \varphi, f \rangle = 0$ implies that

$$(z - \bar{z})\left\langle \frac{1}{\mathcal{A} - \bar{z}}\varphi, \frac{1}{\mathcal{A} - \bar{z}}g \right\rangle = \left\langle \varphi, \frac{1}{\mathcal{A} - z}g \right\rangle - \left\langle \varphi, \frac{1}{\mathcal{A} - \bar{z}}g \right\rangle = \left\langle \frac{1}{\mathcal{A} - \bar{z}}\varphi, g \right\rangle$$

and this leads to the following equality

$$\begin{aligned}
P_{\mathcal{H}}\frac{1}{\mathbf{A} - z}g &= \frac{1}{\mathcal{A} - z}g - q(z)\left\langle \frac{1}{\mathcal{A} - \bar{z}}\varphi, \frac{1}{\mathcal{A} - \bar{z}}g \right\rangle \frac{1}{\mathcal{A} - z}\varphi \\
&= \frac{1}{\mathcal{A} - z}g - \frac{q(z)}{z - \bar{z}}\left\langle \frac{1}{\mathcal{A} - \bar{z}}\varphi, g \right\rangle \frac{1}{\mathcal{A} - z}\varphi.
\end{aligned}$$

Comparison of the latter equality with (2.2) shows that the function $q(z)$ we introduced is related to the function $p(z)$ appearing in (2.2) as follows: $q(z) = (z - \bar{z})p(z)$. Hence the Cayley transform of the operator $\mathcal{A}$ and the restricted Cayley transform of the generalized extension $\mathbf{A}$ are related as follows:

$$P_{\mathcal{H}}\mathbf{U}(\bar{z})f = U(\bar{z})f - (z - \bar{z})p(z)\left\langle \frac{1}{\mathcal{A} - \bar{z}}\varphi, f \right\rangle \frac{1}{\mathcal{A} - z}\varphi. \qquad (2.4)$$

The analytical properties of the function $p(z)$ can be studied in the following way. Krein's formula for the resolvent of rank one perturbations suggests we consider another function $Q^+(z)$ related to the function $p(z)$ as follows

$$p(z) = \frac{1}{Q^+(z) + Q(z)} \Rightarrow Q^+(z) = p^{-1}(z) - Q(z).$$

The function $Q(z)$ in the latter formula is the Q-function corresponding to the self–adjoint operator $\mathcal{A}$ and element $\varphi \in \mathcal{H}_{-2}(\mathcal{A})$. This function describes the analytical properties of the restriction. The analytical properties of the extension itself are characterized by the function $Q^+(z)$.

Theorem 2.1.1 *Every generalized resolvent of the symmetric operator $\mathcal{A}^0$ which is the restriction of the self–adjoint operator $\mathcal{A}$ to the domain*

$$\mathrm{Dom}\,(\mathcal{A}^0) = \{\psi \in \mathrm{Dom}\,(\mathcal{A}) : \langle \psi, \varphi \rangle = 0\}$$

is given by the following formula ($z \in \mathrm{Res}(\mathcal{A})$)

$$\mathbf{R}(z) = \frac{1}{\mathcal{A}-z} - \frac{1}{Q^+(z)+Q(z)} \left\langle \frac{1}{\mathcal{A}-\bar{z}}\varphi, \cdot \right\rangle \frac{1}{\mathcal{A}-z}\varphi. \qquad (2.5)$$

The function $Q(z)$ is the Q-function of the operator $\mathcal{A}$ and element $\varphi \in \mathcal{H}_{-2}(\mathcal{A})$. The function $Q^+(z)$ is a Nevanlinna function.

Proof The fact that representation (2.5) holds has already been proven. We have to show only that the function $Q^+(z)$ is from the Nevanlinna class. Consider formula (2.3) for the Cayley transforms with $z = i$ and $f = (1/(\mathcal{A}+i))\,\varphi$:

$$P_{\mathcal{H}}\mathbf{U}(-i)\frac{1}{\mathcal{A}+i}\varphi = \theta\frac{1}{\mathcal{A}-i}\varphi;$$

$$\theta = 1 - 2ip(i)\left\| \frac{1}{\mathcal{A}-i}\varphi \right\|^2.$$

The Cayley transforms are unitary operators and it follows that $|\theta| \leq 1$. Consider also the parameter α related to the parameter θ as follows

$$\gamma = -i\frac{1+\theta}{1-\theta}\left\| \frac{1}{\mathcal{A}-i}\varphi \right\|^2. \qquad (2.6)$$

The latter formula defines a map of the unit disk in the θ- plane on to the lower half of the γ-plane.

If $|\theta| = 1$ then the range of $\mathbf{U}(i)|_{\mathcal{H}}$ is a subset of $\mathcal{H}$ and the operator $\mathbf{A}$ restricted to the Hilbert space $\mathcal{H}$ coincides with one of the standard extensions of the operator $\mathcal{A}^0$. Using Definition 2.1.3 we conclude that the generalized extension $\mathbf{A}$ is equivalent to one of the self–adjoint extensions of the operator $\mathcal{A}^0$.

Similar calculations can be carried out for an arbitrary point $z = z_0$ instead of $z = i$. It follows that if $p(z_0) = 0$, i.e. if the generalized resolvent and a certain orthogonal resolvent coincide at one point, then the operators $\mathbf{A}$ and

$\mathcal{A}$ are equivalent. To prove that the operators are equivalent it was enough to suppose that the resolvents calculated on the element $(1/(\mathcal{A} - \bar{z}_0))\,\varphi$ are equal. In other words

$$\mathbf{R}(z_0)\frac{1}{\mathcal{A} - \bar{z}_0}\varphi = \frac{1}{\mathcal{A} - z_0}\frac{1}{\mathcal{A} - \bar{z}_0}\varphi$$

implies that $\mathbf{A} = \mathcal{A} \oplus \mathcal{A}^{\perp}$.

Formula (2.5) for the generalized resolvent implies that

$$Q^{+}(z) + Q(z) = \frac{\left\langle \frac{1}{\mathcal{A} - \bar{z}}\varphi, \frac{1}{\mathcal{A} - z}\varphi \right\rangle^{2}}{\left\langle \frac{1}{\mathcal{A} - \bar{z}}\varphi, \left(\frac{1}{\mathcal{A} - z} - \mathbf{R}(z)\right)\frac{1}{\mathcal{A} - z}\varphi \right\rangle}.$$

The denominator on the right hand side of the latter equality does not vanish if the extended operator $\mathbf{A}$ does not coincide with a certain standard extension and it implies that the function $Q^{+}(z)$ is holomorphic in the upper half plane. The function $Q^{+}(z)$ cannot be real at any point in the upper half plane if the extension $\mathbf{A}$ is not a standard extension. In fact, suppose that $Q^{+}(z_0) = -\gamma \in \mathbf{R}, \Im z_0 > 0$. Then the resolvent of the operator $\mathcal{A}^{\gamma}$ and the generalized resolvent coincide at the point $z = z_0$:

$$\mathbf{R}(z_0) = \frac{1}{\mathcal{A}^{\gamma} - z_0}.$$

It follows from our previous calculations that then the operators $\mathbf{A}$ and $\mathcal{A}^{\gamma}$ are also equivalent. Thus there are only two possibilities: either $\mathbf{A}$ is a standard extension or $Q^{+}(z)$ is not real-valued on the upper half plane.

The correspondence (2.6) between the parameters θ and γ is one-to-one and $Q(i)$ has positive imaginary part. It follows that the function $Q^{+}(z)$ has positive imaginary part in the entire upper half plane.

$\square$

The latter theorem describes the set of all generalized resolvents. It can easily be generalized to describe all generalized extensions of a symmetric operator $\mathcal{A}^{0}$ with arbitrary equal finite deficiency indices (see [569, 400]). The case where the symmetric operator $\mathcal{A}^{0}$ has infinite deficiency indices needs more careful consideration. We are going to consider special cases of this formula in the following chapters.

2.2 Model generalized perturbations

2.2.1 Introduction

The generalized extensions of symmetric operators have been studied in the previous section. In this section we are going to construct such extensions explicitly. Let **A** be a self-adjoint operator acting in the extended Hilbert space $\mathbf{H} = \mathcal{H} \oplus \mathcal{H}'$, where $\mathcal{H}$ is the original Hilbert space and $\mathcal{H}' = \mathcal{H}^\perp$ is the extension space. The decomposition of the Hilbert space leads to the natural decomposition of the resolvent of the self-adjoint operator **A** :

$$\frac{1}{\mathbf{A} - z} = \begin{pmatrix} R_{00}(z) & R_{01}(z) \\ R_{10}(z) & R_{11}(z) \end{pmatrix}. \tag{2.7}$$

The operators $R_{01}(z)$ and $R_{10}(z)$ in the latter formula have dimension one, since the deficiency indices of the restricted original operator are equal to $(1,1)$.

Let us study the decomposition of the resolvent operator in the case where the operators $R_{01}(z)$ and $R_{10}(z)$ "are small". Suppose that the coupling vanishes. Then the antidiagonal components of the resolvent $R_{01}(z)$ and $R_{10}(z)$ are equal to zero. The total operator **A** can be decomposed into the orthogonal sum of self-adjoint operators $\mathbf{A} = \mathcal{A} \oplus \mathcal{A}'$. The operators $\mathcal{A}$ and $\mathcal{A}'$ are self-adjoint operators acting in the Hilbert spaces $\mathcal{H}$ and $\mathcal{H}'$ respectively. The operators $R_{00}(z)$ and $R_{11}(z)$ are just the resolvents of the operators $\mathcal{A}$ and $\mathcal{A}'$ in the case of zero coupling. Generalized extensions of the operator $\mathcal{A}^0$ which are constructed starting from the orthogonal sum of the operators $\mathbf{A} = \mathcal{A} \oplus \mathcal{A}'$ have been named in [781, 785] *operators with internal structure*. To underline the physical background the original Hilbert space $\mathcal{H}$ is called **external space**. The extension space $\mathcal{H}'$ has the name **internal space**. The operators $\mathcal{A}$ and $\mathcal{A}'$ acting in the external and internal spaces are called **external** and **internal operators**.

In all examples considered below the interaction between the channels is introduced in the following way. In the first step both operators $\mathcal{A}$ and $\mathcal{A}'$ are restricted to certain symmetric operators $\mathcal{A}^0$ and $\mathcal{A}'^0$ with deficiency indices $(1,1)$. Then the total operator $\mathbf{A}^\Gamma$ is constructed as a self-adjoint extension of the symmetric operator $\mathbf{A}^0 = \mathcal{A}^0 \oplus \mathcal{A}'^0$ with deficiency indices $(2,2)$. Obviously every orthogonal sum of self-adjoint extensions of the operators $\mathcal{A}^0$ and $\mathcal{A}'^0$ inside the spaces $\mathcal{H}$ and $\mathcal{H}'$ can be obtained in this way. The self-adjoint extensions of the operator $\mathbf{A}^0$, which are equal to the orthogonal sum of some self-adjoint extensions of the operators $\mathcal{A}^0$ and $\mathcal{A}'^0$, will be called **separated extensions**. But the set of all self-adjoint extensions of the operator $\mathbf{A}^0$ is much wider and includes the extensions which cannot be decomposed into the orthogonal sum of two self–adjoint operators in $\mathcal{H}$ and $\mathcal{H}'$. The self–adjoint extensions of the operator $\mathbf{A}^0$ which are not separated will be called **connected**.

The case where the internal space $\mathcal{H}'$ has finite dimension is interesting for applications. In that case every restriction of the operator $\mathcal{A}'$ is not densely defined. We use modified von Neumann theory to construct self–adjoint extensions in this case (see Section 2.2.3). We start from the case of densely defined restrictions. The self–adjoint extensions of the operator $\mathbf{A}^0$ can be constructed using von Neumann formulas.

2.2.2   Model perturbations I: densely defined restricted operators

Let us consider a pair of Hilbert spaces $\mathcal{H}, \mathcal{H}'$ and a pair of self–adjoint operators $\mathcal{A}, \mathcal{A}'$ defined respectively in $\mathcal{H}, \mathcal{H}'$. The scalar product in the spaces $\mathcal{H}, \mathcal{H}'$ will be denoted by $\langle \cdot, \cdot \rangle$ and $\langle \cdot, \cdot \rangle'$. The orthogonal sum $\mathbf{A} = \mathcal{A} \oplus \mathcal{A}'$ is a self–adjoint operator acting in the orthogonal sum of the Hilbert spaces $\mathbf{H} = \mathcal{H} \oplus \mathcal{H}'$:

$$\mathbf{A} \begin{pmatrix} \psi_0 \\ \psi' \end{pmatrix} = \begin{pmatrix} \mathcal{A} & 0 \\ 0 & \mathcal{A}' \end{pmatrix} \begin{pmatrix} \psi_0 \\ \psi' \end{pmatrix} = \begin{pmatrix} \mathcal{A}\psi_0 \\ \mathcal{A}'\psi' \end{pmatrix}.$$

In what follows we are going to use the natural embedding of the spaces $\mathcal{H}, \mathcal{H}'$ into the space $\mathbf{H}$. Thus every element $\psi_0 \in \mathcal{H}$ will be identified with the element $(\psi_0, 0) \in \mathbf{H}$. Similarly we identify $\psi' \in \mathcal{H}'$ and $(0, \psi') \in \mathbf{H}$, and a similar identification will be used for the operators. For example the operator $\mathcal{A}$ in the Hilbert space $\mathcal{H}$ and the operator $\mathcal{A}(\psi, \psi') = (\mathcal{A}\psi, 0)$ in $\mathbf{H}$ will be identified.

Consider two arbitrary elements $\varphi \in \mathcal{H}_{-2}(\mathcal{A}) \setminus \mathcal{H}, \varphi' \in \mathcal{H}_{-2}(\mathcal{A}') \setminus \mathcal{H}'$ normalized as follows

$$\left\| \frac{1}{\mathcal{A} - i} \varphi \right\| = 1,$$

$$\left\| \frac{1}{\mathcal{A}' - i} \varphi' \right\|' = 1.$$

$$(2.8)$$

The restrictions $\mathcal{A}^0, \mathcal{A}'^0$ of the operators $\mathcal{A}$ and $\mathcal{A}'$ to the domains

$$\mathrm{Dom}\,(\mathcal{A}^0) = \{\psi \in \mathrm{Dom}\,(\mathcal{A}) : \langle \psi, \varphi \rangle = 0\};$$

$$\mathrm{Dom}\,(\mathcal{A}'^0) = \{\psi' \in \mathrm{Dom}\,(\mathcal{A}') : \langle \psi', \varphi' \rangle' = 0\}$$

are symmetric operators. We consider in this section only the case where φ, φ' are not elements from the Hilbert spaces $\mathcal{H}, \mathcal{H}'$. Then the restricted symmetric operators $\mathcal{A}^0$ and $\mathcal{A}'^0$ are densely defined and closed operators in the Hilbert spaces $\mathcal{H}$ and $\mathcal{H}'$ and the adjoint operators can be determined. The deficiency indices of the operators $\mathcal{A}^0$ and $\mathcal{A}'^0$ are equal to $(1,1)$. Let us denote the deficiency subspaces for the operators $\mathcal{A}^0, \mathcal{A}'^0$ as follows

$$N_z = \text{Ker}\{\mathcal{A}^{0*} - z\},$$

$$N_z' = \text{Ker}\{\mathcal{A}'^{0*} - z\}.$$

We are going to prove that

$$N_z = \mathcal{L}\left\{\frac{1}{A - z}\varphi\right\};$$

$$N_z' = \mathcal{L}\left\{\frac{1}{A' - z}\varphi'\right\}.$$

If $\psi \in N_z, \tilde{u} \in \text{Dom}(\mathcal{A}^0)$, then

$$(\mathcal{A}^{0*} - z)\psi = 0$$

$$\Rightarrow 0 = \langle (\mathcal{A}^{0*} - z)\psi, \tilde{u}\rangle = \langle \psi, (\mathcal{A} - \bar{z})\tilde{u}\rangle$$

$$\Rightarrow \psi \in \mathcal{L}\left\{\frac{1}{A-z}\varphi\right\}.$$

The proof for the operator $\mathcal{A}'$ is similar.

Then the orthogonal sum $\mathbf{A}^0 = \mathcal{A}^0 \oplus \mathcal{A}'^0$ is a dense symmetric restriction of the operator $\mathbf{A}$ acting in the Hilbert space $\mathbf{H}$. The deficiency subspace $\mathbf{N}_z = \text{Ker}\{\mathbf{A}^{0*} - z\}$ is equal to the orthogonal sum of the deficiency subspaces for the symmetric operators $\mathcal{A}^0$ and $\mathcal{A}'^0$

$$\mathbf{N}_z = N_z \oplus N_z'.$$

It follows that the deficiency indices of the operator $\mathbf{A}^0$ are equal to $(2,2)$. In what follows we put $z = i$ to avoid complicated formulas. Every element $\Psi = (\psi, \psi')$ from the domain of the adjoint operator $\mathbf{A}^{0*}$ possesses the following representation

$$\psi = \hat{\psi} + a_+(\Psi)\frac{1}{A - i}\varphi + a_-(\Psi)\frac{1}{A + i}\varphi$$

$$\psi' = \hat{\psi}' + a_+'(\Psi)\frac{1}{A' - i}\varphi' + a_-'(\Psi)\frac{1}{A' + i}\varphi',$$

where $\hat{\psi} \in \text{Dom}(\mathcal{A}^0), \hat{\psi}' \in \text{Dom}(\mathcal{A}'^0), a_\pm, a_\pm' \in \mathbf{C}$.

All self–adjoint extensions of the densely defined symmetric operator $\mathbf{A}^0$ can be constructed using von Neumann theory. Since the vectors φ, φ' are normalized by (2.8), every self–adjoint extension $\mathbf{A}(\mathbf{V})$ of the operator $\mathbf{A}^0$ is described by a certain unitary 2×2 matrix $\mathbf{V}$ such that

$$\mathbf{A}(\mathbf{V}) = \mathbf{A}^{0*}|_{\text{Dom}\,(\mathbf{A}(\mathbf{V}))}; \qquad (2.9)$$

where

$$\text{Dom}\,(\mathbf{A}(\mathbf{V}))$$

$$= \left\{ \Psi = \hat{\Psi} + \frac{a_+(\Psi)}{\mathcal{A} - i}\varphi + \frac{a_-(\Psi)}{\mathcal{A} + i}\varphi + \frac{a'_+(\Psi)}{\mathcal{A}' - i}\varphi' + \frac{a'_-(\Psi)}{\mathcal{A}' + i}\varphi' \in \text{Dom}\,(\mathbf{A}^{0*}) : \right.$$

$$\left. \hat{\Psi} \in \text{Dom}(\mathbf{A}^0), \quad -\mathbf{V}\left(\begin{array}{c} a_- \\ a'_- \end{array} \right) = \left(\begin{array}{c} a_+ \\ a'_+ \end{array} \right) \right\}.$$

The unitary matrices $\mathbf{V}$ describe all possible self–adjoint extensions of the symmetric operator $\mathbf{A}^0$. The von Neumann construction has to be modified in the case where the symmetric operator is not densely defined. It is more convenient to use the Cayley transform of the symmetric operator and its self–adjoint extensions in that case. See Chapter 4 where this approach is used.

The self–adjoint extensions of the operator $\mathbf{A}^0$ can be constructed using another representation for the elements from the domain of the adjoint operator. Consider the two dimensional subspace $\mathbf{M}$ of the Hilbert space $\mathbf{H}$ spanned by the vectors $(1/(\mathcal{A} + i))\,\varphi$ and $(1/(\mathcal{A}' + i))\,\varphi'$

$$\mathbf{M} = \mathcal{L}\left\{ \frac{1}{\mathcal{A} + i}\varphi, \frac{1}{\mathcal{A}' + i}\varphi' \right\}.$$

The orthogonal decomposition of the Hilbert space $\mathbf{H} = \mathcal{H} \oplus \mathcal{H}'$ determines the orthogonal decomposition of the subspace $\mathbf{M} = M \oplus M'$. Every vector $\Xi \in \mathbf{M}$ is equal to the sum of two orthogonal vectors

$$\Xi = \xi + \xi', \quad \xi \in \mathcal{H}, \xi' \in \mathcal{H}'.$$

Then every element $\Psi \in \text{Dom}\,(\mathbf{A}^{0*})$ possesses the following representation

$$\Psi = \hat{\Psi} + \frac{\mathbf{A}}{\mathbf{A} - i}\Xi_+(\Psi) + \frac{1}{\mathbf{A} - i}\Xi_-(\Psi) \qquad (2.10)$$

where $\hat{\Psi} \in \mathrm{Dom}\,(\mathbf{A}^0), \Xi_{\pm}(\Psi) \in \mathbf{M}$. Elementary calculations show that

$$\hat{\Psi} = \hat{\psi} + \hat{\psi}';$$

$$\Xi_+(\Psi) = (a_+(\Psi) + a_-(\Psi))\frac{1}{\mathcal{A}+i}\varphi + (a'_+(\Psi) + a'_-(\Psi))\frac{1}{\mathcal{A}'+i}\varphi';$$

$$\Xi_-(\Psi) = i(a_+(\Psi) - a_-(\Psi))\frac{1}{\mathcal{A}+i}\varphi + i(a'_+(\Psi) - a'_-(\Psi))\frac{1}{\mathcal{A}'+i}\varphi'.$$

$$(2.11)$$

The vectors $\Xi_{\pm}(\Psi)$ will be named *the boundary values* of the element Ψ. The map $L\,:\,\Psi \mapsto (\Xi_+(\Psi), \Xi_-(\Psi))$ acting in the spaces $\mathrm{Dom}\,(\mathbf{A}^{0*}) \to \mathcal{M} = \mathbf{M} \oplus \mathbf{M}$ will be called *boundary map*.

The adjoint operator $\mathbf{A}^{0*}$ acts as follows in the chosen representation

$$\mathbf{A}^{0*}\Psi = \mathbf{A}^{0*}\left(\hat{\Psi} + \frac{\mathbf{A}}{\mathbf{A}-i}\Xi_+(\Psi) + \frac{1}{\mathbf{A}-i}\Xi_-(\Psi)\right)$$

$$(2.12)$$

$$= \mathbf{A}^0\hat{\Psi} - \frac{1}{\mathbf{A}-i}\Xi_+(\Psi) + \frac{\mathbf{A}}{\mathbf{A}-i}\Xi_-(\Psi).$$

Theorem 2.2.1 *The boundary form*

$$B_{\mathbf{A}}[U, V] = \langle U, \mathbf{A}^{0*}V \rangle - \langle \mathbf{A}^{0*}U, V \rangle \qquad (2.13)$$

of the operator $\mathbf{A}^{0*}$ is given by

$$B_{\mathbf{A}}[U, V] = \langle \Xi_+(U), \Xi_-(V) \rangle - \langle \Xi_-(U), \Xi_+(V) \rangle. \qquad (2.14)$$

Proof Consider two arbitrary vectors $U, V \in \mathrm{Dom}\,(\mathbf{A}^{0*})$. Every such vector possesses the representation (2.10) and the action of the adjoint operator is given by the formulas (2.12). Using the fact that the original operator $\mathbf{A}$ is self–adjoint one can carry out the following calculations

$$\langle U, \mathbf{A}^{0*}V \rangle - \langle \mathbf{A}^{0*}U, V \rangle$$

$$= \left\langle \hat{U}_\beta + \frac{\mathbf{A}}{\mathbf{A} - i} \Xi(U) + \frac{1}{\mathbf{A} - i} \Xi_-(U), \right.$$
$$\left. \mathbf{A}^0 \hat{V} - \frac{1}{\mathbf{A} - i} \Xi(V) + \frac{\mathbf{A}}{\mathbf{A} - i} \Xi_-(V) \right\rangle$$

$$- \left\langle \mathbf{A}^0 \hat{U} - \frac{1}{\mathbf{A} - i} \Xi_+(U) + \frac{\mathbf{A}}{\mathbf{A} - i} \Xi_-(U), \right.$$
$$\left. \hat{V} + \frac{\mathbf{A}}{\mathbf{A} - i} \Xi_+(V) + \frac{1}{\mathbf{A} - i} \Xi_-(U) \right\rangle$$

$$= - \left\langle \frac{1}{\mathbf{A} - i} \Xi_-(U), \frac{1}{\mathbf{A} - i} \Xi_+(V) \right\rangle + \left\langle \frac{\mathbf{A}}{\mathbf{A} - i} \Xi_+(U), \frac{\mathbf{A}}{\mathbf{A} - i} \Xi_-(V) \right\rangle$$

$$- \left\langle \frac{\mathbf{A}}{\mathbf{A} - i} \Xi_-(U), \frac{\mathbf{A}}{\mathbf{A} - i} \Xi_+(V) \right\rangle + \left\langle \frac{1}{\mathbf{A} - i} \Xi_+(U), \frac{1}{\mathbf{A} - i} \Xi_-(V) \right\rangle$$

$$= \langle \Xi_+(U), \Xi_-(V) \rangle - \langle \Xi_-(U), \Xi_+(V) \rangle.$$

The theorem is proven.

□

The map
$$L : \ \mathrm{Dom}\,(\mathbf{A}^{0*}) \ \to \ \mathcal{M}$$

$$\Psi \ \mapsto \ (\Xi_+(\Psi), \Xi_-(\Psi))$$

is not invertible. The kernel of the map L coincides with the domain $\mathrm{Dom}\,(\mathbf{A}^0)$ of the restricted operator. Thus the map L can be lifted to the map $\mathbf{L}$ from the factor space $\mathrm{Dom}\,(\mathbf{A}^{0*})/_{\mathrm{Dom}\,(\mathbf{A}^0)}$ onto the space $\mathcal{M}$. The lifted map $\mathbf{L}$ is invertible. Similarly the boundary form $B_{\mathbf{A}}$ defines the following sesquilinear form $\mathbf{B}[\cdot, \cdot]$ on the space $\mathcal{M}$

$$\mathbf{B}[U, V] = B_{\mathbf{A}}[\mathbf{L}^{-1}U, \mathbf{L}^{-1}V], \ \ U, V \in \mathcal{M}.$$

The sesquilinear form $\mathbf{B}$ defines a complex symplectic structure on the four dimensional subspace $\mathcal{M}$.

We proved earlier that the operator $\mathbf{A}^0$ has deficiency indices $(2, 2)$ and that all the self–adjoint extensions of the operator $\mathbf{A}^0$ can be described by certain 2×2 unitary matrices using von Neumann theory. It will be more convenient for us to describe the self–adjoint extensions of the operator $\mathbf{A}^0$ in terms of the Lagrangian planes corresponding to the symplectic structure determined by the form $\mathbf{B}[\cdot, \cdot]$. To every self–adjoint extension $\mathbf{A}$ of $\mathbf{A}^0$ we associate a subspace $E \subset \mathcal{M}$

$$E = L(\mathrm{Dom}\,(\mathbf{A})) \tag{2.15}$$

consisting of all boundary values of elements from the domain of $\mathbf{A}$.

Theorem 2.2.2 *The map* $\mathrm{Dom}\,(\mathbf{A}) \mapsto L(\mathrm{Dom}\,(\mathbf{A})) \subset \mathcal{M}$ *defines a one–to–one correspondence between the set of self–adjoint extensions of* $\mathbf{A}^0$ *and the set of subspaces of* $\mathcal{M}$ *that are Lagrangian with respect to the form* $\mathbf{B}[U, V]$.

Proof Given a self–adjoint extension $\mathbf{A}$ it follows that the form $\mathbf{B}$ vanishes on the subspace (2.15) of $\mathcal{M}$. Suppose that the dimension of E were less than 2. Then the elements from the domain of $\mathbf{A}$ span the subspace of $\mathcal{N}_i \oplus \mathcal{N}_{-i}$ with the dimension less than 2 and this would imply that $\mathbf{A}$ is not self–adjoint. Therefore E must be two dimensional and hence Lagrangian.

Conversely, let E be a Lagrangian subspace of $\mathcal{M}$. Let the domain of $\mathbf{A}$ be defined by (2.15). Then the operator $\mathbf{A}$ is symmetric. Self–adjointness of the operator $\mathbf{A}$ follows from the fact that the map $\mathbf{L}$ is injective.

$\square$

The boundary form $\mathbf{B}$ is equal to zero on every plane in the space $\mathcal{M}$ determined by the boundary conditions connecting the boundary values $(\Xi_+, \Xi_-) \in \mathcal{M}$ by some Hermitian operator Γ acting in M

$$\Xi_- = \Gamma \Xi_+.$$

The orthogonal decomposition of the Hilbert space $\mathbf{H} = \mathcal{H} \oplus \mathcal{H}'$ determines the orthogonal decomposition of the two dimensional space $\mathbf{M} = M \oplus M'$. Then the operator Γ can be presented as follows using the decomposition:

$$\begin{pmatrix} \xi_{-0} \\ \xi_{-1} \end{pmatrix} = \begin{pmatrix} \Gamma_{00} & \Gamma_{01} \\ \Gamma_{10} & \Gamma_{11} \end{pmatrix} \begin{pmatrix} \xi_{+0} \\ \xi_{+1} \end{pmatrix}. \tag{2.16}$$

The operators $\Gamma_{\beta_0 \beta_1}$, $\beta_0, \beta_1 = 0, 1$, in the latter formula are linear operators between the one dimensional vector spaces M, M' respectively. The boundary conditions (2.16) however do not describe all Lagrangian planes of the symplectic form **B**. For example the unperturbed operator corresponds to the boundary conditions

$$\begin{pmatrix} \xi_{+0} \\ \xi_{+1} \end{pmatrix} = \begin{pmatrix} 0 \\ 0 \end{pmatrix},$$

which cannot be described by formula (2.16). The whole family of Lagrangian planes can be obtained using form preserving transformations.

Theorem 2.2.3 *Let the operator* Γ *in* M *be Hermitian. The operator* $\mathbf{A}^{\Gamma}$ *which is the restriction of the operator* $\mathbf{A}^{0*}$ *to the domain of elements* $U \in$ Dom $(\mathbf{A}^{0*})$ *such that the boundary values* LU *satisfy the boundary conditions* *(2.16) is self-adjoint and its resolvent is equal to*

$$\frac{1}{\mathbf{A}^{\Gamma} - \lambda} = \frac{1}{\mathbf{A} - \lambda} + \frac{\mathbf{A} + i}{\mathbf{A} - \lambda} \frac{1}{\Gamma - Q(\lambda)} P_{\mathbf{M}} \frac{\mathbf{A} - i}{\mathbf{A} - \lambda} \qquad (2.17)$$

for any λ, $\Im\lambda \neq 0$. *The operator* $Q(\lambda)$ *is by definition the operator valued* *R-function given on* $\mathbf{M}$ *by the formula*

$$Q(\lambda) = P_{\mathbf{M}} \frac{1 + \lambda \mathcal{A}}{\mathcal{A} - \lambda} P_{\mathbf{M}}. \qquad (2.18)$$

Proof We prove first that the operator $\mathbf{A}^{\Gamma}$ is symmetric. Let $U, V \in$ Dom$(\mathbf{A}^{\Gamma})$. Then Theorem 2.2.1 implies

$$\langle U, \mathbf{A}^{0*}V \rangle - \langle \mathbf{A}^{0*}U, V \rangle = B_{\mathbf{A}}[U, V]$$

$$= \langle \Xi_{+}(U), \Xi_{-}(V) \rangle - \langle \Xi_{-}(U), \Xi_{+}(V) \rangle$$

$$= \langle \Xi_{+}(U), \Gamma\Xi_{+}(V) \rangle - \langle \Gamma\Xi_{+}(U), \Xi_{+}(V) \rangle$$

$$= \langle \Xi_{+}(U), \Gamma\Xi_{+}(V) \rangle_M - \langle \Gamma\Xi_{+}(U), \Xi_{+}(V) \rangle_M$$

$$= 0.$$

$$(2.19)$$

The latter equality holds because the operator Γ is Hermitian in $\mathbf{M}$.

To calculate the resolvent of the operator $\mathbf{A}^{\Gamma}$ one has to solve the following equation

$$(\mathbf{A}^{\Gamma} - \lambda)^{-1}F = U$$

for arbitrary $F \in \mathbf{H}$ and arbitrary λ, $\Im\lambda \neq 0$. The latter equation implies that the vector $U \in$ Dom $(\mathbf{A}^{0*})$ satisfies the following equation

$$F = (\mathbf{A}^{0*} - \lambda) \left(\hat{U} + \frac{\mathbf{A}}{\mathbf{A} - i}\Xi_{+}(U) + \frac{1}{\mathbf{A} - i}\Xi_{-}(U) \right) \qquad (2.20)$$

and the boundary conditions (2.16). Equation (2.20) implies

$$F = (\mathbf{A} - \lambda)\hat{U} + \frac{\mathbf{A} - \lambda}{\mathbf{A} - i}\Xi_-(U) - \frac{1 + \lambda\mathbf{A}}{\mathbf{A} - i}\Xi_+(U).$$

To calculate $\Xi_+(U)$ one applies the operator $(\mathbf{A} - i)/(\mathbf{A} - \lambda)$ to the latter equation

$$\frac{\mathbf{A} - i}{\mathbf{A} - \lambda}F = (\mathbf{A} - i)\hat{U} + \Xi_-(U) - \frac{1 + \lambda\mathbf{A}}{\mathbf{A} - \lambda}\Xi_+(U)$$

and projects onto the subspace $\mathbf{M}$

$$P_{\mathbf{M}}\frac{\mathbf{A} - i}{\mathbf{A} - \lambda}F = \Xi_-(U) - P_{\mathbf{M}}\frac{1 + \lambda\mathbf{A}}{\mathbf{A} - \lambda}\Xi_+(U) = (\Gamma - Q(\lambda))\Xi_+(U)$$

$$\Rightarrow \Xi_+(U) = (\Gamma - Q(\lambda))^{-1}P_{\mathbf{M}}\frac{\mathbf{A} - i}{\mathbf{A} - \lambda}F. \qquad (2.21)$$

Here we used the fact that the element U satisfies the boundary conditions (2.16). The operator $\Gamma - Q(\lambda)$ is invertible, since the operator Γ is self–adjoint and the operator $Q(\lambda)$ has nontrivial imaginary part

$$\Im Q(\lambda) = P_{\mathbf{M}}\frac{\Im\lambda(\mathbf{A}^2 + 1)}{(\mathbf{A} - \Re\lambda)^2 + (\Im\lambda)^2}P_{\mathbf{M}}$$

if $\Im\lambda \neq 0$. Similarly, projection into the orthogonal complement of $\mathbf{M}$ gives

$$\hat{U} = \frac{1}{\mathbf{A} - i}(1 - P_{\mathbf{M}})\frac{\mathbf{A} - i}{\mathbf{A} - \lambda}F + \frac{1}{\mathbf{A} - i}(1 - P_{\mathbf{M}})\frac{1 + \lambda\mathbf{A}}{\mathbf{A} - \lambda}\Xi_+(U). \qquad (2.22)$$

Formulas (2.21),(2.22) give the solution of the problem

$$U = \hat{U} + \frac{\mathbf{A}}{\mathbf{A} - i}\Xi_+(U) + \frac{1}{\mathbf{A} - i}\Xi_-(U)$$

$$= \frac{1}{\mathbf{A} - \lambda}F + \frac{1}{\mathbf{A} - i}\left(-1 + (1 - P_{\mathbf{M}})\frac{1 + \lambda\mathbf{A}}{\mathbf{A} - \lambda}\frac{1}{\Gamma - Q}\right)P_{\mathbf{M}}\frac{\mathbf{A} - i}{\mathbf{A} - \lambda}F$$

$$+ \frac{1}{\mathbf{A} - i}\left(\mathbf{A}\frac{1}{\Gamma - Q} + \Gamma\frac{1}{\Gamma - Q}\right)P_{\mathbf{M}}\frac{\mathbf{A} - i}{\mathbf{A} - \lambda}F$$

$$= \frac{1}{\mathbf{A} - \lambda}F + \frac{1}{\mathbf{A} - i}\left(-1 + \frac{1 + \lambda\mathbf{A}}{\mathbf{A} - \lambda}\frac{1}{\Gamma - Q} - Q\frac{1}{\Gamma - Q}\right)P_{\mathbf{M}}\frac{\mathbf{A} - i}{\mathbf{A} - \lambda}F$$

$$+ \frac{1}{\mathbf{A} - i}\left(\mathbf{A}\frac{1}{\Gamma - Q} + \Gamma\frac{1}{\Gamma - Q}\right)P_{\mathbf{M}}\frac{\mathbf{A} - i}{\mathbf{A} - \lambda}F$$

$$= \frac{1}{\mathbf{A} - \lambda}F + \frac{\mathbf{A} + i}{\mathbf{A} - \lambda}\frac{1}{\Gamma - Q}P_{\mathbf{M}}\frac{\mathbf{A} - i}{\mathbf{A} - \lambda}F.$$

The latter formula coincides with (2.17). Moreover the domain of the resolvent of the operator $\mathbf{A}^{\Gamma}$ coincides with the Hilbert space $\mathbf{H}$. We are going to prove that the kernel of the calculated resolvent is trivial. Suppose that

$$\frac{1}{\mathbf{A} - \lambda}\psi + \frac{\mathbf{A} + i}{\mathbf{A} - \lambda}\frac{1}{\Gamma - Q}P_{\mathbf{M}}\frac{\mathbf{A} - i}{\mathbf{A} - \lambda}\psi = 0.$$

The first term in the latter equality is an element from the domain $\mathrm{Dom}(\mathbf{A})$. The second term is equal to a linear combination of the vectors

$$\frac{1}{\mathcal{A} - \lambda}\varphi, \ \frac{1}{\mathcal{A}' - \lambda}\varphi',$$

which do not belong to the domain $\mathrm{Dom}(\mathbf{A})$. It follows that the kernel of the calculated resolvent is trivial. We have calculated the resolvent of the symmetric operator $\mathbf{A}^{\Gamma}$. The domain of the resolvent coincides with the Hilbert space. This proves that the symmetric operator $\mathbf{A}^{\Gamma}$ is in fact self-adjoint.

$\square$

The operator $\mathbf{A}^{\Gamma}$ is defined on the domain of functions Ψ from $\mathrm{Dom}(\mathbf{A}^{0*})$ possessing the following representation

$$\Psi = \hat{\Psi} + \frac{1}{\mathbf{A} - i}(\Gamma + \mathbf{A})\Xi_{+}(\Psi), \tag{2.23}$$

where $\hat{\Psi} \in \mathrm{Dom}(\mathbf{A}^{0}), \Xi_{+}(\Psi) \in M$. The action of the operator $\mathbf{A}^{\Gamma}$ is given by the formula

$$\mathbf{A}^{\Gamma}\Psi = \mathbf{A}\hat{\Psi} + \frac{1}{\mathbf{A} - i}(-1 + \mathbf{A}\Gamma)\Xi_{+}(\Psi). \tag{2.24}$$

The inverse of the operator $\Gamma - Q(\lambda)$ can be easily calculated, since the orthogonal decomposition $\mathbf{M} = M \oplus M'$ reduces the operator $Q(\lambda) = Q_{0}(\lambda) \oplus Q_{1}(\lambda)$

$$(\Gamma - Q(\lambda))^{-1} =$$

$$\begin{pmatrix} [\Gamma_{00} - Q_{0}(\lambda) - \Gamma_{01}\frac{1}{\Gamma_{11} - Q_{1}(\lambda)}\Gamma_{10}]^{-1} & 0 \\ 0 & [\Gamma_{11} - Q_{1}(\lambda) - \Gamma_{10}\frac{1}{\Gamma_{00} - Q_{0}(\lambda)}\Gamma_{01}]^{-1} \end{pmatrix}$$

$$\times \begin{pmatrix} 1 & -\Gamma_{01}\frac{1}{\Gamma_{11} - Q_{1}(\lambda)} \\ -\Gamma_{10}\frac{1}{\Gamma_{00} - Q_{0}(\lambda)} & 1 \end{pmatrix}. \tag{2.25}$$

The operator Γ acts in the two dimensional subspaces $\Gamma : \mathbf{M} \to \mathbf{M}$. The vectors $e = (1/(\mathcal{A} + i)) \varphi, e' = (1/(\mathcal{A}' + i)) \varphi'$ form an orthonormal basis in the two dimensional subspace $\mathbf{M}$. The operator Γ in this basis is the operator of multiplication by the 2×2 Hermitian matrix

$$\Gamma = \begin{pmatrix} \gamma_{00} & \gamma_{01} \\ \gamma_{10} & \gamma_{11} \end{pmatrix}. \tag{2.26}$$

Similarly, the operator $Q(\lambda)$ is the following diagonal operator in the chosen basis

$$\mathbf{Q} = \begin{pmatrix} Q_0(\lambda) & 0 \\ 0 & Q_1(\lambda) \end{pmatrix} \tag{2.27}$$

$$Q_0(\lambda) = \langle \varphi, \frac{1 + \lambda \mathcal{A}}{\mathcal{A} - \lambda} \frac{1}{\mathcal{A}^2 + 1} \varphi \rangle$$

$$Q_1(\lambda) = \langle \varphi', \frac{1 + \lambda \mathcal{A}'}{\mathcal{A}' - \lambda} \frac{1}{\mathcal{A}'^2 + 1} \varphi' \rangle'.$$

We note that the function $Q_0(\lambda)$ is just the Q-function corresponding to the original operator $\mathcal{A}$ and element φ.

We are going to calculate the resolvent of the operator $\mathbf{A}^\Gamma$ restricted to the original Hilbert space $\mathcal{H}$. Formulas (2.17), (2.25), (2.26) and (2.27) imply

$$P_{\mathcal{H}} \frac{1}{\mathbf{A}^\Gamma - \lambda} P_{\mathcal{H}}$$

$$= \frac{1}{\mathcal{A} - \lambda} - \frac{1}{Q_0(\lambda) + \gamma_{01} \frac{1}{\gamma_{11} - Q_1(\lambda)} \gamma_{10} - \gamma_{00}} \langle \frac{1}{\mathcal{A} - \bar{\lambda}} \varphi, \cdot \rangle \frac{1}{\mathcal{A} - \lambda} \varphi. \tag{2.28}$$

The parameters γ_{00} and γ_{11} are real. Since the parameter $|\gamma_{01}|$ is positive and the function Q_1 is a Nevanlinna function, the function

$$Q_0^+(\lambda) = \frac{1}{\gamma_{11} - Q_1(\lambda)} |\gamma_{10}|^2 - \gamma_{00} \tag{2.29}$$

is a Nevanlinna function also.

The restricted resolvent is determined by the internal operator $\mathcal{A}'$, the vector φ' and real coefficients $\gamma_{00}, \gamma_{11}, |\gamma_{01}|$. The set of functions $Q_0^+(\lambda)$ constructed in the current section does not cover the whole set of Nevanlinna functions. For example rational Nevanlinna functions do not belong to this class. This is related to the fact that the vector φ' which we considered cannot be an element of the Hilbert space $\mathcal{H}'$. To obtain the whole Nevanlinna class of functions one needs to consider non–densely defined restrictions of the internal operator $\mathcal{A}'$, i.e. vectors φ from the Hilbert space. The corresponding construction is presented in the following section.

Changing the roles of the subspaces $\mathcal{H}$ and $\mathcal{H}'$ one can obtain the following formula for the resolvent $1/(\mathbf{A}^{\Gamma} - \lambda)$ restricted to the subspace $\mathcal{H}'$

$$P_{\mathcal{H}'} \frac{1}{\mathbf{A}^{\Gamma} - \lambda} P_{\mathcal{H}'}$$

$$= \frac{1}{\mathcal{A}' - \lambda} - \frac{1}{Q_1(\lambda) + Q_{\bar{1}}^{+}(\lambda)} \left\langle \frac{1}{\mathcal{A}' - \bar{\lambda}} \varphi', \cdot \right\rangle' \frac{1}{\mathcal{A}' - \lambda} \varphi', \qquad (2.30)$$

where

$$Q_{\bar{1}}^{+}(\lambda) = \frac{1}{\gamma_{00} - Q_0(\lambda)} |\gamma_{01}|^2 - \gamma_{11}$$

is a Nevanlinna function.

We have described the self–adjoint extensions of the operator $\mathbf{A}^0$ by Hermitian 2×2 matrices Γ. Von Neumann theory gives the description of all such extensions by unitary matrices $\mathbf{V}$ (see (2.9)). We are going to establish the relation between the matrices Γ and $\mathbf{V}$. Let us suppose that the coordinates $a_{\pm\beta}, \beta = 0, 1$ satisfy the following equation

$$-\mathbf{V} \begin{pmatrix} a_- \\ a'_- \end{pmatrix} = \begin{pmatrix} a_+ \\ a'_+ \end{pmatrix}.$$

Then (2.11) implies that

$$\begin{aligned} -v_{00}(\xi_+ + i\xi_-) - v_{01}(\xi'_+ + i\xi'_-) &= \xi_+ - i\xi_- \\ -v_{10}(\xi_+ + i\xi_-) - v_{11}(\xi'_+ + i\xi'_-) &= \xi'_+ - i\xi'_-. \end{aligned}$$

The latter equations can be written in the matrix form as follows

$$i(\mathbf{I} - \mathbf{V}) \begin{pmatrix} \xi_- \\ \xi'_- \end{pmatrix} = (\mathbf{I} + \mathbf{V}) \begin{pmatrix} \xi_+ \\ \xi'_+ \end{pmatrix},$$

which implies that the Hermitian matrix Γ and unitary matrix $\mathbf{V}$ are related by the Cayley transform

$$\Gamma = -i \frac{\mathbf{I} + \mathbf{V}}{\mathbf{I} - \mathbf{V}}, \quad \mathbf{V} = \frac{\Gamma + i\mathbf{I}}{\Gamma - i\mathbf{I}}. \qquad (2.31)$$

Compare the latter formula with (1.30). The extension of the operator $\mathbf{A}^0$ equal to the unperturbed operator $\mathbf{A}$ is described by the unit matrix $\mathbf{V}$. The Cayley transform of this matrix is not defined. This is another example where the boundary conditions (2.16) do not describe all self–adjoint extensions of $\mathbf{A}^0$.

2.2.3   Model perturbations II: non–densely defined construction

We consider in the present section the case where the vector φ' belongs to the Hilbert space $\mathcal{H}'$. This case needs careful consideration since the operator $\mathcal{A}'$ restricted to the orthogonal complement of φ' is not densely defined. It follows that the adjoint operator is not defined and one cannot use directly the von Neumann theory to construct the self–adjoint extensions of the restricted operator. Self–adjoint extensions of non–densely defined symmetric operators have been considered by M. A. Krasnosel'skiǐ [562, 561]. To avoid complicated constructions we should omit discussion of the adjoint operator and consider directly the perturbed operator. We shall not discuss the case where both restricted operators $\mathcal{A}^0$ and $\mathcal{A}'^0$ are not densely defined. Considering these perturbations we have to take into account that the family of self–adjoint extensions of a nondensely defined symmetric operator contains also self–adjoint operator relations. Thus we are going to prove that the perturbation we are going to define is an operator, not an operator relation.

Let $\varphi \in \mathcal{H}_{-2}(\mathcal{A}) \setminus \mathcal{H}$ and $\varphi' \in \mathcal{H}'$. We denote by $\mathrm{Dom}\,(\mathbf{A}^0)$ the domain of the restricted operator

$$\mathrm{Dom}(\mathbf{A}^0) = \{\Psi = (\psi, \psi') \in \mathrm{Dom}(\mathbf{A}) : \langle \psi, \varphi \rangle = 0, \langle \psi', \varphi' \rangle' = 0\}.$$

The operator $\mathbf{A}^0$ coincides with the operator $\mathbf{A}$ restricted to the domain $\mathrm{Dom}\,(\mathbf{A}^0)$. Let $\mathbf{M}$ be the two dimensional subspace of $\mathbf{H}$ spanned by the vectors $(1/(\mathcal{A}+i))\,\varphi$, $(1/(\mathcal{A}'+i))\,\varphi'$:

$$\mathbf{M} = \mathcal{L}\left\{\frac{1}{\mathcal{A}+i}\varphi, \frac{1}{\mathcal{A}'+i}\varphi'\right\}.$$

Let Γ be a Hermitian operator acting in $\mathbf{M}$. We denote by $\mathrm{Dom}\,(\mathbf{A}^\Gamma)$ the set of elements Ψ from $\mathrm{Dom}\,(\mathcal{A}^{0*}) \oplus \mathrm{Dom}(\mathcal{A}') \subset \mathbf{H}$ possessing the following representation

$$\Psi = \hat{\Psi} + \frac{\mathbf{A}}{\mathbf{A}-i}\Xi_+(\Psi) + \frac{1}{\mathbf{A}-i}\Xi_-(\Psi) \qquad (2.32)$$

with $\hat{\Psi} \in \mathrm{Dom}\,(\mathbf{A}^0)$, $\Xi_\pm(\Psi) \in \mathbf{M}$, $\Xi_- = \Gamma\Xi_+$.

Lemma 2.2.1 *If*

$$\gamma_{01} \neq 0 \qquad (2.33)$$

and/or

$$\gamma_{11} + \left\langle \varphi', \frac{\mathcal{A}'}{\mathcal{A}'^2+1}\varphi' \right\rangle \neq 0, \qquad (2.34)$$

then every element $\Psi \in \mathrm{Dom}\,(\mathbf{A}^{\Gamma})$ *possesses the unique representation*

$$\Psi = \hat{\Psi} + \frac{1}{\mathbf{A} - i}\Gamma\Xi_{+} + \frac{\mathbf{A}}{\mathbf{A} - i}\Xi_{+}, \qquad (2.35)$$

where $\hat{\Psi} \in \mathrm{Dom}\,(\mathbf{A}^{0}), \Xi_{+} \in \mathbf{M}$.

Proof One needs to prove only the uniqueness of representation (2.32) because the domain $\mathrm{Dom}\,(\mathbf{A}^{\Gamma})$ was defined as the set of all vectors possessing representation (2.35). Let Ψ be an element of $\mathrm{Dom}\,(\mathbf{A}^{\Gamma})$. Suppose that there exist vectors $\hat{\Psi}, \hat{\Psi}' \in \mathrm{Dom}\,(\mathbf{A}^{0})$ and $\Xi_{+}, \Xi'_{+} \in \mathbf{M}$ such that

$$\Psi = \hat{\Psi} + \frac{1}{\mathbf{A} - i}(\Gamma + \mathbf{A})\Xi_{+} = \hat{\Psi}' + \frac{1}{\mathbf{A} - i}(\Gamma + \mathbf{A})\Xi'_{+}.$$

This implies that

$$\hat{\Psi} - \hat{\Psi}' = -\frac{1}{\mathbf{A} - i}(\Gamma + \mathbf{A})(\Xi_{+} - \Xi'_{+}). \qquad (2.36)$$

The latter equality implies the equality

$$\hat{\Psi} - \hat{\Psi}' + \frac{1}{\mathbf{A} - i}(\Gamma + i)(\Xi_{+} - \Xi'_{+}) = -(\Xi_{+} - \Xi'_{+}).$$

The left hand side of the latter equality belongs to the domain $\mathrm{Dom}\,(\mathbf{A})$. The right hand side is a linear combination of the vectors $(1/(\mathcal{A} + i))\,\varphi$ and $(1/(\mathcal{A}' + i))\,\varphi'$. The element $(1/(\mathcal{A} + i))\,\varphi$ does not belong to the domain $\mathrm{Dom}\,(\mathcal{A})$. Thus the equality holds only if there exists some constant c such that

$$\Xi_{+} - \Xi'_{+} = c\frac{1}{\mathcal{A}' + i}\varphi'.$$

The difference $\hat{\Psi} - \hat{\Psi}'$ belongs to the domain $\mathrm{Dom}\,(\mathbf{A}^{0})$ and one can apply the operator $\mathbf{A} - i$ to equality (2.36)

$$(\mathbf{A} - i)(\hat{\Psi} - \hat{\Psi}') = -c(\Gamma + \mathbf{A})\frac{1}{\mathcal{A}' + i}\varphi'.$$

Projection into the space $\mathbf{M}$ leads to the equation

$$0 = -cP_{\mathbf{M}}(\Gamma + \mathcal{A}')\frac{1}{\mathcal{A}' - i}\varphi'.$$

The latter equation written in the basis

$$\left\{\frac{1}{\mathcal{A} + i}\varphi, \frac{1}{\mathcal{A}' + i}\varphi'\right\}$$

is equivalent to the following 2×2 linear system

$$
\begin{aligned}
0 &= -c\gamma_{01} \\
0 &= -c(\gamma_{11} + \langle \tfrac{1}{A'+i}\varphi', \tfrac{A'}{A'+i}\varphi' \rangle')
\end{aligned}
$$

Thus the constant c is trivial if one of the equations (2.33) and/or (2.34) is not satisfied. This ends the proof of the lemma.

$\square$

We are going to discuss conditions (2.33) and (2.34) in more detail. If condition (2.33) is not satisfied then the components ψ and ψ' of the elements from the domain $\mathrm{Dom}\,(\mathbf{A}^{\Gamma})$ are independent. This implies that every self–adjoint operator $\mathbf{A}^{\Gamma}$ with domain $\mathrm{Dom}\,(\mathbf{A}^{\Gamma})$ is in fact a separated extension of $\mathbf{A}^{0}$. The resolvent of such an extension restricted to the Hilbert space $\mathcal{H}$ is an orthogonal resolvent. Such resolvents were studied in Chapter 1.

If both conditions (2.33) and (2.34) are satisfied then the resolvent restricted to the space $\mathcal{H}'$ will be a resolvent of a self–adjoint relation, not an operator. Thus $\mathbf{A}^{\Gamma}$ is a self–adjoint relation if and only if it is equal to a direct sum of a self–adjoint operator acting in the Hilbert space $\mathcal{H}$ and a self–adjoint relation in $\mathcal{H}'$. Every such relation is equivalent to a self–adjoint operator in $\mathcal{H}$. Therefore we are going to concentrate our attention on the case where the generalized self–adjoint extension of the operator $\mathbf{A}^{0}$ is a self–adjoint operator, not a self–adjoint relation.

Definition 2.2.1 The Γ-modified operator $\mathbf{A}^{\Gamma}$ *is defined on* $\mathrm{Dom}\,(\mathbf{A}^{\Gamma})$ *using representation (2.35) by the formula*

$$
\begin{aligned}
\mathbf{A}^{\Gamma}\Psi &= \mathbf{A}^{\Gamma}\left(\hat{\Psi} + \frac{1}{\mathbf{A}-i}\Gamma\Xi_{+}(\Psi) + \frac{\mathbf{A}}{\mathbf{A}-i}\Xi_{+}(\psi) \right) \\
&= \mathbf{A}\hat{\Psi} + (\mathbf{A}-i)^{-1}(-1+\mathbf{A}\Gamma)\Xi_{+}(\Psi).
\end{aligned}
\tag{2.37}
$$

Theorem 2.2.4 *Let the operator* Γ *in* $\mathbf{M}$ *be Hermitian. If* $\Gamma_{01} \neq 0$ *then the* Γ-modified operator $\mathbf{A}^{\Gamma}$ *is self–adjoint on the domain* $\mathrm{Dom}\,(\mathbf{A}^{\Gamma})$.

Proof If $\Gamma_{01} \neq 0$, then every element from the domain $\mathrm{Dom}\,(\mathbf{A}^{\Gamma})$ possesses the unique representation (2.35). We show first that the operator $\mathbf{A}^{\Gamma}$ is symmetric. Let $U, V \in \text{Dom} (\mathbf{A}^\Gamma)$. Then

$$\langle U, \mathbf{A}^\Gamma V \rangle - \langle \mathbf{A}^\Gamma U, V \rangle$$

$$= \left\langle \hat{U} + \frac{1}{\mathbf{A} - i}(\Gamma + \mathbf{A})\Xi_+(U), \mathbf{A}^0\hat{V} + \frac{1}{\mathbf{A} - i}(-1 + \mathbf{A}\Gamma)\Xi_+(V) \right\rangle$$

$$- \left\langle \mathbf{A}^0\hat{U} + \frac{1}{\mathbf{A} - i}(-1 + \mathbf{A}\Gamma)\Xi_+(U), \hat{V} + \frac{1}{\mathbf{A} - i}(\Gamma + \mathbf{A})\Xi_+(V) \right\rangle$$

$$= \left\langle \hat{U}, \mathbf{A}^0\hat{V} \right\rangle + \left\langle \frac{1}{\mathbf{A} - i}(\Gamma + \mathbf{A})\Xi_+(U), \mathbf{A}^0\hat{V} \right\rangle$$

$$+ \left\langle \hat{U}, \frac{1}{\mathbf{A} - i}(-1 + \mathbf{A}\Gamma)\Xi_+(V) \right\rangle$$

$$+ \left\langle \frac{1}{\mathbf{A} - i}(\Gamma + \mathbf{A})\Xi_+(U), \frac{1}{\mathbf{A} - i}(-1 + \mathbf{A}\Gamma)\Xi_+(V) \right\rangle$$

$$- \left\langle \mathbf{A}^0\hat{U}, \hat{V} \right\rangle - \left\langle \frac{1}{\mathbf{A} - i}(-1 + \mathbf{A}\Gamma)\Xi_+(U), \hat{V} \right\rangle$$

$$- \left\langle \mathbf{A}^0\hat{U}, \frac{1}{\mathbf{A} - i}(\Gamma + \mathbf{A})\Xi_+(V) \right\rangle$$

$$- \left\langle \frac{1}{\mathbf{A} - i}(-1 + \mathbf{A}\Gamma)\Xi_+(U), \frac{1}{\mathbf{A} - i}(\Gamma + \mathbf{A})\Xi_+(V) \right\rangle$$

$$= \left\langle \Xi_+(U), (\mathbf{A} - i)\hat{V} \right\rangle - \left\langle (\mathbf{A} - i)\hat{U}, \Xi_+(V) \right\rangle$$

$$+ \left\langle (-1 + \Gamma\mathbf{A})\frac{1}{\mathbf{A}^2 + 1}(\Gamma + \mathbf{A})\Xi_+(U), \Xi_+(V) \right\rangle$$

$$- \left\langle (\Gamma + \mathbf{A})\frac{1}{\mathbf{A}^2 + 1}(-1 + \mathbf{A}\Gamma)\Xi_+(U), \Xi_+(V) \right\rangle$$

$$= 0.$$

The first two scalar products in the latter formula are equal to zero because $\Xi_+(U), \Xi_-(V) \in \mathbf{M}$ and $\hat{U}, \hat{V} \in \text{Dom} (\mathbf{A}^0)$. The third scalar product is equal to zero due to the following operator equality

$$(-1 + \Gamma\mathbf{A})\frac{1}{\mathbf{A}^2 + 1}(\Gamma + \mathbf{A})$$

$$= ((\mathbf{A} + \Gamma)\mathbf{A} - \mathbf{A}^2 - 1)\frac{1}{\mathbf{A}^2 + 1}(\Gamma + \mathbf{A})$$

$$= (\mathbf{A} + \Gamma)\frac{\mathbf{A}}{\mathbf{A}^2 + 1}(\Gamma + \mathbf{A}) - (\Gamma + \mathbf{A})$$

$$= (\mathbf{A} + \Gamma)\frac{1}{\mathbf{A}^2 + 1}(-1 + \mathbf{A}\Gamma).$$

Thus we have proved that the operator $\mathbf{A}^\Gamma$ is symmetric.

To prove that the operator $\mathbf{A}^\Gamma$ is self–adjoint we are going to calculate its resolvent, i.e. the solution of the equation

$$(\mathbf{A}^\Gamma - \lambda)^{-1}F = U$$

for arbitrary $F \in \mathbf{H}$ and arbitrary λ, $\Im\lambda \neq 0$. The latter equality implies that

$$F = (\mathbf{A}^\Gamma - \lambda)\left(\hat{U} + \frac{1}{\mathbf{A} - i}(\mathbf{A} + \Gamma)\Xi_+(U)\right)$$

$$= (\mathbf{A} - \lambda)\hat{U} + \frac{1}{\mathbf{A} - i}\left(-1 + \mathbf{A}\Gamma - \lambda(\mathbf{A} + \Gamma)\right)\Xi_+(U).$$

Applying the operator $(\mathbf{A} - i)/(\mathbf{A} - \lambda)$ and projecting into the subspace $\mathbf{M}$ one gets the following equation

$$P_{\mathbf{M}}\frac{\mathbf{A} - i}{\mathbf{A} - \lambda}F = (\Gamma - Q(\lambda))\Xi_+(U),$$

where

$$Q(\lambda) = P_{\mathbf{M}}\frac{1 + \lambda\mathbf{A}}{\mathbf{A} - \lambda}P_{\mathbf{M}}.$$

The vector $\Xi_+(U)$ can be calculated, since the operator $\Gamma - Q(\lambda)$ has non-trivial imaginary part and is therefore invertible:

$$\Xi_+(U) = \frac{1}{\Gamma - Q(\lambda)}P_{\mathbf{M}}\frac{\mathbf{A} - i}{\mathbf{A} - \lambda}F. \tag{2.38}$$

Projection into the orthogonal complement of $\mathbf{M}$ gives the equality

$$\hat{U} = \frac{1}{\mathbf{A} - i}(1 - P_{\mathbf{M}})\frac{\mathbf{A} - i}{\mathbf{A} - \lambda}F + \frac{1}{\mathbf{A} - i}(1 - P_{\mathbf{M}})\frac{1 + \lambda\mathbf{A}}{\mathbf{A} - \lambda}\Xi_+(U). \tag{2.39}$$

Combining formulas (2.38) and (2.39) one gets the solution

$$U = \frac{1}{\mathbf{A} - \lambda}F + \frac{\mathbf{A} + i}{\mathbf{A} - \lambda}\frac{1}{\Gamma - Q(\lambda)}P_{\mathrm{M}}\frac{\mathbf{A} - i}{\mathbf{A} - \lambda}F. \qquad (2.40)$$

The element U belongs to the domain $\mathrm{Dom}\,(\mathbf{A}^{\Gamma})$. The domain of the resolvent coincides with the Hilbert space $\mathbf{H}$. It is necessary to prove that the kernel of the calculated resolvent operator is trivial. Suppose that

$$\frac{1}{\mathbf{A} - \lambda}F + \frac{\mathbf{A} + i}{\mathbf{A} - \lambda}\frac{1}{\Gamma - Q(\lambda)}P_{\mathrm{M}}\frac{\mathbf{A} - i}{\mathbf{A} - \lambda}F = 0.$$

The first term in the latter equality is an element from the domain $\mathrm{Dom}\,(\mathbf{A})$. The second term is equal to a linear combination of the vectors $(1/(\mathcal{A} - \lambda))\,\varphi$ and $(1/(\mathcal{A}' - \lambda))\,\varphi'$. The vector $(1/(\mathcal{A} - \lambda))\,\varphi$ does not belong to the domain $\mathrm{Dom}\,(\mathbf{A})$. Thus there exists a certain constant c such that

$$F = c\varphi'.$$

This implies that

$$\frac{1}{\mathcal{A}' - \lambda}\varphi' + \frac{\mathbf{A} + i}{\mathbf{A} - \lambda}\frac{1}{\Gamma - Q(\lambda)}\left\langle \frac{1}{\mathcal{A}' + i}\varphi', \frac{\mathcal{A}' - i}{\mathcal{A}' - \lambda}\varphi'\right\rangle' \frac{1}{\mathcal{A}' + i}\varphi' = 0.$$

The latter equality holds only if the operator $(\Gamma - Q(\lambda))^{-1}$ is diagonal. This implies that the matrix Γ has to be diagonal, i.e. $\Gamma_{01} = 0$. We thus have a contradiction which proves the statement, ending the proof of the theorem.

□

The resolvent of the constructed self–adjoint operator $\mathbf{A}^{\Gamma}$ is given by formula (2.40). That formula coincides with the formula for the resolvent of the operator $\mathbf{A}^{\Gamma}$ constructed using a densely defined restricted operator. The formula for the resolvent of the operator $\mathbf{A}^{\Gamma}$ restricted to the space $\mathcal{H}$ coincides with the corresponding formula for the restricted resolvent (2.28).

In the case where the restricted internal operator is not densely defined the domain of the modified operator can be described explicitly.

Theorem 2.2.5 *Let φ' be an element from the Hilbert space $\mathcal{H}'$ and let Γ_{01} be different from zero. Then the Γ-modified operator $\mathbf{A}^{\Gamma}$ coincides with the* operator $\mathbf{A}_B$, $B = \begin{pmatrix} \beta_{11} & \beta_{12} \\ \beta_{21} & \beta_{22} \end{pmatrix}$ defined on the domain

$$\mathrm{Dom}\,(\mathbf{A}_B) \;=\; \Big\{ \Psi = (\psi, \psi') \in \mathrm{Dom}\,(\mathcal{A}^{0*}) \oplus \mathrm{Dom}\,(\mathcal{A}') :$$

$$\psi = \hat{\psi} + a_+(\Psi)\frac{\mathcal{A}}{\mathcal{A}^2 + 1}\varphi + a_-(\Psi)\frac{1}{\mathcal{A}^2 + 1}\varphi, \qquad (2.41)$$

$$a_\pm(\Psi) \in \mathbf{C},\ \langle \varphi', \psi' \rangle' = \beta_{21} a_+(\Psi) + \beta_{22} a_-(\Psi) \Big\}$$

by the following formula

$$\mathbf{A}_B \begin{pmatrix} \psi \\ \psi' \end{pmatrix} = \begin{pmatrix} \mathcal{A}^{0*}\psi \\ \mathcal{A}'\psi' + (\beta_{11} a_+(\Psi) + \beta_{12} a_-(\Psi))\,\varphi' \end{pmatrix}, \qquad (2.42)$$

where the coefficients β_{ik} are given by

$$\beta_{11} = \frac{\gamma_{00}}{\gamma_{01}}, \qquad \beta_{21} = -\frac{\gamma_{00}}{\gamma_{01}}c_1 + \frac{|\gamma_{01}|^2 - \gamma_{11}\gamma_{00}}{\gamma_{01}},$$

$$\qquad (2.43)$$

$$\beta_{12} = -\frac{1}{\gamma_{01}}, \qquad \beta_{22} = \frac{c_1}{\gamma_{01}} + \frac{\gamma_{11}}{\gamma_{01}}$$

and $c_1 = \langle \varphi', \frac{\mathcal{A}'}{\mathcal{A}'^2 + 1}\varphi' \rangle'$.

Proof Let $\Psi = (\psi, \psi')$ be an element from $\mathrm{Dom}\,(\mathbf{A}^\Gamma)$. The components of this function possess the following representations

$$\psi = \hat{\psi} + a_+(\Psi)\frac{\mathcal{A}}{\mathcal{A}^2 + 1}\varphi + a_-(\Psi)\frac{1}{\mathcal{A}^2 + 1}\varphi;$$

$$\qquad (2.44)$$

$$\psi' = \hat{\psi}' + a'_+(\Psi)\frac{\mathcal{A}'}{\mathcal{A}'^2 + 1}\varphi' + a'_-(\Psi)\frac{1}{\mathcal{A}'^2 + 1}\varphi',$$

where the coefficients $a_\pm, a'_\pm$ satisfy the following conditions

$$\begin{aligned} a_- &= \gamma_{00} a_+ + \gamma_{01} a'_+ \\ a'_- &= \gamma_{10} a_+ + \gamma_{11} a'_+. \end{aligned} \qquad (2.45)$$

The vector φ' belongs to the Hilbert space $\mathcal{H}'$. Therefore the functions $\frac{1}{\mathcal{A}'^2+1}\varphi'$ and $\frac{\mathcal{A}'}{\mathcal{A}'^2+1}\varphi'$ are elements from the domain of the operator $\mathcal{A}'$

$$\psi' \in \mathrm{Dom}\,(\mathcal{A}').$$

Therefore the domain $\mathrm{Dom}\,(\mathbf{A}^\Gamma)$ is a subset of $\mathrm{Dom}\,(\mathcal{A}^{0*}) \oplus \mathrm{Dom}\,(\mathcal{A}')$. Since Ψ is an element from the domain $\mathrm{Dom}\,(\mathbf{A}^\Gamma)$ the vector $\hat{\psi}'$ is orthogonal to φ'. It follows from (2.44) that

$$
\begin{aligned}
\langle \varphi', \psi' \rangle' &= a'_+(\Psi)\langle\varphi', \frac{\mathcal{A}'}{\mathcal{A}'^2+1}\varphi'\rangle' + a'_-(\Psi)\left\langle \varphi', \frac{1}{\mathcal{A}'^2+1}\varphi'\right\rangle' \\
&= a'_+(\Psi)c_1 + a'_-(\Psi).
\end{aligned}
\tag{2.46}
$$

The coefficients $a'_\pm(\Psi)$ can be calculated from the coefficients $a_\pm(\Psi)$ using conditions (2.45) and taking into account that $\gamma_{01} \neq 0$

$$
\begin{aligned}
a'_+(\Psi) &= -\frac{\gamma_{00}}{\gamma_{01}}a_+(\Psi) &+& \frac{1}{\gamma_{01}}a_-(\Psi) \\
a'_-(\Psi) &= \left(\gamma_{10} - \frac{\gamma_{11}\gamma_{00}}{\gamma_{01}}\right)a_+(\Psi) &+& \frac{\gamma_{11}}{\gamma_{01}}a_-(\Psi).
\end{aligned}
\tag{2.47}
$$

We get from (2.46)

$$
\begin{aligned}
\langle \varphi', \psi' \rangle' &= \left(-\frac{\gamma_{00}}{\gamma_{01}}c_1 + \frac{|\gamma_{01}|^2 - \gamma_{11}\gamma_{00}}{\gamma_{01}}\right)a_+(\Psi) + \left(\frac{c_1}{\gamma_{01}} + \frac{\gamma_{11}}{\gamma_{01}}\right)a_-(\Psi) \\
&= \beta_{21}a_+(\Psi) + \beta_{22}a_-(\Psi).
\end{aligned}
\tag{2.48}
$$

We have proven that $\mathrm{Dom}\,(\mathbf{A}_B) \supset \mathrm{Dom}\,(\mathbf{A}^\Gamma)$. Let us prove the opposite inclusion. Let $\Psi = (\psi, \psi')$ be an arbitrary element from the set $\mathrm{Dom}\,(\mathcal{A}^{0*}) \oplus \mathrm{Dom}\,(\mathcal{A}')$ satisfying the boundary condition (2.48). Every such function possesses the representations (2.44) but such representations may not be unique. To prove that Ψ is from $\mathrm{Dom}\,(\mathbf{A}^\Gamma)$ one has to show that there exist constants $a_\pm, a'_\pm$ satisfying conditions (2.45) and such that the corresponding function $\hat{\psi}'$ is orthogonal to φ'. Consider first the external component ψ. The adjoint operator $\mathcal{A}^{0*}$ is densely defined and therefore the representation (2.44) is unique and the coefficients $a_\pm(\Psi)$ can be calculated. Then the coefficients $a'_\pm(\Psi)$ are determined by (2.47). The boundary condition (2.48) guarantees that the vector

$$
\hat{\psi}' = \psi' - a'_+(\Psi)\frac{\mathcal{A}'}{\mathcal{A}'^2+1}\varphi' - a'_-(\Psi)\frac{1}{\mathcal{A}'^2+1}\varphi'
$$

is orthogonal to φ'. Thus we have proven that the domain $\mathrm{Dom}\,(\mathbf{A}^\Gamma)$ of the operator $\mathbf{A}^\Gamma$ is described by (2.41).

The operator $\mathbf{A}^\Gamma$ is defined by formula (2.37). Let us write the action of the operator explicitly. Let $\Psi \in \mathrm{Dom}\,(\mathbf{A}^\Gamma)$ and $\mathbf{A}^\Gamma\Psi = F = (f, f')$. Then the following equalities hold:

$$f = \mathcal{A}\hat{\psi} - a_+(\Psi)\frac{1}{\mathcal{A}^2+1}\varphi + a_-(\Psi)\frac{\mathcal{A}}{\mathcal{A}^2+1}\varphi$$

$$= \mathcal{A}^{0*}\psi;$$

$$f' = \mathcal{A}'\hat{\psi}' - a'_+(\Psi)\frac{1}{\mathcal{A}'^2+1}\varphi' + a'_-(\Psi)\frac{\mathcal{A}'}{\mathcal{A}'^2+1}\varphi'$$

$$= \mathcal{A}'\psi' - a'_+(\Psi)\varphi'$$

$$= \mathcal{A}'\psi' + \left(\frac{\gamma_{00}}{\gamma_{01}}a_+(\Psi) - \frac{1}{\gamma_{01}}a_-(\Psi)\right)\varphi'$$

$$= \mathcal{A}'\psi' + (\beta_{11}a_+(\Psi) + \beta_{12}a_-(\Psi))\,\varphi'.$$

It follows that the action of the operator $\mathbf{A}^\Gamma$ is defined by (2.42) and the operators $\mathbf{A}^\Gamma$ and $\mathbf{A}_B$ coincide. The theorem is proven.

$\square$

The latter theorem implies that the generalized perturbations of a symmetric operator $\mathbf{A}^0$ can be defined directly using the matrix B. Consider an extension space $\mathcal{H}'$, a self–adjoint operator $\mathcal{A}'$ acting in $\mathcal{H}'$, the element $\varphi' \in \mathcal{H}'$ and a 2×2 matrix

$$B = \begin{pmatrix} \beta_{11} & \beta_{12} \\ \beta_{21} & \beta_{22} \end{pmatrix}.$$

Definition 2.2.2 *The operator $\mathbf{A}_B$ is defined on the domain of functions from*

$$\mathrm{Dom}\,(\mathcal{A}^{0*}) \oplus \mathrm{Dom}\,(\mathcal{A}') \ni \left(\hat{\psi} + a_+(\psi)\frac{\mathcal{A}}{\mathcal{A}^2+1}\varphi + a_-(\psi)\frac{1}{\mathcal{A}^2+1}\varphi, \psi'\right)$$

satisfying the boundary condition

$$\langle \varphi', \psi' \rangle' = \beta_{21}a_+(\Psi) + \beta_{22}a_-(\Psi), \tag{2.49}$$

by the following formula

$$\mathbf{A}_B \begin{pmatrix} \psi \\ \psi' \end{pmatrix} = \begin{pmatrix} \mathcal{A}^{0*}\psi \\ \mathcal{A}'\psi' + (\beta_{11}a_+(\Psi) + \beta_{12}a_-(\Psi))\,\varphi' \end{pmatrix}. \tag{2.50}$$

One can prove that the operator $\mathbf{A}_B$ is self-adjoint if and only if the matrix coefficients β_{ik} satisfy the following conditions

$$
\begin{aligned}
\beta_{00}\overline{\beta}_{11} - \beta_{10}\overline{\beta}_{01} &= 1, \\
\beta_{00}\overline{\beta}_{10} &= \beta_{10}\overline{\beta}_{00}, \\
\beta_{11}\overline{\beta}_{01} &= \beta_{01}\overline{\beta}_{11}.
\end{aligned}
\tag{2.51}
$$

In particular the coefficients β_{ik} defined by (2.43) satisfy the latter conditions if $\gamma_{01} \neq 0$. The operators $\mathbf{A}^\Gamma$ and $\mathbf{A}_B$ coincide if conditions (2.43) are satisfied. This particular construction of generalized extensions was first suggested by K. Makarov [669].

The construction we describe is very useful in different applications when the extension space can be chosen to be finite dimensional and φ' is always an element from the extension Hilbert space $\mathcal{H}'$. The corresponding operators are described by rational Nevanlinna functions only.

2.2.4 Generalized resolvents and model generalized perturbations

The operators with model generalized perturbations lead to the generalized resolvents described by Theorem 2.1.1. We are going to show that every generalized resolvent can be obtained in this way.

Theorem 2.2.6 *Let $\mathcal{A}$ be a self-adjoint operator acting in the Hilbert space $\mathcal{H}$ and let $\mathcal{A}^0$ be its restriction to the domain $\mathrm{Dom}\,(\mathcal{A}^0) = \{\psi \in \mathrm{Dom}\,(\mathcal{A}) : \langle \psi, \varphi \rangle = 0\}$ where $\varphi \in \mathcal{H}_{-2}(\mathcal{A}) \setminus \mathcal{H}$. Let $\mathbf{R}(\lambda)$ be a generalized nonorthogonal resolvent corresponding to a certain generalized extension of the operator $\mathcal{A}^0$. Then there exists a model generalized perturbation of the operator $\mathcal{A}$ such that its resolvent restricted to the Hilbert space $\mathcal{H}$ coincides with the generalized resolvent $\mathbf{R}(\lambda)$.*

Proof To prove the theorem one has to show that there exist a Hilbert space $\mathcal{H}'$, a self-adjoint operator $\mathcal{A}'$ acting in $\mathcal{H}'$, an element $\varphi' \in \mathcal{H}_{-2}(\mathcal{A}')$ and a Hermitian operator Γ acting in the two dimensional space $\mathbf{M} = \mathcal{L}\{\frac{1}{\mathcal{A}+i}\varphi, \frac{1}{\mathcal{A}'+i}\varphi'\}$, such that the restriction of the resolvent

$$
R_{\mathbf{A}\Gamma}(\lambda) = \frac{1}{\mathbf{A} - \lambda} + \frac{\mathbf{A} + i}{\mathbf{A} - \lambda}\frac{1}{\Gamma - Q(\lambda)}P_\mathbf{M}\frac{\mathbf{A} - i}{\mathbf{A} - \lambda}
$$

to the space $\mathcal{H}$ coincides with the generalized resolvent $\mathbf{R}(\lambda)$. The restriction of the resolvent $R_{\mathbf{A}\Gamma}$ is given by the formula (2.28)

$$
P_\mathcal{H} R_{\mathbf{A}\Gamma} P_\mathcal{H} = \frac{1}{\mathcal{A} - \lambda} - \frac{1}{Q_0(\lambda) + Q_0^+(\lambda)}\left\langle \frac{1}{\mathcal{A} - \overline{\lambda}}\varphi, \cdot \right\rangle \frac{1}{\mathcal{A} - \lambda}\varphi,
\tag{2.52}
$$

where $Q_0^+(\lambda) = \frac{1}{\gamma_{11}-Q_1(\lambda)}|\gamma_{01}|^2 - \gamma_{00}$. The function $Q_1(\lambda)$ is the Nevanlinna function defined by the internal operator $\mathcal{A}'$ and the parameter φ' :

$$Q_1(\lambda) = \left\langle \varphi', \frac{1+\lambda\mathcal{A}'}{\mathcal{A}'-\lambda}\frac{1}{\mathcal{A}'^2+1}\varphi' \right\rangle.$$

Every such function $Q_1(\lambda)$ can be presented by the following integral

$$Q_1(z) = \int_{-\infty}^{\infty} \frac{1+\lambda x}{x-\lambda}\frac{d\mu(x)}{x^2+1}, \tag{2.53}$$

where the Borel measure $d\mu(x)$ satisfies the condition $\int_{-\infty}^{\infty}(x^2+1)^{-1}d\mu(x) < \infty$. One can consider the spectral measure associated with the self-adjoint operator $\mathcal{A}'$ and the vector $\frac{1}{\mathcal{A}'+i}\varphi' \in \mathcal{H}'$ to get the latter representation. Moreover every function with the representation (2.53) corresponds to a certain self-adjoint operator $\mathcal{A}'$ and element φ'.

Every generalized resolvent of the operator $\mathcal{A}^0$ is given by the same formula (2.52) with arbitrary Nevanlinna function $Q_0^+(\lambda)$. To prove the theorem it is enough to show that for any Nevanlinna function $Q_0^+(\lambda)$ there exist certain parameters $\gamma_{00}, \gamma_{11}, \gamma_{01}$ such that the function

$$\tilde{Q}(\lambda) = \gamma_{11} - \frac{|\gamma_{10}|^2}{Q_0^+(\lambda)+\gamma_{00}}$$

possesses the representation (2.53). If the function $Q_0^+(\lambda)$ is equal to a real constant c^+ then the generalized resolvent is orthogonal and does not satisfy the conditions of the theorem.

Consider now the case where the function $Q_0^+(\lambda)$ is not equal to a real constant. The function $\tilde{Q}(\lambda)$ is a Nevanlinna function and thus possesses representation (1.7)

$$\tilde{Q}(\lambda) = a + b\lambda + \int_{-\infty}^{\infty} \frac{1+\lambda x}{x-\lambda}\frac{d\nu(x)}{x^2+1}. \tag{2.54}$$

The function $\tilde{Q}(\lambda)$ can be presented by the integral (2.53) if the parameters a and b in the latter representation are equal to zero. The parameter b is given by the following formula

$$b = \lim_{y\to\infty} \frac{\Im\tilde{Q}(iy)}{y}.$$

The imaginary part of the quotient $-\dfrac{|\gamma_{01}|^2}{Q_0^+(\lambda)+\gamma_{00}}$

$$\Im\left(-\frac{|\gamma_{01}|^2}{Q_0^+(iy)-\gamma_{00}}\right) = \frac{\Im Q_0^+(iy)|\gamma_{01}|^2}{(\Re Q_0^+(iy)-\gamma_{00})^2+(\Im Q_0^+(iy))^2}$$

tends to infinity when $y \to \infty$ only if the denominator has limit zero there. The parameter γ_{00} can always be chosen in such a way that the limit is not equal to zero. This implies that the parameter γ_{11} can be chosen in such a way that the function $\tilde{Q}(\lambda)$ possesses the representation (2.54) with $a = b = 0$. Thus for the chosen values of the parameters the function $\tilde{Q}$ possesses the representation (2.53). The theorem is proven.

□

The case where the generalized resolvent is orthogonal corresponds to the standard extensions of the operator $\mathcal{A}^0$ and was studied in Chapter 1. Thus every generalized resolvent of any symmetric operator with deficiency indices $(1,1)$ is orthogonal or equal to the restricted resolvent of a certain operator with model generalized perturbation. Examples of such generalized perturbations of differential operators are considered in the following section.

2.3   Generalized point interactions for differential operators

2.3.1   Generalized point interaction in dimension three

We consider in this section the Laplace operator in dimension three with the interaction determined using the model of generalized perturbations. This operator is a generalization of the Laplace operator with delta potential considered in detail in Section 1.5.1. We are going to follow the general scheme developed in the previous sections. This model was first investigated by B.S.Pavlov [781].

Let us consider the unperturbed operator $\mathcal{L} = L \oplus L'$ acting in the orthogonal sum of the Hilbert spaces $\mathbf{H} = \mathcal{H} \oplus \mathcal{H}'$. The first operator in the latter decomposition is the unperturbed Laplace operator in dimension three:

$$L = -\Delta, \quad \mathcal{H} = L_2(\mathbf{R}^3).$$

The second operator is a finite dimensional Hermitian matrix which determines a certain linear operator in a finite dimensional complex space $\mathcal{H}' = \mathbf{C}^n$. The modified operator is constructed using the following parameters

- vector $\varphi \in \mathcal{H}_{-2}(L) \setminus \mathcal{H}$;

- vector $\varphi' \in \mathcal{H}_{-2}(L')$;

• Hermitian operator Γ (acting in a subspace of H).

The vector φ should be chosen having support at one point in order to model a point interaction in the external space. The set of bounded functionals with support at the origin is formed by the delta function and its derivatives but only the delta function itself is an element from the Hilbert space $\mathcal{H}_{-2}(-\Delta)$. Therefore the vector φ can be chosen equal to the delta function

$$\varphi = c\delta,$$

where $c = 2\sqrt[4]{2}\sqrt{\pi}$ is a normalizing constant. The space $\mathcal{H}_{-2}(L')$ coincides with the original internal Hilbert space $\mathcal{H}'$, since the space $\mathcal{H}'$ has finite dimension. The vector φ' from the internal space $\mathcal{H}'$ can be chosen arbitrarily but we suppose that this vector is generating for the operator L' and the space $\mathcal{H}'$. We also suppose that the vector φ' is normalized $\| \frac{1}{L'+i}\varphi' \|' = 1$. Then the restricted operator $\mathcal{L}^0$ coincides with the operator $\mathcal{L}$ restricted to the domain

$$\mathrm{Dom}\,(\mathcal{L}^0) = \{\Psi = (\psi, \psi') \in \mathrm{Dom}(\mathcal{L}) : \psi(0) = 0, \langle \psi', \varphi' \rangle' = 0\}.$$

The subspace $\mathbf{M}$ of $\mathbf{H}$ is spanned by the vectors $(c\frac{e^{-\sqrt{i}|x|}}{4\pi|x|}, 0)$ and $(0, \frac{1}{L'+i}\varphi')$ and the domain $\mathrm{Dom}\,(\mathcal{A}^\Gamma)$ is given by

$$\mathrm{Dom}\,(\mathcal{A}^\Gamma) \;=\; \left\{\Psi = \hat{\Psi} + \frac{\mathcal{L}}{\mathcal{L}-i}\Xi_+(\Psi) + \frac{1}{\mathcal{L}-i}\Gamma\Xi_+(\Psi) : \right.$$

$$\left. \hat{\Psi} \in \mathrm{Dom}\,(\mathcal{L}^0), \Xi_+(\Psi) \in \mathbf{M}\right\}. \tag{2.55}$$

We suppose that the Hermitian operator Γ acting in $\mathbf{M}$ satisfies the condition $\Gamma_{01} \neq 0$. The latter condition guarantees that the modified operator cannot be decomposed into the orthogonal sum of the operators acting in the external and internal spaces. The same condition implies that the representation (2.55) is unique and the Γ-modified operator can be defined by the formula

$$\mathcal{L}^\Gamma\Psi \;=\; \mathcal{L}^\Gamma\left(\hat{\Psi} + \frac{1}{\mathcal{L}-i}\Gamma\Xi_+(\Psi) + \frac{\mathcal{L}}{\mathcal{L}-i}\Xi_+(\Psi)\right)$$

$$= \mathcal{L}\hat{\Psi} + (\mathcal{L}-i)^{-1}(-1 + \mathcal{L}\Gamma)\Xi_+(\Psi). \tag{2.56}$$

The resolvent of the modified self–adjoint operator is given by the formula

$$\frac{1}{\mathcal{L}^\Gamma - \lambda} = \frac{1}{\mathcal{L}-\lambda} + \frac{\mathcal{L}+i}{\mathcal{L}-\lambda}\frac{1}{\Gamma - Q(\lambda)}P_\mathbf{M}\frac{\mathcal{L}-i}{\mathcal{L}-\lambda}, \tag{2.57}$$

where $Q(\lambda) = P_{\mathbf{M}} \frac{1+\lambda\mathcal{L}}{\mathcal{L}-\lambda} P_{\mathbf{M}}$ is an operator acting in $\mathbf{M}$. The Q-function of the external operator can be easily calculated:

$$Q_0(\lambda) = \left\langle \varphi, \frac{1+L\lambda}{L-\lambda} \frac{1}{L^2+1} \varphi \right\rangle = 1 + c^2 \frac{i\sqrt{\lambda}}{4\pi}.$$

Thus the resolvent restricted to the external space $\mathcal{H}$ is given by

$$P_{\mathcal{H}} \frac{1}{\mathcal{L}^\Gamma - \lambda} P_{\mathcal{H}} = \frac{1}{L-\lambda} - \frac{c^2}{1 + c^2 \frac{i\sqrt{\lambda}}{4\pi} + Q_0^+(\lambda)} \left\langle \frac{1}{L-\bar\lambda} \delta, \cdot \right\rangle \frac{1}{L-\lambda} \delta, \quad (2.58)$$

where $Q_0^+(\lambda) = \frac{1}{\gamma_{11}-Q_1(\lambda)} |\gamma_{01}|^2 - \gamma_{00}$, $Q_1(\lambda) = \langle \varphi', \frac{1+\lambda L'}{L'-\lambda} \frac{1}{L'^2+1} \varphi' \rangle'$. The latter formula for the restricted resolvent can be simplified if one introduces another function with positive imaginary part in the upper half plane $\tilde{Q}_0^+(\lambda) = \frac{1}{c^2}\left(1 + Q_0^+(\lambda)\right)$:

$$P_{\mathcal{H}} \frac{1}{\mathcal{L}^\Gamma - \lambda} P_{\mathcal{H}} = \frac{1}{L-\lambda} - \frac{1}{\frac{i\sqrt{\lambda}}{4\pi} + \tilde{Q}_0^+(\lambda)} \left\langle \frac{1}{L-\bar\lambda} \delta, \cdot \right\rangle \frac{1}{L-\lambda} \delta. \quad (2.59)$$

This resolvent coincides with the resolvent of the formal Laplace operator in $L_2(\mathbf{R}^3)$ with energy dependent boundary condition at the origin. We note that the restricted resolvent does not coincide with the resolvent of any operator. Every function ψ from the domain $\text{Dom}(L^{0*}) = W_2^2(\mathbf{R}^3 \setminus \{0\})$ of the operator adjoint to the restricted operator L^0 possesses the following asymptotic representation at the origin

$$\psi(x) = \frac{\psi_s}{4\pi|x|} + \psi_0 + o(|x|), \quad x \to 0. \quad (2.60)$$

In fact every function from the domain of the adjoint operator possesses the following representation

$$\begin{aligned}
\psi(x) &= \hat\psi(x) + \frac{L}{L-i} a_+(\Psi) \frac{1}{L+i}\varphi + \frac{1}{L-i} a_-(\Psi) \frac{1}{L+i}\varphi \\
&= \hat\psi + a_+(\Psi)\frac{c}{2}\left(\frac{e^{i\sqrt{i}|x|}}{4\pi|x|} + \frac{e^{-\sqrt{i}|x|}}{4\pi|x|}\right) \\
&\quad + a_-(\Psi)\frac{c}{2i}\left(\frac{e^{i\sqrt{i}|x|}}{4\pi|x|} - \frac{e^{-\sqrt{i}|x|}}{4\pi|x|}\right) \\
&= c a_+(\Psi)\frac{1}{4\pi|x|} + \frac{1}{c}(-a_+(\Psi) + a_-(\Psi)) + o(1).
\end{aligned}$$

It follows that the coefficients ψ_s, ψ_0 and $a_\pm(\Psi)$ are related as follows

$$\psi_s = ca_+(\Psi)$$

$$\psi_0 = \frac{1}{c}(-a_+(\Psi) + a_-(\Psi)).$$

Therefore the restricted resolvent gives the solution to the following boundary problem

$$\begin{cases} -\Delta\psi - \lambda\psi = 0; \\ \tilde{Q}_0^+(\lambda)\psi_s + \psi_0 = 0 \end{cases} \tag{2.61}$$

where $\psi \in W_2^2(\mathbf{R}^3 \setminus \{0\})$ and the coefficients ψ_s, ψ_0 are determined by (2.60). The symbol $-\Delta$ in the latter expression denotes the differential Laplace operator which is defined on the Sobolev space $W_2^2(\mathbf{R}^3 \setminus \{0\})$ in the generalized sense with the set of test–functions $C_0^\infty(\mathbf{R}^3 \setminus \{0\})$. The spectral problem determined by (2.61) is not a spectral problem for any self–adjoint operator. To study the latter problem it is useful to consider the total self–adjoint operator $\mathcal{L}^\Gamma$ and use its spectral representation. All the eigenfunctions of the operator $\mathcal{L}^\Gamma$ can be calculated explicitly. We are going to concentrate our attention on the properties of the zero component.

The modified operator $\mathcal{L}^\Gamma$ is a rank two perturbation of the original operator $\mathcal{L}$. The spectrum of the original operator consists of the branch of the continuous spectrum $[0, \infty)$ and n eigenvalues. The absolutely continuous spectrum is invariant with respect to finite rank perturbations (see Chapter 4). Therefore the spectrum of the operator $\mathcal{L}^\Gamma$ consists of the branch of the absolutely continuous spectrum $[0, \infty)$ and perhaps several eigenvalues. The eigenvalues coincide with the singularities of the resolvent which are situated on the negative part of the real line.

The discrete spectrum eigenfunction $\Psi = (\psi, \psi')$ corresponding to the energy E is a solution to the equation

$$(\mathcal{L}^\Gamma - E)\Psi = 0. \tag{2.62}$$

The latter equation can be written as follows using the representation (2.55)

$$(\mathcal{L} - E)\hat{\Psi} + \frac{1}{\mathcal{L} - i}(-1 + \mathcal{L}\Gamma - E\Gamma - E\mathcal{L})\Xi_+(\Psi) = 0. \tag{2.63}$$

Applying the operator $\frac{\mathcal{L} - i}{\mathcal{L} - E}$ and projecting into the subspace M one gets the following equation

$$(\Gamma - Q(\lambda))\,\Xi_+(\Psi) = 0.$$

Thus the discrete spectrum can be situated only at the points where the operator $\Gamma - Q(\lambda)$ in M is not invertible. Considering this operator in the standard basis in $\mathbf{M}$ one gets the equation for the discrete spectrum of the operator $\mathcal{L}^\Gamma$

$$\det\left(\begin{pmatrix} \gamma_{00} & \gamma_{01} \\ \gamma_{10} & \gamma_{11} \end{pmatrix} - \begin{pmatrix} 1 + c^2\frac{i\sqrt{\lambda}}{4\pi} & 0 \\ 0 & Q_1(\lambda) \end{pmatrix}\right) = 0$$

$$\Rightarrow 1 + c^2\frac{i\sqrt{\lambda}}{4\pi} + Q_0^+(\lambda) = 0 \qquad (2.64)$$

$$\Rightarrow \frac{i\sqrt{\lambda}}{4\pi} + \tilde{Q}_0^+(\lambda) = 0.$$

The latter equation has solutions on the physical sheet of the energy parameter λ only on the negative part of the real axis, since the functions $Q_0(\lambda) = 1 + c^2\frac{i\sqrt{\lambda}}{4\pi}$ and $Q_0^+(\lambda)$ have imaginary parts with the same sign in the lower and upper half planes and the function $\frac{i\sqrt{\lambda}}{4\pi}$ is pure imaginary considered on the positive part of the real axis. The function $\tilde{Q}_0^+(\lambda)$ is real on the whole real line. The latter equation has no more than $N'+1$ solutions on the negative part of the real axis where N' is the dimension of the internal space $\mathcal{H}'$. The function $\frac{i\sqrt{\lambda}}{4\pi}$ increases monotonically on the interval $\lambda \in (-\infty, 0]$ from $-\infty$ to 0. The functions $Q_1(\lambda)$, $Q_0^+(\lambda)$ and $\tilde{Q}_0^+(\lambda)$ are rational Nevanlinna functions. The singularities of the functions are situated on the real axis. These functions are increasing functions on each interval which does not contain the singularities. Each function has no more than N' singularities and it follows that equation (2.64) has no more than $N'+1$ solutions on the physical sheet. This equation has no more than $2N'+1$ solutions in total. The other solutions are situated on the nonphysical sheet of the energy parameter λ ($\Im k < 0$.)

We have proven that the discrete spectrum can be situated only at the points which are solutions of (2.64). The eigenfunctions corresponding to the solution $\lambda = E < 0$ of (2.64) can be calculated as follows. Let $\Xi_+(\Psi) = a_+\frac{1}{L+i}\varphi + a'_+\frac{1}{L'+i}\varphi'$ be an element from the kernel of the operator $\Gamma - Q(E)$. Then the coordinates a_+, a'_+ satisfy the following 2×2 linear system

$$\begin{pmatrix} \gamma_{00} - Q_0(E) & \gamma_{01} \\ \gamma_{10} & \gamma_{11} - Q_1(E) \end{pmatrix} \begin{pmatrix} a_+ \\ a'_+ \end{pmatrix} = 0.$$

It follows that the coordinates a'_+ and a_+ are related as follows

$$a'_+ = -\frac{\gamma_{10}}{\gamma_{11} - Q_1(E)}a_+. \qquad (2.65)$$

The elements from the domain of the modified operator $\mathcal{L}^\Gamma$ satisfy the boundary conditions

$$\begin{pmatrix} a_- \\ a'_- \end{pmatrix} = \begin{pmatrix} \gamma_{00} & \gamma_{01} \\ \gamma_{10} & \gamma_{11} \end{pmatrix} \begin{pmatrix} a_+ \\ a'_+ \end{pmatrix}.$$

Thus the boundary values of the external component satisfy the energy dependent generalized boundary conditions at the origin

$$a_- = -Q_0^+(E)a_+, \tag{2.66}$$

where the function $Q_0^+(E)$ is given by formula (2.29).

Applying the resolvent of the original operator $\frac{1}{\mathcal{L}-E}$ to equation (2.63) we get the following expression for $\hat{\Psi}$:

$$\hat{\Psi} = -\frac{1}{\mathcal{L}-i}\left(\Gamma - \frac{1+E\mathcal{L}}{\mathcal{L}-E}\right)\Xi_+(\Psi).$$

Then the element Ψ is given by

$$\begin{aligned} \Psi &= \hat{\Psi} + \frac{1}{\mathcal{L}-i}(\Gamma+\mathcal{L})\Xi_+(\Psi) \\ &= \frac{1}{\mathcal{L}-i}\left(-\Gamma + \frac{1+E\mathcal{L}}{\mathcal{L}-E} + \Gamma + \mathcal{L}\right)\Xi_+(\Psi) \tag{2.67} \\ &= \frac{\mathcal{L}+i}{\mathcal{L}-E}\Xi_+(\Psi). \end{aligned}$$

The external component of the eigenfunction is given by the following expression

$$\psi(x) = c\frac{e^{-\sqrt{-E}|x|}}{4\pi|x|}a_+. \tag{2.68}$$

The internal component is given by

$$\psi'(x) = \frac{1}{L'-E}\varphi'a'_+ = -\frac{\gamma_{10}}{\gamma_{11}-Q_1(E)}a_+\frac{1}{L'-E}\varphi'. \tag{2.69}$$

The constant a_+ can be fixed from the normalizing condition $\|\Psi\| = 1$. The norm of the calculated eigenfunction is given by

$$\begin{aligned} \|\Psi\|^2 &= \|\psi\|^2 + \|\psi'\|'^2 \\ &= \left(c^2\int_0^\infty \frac{\exp(-2\sqrt{-E}r)}{4\pi r^2}r^2 dr + \left\|\frac{1}{L'-E}\varphi'\right\|'^2\left|\frac{\gamma_{10}}{\gamma_{11}-Q_1(E)}\right|^2\right)|a_+|^2 \\ &= \left(\frac{1}{\sqrt{-2E}} + \left\|\frac{1}{L'-E}\varphi'\right\|'^2\left|\frac{\gamma_{10}}{\gamma_{11}-Q_1(E)}\right|^2\right)|a_+|^2. \end{aligned}$$

Thus the normalized eigenfunction $\Psi(E)$ is given by the following expression

$$\Psi(E) = \frac{1}{\sqrt{\frac{1}{\sqrt{-2E}} + \| \frac{1}{L'-E}\varphi' \|^2 |\frac{\gamma_{10}}{\gamma_{11}-Q_1(E)}|^2}} \begin{pmatrix} c\dfrac{e^{-\sqrt{-E}|x|}}{4\pi|x|} \\[2mm] -\dfrac{\gamma_{10}}{\gamma_{11}-Q_1(E)}\dfrac{1}{L'-E}\varphi' \end{pmatrix}.$$

(2.70)

The continuous spectrum eigenfunctions can be calculated as follows. Consider the kernel $r_{00}(\lambda,x,y)$ of the resolvent restricted to the external space

$$r_{00}(\lambda,x,y) = \frac{e^{i\sqrt{\lambda}|x-y|}}{4\pi|x-y|} - \frac{e^{i\sqrt{\lambda}|x|}}{4\pi|x|}\frac{1}{\frac{i\sqrt{\lambda}}{4\pi}+\tilde{Q}_0^+(\lambda)}\frac{e^{i\sqrt{\lambda}|y|}}{4\pi|y|}.$$

Considering the limit of the latter expression where $y \to \nu\infty$, $\nu \in \mathbf{R}^3$, $|\nu| = 1$ and $\lambda \to E \in \mathbf{R}_+$ one can calculate the external component of the continuous spectrum eigenfunction

$$\psi(E,\nu,x) = e^{-i\sqrt{E}\langle x,\nu\rangle} - \frac{e^{i\sqrt{E}|x|}}{4\pi|x|}\frac{1}{\frac{i\sqrt{E}}{4\pi}+\tilde{Q}_0^+(E)}.$$

(2.71)

The continuous spectrum eigenfunctions are parameterized by the energy parameter $E \in [0,\infty)$ and the direction of the incoming plane wave $\nu \in \mathbf{R}^3$, $\| \nu \|= 1$. The external component of the eigenfunction satisfies the energy dependent boundary conditions (2.66). The internal component of the eigenfunction is given by

$$\begin{aligned} \psi'(E,\nu) &= \frac{c}{L'-E}\varphi'\frac{\gamma_{10}}{\gamma_{11}-Q_1(E)}\frac{1}{Q_0(E)+Q_0^+(E)} \\[2mm] &= \frac{c}{L'-E}\varphi'\frac{\gamma_{10}}{(Q_0(E)-\gamma_{00})(\gamma_{11}-Q_1(E))+|\gamma_{01}|^2}. \end{aligned}$$

(2.72)

The scattering amplitude can be calculated from the asymptotics of the external component

$$\rho(E) = \frac{1}{\frac{i\sqrt{E}}{4\pi}+\tilde{Q}_0^+(E)}.$$

It does not depend on the direction of the incoming plane wave and this shows that the scattering is nontrivial in the S-channel only. The kernel of the corresponding scattering matrix is given by

$$
\begin{aligned}
s(E,\omega,\nu) &= \delta(\omega+\nu) - \frac{i\sqrt{E}}{2\pi}\frac{1}{\frac{i\sqrt{E}}{4\pi}+\tilde{Q}_0^+(E)} \\
&= \delta(\omega+\nu) - \frac{i\sqrt{E}}{2\pi}\frac{c^2}{Q_0(E)+Q_0^+(E)}.
\end{aligned}
\tag{2.73}
$$

The vectors ω and ν denote the directions of the outgoing and incoming plane waves in the latter formula.

The singularities of the scattering matrix are determined by equation (2.64). These singularities are situated on the negative part of the real axis and on the nonphysical sheet. The singularities of the first kind coincide with the bound states. The other singularities are the resonances. We are going to study the scattering matrices for generalized perturbations in more detail in Chapter 4. This model of the generalized point interaction in dimension three was first considered by B.Pavlov [781].

2.3.2 Generalized delta interaction in dimension one

This section is devoted to the construction of the model operator describing the two-body problem on the line. This problem is usually described by the Schrödinger operator with the potential depending on the distance between the particles. Our aim in this section is to construct the corresponding model operator with the generalized singular interaction. This operator will be used in Chapter 7 to construct the model operator discribing the system of three one-dimensional quantum particles. We confine our consideration to the case where the dimension of the extension space is finite.

Schrödinger operator for two one-dimensional particles

In this section we recall some facts concerning the Schrödinger operator describing system of two one-dimensional particles. We confine our consideration to the case of particles having equal masses. The Schrödinger operator can be written in the centre of mass system as follows

$$
\mathcal{A}_V = -\frac{d^2}{dx^2} + V(|\,x\,|),
\tag{2.74}
$$

where x is the distance between the particles and V is the interaction potential. The operator $\mathcal{A}_V$ has been studied as a self-adjoint operator in the

Hilbert space $L_2(\mathbf{R})$ for potentials with finite first momentum [368, 365, 363, 367, 7, 7, 237, 707]:

$$\int_{-\infty}^{+\infty} |x\, V(x)|dx < \infty. \qquad (2.75)$$

The scattering problem for the two-body Schrödinger operator is formulated with the unperturbed operator equal to the second derivative operator

$$\mathcal{A} = -\frac{d^2}{dx_{12}^2}, \qquad (2.76)$$

defined on the standard domain $\text{Dom}\,(\mathcal{A}) = W_2^2(\mathbf{R})$. The unperturbed and perturbed operators have the same branch of absolutely continuous spectrum $[0, \infty)$. The perturbed operator $\mathcal{A}_V$ can have some additional negative eigenvalues – two-body bound states. Let $f_-(x, k), f_+(x, k)$, $k \in \mathbf{R} \setminus \{0\}$, be the solutions of the equation

$$-\frac{d^2 f(x, k)}{dx^2} + V(x)f(x, k) = k^2 f(x, k)$$

in the generalized sense having the following asymptotics

$$f_-(x, k) \sim e^{ikx}, \quad x \to +\infty,$$

$$f_+(x, k) \sim e^{-ikx}, \quad x \to -\infty. \qquad (2.77)$$

The solutions $f_j(k, x)$ are asymptotic to sums of exponentials as $x \to \mp\infty$

$$f_-(x, k) \sim \frac{1}{T_-(k)}e^{ikx} + \frac{R_-(k)}{T_-(k)}e^{-ikx}, \quad x \to -\infty,$$

$$f_+(x, k) \sim \frac{1}{T_+(k)}e^{-ikx} + \frac{R_+(k)}{T_+(k)}e^{ikx}, \quad x \to +\infty. \qquad (2.78)$$

The matrix

$$S(k) = \begin{pmatrix} T_+(k) & R_-(k) \\ R_+(k) & T_-(k) \end{pmatrix}$$

is called the scattering matrix. This matrix is unitary

$$|T_-|^2 + |R_-|^2 = 1 = |T_+|^2 + |R_+|^2,$$

$$T_-(k)R_+(-k) + R_-(k)T_+(-k) = 0.$$

One can prove that the transition coefficients coincide and that the following asymptotics for the coefficients of the scattering matrix are valid [363] when

$k \to \infty$

$$T_+(k) = T_-(k) = 1 + O\left(\frac{1}{|k|}\right);$$

$$R_-(k) = O\left(\frac{1}{|k|}\right); \quad R_+(k) = O\left(\frac{1}{|k|}\right).$$

(2.79)

The following estimates are valid in the low-energy domain

$$T_\pm(k) = O(k), \quad R_\pm(k) = -1 + O(k), \quad k \to 0. \qquad (2.80)$$

These asymptotics will be called "standard" in what follows. The model operators which will be constructed in the following section define unitary scattering matrices. But the coefficients of these matrices do not necessarily have standard asymptotics. In order to make the model realistic we are going to confine our consideration to model operators with standard asymptotics of the scattering matrix.

Model operator

Consider the unpertubed operator $\mathbf{A} = A \oplus A'$ defined in the orthogonal sum of the Hilbert spaces $\mathbf{H} = L_2(\mathbf{R}) \oplus \mathcal{H}'$, where

- A is the second derivative operator $A = -\frac{d^2}{dx^2}$ defined on the domain $\mathrm{Dom}\,(A) = W_2^2(\mathbf{R})$ in the Hilbert space $\mathcal{H} = L_2(\mathbf{R})$;

- A' is a Hermitian operator (matrix) acting in the finite dimensional Hilbert space $\mathcal{H}' = \mathbf{C}^n$.

Our aim is to construct the second derivative operator with the generalized δ-interaction at the origin. Therefore to define the restricted operator we use a vector $\varphi = c\delta$, where δ is the Dirac delta function and the normalizing constant $c = 2^{3/4}$ is chosen in order to fulfil the normalizing condition $\langle \varphi, \frac{1}{A^2+1}\varphi \rangle' = 1$. Then the restricted operator A^0 is defined on the domain $\mathrm{Dom}\,(A^0) = \{\psi \in W_2^2(\mathbf{R}) : \psi(0) = 0\}$. Every function ψ from the domain of the adjoint operator can be written using the standard representation (2.32)

$$
\begin{aligned}
\psi(x) = \ & \hat{\psi}(x) + a_+(\Psi)\frac{e^{-|x|/\sqrt{2}}}{2^{3/4}}\left(\sin(|x|/\sqrt{2}) - \cos(|x|/\sqrt{2})\right) \\
& -a_-(\Psi)\frac{e^{-|x|/\sqrt{2}}}{2^{3/4}}\left(\sin(|x|/\sqrt{2}) + \cos(|x|/\sqrt{2})\right).
\end{aligned}
$$

(2.81)

Let u, v be two functions from the domain of the adjoint operator. Then the boundary form of the adjoint operator is given by

$$\langle u, A^{0*}v\rangle - \langle A^{0*}u, v\rangle = \overline{a_+(U)}a_-(V) - \overline{a_-(U)}a_+(V), \qquad (2.82)$$

since the element φ is normalized.

Every function from the domain of the adjoint operator is continuous at the origin

$$\psi(-0) = \psi(+0) \equiv \psi(0),$$

but the first derivative can have a nontrivial jump discontinuity $[d\psi/dx]$ at the origin

$$[d\psi/dx](0) = \frac{d\psi}{dx}(+0) - \frac{d\psi}{dx}(-0).$$

The boundary values $\psi(0), [d\psi/dx](0)$ can easily be calculated from $a_\pm(\Psi)$

$$\begin{cases} \psi(0) & = \quad -2^{-3/4}(a_+(\Psi) + a_-(\Psi)) \\ [d\psi/dx](0) & = \quad 2^{3/4}a_+(\Psi). \end{cases} \qquad (2.83)$$

Then the domain of the adjoint operator A^{0*} is given by $\mathrm{Dom}\,(A^{0*}) = \{\psi \in W_2^2(\mathbf{R}\setminus\{0\}) : \psi(-0) = \psi(+0)\}$. The boundary form of the operator A^{0*} can be written in terms of $\psi(0)$ and $[d\psi/dx](0)$ as follows

$$\langle u, A^{0*}v\rangle - \langle A^{0*}u, v\rangle = \overline{u(0)}[dv/dx](0) - \overline{[du/dx](0)}v(0). \qquad (2.84)$$

In what follows we are going to use the boundary values $\psi(0), [d\psi/dx](0)$ instead of $a_+(\Psi), a_-(\Psi)$, since the formulas for the boundary form of the adjoint operator written in terms of the boundary values $a_+(\Psi), a_-(\Psi)$ and $\psi(0), [d\psi/dx](0)$ are identical (compare (2.82) and (2.84)).

To define the perturbed operator we choose an arbitrary matrix B satisfying the conditions (2.51) and the vector $\varphi' \in \mathcal{H}', \|\frac{1}{A'+i}\varphi'\|'= 1$. Then the perturbed operator $\mathbf{A}_B$ is defined on the domain of functions from

$$\mathrm{Dom}\,(A^{0*}) \oplus \mathcal{H}' = \{\psi \in W_2^2(\mathbf{R}\setminus\{0\}) : \psi(-0) = \psi(+0)\} \oplus \mathcal{H}' \ni (\psi, \psi')$$

satisfying the boundary conditions

$$\langle \varphi', \psi'\rangle' = \beta_{21}\psi(0) + \beta_{22}[d\psi/dx](0) \qquad (2.85)$$

by the formula

$$\mathbf{A}_B \begin{pmatrix} \psi \\ \psi' \end{pmatrix} = \begin{pmatrix} A^{0*}\psi \\ A'\psi' + (\beta_{11}\psi(0) + \beta_{12}[d\psi/dx](0))\,\varphi' \end{pmatrix}. \qquad (2.86)$$

The operator $\mathbf{A}_B$ is a rank two perturbation of the operator $\mathbf{A}$ and its resolvent can be calculated using Theorem 2.2.3. We are going to calculate this resolvent explicitly. The following function will play an important role in the sequel:

$$D(\lambda) = \frac{\beta_{11}F_1(\lambda) + \beta_{21}}{\beta_{12}F_1(\lambda) + \beta_{22}}, \qquad (2.87)$$

where the function $F_1(\lambda) = \langle \varphi', \frac{1}{A'-\lambda}\varphi'\rangle'$ is analytic in the upper half plane $\Im\lambda > 0$ and has positive imaginary part there. Conditions (2.51) imply that the function $D(\lambda)$ is analytic in the upper half plane and has positive imaginary part there. Moreover the function $F_1(\lambda)$ is rational and therefore the function $D(\lambda)$ is also rational.

Lemma 2.3.1 *The resolvent of the perturbed operator $\mathbf{A}_B$ is given by the following formula*

$$\frac{1}{\mathbf{A}_B - \lambda} = \frac{1}{\mathbf{A} - \lambda} - \begin{pmatrix} \Delta R_{00}(\lambda) & \Delta R_{01}(\lambda) \\ \Delta R_{10}(\lambda) & \Delta R_{11}(\lambda) \end{pmatrix},$$

where the operators ΔR_{ik} are defined as follows

- $$\Delta R_{00}(\lambda) = \frac{D(\lambda)}{D(\lambda) + 2i\sqrt{\lambda}} \left(\int_{-\infty}^{+\infty} \frac{e^{i\sqrt{\lambda}|y|}}{2i\sqrt{\lambda}} \cdot dy \right) e^{i\sqrt{\lambda}|x|}, \qquad (2.88)$$

- $$\Delta R_{10}(\lambda) = \frac{\beta_{11} - \beta_{12}D(\lambda)}{D(\lambda) + 2i\sqrt{\lambda}} \left(\int_{-\infty}^{+\infty} e^{i\sqrt{\lambda}|y|} \cdot dy \right) \frac{1}{A'-\lambda}\varphi', \qquad (2.89)$$

- $$\Delta R_{01}(\lambda) = \frac{-1}{(2i\sqrt{\lambda} + D(\lambda))(\beta_{22} + \beta_{12}F_1(\lambda))} \left\langle \varphi', \frac{1}{A'-\lambda} \cdot \right\rangle e^{i\sqrt{\lambda}|x|}, \qquad (2.90)$$

- $$\Delta R_{11}(\lambda) = \frac{2\beta_{12}i\sqrt{\lambda} + \beta_{11}}{(2\beta_{12}i\sqrt{\lambda} + \beta_{11})F_1(\lambda) + 2\beta_{22}i\sqrt{\lambda} + \beta_{21}}$$
$$\left\langle \varphi', \frac{1}{A'-\lambda} \cdot \right\rangle \frac{1}{A'-\lambda}\varphi'. \qquad (2.91)$$

Proof Consider an arbitrary $\Psi = (\psi, \psi') \in \mathbf{H} = \mathcal{H} \oplus \mathcal{H}'$. Let $\dfrac{1}{\mathbf{A}_B - \lambda} \Psi = G$. This implies that $G \in \mathrm{Dom}\,(\mathbf{A}_B)$ and $\Psi = (\mathbf{A}_B - \lambda)G$. The latter equation can be written for the components as follows

$$\left(-\frac{d^2}{dx^2} - \lambda\right) g(x) = \psi(x), \quad x \neq 0; \tag{2.92}$$

$$A'g' + \{\beta_{11}g(0) + \beta_{12}[dg/dx](0)\}\,\varphi' - \lambda g' = \psi'.$$

We apply the operator $\dfrac{1}{A' - \lambda}$ to the left and right hand sides of the latter equation

$$g' = \frac{1}{A' - \lambda}\psi' - \{\beta_{11}g(0) + \beta_{12}[dg/dx](0)\}\,\frac{1}{A' - \lambda}\varphi'.$$

The projection on the element φ' gives the following relation

$$\langle \varphi', g' \rangle' = \left\langle \varphi', \frac{1}{A' - \lambda}\psi' \right\rangle' - \{\beta_{11}g(0) + \beta_{12}[dg/dx](0)\}\, F_1(\lambda). \tag{2.93}$$

Every solution to (2.92) which is continuous at the origin is given by

$$g(x) \;=\; \left(\frac{1}{A - \lambda}\psi\right)(x) + q e^{i\sqrt{\lambda}|x|}$$

$$=\; \int_{-\infty}^{+\infty} \frac{e^{i\sqrt{\lambda}|y-x|}}{2i\sqrt{\lambda}}\psi(y)dy + q e^{i\sqrt{\lambda}|x|},$$

where $q \in \mathbf{C}$ is a parameter which will be fixed later. The boundary values of the function $g(x)$ at the origin are equal to

$$\begin{cases} g(0) & = & \displaystyle\int_{-\infty}^{\infty} \frac{e^{i\sqrt{\lambda}|y|}}{2i\sqrt{\lambda}}\psi(y)dy + q; \\[3mm] [dg/dx](0) & = & 2i\sqrt{\lambda}q. \end{cases}$$

The element G belongs to the domain of the operator $\mathbf{A}_B$ and satisfies the boundary conditions (2.85). It follows from (2.93) that the boundary values of $g(x)$ should satisfy the following equation

$$\{\beta_{21} + \beta_{11}F_1(\lambda)\}\, g(0) + \{\beta_{22} + \beta_{12}F_1(\lambda)\}\,[dg/dx](0)$$

$$= \langle \varphi', \frac{1}{A' - \lambda}\psi' \rangle'.$$

The parameter q can now be calculated

$$q = \frac{1}{2i\sqrt{\lambda} + D(\lambda)}$$
$$\times \left(\frac{\langle \varphi', \frac{1}{A'-\lambda}\psi' \rangle'}{\beta_{22} + \beta_{12}F_1(\lambda)} - D(\lambda) \int_{-\infty}^{+\infty} \frac{e^{i\sqrt{\lambda}|y|}}{2i\sqrt{\lambda}} \psi(y) dy \right).$$

It follows that

$$g(x) = \left(\frac{1}{A-\lambda}\psi \right)(x) + \frac{e^{i\sqrt{\lambda}|x|}}{2i\sqrt{\lambda} + D(\lambda)} \left(\frac{\langle \varphi', \frac{1}{A'-\lambda}\psi' \rangle'}{\beta_{22} + \beta_{12}F_1(\lambda)} \right.$$
$$\left. -D(\lambda) \int_{-\infty}^{+\infty} \frac{e^{i\sqrt{\lambda}|y|}}{2i\sqrt{\lambda}} \psi(y) dy \right),$$

$$g' = \frac{1}{A'-\lambda}\psi'$$

$$-\frac{1}{A'-\lambda}\varphi' \left(\frac{2i\sqrt{\lambda}(\beta_{11} - \beta_{12}D(\lambda))}{2i\sqrt{\lambda} + D(\lambda)} \int_{-\infty}^{+\infty} \frac{e^{i\sqrt{\lambda}|y|}}{2i\sqrt{\lambda}} \psi(y) dy \right.$$
$$\left. + \frac{\beta_{11} + 2i\sqrt{\lambda}\beta_{12}}{2i\sqrt{\lambda}(\beta_{22} + \beta_{12}F_1(\lambda)) + \beta_{21} + \beta_{11}F_1(\lambda)} \langle \varphi', \frac{1}{A'-\lambda}\psi' \rangle' \right).$$

Formulas (2.88),(2.89),(2.90),(2.91) follow from the last two equations. The lemma is proven.

□

The resolvent of the operator $\mathbf{A}_B$ restricted to the original Hilbert space $\mathcal{H} = L_2(\mathbf{R})$ is given by

$$P_{\mathcal{H}} \frac{1}{\mathbf{A}_B - \lambda}|_{\mathcal{H}} = \int_{-\infty}^{+\infty} \frac{e^{i\sqrt{\lambda}|y-x|}}{2i\sqrt{\lambda}} \cdot dy$$

$$-\frac{D(\lambda)}{D(\lambda) + 2i\sqrt{\lambda}} \left(\int_{-\infty}^{+\infty} \frac{e^{i\sqrt{\lambda}|y|}}{2i\sqrt{\lambda}} \cdot dy \right) e^{i\sqrt{\lambda}|x|}. \tag{2.94}$$

This generalized resolvent coincides with the resolvent of the differential expression $-d^2/dx^2$ defined on functions from $W_2^2(\mathbf{R}\backslash\{0\})$ satisfying the energy dependent boundary conditions at the origin:

$$\begin{cases} \psi(-0) = \psi(+0); \\ [d\psi/dx](0) = -D(\lambda)\psi(\pm 0). \end{cases} \tag{2.95}$$

The latter differential expression does not define any operator in the Hilbert space $L_2(\mathbf{R}) = \mathcal{H}$, since the boundary conditions depend on the energy parameter λ. In the physics literature the latter expression is used to define the second derivative operator with the energy dependent delta interaction. The advantage of the approach considered here is that the differential expression with energy dependent boundary conditions appears as the restriction of the self–adjoint operator $\mathbf{A}_B$ to the original Hilbert space $\mathcal{H}$.

The singularities of the resolvent $\dfrac{1}{\mathbf{A}_B - \lambda}$ on the physical sheet are situated at the points which satisfy the equation $D(\lambda) + 2i\sqrt{\lambda} = 0$. They correspond to the eigenvalues of the operator $\mathbf{A}_B$. The absolutely continuous spectrum of the operator is determined by the discontinuity of the resolvent on the positive part of the real axis due to the discontinuity of the function $\sqrt{\lambda}$ there.

The discrete spectrum eigenfunctions are solutions of the equation $\mathbf{A}_B \Psi_s = \lambda_s \Psi_s$, where $\lambda_s = -\chi^2 < 0$, $\chi > 0$, $s = 1, 2, ..., N_{bs}^2$, are the negative real solutions of the equation

$$2\chi_s = D(-\chi_s^2).$$

The eigenfunctions can be explicitly calculated

$$\Psi_s = c_s \begin{pmatrix} \psi_{s0} \\ \psi_{s1} \end{pmatrix},$$

$$\psi_{s0}(x) = e^{-\chi_s |x|},$$

$$\psi_{s1} = (2\beta_{12}\chi_s - \beta_{11})\frac{1}{A' + \chi^2}\varphi'. \tag{2.96}$$

The constant c_s can be determined from the normalizing condition $\| \Psi_s \|_{\mathcal{H}} = 1$:

$$c_s = \left(\frac{1}{\chi_s} + |2\beta_{12}\chi_s - \beta_{11}|^2 \left\|\frac{1}{A' + \chi_s^2}\varphi'\right\|^2\right)^{-1/2}. \tag{2.97}$$

The continuous spectrum eigenfunctions $\Psi(\lambda) = (\psi(\lambda), \psi'(\lambda))$, $\lambda \geq 0$, are generalized solutions of the following equation

$$\begin{pmatrix} -\frac{d^2}{dx^2}\psi \\ A'\psi' + (\beta_{11}\psi(0) + \beta_{12}[d\psi/dx](0))\,\varphi' \end{pmatrix} = \lambda \begin{pmatrix} \psi \\ \psi' \end{pmatrix}, \tag{2.98}$$

satisfying the boundary conditions (2.85). The latter equation can be reduced to the usual one dimensional Schrödinger equation on the axis with energy dependent boundary conditions at the origin. This reduction is similar to the one which we have carried out calculating the resolvent of the total operator. The second equation (2.98) implies that

$$\psi' = -\{\beta_{11}\psi(0) + \beta_{12}[d\psi/dx](0)\}\frac{1}{A'-\lambda}\varphi'.$$

Substitution into the boundary conditions (2.85) gives the energy dependent boundary conditions (2.95) for the component ψ of the eigenfunction.

The multiplicity of the continuous spectrum is equal to 2. As in (2.77) the following representations for the eigenfunctions can be used

$$\Psi_{\pm}(\lambda) = \frac{1}{2\sqrt{\pi k}}\begin{pmatrix}\psi_{\pm 0}(\lambda)\\ \psi_{\pm 1}(\lambda)\end{pmatrix}, \quad \lambda = k^2$$

$$\psi_{-0}(\lambda, x) = \begin{cases} e^{ikx} + R_-(k)e^{-ikx}; & x < 0 \\ T_-(k)e^{ikx}; & x > 0 \end{cases}$$

$$\psi_{+0}(\lambda, x) = \begin{cases} T_+(k)e^{-ikx}; & x < 0 \\ e^{-ikx} + R_+(k)e^{ikx}; & x > 0 \end{cases}$$

The left and right reflection and transition coefficients are identical due to the symmetry of the problem

$$R_-(k) = R_+(k) \equiv R(k), \quad T_-(k) = T_+(k) \equiv T(k).$$

The transition and reflection coefficients are calculated from the energy dependent boundary conditions (2.95)

$$\begin{aligned} T(k) &= \frac{2ik}{D(\lambda) + 2ik} \\ R(k) &= \frac{-D(\lambda)}{D(\lambda) + 2ik}. \end{aligned} \tag{2.99}$$

The components $\psi_{\pm 1}(\lambda)$ of the eigenfunctions are identical

$$\psi_{\pm 1}(\lambda) = \psi'(\lambda) = i\frac{\sqrt{k}}{\sqrt{\pi}}\frac{\beta_{12}D(k^2) - \beta_{11}}{D(k^2) + 2ik}\frac{1}{A'-k^2}\varphi'. \tag{2.100}$$

The reflection and transition coefficients form the unitary scattering matrix

$$S(k) = \begin{pmatrix} T(k) & R(k) \\ R(k) & T(k) \end{pmatrix}. \tag{2.101}$$

This is the scattering matrix for the operators $\mathbf{A}_B$ and $\mathbf{A}$. The unitarity of the scattering matrix we calculated can be proven directly using the fact that the function $D(k^2)$ is real for the real values of the parameter k.

The discrete spectrum eigenfunctions Ψ_s and continuous spectrum eigenfunctions $\Psi_{\pm}(\lambda)$ define the spectral decomposition of the operator $\mathbf{A}_B$:

Theorem 2.3.1 *Let $U = (u, u'), V = (v, v') \in H$ have components U, V infinitely differentiable outside the origin with compact support. Then the following formula is valid*

$$< U, V >_H = \sum_{s=1}^{N_{bs}^2} \langle U, \Psi_s \rangle_H \langle \Psi_s, V \rangle_H$$

$$+ \sum_{\alpha=\pm} \int_0^\infty d\lambda \left(\int_{-\infty}^{+\infty} \overline{u(x)} \psi_{\alpha 0}(x) dx + \langle u', \psi_{\alpha 1} \rangle' \right) \quad (2.102)$$

$$\times \left(\int_{-\infty}^{+\infty} \overline{\psi_{\alpha 0}(x)} v(x) dx + \langle \psi_{\alpha 1}, v' \rangle' \right).$$

Moreover if $V \in \text{Dom}(A_B)$ then

$$\langle U, A_B V \rangle_H = \sum_{s=1}^{N_{bs}^2} \lambda_s \langle U, \Psi_s \rangle_H \langle \Psi_s, V \rangle_H$$

$$+ \sum_{\alpha=\pm} \int_0^\infty \lambda d\lambda \left(\int_{-\infty}^{+\infty} \overline{u(x)} \psi_{\alpha 0}(x) dx + \langle u', \psi_{\alpha 1} \rangle' \right) \quad (2.103)$$

$$\times \left(\int_{-\infty}^{+\infty} \overline{\psi_{\alpha 0}(x)} v(x) dx + \langle \psi_{\alpha 1}, v' \rangle' \right).$$

Proof The theorem can be proven by integrating the resolvent of the operator A_B over the contour surrounding the discrete and continuous spectra.

□

Restrictions on the model

Only the model operators with the standard behavior (2.79) of the scattering matrix will be considered in what follows. The function $D(\lambda)$ is a rational function. It is analytic in the upper half plane and has positive imaginary part there. It is real on the real axis. Every such function has the following asymptotics at infinity: $D(\lambda) = c_1 \lambda + c_0 + O(\frac{1}{\lambda}), c_1, c_0 \in \mathbf{R}, c_1 \geq 0$. The transition coefficient $T(k)$ tends to one at infinity if and only if the linear term in the asymptotics is absent ($c_1 = 0$). The function $F_1(\lambda)$ decreases at infinity, since $\varphi' \in \text{Dom}(A')$. Therefore the function $D(\lambda)$ given by (2.87) is bounded at infinity if and only if the parameter β_{22} is different from zero: $\beta_{22} \neq 0$.

The reflection coefficient tends to zero at infinity in this case and the scattering matrix has the standard behavior at infinity (2.79). The scattering matrix for the model operator has standard zero energy behavior if no zero energy bound state is present

$$D(0) \neq 0. \tag{2.104}$$

In the sequel we are going to consider only the model operators with standard behavior of the scattering matrix.

The singularities of the scattering matrix are situated on the positive part of the imaginary axis on the k–plane and in the lower half plane ($k = \sqrt{\lambda}$). These singularities correspond to the bound states and resonances respectively. We are going to consider the case $\beta_{12} = \beta_{21} = 0$. The number of negative eigenvalues of the perturbed and unperturbed operators coincide in this case.

Lemma 2.3.2 *Let $\beta_{12} = \beta_{21} = 0$ and let all the eigenvalues of A' be negative. Then the equation*

$$D(k^2) + 2ik = 0 \tag{2.105}$$

has exactly $N' = \dim \mathcal{H}'$ solutions in the upper half plane $\Im k > 0$. All these solutions are situated on the imaginary axis.

Proof The solutions of the equation on the physical sheet are situated on the imaginary axis because the functions $D(\lambda)$ and $2i\sqrt{\lambda}$ have imaginary parts with the same sign on the λ-plane outside the real axis, where these functions are real. The function $D(\lambda) = \beta_{11}^2 F_1(\lambda)$ considered on the real axis is a continuous increasing function on each interval not containing the singularities which coincide with the eigenvalues of the operator A'. The number of singularities on the negative half axis coincides with N', since all the eigenvalues of the operator A' are negative. The function $2i\sqrt{\lambda}$ is a negative increasing function. It follows that the equation has exactly N' solutions in the upper half plane $\Im k > 0$ and all these solutions are situated on the imaginary axis. $\square$

The constant β_{11} can be considered as a perturbation parameter. The eigenvalues of the operator $\mathbf{A}_B$ tend to the eigenvalues of the operator A' in the limit $\beta_{11} \to 0$.

Equation (2.105) has exactly $2N' + 1$ solutions since all these solutions are roots of a polynomial in k of degree $2N' + 1$. The number of the solutions on the nonphysical sheet $\Im k < 0$ is equal to $N' + 1$. We suppose that all these solutions are situated on the imaginary axis. This is true if the parameter β_{11} is small. In the general situation only $N' - 1$ solutions are situated on the imaginary axis in the lower half plane and two solutions can have nontrivial real part.

The model operator we considered will be used in Chapter 7 to construct the operator describing a system of three one-dimensional quantum particles interacting by two-body delta interactions.

Chapter 3

Finite rank perturbations and distribution theory

3.1 Finite rank perturbations

3.1.1 Preliminaries

Finite rank perturbations of self–adjoint operators are studied in this chapter. Our attention will be concentrated on singular perturbations. We follow the main ideas developed in Chapter 1 for perturbations of rank one.

Additive finite rank perturbations of a self–adjoint operator $\mathcal{A}$ acting in the Hilbert space $\mathcal{H}$ are given formally by

$$\mathcal{A}_T = \mathcal{A} + T, \tag{3.1}$$

where the operator T is a finite dimensional operator acting from the Hilbert space $\mathcal{H}_2(\mathcal{A})$ to the Hilbert space $\mathcal{H}_{-2}(\mathcal{A})$. We consider the case where the operator T is a symmetric operator acting in these Hilbert spaces, i.e. for every two functions $u, v \in \mathcal{H}_2(\mathcal{A})$ the following equality holds

$$\langle u, Tv \rangle = \langle Tu, v \rangle. \tag{3.2}$$

The fact that the image of the operator T has finite dimension implies that there exist linearly independent elements $\varphi_j, j = 1, 2, ..., n$, from the space $\mathcal{H}_{-2}(\mathcal{A})$ which span the image space, i.e. the following formula holds

$$Tu = \sum_{j=1}^{n} \psi_j(u)\varphi_j, \tag{3.3}$$

where ψ_j are certain linear bounded functionals on $\mathcal{H}_2(\mathcal{A})$. Formula (3.2) implies that the functionals ψ_j are equal to the linear combinations of the functionals φ_j, i.e. there exists a matrix $\mathbf{T} = \{t_{ij}\}_{i,j=1}^{n}$, $t_{ij} \in \mathbf{C}$, such that the following equality holds

$$\psi_j = \sum_{i=1}^{n} \overline{t_{ji}}\varphi_i. \qquad (3.4)$$

Moreover the latter representation and formula (3.2) imply that

$$0 = \langle u, Tv \rangle - \langle Tu, v \rangle$$

$$= \sum_{j,i=1}^{n} (t_{ji} - \overline{t_{ij}}) \langle \varphi_j, v \rangle \overline{\langle \varphi_j, u \rangle}.$$

Therefore the matrix $\mathbf{T}$ must in fact be Hermitian $t_{ji} = \overline{t_{ij}}$. Thus every symmetric finite rank perturbation of a self–adjoint operator is determined by a finite set of elements $\varphi_j, j = 1, 2, ..., n$, from the Hilbert space $\mathcal{H}_{-2}(\mathcal{A})$ and a Hermitian $n \times n$ matrix $\mathbf{T} = \{t_{ij}\}_{i,j=1}^{n}$. We suppose in addition that the matrix $\mathbf{T}$ is invertible. The latter assumption does not restrict the set of perturbations considered. If the matrix $\mathbf{T}$ is not invertible, i.e. has zero determinant, then let us denote by $\mathrm{Ker}\,(T)$ the kernel of the operator T. Considering the orthogonal complement to the subspace $\mathrm{Ker}(T)$ we get a finite rank operator of order less than n determined by a nondegenerate Hermitian matrix. Thus every additive symmetric finite rank perturbation of the operator $\mathcal{A}$ is given on the domain of $\mathcal{A}$ by

$$\mathcal{A}_T = \mathcal{A} + \sum_{i,j=1}^{n} t_{ji}\langle \varphi_i, \cdot \rangle \varphi_j, \qquad (3.5)$$

where the matrix $\mathbf{T}$ is invertible and Hermitian and the vectors $\varphi_j, j = 1, ..., n$, are elements from the Hilbert space $\mathcal{H}_{-2}(\mathcal{A})$. We suppose that the vectors φ_j form an orthonormal system in the Hilbert space $\mathcal{H}_{-2}(\mathcal{A})$:

$$\langle \frac{1}{\mathcal{A}-i}\varphi_i, \frac{1}{\mathcal{A}-i}\varphi_j \rangle = \delta_{i,j},$$

where $\delta_{i,j}$ is the Kronecker symbol. The following definition will be used

Definition 3.1.1 *The set of vectors $\varphi_j \in \mathcal{H}_{-2}(\mathcal{A}) \backslash \mathcal{H}, j = 1, 2, ..., n$, is called $\mathcal{H}$-independent if and only if the equality*

$$\sum_{j=1}^{n} c_j \varphi_j \in \mathcal{H}, \quad c_j \in \mathbf{C}$$

always implies

$$c_1 = c_2 = ... = c_n = 0.$$

In this section we are going to study H-independent perturbations only.

The operator $\mathcal{A}_T$ on the domain Dom $(\mathcal{A})$ is not an operator acting in the Hilbert space $\mathcal{H}$. But this operator is symmetric as an operator acting from $\mathcal{H}_2(\mathcal{A}) = $ Dom $(\mathcal{A})$ to $\mathcal{H}_{-2}(\mathcal{A})$. The self-adjoint operator corresponding to the formally symmetric operator (3.5) coincides with one of the self-adjoint extensions of the operator $\mathcal{A}^0$ equal to the operator $\mathcal{A}$ restricted to the following domain

$$\text{Dom}(\mathcal{A}^0) = \text{Dom}(\mathcal{A}) \cap \text{Ker}(T).$$

Lemma 3.1.1 *Suppose that the vectors $\varphi_j \in \mathcal{H}_{-2}(\mathcal{A}) \setminus \mathcal{H}$, $j = 1, 2, ..., n$, are $\mathcal{H}$-independent and form an orthonormal system in $\mathcal{H}_{-2}(\mathcal{A})$. Then the restriction $\mathcal{A}^0$ of the operator $\mathcal{A}$ to the domain $\text{Dom}(\mathcal{A}^0) = \{\psi \in \text{Dom}(\mathcal{A}) : \langle \varphi_j, \psi \rangle = 0, j = 1, 2, ..., n\}$ is a densely defined symmetric operator with the deficiency indices (n, n).*

Proof To prove the lemma we consider the sequence of operators $\mathcal{A}_l^0, l = 0, 1, 2, ..., n$, which are the restrictions of the operator $\mathcal{A}$ to the domains

$$\text{Dom}(\mathcal{A}_l^0) = \{\psi \in \text{Dom}(\mathcal{A}) : \langle \varphi_j, \psi \rangle = 0, j = 1, 2, ..., l\}$$

respectively. Each operator $\mathcal{A}_{l+1}^0$ is a restriction of the operator $\mathcal{A}_l^0$. The first operator in the chain $\mathcal{A}_0^0$ coincides with the operator $\mathcal{A}$. The operator $\mathcal{A}_n^0$ coincides with the operator $\mathcal{A}^0$. The following equality holds for the domains of the operators

$$\text{Dom}(\mathcal{A}_{l+1}^0) = \{\psi \in \text{Dom}(\mathcal{A}_l^0) : \langle \varphi_{l+1}, \psi \rangle = 0\}.$$

Using mathematical induction one can prove that each operator $\mathcal{A}_l^0$ is densely defined. It follows from Lemma 1.2.3 that the operator $\mathcal{A}_1^0$ is densely defined and has deficiency indices $(1, 1)$. Suppose that the operator $\mathcal{A}_l^0$ is densely defined. Then there exists a sequence $f_m \in \text{Dom}(\mathcal{A}_n^0)$ converging to f in the original norm

$$\lim_{m \to \infty} \| f - f_m \| = 0.$$

The functional φ_{l+1} is not a bounded functional on Dom $(\mathcal{A}_l^0)$ with norm equal to the norm in the original Hilbert space $\mathcal{H}$. To prove the latter statement let us suppose that φ_{l+1} is a bounded functional on Dom $(\mathcal{A}_l^0)$. Then there exists a vector $\varphi \in \mathcal{H}$ such that for any $\psi \in \text{Dom}(\mathcal{A}_l^0)$ the following as valid

$$\langle \varphi_{l+1}, \psi \rangle = \langle \varphi, \psi \rangle.$$

This implies that there exist constants $c_1, c_2, ..., c_l$ such that

$$c_1 \varphi_1 + c_2 \varphi_2 + ... + c_l \varphi_l - \varphi_{l+1} = \varphi \in \mathcal{H},$$

which is impossible, since the vectors φ_j are $\mathcal{H}$-independent.

Therefore there exists a sequence $\psi_m \in \text{Dom}\,(\mathcal{A}_n^0)$ with unit norm $\|\,\psi_m\,\|$ $= 1$ such that the sequence $\langle \varphi_{l+1}, \psi_m \rangle$ diverges to infinity when $m \to \infty$. The same arguments as those used in the proof of Lemma 1.2.3 imply that the operator $\mathcal{A}_{l+1}^0$ is also densely defined. Thus the operator $\mathcal{A}^0 = \mathcal{A}_n^0$ is densely defined. Moreover the same arguments imply that the deficiency indices of the symmetric operator $\mathcal{A}^0$ are no larger than (n, n).

To complete the proof it is enough to show that the operator $\mathcal{A}^0$ has at least n linearly independent deficiency elements corresponding to the complex number i. The vectors $\frac{1}{\mathcal{A}-i}\varphi_j, j = 1, 2, ..., n$, defined in the generalized sense are deficiency elements for the operator $\mathcal{A}^0$. These elements are linearly independent, since they form an orthogonal system. This completes the proof of the lemma.

$$\square$$

In the proof of the latter lemma we have not used the fact that the vectors φ_j have unit norm in $\mathcal{H}_{-2}(\mathcal{A})$. In fact every $\mathcal{H}$-independent system $\{\varphi_j\}$ can be orthonormalized. Therefore we are going to use only orthonormal $\mathcal{H}$-independent systems $\{\varphi_j\}_{j=1}^n$ in what follows.

Every element ψ from the domain of the adjoint operator $\text{Dom}(\mathcal{A}^{0*})$ can be presented in the following form

$$\psi = \hat{\psi} + \sum_{j=1}^{n}\left(a_{+j}(\psi)\frac{1}{\mathcal{A}-i}\varphi_j + a_{-j}(\psi)\frac{1}{\mathcal{A}+i}\varphi_j\right),\qquad(3.6)$$

where $\hat{\psi} \in \text{Dom}(\mathcal{A}^0)$, $a_{\pm j}(\psi) \in \mathbf{C}$. We are going to use the following vector notation

$$\vec{a}_{\pm} \equiv \{a_{\pm j}\}_{j=1}^{n}.\qquad(3.7)$$

The adjoint operator $\mathcal{A}^{0*}$ acts as follows on every $\psi \in \text{Dom}(\mathcal{A}^{0*})$:

$$\mathcal{A}^{0*}\left(\hat{\psi} + \sum_{j=1}^{n}\left(a_{+j}(\psi)\frac{1}{\mathcal{A}-i}\varphi_j + a_{-j}(\psi)\frac{1}{\mathcal{A}+i}\varphi_j\right)\right)$$

$$= \mathcal{A}\hat{\psi} + \sum_{j=1}^{n}\left(a_{+j}(\psi)\frac{i}{\mathcal{A}-i}\varphi_j + a_{-j}(\psi)\frac{-i}{\mathcal{A}+i}\varphi_j\right).$$

The boundary form of the adjoint operator is given by

$$\langle \mathcal{A}^{0*}\psi, \eta \rangle - \langle \psi, \mathcal{A}^{0*}\eta \rangle$$

$$= 2i \sum_{j,k=1}^{n} \left(\overline{a_{-j}(\psi)}a_{-k}(\eta) - \overline{a_{+j}(\psi)}a_{+k}(\eta) \right) \left\langle \frac{1}{\mathcal{A}-i}\varphi_j, \frac{1}{\mathcal{A}-i}\varphi_k \right\rangle$$

$$= 2i \sum_{j=1}^{n} \left(\overline{a_{-j}(\psi)}a_{-j}(\eta) - \overline{a_{+j}(\psi)}a_{+j}(\eta) \right)$$

$$= 2i \left(\langle \vec{a}_-(\psi), \vec{a}_-(\eta) \rangle_{\mathbf{C}^n} - \langle \vec{a}_+(\psi), \vec{a}_+(\eta) \rangle_{\mathbf{C}^n} \right),$$

where $\psi, \eta \in \mathrm{Dom}\,(\mathcal{A}^{0*})$ and we have taken into account the fact that the vectors $\frac{1}{\mathcal{A}-i}\varphi_j$ form an orthonormal system.

The self–adjoint extensions of the operator $\mathcal{A}^0$ can be parameterized by $n \times n$ unitary matrices using the von Neumann theory. Let $\mathbf{V} = \{v_{jk}\}_{j,k=1}^{n}$ be such a matrix. The corresponding self–adjoint operator $\mathcal{A}(\mathbf{V})$ coincides with the restriction of the operator $\mathcal{A}^{0*}$ to the domain $\mathrm{Dom}\,(\mathcal{A}(\mathbf{V})) = \{\psi \in \mathrm{Dom}(\mathcal{A}^{0*}): -\mathbf{V}\vec{a}_-(\psi) = \vec{a}_+(\psi)\}$. The extension corresponding to the matrix $\mathbf{V} = I$ coincides with the original operator $\mathcal{A}$.

To define the finite rank perturbations of $\mathcal{A}$ we consider again two scales of Banach spaces. The first set of spaces is the standard scale of Hilbert spaces associated with the operator $\mathcal{A}$

$$\mathcal{H}_2(\mathcal{A}) \subset \mathcal{H}_1(\mathcal{A}) \subset \mathcal{H} \subset \mathcal{H}_{-1}(\mathcal{A}) \subset \mathcal{H}_{-2}(\mathcal{A}).$$

Let the vectors $\varphi_j, j = 1, 2, ..., n$, be $\mathcal{H}$-independent. Then the second scale of Hilbert spaces

$$\mathcal{H}_2(\mathcal{A}) = \mathrm{Dom}(\mathcal{A}) \subset \mathcal{H}_{\{\varphi\}}(\mathcal{A}) \subset \mathcal{H} \subset \mathcal{H}_{\{\varphi\}}(\mathcal{A})^* \subset \mathrm{Dom}(\mathcal{A})^* = \mathcal{H}_{-2}(\mathcal{A}) \tag{3.8}$$

is constructed using the operators $\mathcal{A}$ and T. $\mathcal{H}_{\{\varphi\}}(\mathcal{A})$ denotes in the latter formula the domain of the adjoint operator $\mathcal{A}^{0*}$. The norms in the spaces $\mathcal{H}_2(\mathcal{A})$ and $\mathcal{H}_{-2}(\mathcal{A})$ are equal to the standard norms in these Hilbert spaces. The norm in the space $\mathcal{H}_{\{\varphi\}}(\mathcal{A})$ is defined using the orthonormal basis in the deficiency subspace. Let ψ be an element from the space $\mathcal{H}_{\{\varphi\}}(\mathcal{A}) = \mathrm{Dom}(\mathcal{A}^{0*})$. Then ψ possesses the representation (3.6). This representation can be written as

$$\psi = \hat{\psi} + \sum_{j=1}^{n} a_{+j}(\psi) \left(\frac{i}{\mathcal{A}^2 + 1} + \frac{\mathcal{A}}{\mathcal{A}^2 + 1} \right) \varphi_j$$

$$+ \sum_{j=1}^{n} a_{-j}(\psi) \left(\frac{-i}{\mathcal{A}^2 + 1} + \frac{\mathcal{A}}{\mathcal{A}^2 + 1} \right) \varphi_j \qquad (3.9)$$

$$= \tilde{\psi} + \sum_{j=1}^{n} b_j(\psi) \frac{\mathcal{A}}{\mathcal{A}^2 + 1} \varphi_j,$$

where

$$\tilde{\psi} = \hat{\psi} + \sum_{j=1}^{n} (a_{+j}(\psi) - a_{-j}(\psi)) \frac{i}{\mathcal{A}^2 + 1} \varphi_j \in \text{Dom}(\mathcal{A});$$

$$b_j(\psi) = a_{+j}(\psi) + a_{-j}(\psi) \in \mathbf{C}. \qquad (3.10)$$

The norm in the space $\mathcal{H}_{\{\varphi\}}(\mathcal{A})$ is then defined by the formula

$$\| \psi \|_{\mathcal{H}_{\{\varphi\}}(\mathcal{A})} = \| \tilde{\psi} \|_2 + \sqrt{\sum_{j=1}^{n} |b_j(\psi)|^2}. \qquad (3.11)$$

The space $\mathcal{H}_{\{\varphi\}}(\mathcal{A})$ can be considered as a finite dimensional extension of the space $\mathcal{H}_2(\mathcal{A})$ in the sense that $\mathcal{H}_{\{\varphi\}}(\mathcal{A})$ is isomorphic with $\mathcal{H}_2(\mathcal{A}) \dotplus \mathbf{C}^n$. The natural norm preserving embedding ρ_n of $\mathcal{H}_2(\mathcal{A}) \dotplus \mathbf{C}^n$ into $\mathcal{H}_{\{\varphi\}}(\mathcal{A})$ is defined by the formula (3.9). The norms in the dual spaces are defined correspondingly.

3.1.2   Form bounded finite rank perturbations

Let the operator T be form bounded with respect to the operator $\mathcal{A}$, i.e. the following inequality holds

$$\langle \psi, T\psi \rangle \le a \| \psi \|_1^2 + b \| \psi \|^2 \qquad (3.12)$$

for every ψ from the domain $\text{Dom}(\mathcal{A})$ and some $a, b \ge 0$. This assumption holds if and only if all the vectors $\varphi_j, j = 1, 2, ..., n$, are from the Hilbert space $\mathcal{H}_{-1}(\mathcal{A})$. In fact, let $\varphi_j \in \mathcal{H}_{-1}(\mathcal{A})$. Then the following inequality holds

$$
\begin{aligned}
|\langle \psi, T\psi \rangle| &= \left| \sum_{i,j=1}^n t_{ji} \langle \psi, \varphi_j \rangle \langle \varphi_i, \psi \rangle \right| \\
&\leq \left(\sum_{i,j=1}^n |t_{ij}| \left\| \frac{1}{\sqrt{|\mathcal{A}|} + i} \varphi_j \right\| \left\| \frac{1}{\sqrt{|\mathcal{A}|} + i} \varphi_i \right\| \right) \| (\sqrt{|\mathcal{A}|} - i)\psi \|^2 \\
&= \left(\sum_{i,j=1}^n |t_{ij}| \, \| \varphi_j \|_{-1} \, \| \varphi_i \|_{-1} \right) \| \psi \|_1^2,
\end{aligned}
$$

since the norms of the vectors $\frac{1}{\sqrt{|\mathcal{A}|}+i} \varphi_j$ in the Hilbert space $\mathcal{H}$ can be calculated as follows

$$
\left\| \frac{1}{\sqrt{|\mathcal{A}|} + i} \varphi_j \right\|^2 = \left\langle \varphi_j, \frac{1}{|\mathcal{A}| + 1} \varphi_j \right\rangle = \| \varphi_j \|_{-1}^2 .
$$

Moreover Lemma 1.2.1 can be easily generalized to prove that every finite rank $\mathcal{H}_{-1}$ perturbation is in fact infinitesimally form bounded.

Theorem 3.1.1 *Let $\varphi_j \in \mathcal{H}_{-1}(\mathcal{A}) \setminus \mathcal{H}$ be $\mathcal{H}$-independent and form an orthonormal basis in $\mathcal{H}_{-2}(\mathcal{A})$, i.e. $\langle \frac{1}{\mathcal{A}-i}\varphi_j, \frac{1}{\mathcal{A}-i}\varphi_k \rangle = \delta_{jk}$ and let $\mathbf{T} = \{t_{ij}\}_{i,j=1}^n$ be a Hermitian invertible matrix. Then the self-adjoint operator $\mathcal{A}_T = \mathcal{A} + \sum_{i,j=1}^n t_{ij} \langle \varphi_j, \cdot \rangle \varphi_i$ is the self-adjoint restriction of the operator $\mathcal{A}^{0*}$ to the following domain*

$$
\mathrm{Dom}(\mathcal{A}_T) = \{ \psi \in \mathrm{Dom}(\mathcal{A}^{0*}) : \vec{a}_+(\psi) = -(\mathbf{T}^{-1} + \mathbf{\Phi})^{-1}(\mathbf{T}^{-1} + \mathbf{\Phi}^*)\vec{a}_-(\psi) \},
$$
$$
(3.13)
$$

where $\mathbf{\Phi}$ is the $n \times n$ matrix $\mathbf{\Phi}_{ij} = \langle \varphi_i, \frac{1}{\mathcal{A}-i}\varphi_j \rangle$, $i, j = 1, 2, ..., n$.

Comment The matrix $\mathbf{T}$ is supposed to be invertible. For $\mathbf{T} = 0$ we have therefore $\mathcal{A}_0 = \mathcal{A}$. The matrix $\mathbf{V} = (\mathbf{T}^{-1} + \mathbf{\Phi})^{-1}(\mathbf{T}^{-1} + \mathbf{\Phi}^*)$ is unitary.

Proof The proof is quite similar to the one of Theorem 1.3.1. The operator $\mathcal{A}_T$ is defined as a linear operator acting in the Hilbert spaces $\mathcal{H}_{\{\varphi\}}(\mathcal{A}) = \mathrm{Dom}(\mathcal{A}^{0*}) \to \mathrm{Dom}(\mathcal{A})^*$. Let $\psi \in \mathcal{H}_{\{\varphi\}}(\mathcal{A})$. Then the linear operator $\mathcal{A}_T$ acts as follows

$$\mathcal{A}_T\psi = \left(\mathcal{A} + \sum_{i,j=1}^{n} t_{ij}\langle\varphi_j, \cdot\rangle\varphi_i\right)$$

$$\left(\hat{\psi} + \sum_{k=1}^{n}\left(a_{+k}(\psi)\frac{1}{\mathcal{A}-i}\varphi_k + a_{-k}(\psi)\frac{1}{\mathcal{A}+i}\varphi_k\right)\right)$$

$$= \mathcal{A}\hat{\psi} + \sum_{i,j=1}^{n} t_{ij}\langle\varphi_j, \hat{\psi}\rangle\varphi_i$$

$$+ \sum_{k=1}^{n}\left(a_{+k}(\psi)(\varphi_k + \frac{i}{\mathcal{A}-i}\varphi_k) + a_{-k}(\psi)(\varphi_k - \frac{i}{\mathcal{A}+i}\varphi_k)\right)$$

$$+ \sum_{i,j,k=1}^{n}\left(a_{+k}t_{ij}\left\langle\varphi_j, \frac{1}{\mathcal{A}-i}\varphi_k\right\rangle\varphi_i + a_{-k}t_{ij}\left\langle\varphi_j, \frac{1}{\mathcal{A}+i}\varphi_k\right\rangle\varphi_i\right)$$

$$= \mathcal{A}^{0*}\psi + \sum_{k=1}^{n}\{a_{+k} + a_{-k}$$

$$+ \sum_{i,j=1}^{n}\left(t_{kj}\left\langle\varphi_j, \frac{1}{\mathcal{A}-i}\varphi_i\right\rangle a_{+i} + t_{kj}\left\langle\varphi_j, \frac{1}{\mathcal{A}+i}\varphi_i\right\rangle a_{-i}\right)\}\varphi_k.$$

In the calculations we used the fact that $\langle\varphi_j, \hat{\psi}\rangle = 0$. The domain $\text{Dom}(\mathcal{A}_T)$ of the self–adjoint operator $\mathcal{A}_T$ coincides with the following set: $\{\psi \in \text{Dom}(\mathcal{A}^{0*}) : \mathcal{A}_T\psi \in \mathcal{H}\}$. The element $\mathcal{A}_T\psi$ belongs to the Hilbert space $\mathcal{H}$ if and only if the expression in the braces { } is equal to zero, i.e. the following equation is satisfied

$$a_{+k} + a_{-k} + \sum_{i,j=1}^{n}\left(t_{kj}\left\langle\varphi_j, \frac{1}{\mathcal{A}-i}\varphi_i\right\rangle a_{+i} + t_{kj}\left\langle\varphi_j, \frac{1}{\mathcal{A}+i}\varphi_i\right\rangle a_{-i}\right) = 0 \quad (3.14)$$

for every $k = 1, 2, ..., n$. The latter equation can be written in the matrix form using (3.7)

$$(I + \mathbf{T}\Phi)\vec{a}_+ = -(I + \mathbf{T}\Phi^*)\vec{a}_-, \quad (3.15)$$

where I denotes the unit $n \times n$ matrix. The matrix $I + \mathbf{T}\Phi$ is invertible, since the matrix $\mathbf{T}$ is an invertible Hermitian matrix and the imaginary part of the matrix Φ is equal to the unit matrix. The latter statement follows from the following formulas

$$\Phi = \Re\Phi + i\Im\Phi,$$

$$(\Re\Phi)_{ij} = \langle\varphi_i, \frac{\mathcal{A}}{\mathcal{A}^2+1}\varphi_j\rangle, \quad (\Im\Phi)_{ij} = \langle\varphi_j, \frac{1}{\mathcal{A}^2+1}\varphi_j\rangle = \delta_{ij}, \qquad (3.16)$$

which are valid, since the vectors φ_j form an orthonormal system in the space $\mathcal{H}_{-2}(\mathcal{A})$. Note that the matrix $\Re\Phi$ is Hermitian but does not necessarily have real coefficients outside the diagonal. Thus condition (3.15) implies

$$\vec{a}_+(\psi) = -\left(\mathbf{T}^{-1}+\Phi\right)^{-1}\left(\mathbf{T}^{-1}+\Phi^*\right)\vec{a}_-(\psi). \qquad (3.17)$$

The matrix $\left(\mathbf{T}^{-1}+\Phi\right)^{-1}\left(\mathbf{T}^{-1}+\Phi^*\right)$ is unitary and the adjoint operator restricted to the domain of functions satisfying the conditions (3.17) is self-adjoint.

If the matrix $\mathbf{T}$ is equal to zero, then (3.14) implies that $\vec{a}_- = -\vec{a}_+$ and the extended operator is just the original operator $\mathcal{A}$. This completes the proof of the theorem.

$\square$

We consider now the set of all self-adjoint extensions of the operator $\mathcal{A}^0$. Every such operator coincides with the restriction of the adjoint operator $\mathcal{A}^{0*}$ to the domain of functions satisfying boundary conditions of the form

$$\vec{a}_+(\psi) = -\mathbf{V}\vec{a}_-(\psi),$$

where $\mathbf{V}$ is a certain unitary $n \times n$ matrix. The domain of the extended self-adjoint operator coincides with the domain of a certain operator $\mathcal{A}_T$ if and only if the following equality holds

$$\mathbf{V} = (\mathbf{T}^{-1}+\Phi)^{-1}(\mathbf{T}^{-1}+\Phi^*). \qquad (3.18)$$

The following lemma describes the set of all self-adjoint extensions of the operator $\mathcal{A}^0$ which can be obtained as finite rank additive perturbations of the operator $\mathcal{A}$ providing that the family of singular vectors φ_j from $\mathcal{H}_{-1}(\mathcal{A})$ remains unchanged.

Lemma 3.1.2 *Let $\varphi_j \in \mathcal{H}_{-1}(\mathcal{A}) \setminus \mathcal{H}, j = 1, 2, ..., n$, be $\mathcal{H}$-independent and form an orthogonal system in $\mathcal{H}_{-2}(\mathcal{A})$. If $\det(\mathbf{V}+\frac{i-\Re\Phi}{i+\Re\Phi}) \neq 0$ then the operator $\mathcal{A}^{0*}$ restricted to the domain of functions $\{\psi \in \mathrm{Dom}(\mathcal{A}^{0*}) : -\mathbf{V}\vec{a}_-(\psi) = \vec{a}_+(\psi)\}$ is a finite dimensional additive perturbation of the operator $\mathcal{A}$.*

Proof Suppose that the matrix $I - \mathbf{V}$ is invertible ($\det(I - \mathbf{V}) \neq 0$). Then the matrix $\mathbf{T}^{-1}$ can be calculated as follows using the representation (3.16)

$$\mathbf{T}^{-1} = -\Phi(I-\mathbf{V}^*)^{-1} - \Phi^*(I-\mathbf{V})^{-1} = -\Re\Phi + i\frac{I+\mathbf{V}}{I-\mathbf{V}}. \qquad (3.19)$$

The determinant of the matrix in the right hand side of the latter equality
is not equal to zero

$$\det\left(-\Re\Phi + i\frac{I+V}{I-V}\right) = (\det(I-V))^{-1}\det(\Re\Phi + i)\det\left(V + \frac{i-\Re\Phi}{i+\Re\Phi}\right).$$

Then the Hermitian matrix $-\Re\Phi + i\dfrac{I+V}{I-V}$ is invertible. The matrix $\mathbf{T}$ can
be reconstructed as follows

$$\mathbf{T} = (-\Re\Phi + i\frac{I+V}{I-V})^{-1}. \tag{3.20}$$

$\mathbf{T}$ is then Hermitian.

Consider now the case where $\det(I-V) = 0$. This equality implies that
the matrix V has a nontrivial eigensubspace N_1 corresponding to the eigenvalue 1. The restriction of the adjoint operator $\mathcal{A}^{0*}$ to the subspace

$$\{\psi \in \mathrm{Dom}(\mathcal{A}^{0*}) : P_{N_1}\vec{a}_+(\psi) = -P_{N_1}\vec{a}_-(\psi)\}$$

coincides with the restriction of the operator $\mathcal{A}$ to the same subspace. Thus
the set of vectors $\{\varphi_j\}_{j=1}^n$ contains extra elements in the following sense: one
can find some new set of elements $\{\varphi_j^*\}_{j=1}^{n^*}$, $n^* < n$, such that the corresponding matrix V^* has a trivial eigensubspace N_1^*. We have already proven that
every unitary matrix which does not have eigenvalue 1 describes a certain
finite rank perturbation of the original operator. The lemma is proven.

□

The latter lemma characterizes the set of self–adjoint extensions of the operator $\mathcal{A}^0$ which are finite rank perturbations of the original operator. The
self–adjoint extensions corresponding to the matrices V satisfying the equality $\det(V + \frac{i-\Re\Phi}{i+\Re\Phi}) = 0$ can be described by finite rank perturbations with
infinite strength using the projective space formalism. This approach is developed in Section 3.2.4 where point interactions for the second derivative
operator in one dimension are studied.

3.1.3   Form unbounded finite rank perturbations

Form unbounded finite rank perturbations can be described following the
main ideas developed in Chapter 1 for form unbounded rank one perturbations. We suppose in this section that the vectors φ_j in the representation (3.5) are elements from the Hilbert space $\mathcal{H}_{-2}(\mathcal{A})$ and that the matrix $\mathbf{T}$ is Hermitian. We consider again only systems of $\mathcal{H}$-independent

vectors. The quadratic form defined by the perturbation is not necessarily bounded with respect to the quadratic form of the operator $\mathcal{A}$. The unique self–adjoint extension of the operator $\mathcal{A}^0$ corresponding to the expression (3.5) defined on the domain $\mathrm{Dom}(\mathcal{A})$ can be determined only if all elements φ_j can be extended as bounded linear functionals to the domain $\mathcal{H}_{\{\varphi\}}(\mathcal{A}) = \mathrm{Dom}(\mathcal{A}^{0*})$. The space $\mathcal{H}_{\{\varphi\}}(\mathcal{A})$ is a finite dimensional extension of the space $\mathcal{H}_2(\mathcal{A}) = \mathrm{Dom}(\mathcal{A})$. Thus the extensions are defined if the following coefficients are determined

$$\Phi_{ij} = \left\langle \varphi_i, \frac{1}{\mathcal{A} - i}\varphi_j \right\rangle = \left\langle \varphi_i, \frac{\mathcal{A}}{\mathcal{A}^2 + 1}\varphi_j \right\rangle + i\left\langle \varphi_i, \frac{1}{\mathcal{A}^2 + 1}\varphi_j \right\rangle. \qquad (3.21)$$

The second scalar product in the right hand side of the latter equality is well defined, since $\varphi_j, \varphi_i \in \mathcal{H}_{-2}(\mathcal{A})$. Therefore to determine the matrix Φ it is enough to find a Hermitian matrix $\mathbf{R}$ with the coefficients

$$\mathbf{R}_{ij} = \left\langle \varphi_i, \frac{\mathcal{A}}{\mathcal{A}^2 + 1}\varphi_j \right\rangle, \qquad (3.22)$$

provided these are well defined.

Let us denote by $\mu_{\varphi_i,\varphi_j}(\lambda)$ the spectral measure (in general complex valued) corresponding to the elements φ_i, φ_j and the operator $\mathcal{A}$:

$$\left\langle \frac{1}{\mathcal{A} - z_i}\varphi_i, \frac{1}{\mathcal{A} - z_j}\varphi_j \right\rangle = \int_{-\infty}^{\infty} \frac{1}{\lambda - \bar{z}_i}\frac{1}{\lambda - z_j}d\mu_{\varphi_i,\varphi_j}(\lambda);$$

$$\int_{-\infty}^{\infty} \frac{1}{\lambda^2 + 1}d|\mu_{\varphi_i,\varphi_j}(\lambda)| \leq \| \varphi_i \|_{\mathcal{H}_{-2}(\mathcal{A})} \| \varphi_j \|_{\mathcal{H}_{-2}(\mathcal{A})} < \infty.$$

If the integral $\int_{-\infty}^{\infty} \frac{\lambda}{\lambda^2+1}|d\mu_{\varphi_i,\varphi_j}(\lambda)|$ is absolutely convergent then the scalar product $\langle \varphi_i, \frac{\mathcal{A}}{\mathcal{A}^2+1}\varphi_j \rangle$ is indeed well defined. The latter integral converges absolutely if both vectors φ_i, φ_j are elements from the space $\mathcal{H}_{-1}(\mathcal{A})$. The same is true if the following inclusions hold: $\varphi_i \in \mathcal{H}_{-2,0}(\mathcal{A}), \varphi_j \in \mathcal{H}_{0,-2}(\mathcal{A})$. (The spaces $\mathcal{H}_{n,m}(\mathcal{A})$ were defined in Section 1.4.)

Definition 3.1.2 *The Hermitian matrix* $\mathbf{R}$ *corresponding to the set of vectors* $\varphi_j \in \mathcal{H}_{-2}(\mathcal{A}), j = 1, ..., n$, *and self–adjoint operator* $\mathcal{A}$ *is called* **admissible** *if the following equality holds*

$$\langle f, \frac{\mathcal{A}}{\mathcal{A}^2 + 1}g \rangle = \sum_{i,j=1}^{n} \bar{f}_j \mathbf{R}_{ji} g_i \qquad (3.23)$$

for every two functions f *and* g *provided that the following conditions are fulfilled:*

- *the functions f and g are elements from the linear hull of the vectors φ_j, i.e. the following representations hold*

$$f = \sum_{j=1}^{n} f_j \varphi_j, \quad g = \sum_{j=1}^{n} g_j \varphi_j;$$

- *the spectral measure $\mu_{f,g}$ corresponding to the functions f and g determines the absolutely convergent integral*

$$\int_{-\infty}^{\infty} \frac{|\lambda|}{\lambda^2 + 1} d|\mu_{f,g}(\lambda)| < \infty. \tag{3.24}$$

There exist sets of vectors φ_j such that for no two nontrivial linear combinations of them the integral (3.24) converges. Every Hermitian matrix $\mathbf{R}$ is admissible in this case. If all the vectors φ_j are elements from the Hilbert space $\mathcal{H}_{-1}(\mathcal{A})$ then the admissible matrix $\mathbf{R}$ is unique.

The choice of the admissible matrix $\mathbf{R}$ determines the extensions of all functionals $\varphi_j \in \mathcal{H}_{-2}(\mathcal{A}) = \mathcal{H}_2(\mathcal{A})^*$ to linear bounded functionals on $\mathcal{H}_{\{\varphi\}}(\mathcal{A})$. In fact, let

$$\psi = \tilde{\psi} + \sum_{j=1}^{n} b_j(\psi) \frac{\mathcal{A}}{\mathcal{A}^2 + 1} \varphi_j \in \mathcal{H}_{\{\varphi\}}.$$

Then the scalar product $\langle \varphi_k, \psi \rangle$ can be calculated as follows

$$
\begin{aligned}
\langle \varphi_k, \psi \rangle &= \langle \varphi_k, \tilde{\psi} \rangle + \sum_{j=1}^{n} b_j(\psi) \left\langle \varphi_k, \frac{\mathcal{A}}{\mathcal{A}^2 + 1} \varphi_j \right\rangle \\
&= \langle \varphi_k, \tilde{\psi} \rangle + \sum_{j=1}^{n} \mathbf{R}_{kj} b_j(\psi).
\end{aligned}
$$

This definition is correct and defines a linear functional, since the matrix $\mathbf{R}$ is admissible. Obviously this extension is a bounded functional.

We shall consider in what follows only admissible Hermitian matrices $\mathbf{R}$. The following theorem can be proven.

Theorem 3.1.2 *Let $\varphi_j \in \mathcal{H}_{-2}(\mathcal{A}) \setminus \mathcal{H}$ be $\mathcal{H}$-independent and form an orthonormal basis in $\mathcal{H}_{-2}(\mathcal{A})$. Let $\mathbf{T} = \{t_{ij}\}_{i,j=1}^{n}$ be a Hermitian invertible matrix. Suppose that the Hermitian matrix $\mathbf{R} = \{\mathbf{R}_{ij}\}_{i,j=1}^{n}$ is admissible for* the set of vectors $\varphi_j, j = 1, ..., n$, and operator $\mathcal{A}$. Then the self–adjoint operator $\mathcal{A}_T = \mathcal{A} + \sum_{i,j=1}^{n} t_{ij} \langle \varphi_j, \cdot \rangle \varphi_i$ is the self–adjoint restriction of the operator $\mathcal{A}^{0*}$ to the following domain

$$\text{Dom}(\mathcal{A}_T) = \{ \psi \in \text{Dom}(\mathcal{A}^{0*}) : \vec{a}_+(\psi) = -\mathbf{V}\vec{a}_-(\psi) \}, \qquad (3.25)$$

where $\Phi = \mathbf{R} + iI$;

$$\mathbf{V} = \frac{\mathbf{T}^{-1} + \Phi^*}{\mathbf{T}^{-1} + \Phi} = \frac{\mathbf{T}^{-1} + \mathbf{R} - iI}{\mathbf{T}^{-1} + \mathbf{R} + iI}.$$

Proof The proof is very similar to that of Theorem 3.1.1.

$\square$

We note that the domain of the operator $\mathcal{A}_T$ depends on the choice of the admissible matrix $\mathbf{R}$. If all the vectors φ_j are elements from the Hilbert space $\mathcal{H}_{-1}(\mathcal{A})$ then the admissible matrix $\mathbf{R}$ is unique and there is a unique self–adjoint operator corresponding to the formal expression (3.5). This case has been studied in the previous section.

The unique admissible matrix $\mathbf{R}$ can be determined by some extra conditions even if the vectors φ_j are not elements from the Hilbert space $\mathcal{H}_{-1}(\mathcal{A})$. For example such a unique matrix could exist if all the vectors φ_j and the operator $\mathcal{A}$ are homogeneous with respect to a certain group of unitary transformations. We develop this approach in Section 3.2.3 for differential operators with point interactions.

Let us introduce two n-dimensional vectors

$$\left(\langle \vec{\varphi}, \tilde{\psi} \rangle \right)_j = \langle \varphi_j, \tilde{\psi} \rangle; \quad j = 1, 2, ..., n;$$

$$\left(\vec{b}(\psi) \right)_j = b_j(\psi),$$

where the function $\tilde{\psi}$ and the coefficients $b_j(\psi)$ are defined by (3.10). Then the self–adjoint operator $\mathcal{A}_T$ can be characterized by the boundary conditions connecting the boundary values $\langle \vec{\varphi}, \tilde{\psi} \rangle$ and $\vec{b}(\psi)$

$$\langle \vec{\varphi}, \tilde{\psi} \rangle = \Gamma \, \vec{b}(\psi), \qquad (3.26)$$

where Γ is the Hermitian $n \times n$ matrix

$$\Gamma = -(\mathbf{T}^{-1} + \mathbf{R}).$$

This representation is similar to that used in Chapter 1. In the case of form bounded perturbations the matrix Γ is uniquelly defined. The matrices Γ and V defining the same self–adjoint extension are related by the Cayley transform

$$\Gamma = -i\frac{I + V}{I - V}, \quad V = \frac{\Gamma + I}{\Gamma - I}. \tag{3.27}$$

The resolvent of the operator $\mathcal{A}_T$ can be calculated using Krein's formula first obtained in [569].

Theorem 3.1.3 *Let $\mathcal{A}_T = \mathcal{A} + \sum_{i,j=1}^{n} t_{ji}\langle\varphi_i, \cdot\rangle\varphi_j$ be the self–adjoint restriction of the operator $\mathcal{A}^{0*}$ to the domain of functions from $\mathrm{Dom}\,(\mathcal{A}^0)$ satisfying the boundary conditions (3.26). Let $\mathbf{Q}(\lambda)$ be the Q–matrix function corresponding to the operator $\mathcal{A}$ and vectors φ_j*

$$(\mathbf{Q}(\lambda))_{jk} = \left\langle \varphi_j, \frac{1 + \lambda\mathcal{A}}{\mathcal{A} - \lambda}\frac{1}{\mathcal{A}^2 + 1}\varphi_k \right\rangle. \tag{3.28}$$

Then the resolvent of the operator $\mathcal{A}_T$ is given by

$$\frac{1}{\mathcal{A}_T - \lambda} = \frac{1}{\mathcal{A} - \lambda} + \sum_{j,k=1}^{n} \frac{1}{\mathcal{A} - \lambda}\varphi_j\,(\Gamma - Q(\lambda))_{jk}^{-1}\left\langle \frac{1}{\mathcal{A} - \bar{\lambda}}\varphi_k, \cdot \right\rangle. \tag{3.29}$$

Proof The resolvent equality

$$\frac{1}{\mathcal{A}_T - \lambda}f = \psi = \tilde{\psi} + \sum_{j=1}^{n} b_j(\psi)\frac{\mathcal{A}}{\mathcal{A}^2 + 1}\varphi_j$$

implies

$$f = (\mathcal{A} - \lambda)\tilde{\psi} - \sum_{j=1}^{n} b_j(\psi)\frac{1 + \lambda\mathcal{A}}{\mathcal{A}^2 + 1}\varphi_j.$$

Appliying the resolvent of the original operator and projecting to the elements φ_k we get the system of linear equations

$$\left\langle \varphi_k, \frac{1}{\mathcal{A} - \lambda}f \right\rangle = \langle\varphi_k, \tilde{\psi}\rangle - \sum_{j=1}^{n} b_j(\psi)\left\langle \varphi_k, \frac{1 + \lambda\mathcal{A}}{\mathcal{A} - \lambda}\frac{1}{\mathcal{A}^2 + 1}\varphi_j \right\rangle, \quad k = 1, 2, ..., n,$$

which can be written as follows using matrix notation

$$\left\langle \varphi, \frac{\vec{1}}{\mathcal{A} - \lambda}f \right\rangle = \langle\vec{\varphi}, \tilde{\psi}\rangle - \mathbf{Q}\vec{b}(\psi)$$

$$= (\Gamma - Q)\vec{b}(\psi), \tag{3.30}$$

since the function ψ satisfies the boundary conditions. The function ψ is given by the following equality

$$\frac{1}{\mathcal{A}_T - \lambda} f = \psi = \frac{1}{\mathcal{A} - \lambda} f + \sum_{j=1}^{n} \frac{1}{\mathcal{A} - \lambda} \varphi_j b_j(\psi),$$

which implies (3.29). The Theorem is proven.

$\square$

Note that the matrix $\Gamma - \mathbf{Q}$ is invertible, since the matrix Γ is Hermitian and the matrix $\mathbf{Q}$ has nontrivial imaginary part $\mathbf{Q} = \Re\mathbf{Q} + iI$.

3.1.4 Generalized finite rank perturbations

Generalized finite rank perturbations of any self–adjoint operator $\mathcal{A}$ acting in $\mathcal{H}$ coincide with the generalized extensions of a certain restriction $\mathcal{A}^0$ of the operator $\mathcal{A}$ having finite deficiency indices. The rank of the generalized perturbation is equal to the deficiency index of the symmetric operator $\mathcal{A}^0$. Every generalized extension $\mathbf{A}^+$ of $\mathcal{A}^0$ will be called **a generalized perturbation** of the operator $\mathcal{A}$. We consider in this chapter only symmetric restrictions with finite deficiency indices, i.e. only finite rank generalized perturbations. Without loss of generality we can suppose that the operator $\mathbf{A}^+$ coincides with a self–adjoint perturbation of a certain unperturbed operator

$$\mathbf{A} = \mathcal{A} \oplus \mathcal{A}' \qquad (3.31)$$

acting in the extended Hilbert space $\mathbf{H} = \mathcal{H} \oplus \mathcal{H}'$. The operators $\mathcal{A}$ and $\mathcal{A}'$ are two self–adjoint operators acting in the Hilbert spaces $\mathcal{H}$ and $\mathcal{H}'$ respectively. We no longer limit the rank of the self–adjoint perturbation of $\mathbf{A}$ as was done in Chapter 2 where only rank two perturbations were considered. The generalized extensions described will be called **generalized finite rank perturbations with internal structure** in order to underline the role of the operator $\mathcal{A}'$, which will be called **an internal operator** in what follows.

To define symmetric restrictions of the operator $\mathbf{A}$ consider two finite sets of vectors $\{\varphi_j\}_{j=1}^{n} \subset \mathcal{H}_{-2}(\mathcal{A})$ and $\{\varphi_j'\}_{j=1}^{n'} \subset \mathcal{H}_{-2}(\mathcal{A}')$. Without loss of generality we suppose that the vectors form orthonormal systems as elements of $\mathcal{H}_{-2}(\mathcal{A})$, $\mathcal{H}_{-2}(\mathcal{A}')$. Let the operators $\mathcal{A}^0$ and $\mathcal{A}'^0$ be the restrictions of the operators $\mathcal{A}$ and $\mathcal{A}'$ to the following domains

$$\begin{aligned}
\text{Dom}\,(\mathcal{A}^0) &= \{\psi \in \text{Dom}\,(\mathcal{A}) : \langle \varphi_j, \psi \rangle_H = 0, j = 1, 2, ..., n\}; \\
\text{Dom}\,(\mathcal{A}'^0) &= \{\psi \in \text{Dom}\,(\mathcal{A}') : \langle \varphi_j', \psi \rangle_{H'} = 0, j = 1, 2, ..., n'\}.
\end{aligned} \qquad (3.32)$$

Let the vectors φ_j, φ'_j be $\mathcal{H}$- and $\mathcal{H}'$-independent respectively. Then Lemma 3.1.1 implies that the symmetric operators $\mathcal{A}^0$ and $\mathcal{A}'^0$ have deficiency indices (n, n) and (n', n') respectively. The corresponding deficiency subspaces at the point z are given by

$$
\begin{aligned}
N(z) &= \tfrac{1}{A-z}\mathcal{L}\{\varphi_j\}_{j=1}^n; \\
N'(z) &= \tfrac{1}{A'-z}\mathcal{L}\{\varphi'_j\}_{j=1}^{n'}.
\end{aligned}
\tag{3.33}
$$

Then the deficiency subspace at the point z for the symmetric operator $\mathbf{A}^0 = \mathcal{A}^0 \oplus \mathcal{A}'^0$ is equal to

$$
\mathcal{N}(z) = N(z) \oplus N'(z).
\tag{3.34}
$$

The deficiency indices for the operator $\mathbf{A}^0$ are equal to the sum of the deficiency indices for the operators $\mathcal{A}^0$ and $\mathcal{A}'^0$. Let us denote by M the deficiency subspace at point $-i$

$$
M = \mathcal{N}(-i).
$$

Then every element $\Psi = (\psi, \psi')$ from the domain $\mathrm{Dom}\,(\mathbf{A}^{0*})$ of the adjoint operator $\mathcal{A}^0$ possesses the following representation

$$
\Psi = \hat{\Psi} + \frac{\mathbf{A}}{\mathbf{A} - i}\Xi_+(\Psi) + \frac{1}{\mathbf{A} - i}\Xi_-(\Psi),
\tag{3.35}
$$

where $\hat{\Psi} = (\hat{\psi}, \hat{\psi}') \in \mathrm{Dom}\,(\mathbf{A}^0)$, $\Xi_\pm(\Psi) = (\xi_\pm(\Psi), \xi'_\pm(\Psi)) \in M$.
   The adjoint operator $\mathbf{A}^{0*}$ acts as follows

$$
\begin{aligned}
\mathbf{A}^{0*}\Psi &= \mathbf{A}^{0*}\left(\hat{\Psi} + \frac{\mathbf{A}}{\mathbf{A} - i}\Xi_+(\Psi) + \frac{1}{\mathbf{A} - i}\Xi_-(\Psi)\right) \\
&= \mathbf{A}\hat{\Psi} - \frac{1}{\mathbf{A} - i}\Xi_+(\Psi) + \frac{\mathbf{A}}{\mathbf{A} - i}\Xi_-(\Psi).
\end{aligned}
\tag{3.36}
$$

The boundary form of the adjoint operator can be calculated following the main lines of the proof of Theorem 2.2.1

$$
\begin{aligned}
B_{\mathbf{A}^{0*}}[U, V] &= \langle U, \mathbf{A}^{0*}V\rangle_{\mathbf{H}} - \langle \mathbf{A}^{0*}U, V\rangle_{\mathbf{H}} \\
&= \langle \Xi_+(U), \Xi_-(V)\rangle_M - \langle \Xi_-(U), \Xi_+(V)\rangle_M.
\end{aligned}
\tag{3.37}
$$

   Similar to the case of rank one generalized interactions the self–adjoint extensions of the operator $\mathbf{A}^0$ can be described by subspaces of M that are Lagrangian with respect to the form (3.37) (see Theorem 2.2.2). This subspace can be determined in the following way by a Hermitian operator Γ acting in M.

Definition 3.1.3 *Let Γ be a certain Hermitian (self–adjoint) operator acting in M. Then the operator $\mathbf{A}^\Gamma$ - the extension of the operator $\mathbf{A}^0$ determined by Γ - is the restriction of the operator $\mathbf{A}^{0*}$ to the set of functions*

$$\Psi = \hat{\Psi} + \frac{\mathbf{A}}{\mathbf{A} - i}\Xi_+(\Psi) + \frac{1}{\mathbf{A} - i}\Xi_-(\Psi) \in \mathrm{Dom}\,(\mathbf{A}^{0*}) \qquad (3.38)$$

satisfying the following conditions

$$\Xi_-(\Psi) = \Gamma\Xi_+(\Psi). \qquad (3.39)$$

Note that the same boundary conditions can be written as follows

$$P_M\tilde{\Psi} = \Gamma\Xi_+(\Psi), \qquad (3.40)$$

where

$$\tilde{\Psi} = \hat{\Psi} + \frac{1}{\mathbf{A} - i}\Xi_-(\Psi) \in \mathrm{Dom}\,(\mathbf{A})$$

and P_M denotes the orthogonal projector to M. In fact $P_M(\mathbf{A} - i)\hat{\Psi} = 0$ and $P_M\Xi_-(\Psi) = \Xi_-(\Psi)$.

The operator Γ acting in the space $M = N(-i) \oplus N'(-i)$ possesses the natural decomposition

$$\Gamma = \begin{pmatrix} \Gamma_{00} & \Gamma_{01} \\ \Gamma_{10} & \Gamma_{11} \end{pmatrix}. \qquad (3.41)$$

Let us consider now the case where the vectors φ'_j are not necessarily $\mathcal{H}'$-independent. In this case the restricted operator $\mathcal{A}'^0$ is not densely defined. To avoid the discussion of the adjoint operator let us define the perturbed operator directly on the domain $\mathrm{Dom}\,(\mathbf{A}^\Gamma)$ of functions from the Hilbert space $\mathbf{H}$ possessing the representation

$$\Psi = \hat{\Psi} + \frac{1}{\mathbf{A} - i}(\mathbf{A} + \Gamma)\Xi_+(\Psi), \qquad (3.42)$$

where $\hat{\Psi} \in \mathrm{Dom}\,(\mathbf{A}^0)$ and $\Xi_+(\Psi) \in M = \mathcal{L}\{\frac{1}{A+i}\varphi_j, \frac{1}{A'+i}\varphi'_j\}$.

Lemma 3.1.3 *Let the set of vectors $\{\varphi_j\}_{j=1}^n$ be $\mathcal{H}$-independent. If*

$$\mathrm{Ker}(\Gamma_{01}) \cap N'(-i) = \{0\}, \qquad (3.43)$$

then every element $\Psi \in \mathrm{Dom}\,(\mathbf{A}^\Gamma)$ possesses the unique representation

$$\Psi = \hat{\Psi} + \frac{1}{\mathbf{A} - i}(\mathbf{A} + \Gamma)\Xi_+(\Psi), \qquad (3.44)$$

where $\hat{\Psi} \in \mathrm{Dom}(\mathbf{A}^0)$, $\Xi_+(\Psi) \in M$.

Proof The proof of this lemma follows the same lines as that of Lemma 2.2.1. Suppose that this representation is not unique. It follows that there exist $\hat{\Psi}, \hat{\Psi}^* \in \mathrm{Dom}\,(\mathbf{A}^0), \Xi_+, \Xi_+^* \in M$, such that the following equality holds

$$\hat{\Psi} + \frac{1}{\mathbf{A} - i}(\mathbf{A} + \Gamma)\Xi_+ = \hat{\Psi}^* + \frac{1}{\mathbf{A} - i}(\mathbf{A} + \Gamma)\Xi_+^*. \tag{3.45}$$

It follows that

$$\hat{\Psi} - \hat{\Psi}^* + \frac{1}{\mathbf{A} - i}(\Gamma + i)(\Xi_+ - \Xi_+^*) = -(\Xi_+ - \Xi_+^*).$$

The left hand side of the latter equality belongs to the domain $\mathrm{Dom}\,(\mathbf{A})$, hence the vector $\Xi \equiv \Xi_+ - \Xi_+^*$ belongs to the subspace $N'(-i)$, since the vectors φ_j are $\mathcal{H}$-independent. Applying the operator $\mathbf{A} - i$ to (3.45) and projecting to M we get

$$0 = -P_M(\Gamma + \mathbf{A})\Xi.$$

The latter equality implies in particular that

$$\Gamma_{01}\Xi = 0$$

and it follows that $\Xi = 0$ and therefore $\hat{\Psi} = \hat{\Psi}^*$. The lemma is proven.

$\square$

Definition 3.1.4 *The Γ-modified operator $\mathbf{A}^\Gamma$ is defined on the domain* $\mathrm{Dom}\,(\mathbf{A}^\Gamma)$ *by*

$$\begin{aligned}
\mathbf{A}^\Gamma \Psi &= \mathbf{A}^\Gamma \left(\hat{\Psi} + \frac{1}{\mathbf{A} - i}(\mathbf{A} + \Gamma)\Xi_+(\Psi) \right) \\
&= \mathbf{A}\hat{\Psi} + \frac{1}{\mathbf{A} - i}(-1 + \mathbf{A}\Gamma)\Xi_+(\Psi).
\end{aligned} \tag{3.46}$$

The resolvent of the Γ-modified operator can be easily calculated.

Theorem 3.1.4 *Let the set of vectors $\{\varphi_j\}_{j=1}^n$ be $\mathcal{H}$-independent. Let the operator Γ in M be Hermitian and let (3.43) hold. Then the operator $\mathbf{A}^\Gamma$ is self-adjoint and its resolvent is given by the formula*

$$\frac{1}{\mathbf{A}^\Gamma - \lambda} = \frac{1}{\mathbf{A} - \lambda} + \frac{\mathbf{A} + i}{\mathbf{A} - \lambda} \frac{1}{\Gamma - Q(\lambda)} P_M \frac{\mathbf{A} - i}{\mathbf{A} - \lambda} \tag{3.47}$$

for any $\lambda, \Im\lambda \neq 0$, where

$$Q(\lambda) = P_M \frac{1 + \lambda \mathbf{A}}{\mathbf{A} - \lambda} P_M. \tag{3.48}$$

Proof The fact that the operator $\mathbf{A}^\Gamma$ is symmetric and the formula for its resolvent can be proven following the main lines of the proof of Theorem 2.2.4. The latter formula defines the resolvent operator with domain equal to the Hilbery space $\mathbf{H}$. We are now going to show that the kernel of the calculated resolvent operator is trivial. Let

$$\frac{1}{\mathbf{A}^\Gamma - \lambda} F = \frac{1}{\mathbf{A} - \lambda} F + \frac{\mathbf{A} + i}{\mathbf{A} - \lambda} \frac{1}{\Gamma - Q(\lambda)} P_M \frac{\mathbf{A} - i}{\mathbf{A} - \lambda} F = 0,$$

which implies

$$F = -(\mathbf{A} + i) \frac{1}{\Gamma - Q(\lambda)} P_M \frac{\mathbf{A} - i}{\mathbf{A} - \lambda} F.$$

The left hand side of the latter equality belongs to the Hilbert space $\mathbf{H}$. The right hand side is a linear combination of the vectors φ_j and φ'_j. Therefore $\frac{1}{\mathbf{A} - i} F \in N'(-i)$, since the vectors φ_j are $\mathcal{H}$-independent. It follows then that

$$\Gamma_{01} \frac{1}{\mathcal{A}' + i} F = 0$$

and (3.43) implies that $F = 0$. The theorem is proven.

□

The resolvent of the operator $\mathbf{A}^\Gamma$ restricted to the external space $\mathcal{H}$ is given by

$$P_{\mathcal{H}} \frac{1}{\mathbf{A}^\Gamma - \lambda} P_{\mathcal{H}} = \frac{1}{\mathcal{A} - \lambda}$$

$$-\frac{\mathcal{A} + i}{\mathcal{A} - \lambda} \frac{1}{Q_0(\lambda) + Q_0^+(\lambda)} P_{N(-i)} \frac{\mathcal{A} - i}{\mathcal{A} - \lambda}, \tag{3.49}$$

where

$$Q_0(\lambda) = P_{N(-i)} \frac{1 + \lambda \mathcal{A}}{\mathcal{A} - \lambda} P_{N(-i)},$$

$$Q_1(\lambda) = P_{N'(-i)} \frac{1 + \lambda \mathcal{A}'}{\mathcal{A}' - \lambda} P_{N'(-i)},$$

$$Q_0^+(\lambda) = -\Gamma_{00} + \Gamma_{01} \frac{1}{\Gamma_{11} - Q_1(\lambda)} \Gamma_{10}.$$

The operators $Q_0(\lambda)$, $Q_1(\lambda)$ and $Q_0^+(\lambda)$ have nonnegative imaginary parts in the upper half plane $\Im \lambda \geq 0$.

The resolvent formula we obtained is a particular case of the following formula originally proven by M.Krein [569].

Theorem 3.1.5 *Every generalized resolvents* $\mathbf{R}(z)$ *of the operator* $\mathcal{A}^0$ *is given by the formula*

$$\mathbf{R}(\lambda) = \frac{1}{\mathcal{A} - \lambda} - \sum_{j,k=1}^{n} \frac{1}{\mathcal{A} - \lambda} \varphi_k \left(\mathbf{Q}_0(\lambda) + \mathbf{Q}_0^+(\lambda) \right)_{kj}^{-1} \left\langle \frac{1}{\mathcal{A} - \bar{\lambda}} \varphi_j, \cdot \right\rangle,$$

$$\Im \lambda \geq 0, \tag{3.50}$$

where $\mathbf{Q}_0(\lambda)$ *is the* $n \times n$ *matrix with elements*

$$(\mathbf{Q}_0(\lambda))_{jk} = \left\langle \varphi_j, \frac{1 + \lambda A}{A - \lambda} \frac{1}{A^2 + 1} \varphi_k \right\rangle$$

and $\mathbf{Q}_0^+(\lambda)$ *is an arbitrary* $n \times n$ *function, holomorphic in the upper half plane with nonnegative imaginary part, i.e.*

$$\Im \sum_{j,k=1}^{n} \bar{\xi}_j (\mathbf{Q}_0^+(\lambda))_{jk} \xi_k \geq 0 \tag{3.51}$$

for arbitrary $(\xi_1, \xi_2, ..., \xi_n) \in \mathbf{C}^n$ *and* $\Im \lambda \geq 0$.

3.2   Point interactions for differential operators and distribution theory

3.2.1   Point interactions for differential operators as finite rank perturbations

Finite rank perturbations of differential operators are studied in the current section. We restrict our consideration to the case of perturbations with support at the origin. Let L be the self–adjoint operator $(iD_x)^n$ with domain equal to $W_2^n(\mathbf{R})$. We are going to consider additive perturbations of L corresponding to the formal expression $L_T = L + T$, where T is an operator from $W_2^n(\mathbf{R})$ into $W_2^{-n}(\mathbf{R})$. Such an operator is said to be supported at the origin if its Schwartz distributional kernel is supported at the origin in $\mathbf{R} \times \mathbf{R}$, or, which is the same, supp $Tf \subset \{0\}$ for all $f \in \mathrm{Dom}(T)$, and $Tf = 0$ whenever $0 \notin \mathrm{supp}\, f$. We shall confine ourselves to operators T that act continuously from $W_2^n(\mathbf{R})$ into $W_2^{-n}(\mathbf{R})$, or, equivalently, satisfy the estimate

$$|(Tf)(\varphi)| \leq C \parallel f \parallel_{W_2^n} \parallel \varphi \parallel_{W_2^n} \tag{3.52}$$

for all $f, \varphi \in W_2^n(\mathbf{R})$. The Schwartz kernel of T must consist of a linear combination of derivatives $D^k \delta \otimes D^l \delta$ at $(0,0) \in \mathbf{R}^2$. Because of (3.52) the possible values of k and l are $0 \le k \le n-1$, $0 \le l \le n-1$. Thus T must have the form

$$T\psi = \sum_{l,m=0}^{n-1} t_{lm}\delta^{(m)}(\psi)\delta^{(l)}.$$

We suppose that the matrix $\{v_{lm}\}$ is Hermitian. Thus our perturbed operator will have the form

$$L_T = (iD_x)^n + \sum_{l,m=0}^{n-1} t_{lm}\delta^{(m)}(\cdot)\delta^{(l)} \qquad (3.53)$$

where $\{t_{lm}\}$ is a Hermitian matrix. The distributions $\delta^{(m)}, m = 1, ..., n-1$, are elements from the Hilbert space $\mathcal{H}_{-2}(L)$ but not all these functions belong to the space $\mathcal{H}_{-1}(L)$. Therefore finite rank perturbations given formally by (3.53) constitute an important example of form unbounded finite rank perturbations.

The formal operator L_T given by formula (3.53) is a finite rank perturbation of the operator L. The linear operator defined by the formal differential expression (3.53) restricted to $C_0^\infty(\mathbf{R} \setminus \{0\})$ coincides with the operator $L^0 = L|_{C_0^\infty(\mathbf{R}\setminus\{0\})}$. It follows that any self–adjoint operator corresponding to the formal expression (3.53) must coincide with some self–adjoint extension of L^0. Every such extension can be described by boundary conditions on the functions from the domain of the extended operator.

The domain of the adjoint operator L^{0*} is equal to $W_2^n(\mathbf{R} \setminus \{0\})$. The deficiency indices of the operator L^0 are equal to (n,n). For given z with $\Im z \ne 0$ the deficiency subspace N_z consists of the set of solutions $g \in W_2^n(\mathbf{R}\setminus\{0\})$ of the equation

$$(iD_x)^n g(x) = zg(x). \qquad (3.54)$$

Any element of N_z can be written as a linear combination of functions of the form

$$g(x) = \Theta(-\Im\lambda x)\exp(-i\lambda x), \qquad (3.55)$$

where λ is some complex root of the equation $\lambda^n = z$.

For $f \in W_2^n(\mathbf{R} \setminus \{0\})$ let $J_n^+ f$ be the vector of boundary values at the origin

$$J_n^+ f = \left(f(+0), f'(+0), ..., f^{(n-1)}(+0) \right) \in \mathbf{C}^n,$$

and let $J_n^- f$ be defined similarly; furthermore set

$$J_n f = (J_n^- f, J_n^+ f) \in \mathbf{C}^{2n}.$$

Lemma 3.2.1 *For any z with $\Im z \ne 0$ the map J_n is a bijection from $N_z \oplus N_{\bar{z}}$ onto $\mathbf{C}^{2n}$.*

Proof Since J_n is a mapping between spaces of the same dimension, it is enough to prove that J_n is injective. Thus assume that $u \in N_z, v \in N_{\bar z}$, and $J_n(u + v) = 0$. Then

$$\int_0^\infty (iD_x)^n(u(x) + v(x))\overline{u(x)}dx = \int_0^\infty (u(x) + v(x))\overline{(iD_x)^n u(x)}dx,$$

since the boundary terms vanish by assumption. Using (3.54) we obtain

$$\int_0^\infty (zu(x) + \bar z v(x))\overline{u(x)}dx = \int_0^\infty (u(x) + v(x))\overline{zu(x)}dx,$$

or

$$(z - \bar z)\int_0^\infty |u(x)|^2 dx = 0.$$

Since $\Im z \neq 0$, it follows that $u(x) = 0$ for $x > 0$. In the same way we can prove that u and v both vanish on all of $\mathbf{R}$.

$\square$

Let us introduce the $2n$-dimensional vectors

$$\begin{aligned}\vec\delta(f) &= I(J_n^+ f + J_n^- f)/2,\\ \vec\beta(f) &= I(J_n^+ f - J_n^- f)/2,\end{aligned}$$

where I is the $n \times n$ diagonal matrix with diagonal entries $I_{ll} = (-1)^l, l = 0, 1, ..., n-1$. It follows that

$$J_n^\pm f = I\left(\vec\delta(f) \pm \vec\beta(f)\right). \tag{3.56}$$

Lemma 3.2.2 *The boundary form of the adjoint operator L^{0*} is equal to*

$$\begin{aligned}\langle L^{0*}u, v\rangle - \langle u, L^{0*}v\rangle &= i^n\left\{-\langle IAJ_n^+ u, J_n^+ v\rangle_{\mathbf{C}^n} + \langle IAJ_n^- u, J_n^- v\rangle_{\mathbf{C}^n}\right\}\\ &= 2(-i)^n\left\{\langle IA\vec\delta(u), \vec\beta(v)\rangle_{\mathbf{C}^n} + \langle IA\vec\beta(u), \vec\delta(v)\rangle_{\mathbf{C}^n}\right\},\end{aligned} \tag{3.57}$$

where A is the $n \times n$ antidiagonal matrix with coefficients

$$A_{lm} = \delta_{l+m}^{n-1}, \quad l, m = 0, 1, ..., n-1.$$

Proof The first formula (3.57) is obtained by partial integrations. Inserting the expressions (3.56) for $J_n^\pm$ gives

$$\langle L^{0*}u, v\rangle - \langle u, L^{0*}v\rangle = -2i^n\left\{\langle AI\vec\delta(u), \vec\beta(v)\rangle_{\mathbf{C}^n} + \langle AI\vec\beta(u), \vec\delta(v)\rangle_{\mathbf{C}^n}\right\}.$$

The second formula (3.57) now follows from the matrix identity $AI = (-1)^{n-1}IA$.

$\square$

To every self–adjoint extension L of L^0 we can associate the subspace $E \subset \mathbb{C}^{2n}$

$$E = J_n(\mathrm{Dom}(L)), \qquad (3.58)$$

consisting of all boundary values of elements in the domain of L.

Using the expression (3.57) for the boundary form of L^{0*} we can define a sesquilinear form on $\mathbb{C}^{2n}$ as follows. For $s = (s_-, s_+)$ and $t = (t_-, t_+) \in \mathbb{C}^{2n}$ we set

$$\mathbf{Q}[s, t] = \langle As_+, t_+ \rangle_{\mathbb{C}^n} - \langle As_-, t_- \rangle_{\mathbb{C}^n}.$$

Then (3.57) can be written as

$$\langle L^{0*}u, v \rangle - \langle u, L^{0*}v \rangle = i^n \mathbf{Q}[J_n u, J_n v].$$

Theorem 3.2.1 *The map* $\mathrm{Dom}(L) \mapsto J_n(\mathrm{Dom}(L)) \subset \mathbb{C}^{2n}$ *defines a 1-1 correspondence between the set of self–adjoint extensions of* L^0 *and the set of subspaces of* $\mathbb{C}^{2n}$ *that are Lagrangian with respect to the form* $\mathbf{Q}$.

Proof Given a self–adjoint extension L it follows from Lemma 3.2.2 that the form $\mathbf{Q}$ vanishes on the subspace (3.58) of $\mathbb{C}^{2n}$. If the dimension of E were less than n, then the elements of the domain of L span a subspace of $N_z \oplus N_{\bar{z}}$ with dimension less than n, and this would imply that L is not self–adjoint. Therefore E must be n-dimensional and hence Lagrangian.

Conversely, let E be Lagrangian. Let the domain of L be defined by (3.58). Then the operator L is symmetric. self–adjointness of the operator L follows from the fact that the map J_n restricted to $N_z \oplus N_{\bar{z}}$ is injective.

$\square$

We have shown that the self–adjoint extensions of the operator L^0 can be described by Lagrangian planes of the boundary form. To define which self–adjoint extension corresponds to the formal expression (3.53) the matrix Φ introduced in Section 3.1.3 has to be determined. One needs to define the following formal expressions

$$\Phi_{ij} = \langle \delta^{(i)}, (L - i)^{-1}\delta^{(j)} \rangle. \qquad (3.59)$$

The latter expressions are well defined if and only if $i + j + 2 \leq n$. This follows from the fact that the i-th derivative of the delta function is defined only on the functions having i first continuous derivatives. Then the scalar product $\langle \delta^{(i)}, (L-i)^{-1}\delta^{(j)} \rangle$ is equal to the value of the distribution $\delta^{(i)}$ on the test function $(L-i)^{(-1)}\delta^{(j)}$. The scalar product which appears in (3.59) can be defined for all values of i and j only in the framework of distribution theory with discontinuous test functions. This theory is developed in the next section. We use the fact that the distributions $\delta^{(j)}$ are homogeneous with respect to the scaling transformations to extend these distributions to the set of test functions which are discontinuous at the origin. One can use the homogeneity property directly in order to determine the scalar product $\langle \delta^{(i)}, \frac{1}{L-i}\delta^{(j)} \rangle$ using the methods developed in Section 3.1.3. The approach based on distribution theory with discontinuous test functions has the following advantage: it is not necessary to prove that the resulting Φ-matrix is admissible. This property is satisfied automatically.

3.2.2   Distribution theory for discontinuous test functions

We introduce the space K of test functions as follows.

Definition 3.2.1 *The set of test functions K is the set of all functions defined on $\mathbf{R} \setminus \{0\}$ with support contained in some bounded interval and having bounded derivatives of all orders.*

Thus the functions in K can be discontinuous at the origin, but the limits of the functions and all derivatives from the left and from the right of the point zero exist and are finite. Convergence in this space is defined as follows.

Definition 3.2.2 *A sequence $\{\varphi_n\}$ of functions in K is said to converge to a function $\varphi \in K$ if and only if*

1. *there exists a compact interval outside which all the functions φ_n vanish;*

2. *for any k the sequence $\{\varphi_n^{(k)}\}$ of derivatives of order k converges uniformly to $\varphi^{(k)}$.*

Let $\mathcal{D}(\mathbf{R})$ denote the set of $C_0^\infty(\mathbf{R})$ test functions. Let $\mathbf{R}_+$ denote the closed positive half axis, and let $\mathcal{D}(\mathbf{R}_+)$ denote the set of restrictions to $\mathbf{R}_+$ of functions from $\mathcal{D}(\mathbf{R})$; the expressions $\mathbf{R}_-$ and $\mathcal{D}(\mathbf{R}_-)$ will have the analogous meaning. If $\varphi \in K$, let φ_+ be the function on $\mathbf{R}_+$ which is equal to φ on $\mathbf{R}_+ \setminus \{0\}$ and is continuous on $\mathbf{R}_+$. Then φ_+ can be extended to a function

$\tilde{\varphi}_+ \in \mathcal{D}(\mathbf{R})$, hence $\varphi_+ \in \mathcal{D}(\mathbf{R}_+)$. Defining $\mathcal{D}(\mathbf{R}_-)$ and φ_- similarly we see that the map

$$K \ni \varphi \mapsto (\varphi_-, \varphi_+) \in \mathcal{D}(\mathbf{R}_-) \times \mathcal{D}(\mathbf{R}_+) \qquad (3.60)$$

is a bijection which preserves convergence of sequences.

Definition 3.2.3 *A distribution f in K' is a linear form on K which is continuous in the sense that $f(\varphi_n) \to f(\varphi)$ whenever φ_n tends to φ in K. A sequence $f_n \in K'$ is said to converge to $f \in K'$ as $n \to \infty$, if $f_n(\varphi) \to f(\varphi)$ as $n \to \infty$ for every $\varphi \in K$.*

Denote the space of continuous linear forms on $\mathcal{D}(\mathbf{R}_+)$ by $\mathcal{D}'(\mathbf{R}_+)$. Let $\mathcal{D}_{\mathbf{R}_-}(\mathbf{R})$ denote the space of functions in $\mathcal{D}(\mathbf{R})$ that are supported in $\mathbf{R}_-$. Viewing $\mathcal{D}(\mathbf{R}_+)$ as a factor space $\mathcal{D}(\mathbf{R})/\mathcal{D}_{\mathbf{R}_-}(\mathbf{R})$ we see that $\mathcal{D}'(\mathbf{R}_+)$ can be identified with the set of distributions in $\mathcal{D}'(\mathbf{R})$ that annihilate all test functions in $\mathcal{D}_{\mathbf{R}_-}(\mathbf{R})$; this is of course the set $\mathcal{D}'_{\mathbf{R}_+}(\mathbf{R})$ of distributions supported in $\mathbf{R}_+$. We introduce the analogous notation with $\mathbf{R}_+$ replaced by $\mathbf{R}_-$ and vice versa. The space K' of generalized distributions can now be identified with the space $\mathcal{D}'(\mathbf{R}_-) \times \mathcal{D}'(\mathbf{R}_+)$ of pairs (f_-, f_+) of elements $f_- \in \mathcal{D}'(\mathbf{R}_-)$, $f_+ \in D'(\mathbf{R}_+)$. Inдеed, the identification

$$\mathcal{D}'(\mathbf{R}_-) \times \mathcal{D}'(\mathbf{R}_+) \ni (f_-, f_+) \mapsto f \in K' \qquad (3.61)$$

is defined by

$$f(\varphi) = f_-(\varphi_-) + f_+(\varphi_+), \quad \varphi \in K.$$

Using the observations above it is easy to verify that the identification (3.61) is a vector space isomorphism preserving (weak) convergence of sequences.

Since the differentiation operator d/dx preserves the space $\mathcal{D}_{\mathbf{R}_-}(\mathbf{R})$, it operates in a natural way in the factor space $\mathcal{D}(\mathbf{R})/\mathcal{D}_{\mathbf{R}_-}(\mathbf{R}) \sim \mathcal{D}(\mathbf{R}_+)$, and the same remark applies of course to $\mathcal{D}(\mathbf{R}_-)$, hence to the product space $\mathcal{D}(\mathbf{R}_-) \times \mathcal{D}(\mathbf{R}_+) \sim K$. This differentiation operator corresponds to pointwise differentiation of functions in K. Later we shall sometimes view the elements of K as distributions in K' and apply the distributional derivative to them. To distinguish those two operations we shall denote the pointwise differentiation operator by d/dx and the distributional derivative in K' by D_x. The distribution derivative in the space of Schwartz distributions $\mathcal{D}'(\mathbf{R})$ will also be denoted D_x.

Definition 3.2.4 *The generalized derivative $D_x f$ of a distribution $f \in K'$ is defined by*

$$(D_x f)(\varphi) = -f(\frac{d}{dx}\varphi) \quad \text{for all} \quad \varphi \in K.$$

Using the identification (3.61) and the definition of derivative for standard (Schwartz) distributions we see immediately that

$$D_x f = (D_x f_-, D_x f_+), \qquad (3.62)$$

where the derivatives applied to f_- and f_+ are the distributional derivatives in the sense of standard distribution theory.

To any function $f \in L^1_{loc}$ we can associate the linear form $K \ni \varphi \mapsto \int f\varphi dx$, which is an element of K'. Thus we obtain an embedding $j : L^1_{loc} \to K'$, and in particular the space of test functions K becomes embedded in K'. Using the identification (3.61) we can represent $j(f)$ as the pair $(\chi_- f, \chi_+ f)$, where χ_- and χ_+ are the characteristic functions of $\mathbf{R}_-$ and $\mathbf{R}_+$, respectively.

If f is absolutely continuous, the Schwartz distributional derivative d/dx and the classical pointwise derivative (defined a.e.) of f coincide and are in L^1_{loc}, hence are elements of K', and can be compared to $D_x f$, the derivative of f considered as an element of K'. It turns out that those derivatives do not coincide unless $f(0) = 0$.

Example. The derivative of the constant distribution $c \in K'$ is equal to the distribution $(-c\delta, c\delta) \in K'$. In fact, using the identification $c = (c\chi_-, c\chi_+)$ and formula (3.62) we obtain the result.

On the other hand, if f is absolutely continuous and $f(0) = 0$, then $D_x f$ (in the sense of K') is equal to the pointwise derivative df/dx. Indeed, $f = (\chi_- f, \chi_+ f)$, and since $\chi_- f$ and $\chi_+ f$ are absolutely continuous, the Schwartz distributional derivatives of those functions coincide with the pointwise derivatives.

Lemma 3.2.3 *Assume $f \in K'$ and $D_x f = 0$. Then $f = 0$.*

Proof Let $f = (f_-, f_+)$. The assumption that $D_x f = 0$ in K' implies that $D_x f_- = D_x f_+ = 0$ in the sense of standard distribution theory, hence f_- and f_+ are constant functions. But f_- and f_+ vanish on half lines, hence $f_- = f_+ = 0$.

$\square$

Corollary 3.2.1 *If $f \in L^1_{loc} \subset K'$, there is a unique $F \in K'$ such that $D_x F = f$.*

Proof The function $F(x) = \int_0^x f(t)dt$ is absolutely continuous and $F(0) = 0$, hence $D_x F$ (in the sense of K') is equal to $df/dx = F(x)$ (defined a.e.). This proves existence, and the uniqueness follows from the lemma.

□

Considering an element $f \in K'$ as a linear form on K we can construct an element in $\mathcal{D}'(\mathbf{R})$ by restricting the linear form to $\mathcal{D}(\mathbf{R})$, which is a subset of K. Denote the map from K' to $\mathcal{D}'(\mathbf{R})$ obtained in this way by η. If $f = (f_-, f_+)$, it is easy to see that $\eta(f)$ can be written as $\eta(f) = f_- + f_+$. The map η is obviously not injective, because the element $(h, -h) \in K'$ is mapped to zero by η for any element $h \in \mathcal{D}'(\mathbf{R})$ supported at the origin. That η is surjective onto $\mathcal{D}'(\mathbf{R})$ can be seen as follows. Let g be arbitrary in $\mathcal{D}'(\mathbf{R})$. Write $g = g_0 + g_1$, where g_0 has compact support and g_1 vanishes in some neighborhood of the origin. Then η maps $(\chi_- g_1, \chi_+ g_1)$ to g_1, so it remains only to consider g_0. Choose a continuous function G and an integer k such that $D^k G = g$, and define $F \in K'$ by $F = (F_-, F_+) = (\chi_- G, \chi_+ G)$. Then $\eta(F) = F_- + F_+ = G$ and

$$\eta(D^k F) = \eta(D^k(F_-, F_+)) = \eta((D^k F_-, D^k F_+))$$

$$= D^k F_- + D^k F_+ = D^k G = g_0,$$

which proves that $g \in \eta(K')$.

The reflection operator $\mathbf{I}$ is defined for test functions $\varphi \in K$ by $\mathbf{I}\varphi(x) = \varphi(-x)$ and for distributions in K' by

$$(\mathbf{I}f)(\varphi) = f(\mathbf{I}\varphi), \quad \varphi \in K. \tag{3.63}$$

The definition of $\mathbf{I}f$ for Schwartz distributions f is analogous to (3.63). If $\varphi \in K$ is equal to $(\varphi_-, \varphi_+) \in \mathcal{D}_{\mathbf{R}_-}(\mathbf{R}) \times \mathcal{D}_{\mathbf{R}_+}(\mathbf{R})$ under the identification (3.60), then clearly $\mathbf{I}\varphi = (\mathbf{I}\varphi_+, \mathbf{I}\varphi_-)$. From this we immediately obtain a similar formula for distributions in K': if $f = (f_-, f_+) \in K'$, then

$$\mathbf{I}f = (\mathbf{I}f_+, \mathbf{I}f_-). \tag{3.64}$$

This shows that η commutes with $\mathbf{I}$:

$$\eta(\mathbf{I}f) = \eta((\mathbf{I}f_+, \mathbf{I}f_-)) = \mathbf{I}f_+ + \mathbf{I}f_- = \mathbf{I}(f_- + f_+) = \mathbf{I}(\eta(f)), \quad f \in K'.$$

We say that a distribution $f \in K'$ is *even* if $\mathbf{I}f = f$, and that f is *odd* if $\mathbf{I}f = -f$.

Examples. The distributions (δ, δ) and $(\delta', -\delta')$ are even, and the distributions $(\delta, -\delta)$ and (δ', δ') are odd.

The scaling transformation S_c, $c > 0$, is defined for test functions $\varphi \in K$ in the usual way:

$$S_c\varphi(x) = \varphi(cx),$$

and for distributions $f \in K'$ by

$$S_c f(\varphi) = \frac{1}{c} f(S_c \varphi), \quad \varphi \in K, \quad c > 0. \tag{3.65}$$

Since $S_c \varphi = (S_c \varphi_-, S_c \varphi_+)$ if $\varphi = (\varphi_-, \varphi_+)$, it is easily seen that

$$S_c f = (S_c f_-, S_c f_+), \quad f = (f_-, f_+) \in K',$$

where S_c operates on Schwartz distributions by analogy with (3.65). We shall say that $f \in K'$ is (positively) homogeneous of order $r \in \mathbf{R}$, if $S_c f = c^r f$ for all $c > 0$. It is clear then that $f = (f_-, f_+) \in K'$ is homogeneous of degree r if and only if f_- and f_+ are homogeneous distributions of degree r in the usual sense. Since $(\delta, \delta) \in K'$ is homogeneous of degree -1 we conclude that its n-th derivative (in K') is homogeneous of degree $-n - 1$.

We introduce two elements in K' by

$$\begin{aligned}
\delta(\psi) &= (\psi(-0) + \psi(+0))/2 \\
\beta(\psi) &= (-\psi(-0) + \psi(+0))/2
\end{aligned} \tag{3.66}$$

or

$$\delta = (\delta/2, \delta/2), \quad \beta = (-\delta/2, \delta/2).$$

Denote by $\mathbf{1}$ the function with constant value 1 and by $\mathbf{sg}$ the function $x/|x|$. Considering those functions as elements of K' we can compute their derivatives:

$$\begin{aligned}
D_x\mathbf{1} &= D_x(\chi_-, \chi_+) = (-\delta, \delta) = 2\beta \\
D_x\mathbf{sg} &= D_x(-\chi_-, \chi_+) = (\delta, \delta) = 2\delta.
\end{aligned} \tag{3.67}$$

For arbitrary $\psi \in K$ we now look for constants A and B such that

$$\psi = A\mathbf{1} + B\mathbf{sg} + \psi_1, \tag{3.68}$$

where $\psi_1 \in K$ satisfies $\psi_1(-0) = \psi_1(+0) = 0$. It is clear that we must take

$$\begin{aligned}
A &= (\psi(-0) + \psi(+0))/2 = \delta(\psi) \\
B &= (-\psi(-0) + \psi(+0))/2 = \beta(\psi).
\end{aligned}$$

Since ψ_1 is continuous at the origin we have $D_x\psi_1 = (d/dx)\psi$, hence we obtain by differentiating (3.68)

$$D_x\psi = D_x\psi + AD_x\mathbf{1} + BD_x\mathbf{sg} = \frac{d}{dx}\psi + \delta(\psi) \cdot 2\beta + \beta(\psi) \cdot 2\delta.$$

Iterating this formula we obtain

$$D_x^2 \psi = \left(\frac{d}{dx}\right)^2 \psi + 2\delta \left(\frac{d\psi}{dx}\right) \beta + 2\beta \left(\frac{d\psi}{dx}\right) \delta + 2\delta(\psi)\beta^{(1)} + 2\beta(\psi)\delta^{(1)}.$$

Continuing in this way we finally obtain

$$D_x^n \psi = \left(\frac{d}{dx}\right)^n \psi + 2\sum_{j=0}^{n-1}(-1)^j \left(\delta^{(j)}(\psi)\beta^{(n-j-1)} + \beta^{(j)}(\psi)\delta^{(n-j-1)}\right). \quad (3.69)$$

The following lemma will be needed below.

Lemma 3.2.4 *Assume that the distribution $f \in K'$ satisfies*

$$f(\varphi) = D_x^n \delta(\varphi) \quad \text{for all} \quad \varphi \in \mathcal{D}(\mathbf{R}),$$

and that f is homogeneous of order $-n-1$. Assume moreover that f is an even distribution if n is an even number and that f is an odd distribution if n is odd. Then $f = D_x^n \delta$.

Proof The assumption implies that $f(\varphi) = 0$ for all $\varphi \in \mathcal{D}(\mathbf{R} \setminus \{0\})$, hence f must be supported at the origin. Then f must be a linear combination of derivatives of δ and β, and since f is also homogeneous of degree $-n-1$, we know that $f = aD_x^n\delta + bD_x^n\beta$ for suitable constants a and b. But $(D_x^n\beta)(\varphi) = 0$ for all $\varphi \in \mathcal{D}(\mathbf{R})$, whereas $(D_x^n\delta)(\varphi) = (-1)^n\varphi^{(n)}(0)$ for such φ, hence $a = 1$. The distribution $D_x^n\delta$ is even if n is even and odd if n is odd, and f has the same property by assumption, while β has the opposite parity; this implies that $b = 0$, which completes the proof.

$\square$

We are now going to discuss multiplication by test functions. Let $K_{\text{loc}} \subset L_{loc}^1$ denote the set of functions that are C^∞ outside the origin and are equal to some element of K in a neighborhood of the origin. Considering the elements of K_{loc} as distributions in K' we want to define the product of $f \in K'$ and $\psi \in K_{\text{loc}}$.

Definition 3.2.5 *The product of $f \in K'$ and $\psi \in K_{\text{loc}}$ is the element $f\psi \in K'$ defined as*

$$(f\psi)(\varphi) = f(\psi\varphi), \quad \varphi \in K.$$

It follows immediately from the definition that $f\psi = (f_-\psi_-, f_+\psi_+)$. Note that $f_+ \in \mathcal{D}'_{\mathbf{R}_+}(\mathbf{R})$, and $\psi_+ \in \mathcal{D}(\mathbf{R}_+) \sim \mathcal{D}(\mathbf{R})/\mathcal{D}_{\mathbf{R}_-}(\mathbf{R})$. Therefore the product of f_+ and ψ_+ is well defined as a Schwartz distribution and the result is an element of $\mathcal{D}'_{\mathbf{R}_+}(\mathbf{R})$.

As an example let us compute the product $\delta\psi$, where $\delta = (\delta/2, \delta/2) \in K'$ and $\psi \in K_{\mathrm{loc}}$. It is clear that $\delta\psi = 0$ if $\psi(-0) = \psi(+0) = 0$. Using the representation (2.10) and observing that $\delta \cdot \mathbf{1} = \delta$ and $\delta \cdot \mathbf{sg} = \beta$ we therefore obtain

$$\delta\psi = \delta(\psi)\delta + \beta(\psi)\beta. \tag{3.70}$$

In the same way we note that $\beta \cdot \mathbf{1} = \beta$ and $\beta \cdot \mathbf{sg} = \delta$, and hence

$$\beta\psi = A\beta + B\delta = \delta(\psi)\beta + \beta(\psi)\delta. \tag{3.71}$$

3.2.3 Differential operator of order n in one dimension

We shall now apply the distribution theory developed in the previous section to determine the domain of the operator L_T in (3.53). If the Dirac distributions in (3.53) are interpreted as elements of K' and $W_2^n(\mathbf{R} \setminus \{0\})$ is considered as a subset of K', then the operator L_T maps $W_2^n(\mathbf{R} \setminus \{0\})$ into K'. Note that the differential operator D_x then has to be understood according to Definition 3.2.4. Composing with the map $\eta : K' \to \mathcal{D}'(\mathbf{R})$ we get an operator $u \mapsto \eta(L_T u)$ from $W_2^n(\mathbf{R} \setminus \{0\})$ into $\mathcal{D}'(\mathbf{R})$. We can then define a new operator $\mathbf{L}_T$ in $L_2(\mathbf{R})$ as $\mathbf{L}_T u = \eta(L_T u)$ with domain

$$\mathrm{Dom}\,(\mathbf{L}_T) = \{u \in W_2^n(\mathbf{R} \setminus \{0\}); L_T u \in L_2(\mathbf{R})\}. \tag{3.72}$$

Theorem 3.2.2 *The operator*

$$\mathbf{L}_T = \eta\left((iD_x)^n + \sum_{l,m=0}^{n-1} t_{lm}\delta^{(m)}(\cdot)\delta^{(l)} \right)$$

defined by the Hermitian matrix $\mathbf{T} = \{t_{lm}\}_{l,m=0}^{n-1}$ *is self–adjoint on the domain* (3.72). *The operator* $\mathbf{L}_T$ *coincides with the operator* L^{0*} *restricted to the domain of functions* $u \in W_2^n(\mathbf{R} \setminus \{0\})$ *satisfying the following boundary condition at the origin*

$$2\vec{\beta}(u) = (-i)^n A\mathbf{T}\vec{\delta}(u). \tag{3.73}$$

Remark The matrix A was introduced in Lemma 3.2.2.

Proof If L_T acts from $W_2^n(\mathbf{R} \setminus \{0\}) \subset K'$ into K' as explained above, then by (3.69)

$$
L_T u = i^n v + 2i^n \sum_{l=0}^{n-1} (-1)^{n-l-1} \left(\delta^{(n-l-1)}(u)\beta^{(l)} + \beta^{(n-l-1)}(u)\delta^{(l)} \right)
$$

$$
+ \sum_{l,m=0}^{n-1} t_{lm}\delta^{(m)}(u)\delta^{(l)},
$$

(3.74)

where $v = (d/dx)^n u \in L_2(\mathbf{R})$. Note that here $\delta^{(j)} \in K'$, and hence

$$
\delta^{(j)}(u) = \frac{(-1)^j}{2}(u^{(j)}(+0) + u^{(j)}(-0)).
$$

Since $\eta(\beta^{(j)}) = 0$ and $\eta(\delta^{(j)}) = \delta^{(j)} \in \mathcal{D}'(\mathbf{R})$, this gives

$$
\mathbf{L}_T u = i^n v + 2i^n \sum_{l=0}^{n-1} (-1)^{n-l-1}\beta^{(n-l-1)}(u)\delta^{(l)} + \sum_{l,m=0}^{n-1} c_{lm}\delta^{(m)}(u)\delta^{(l)}. \quad (3.75)
$$

The requirement that $\mathbf{L}_T u \in L_2(\mathbf{R})$ then means that for each l

$$
2i^n(-1)^{n-l-1}\beta^{(n-l-1)}(u) + \sum_{m=0}^{n-1} t_{lm}\delta^{(m)}(u) = 0,
$$

which is formula (3.73). This boundary condition defines a certain Lagrangian plane in $\mathbf{C}^{2n}$ and it follows from Theorem 3.2.1 that the restricted operator is self–adjoint.

$\square$

The operators with singular interaction form a proper subset of the set of all self–adjoint perturbations of the operator L. This is related to the fact that not every Lagrangian subspace of $\mathbf{C}^{2n}$ can be described by a relation of the form (3.73). Let $\mathbf{T}$ be presented in the form $\mathbf{T} = \mathbf{T}_0^{-1}\mathbf{T}_1$, where $\mathbf{T}_0, \mathbf{T}_1$ are Hermitian $n \times n$ matrices. Then the boundary conditions (3.9) can be written as

$$
2\mathbf{T}_0 A \vec{\beta}(\psi) = -i^n \mathbf{T}_1 \vec{\delta}(\psi). \quad (3.76)
$$

The latter equation can be considered even if the matrix $\mathbf{T}_0$ is not invertible. This situation corresponds formally to the matrices $\mathbf{T}$ with infinite determinant and can be considered as a singular interaction with infinite coupling. Every Lagrangian subspace of $\mathbf{C}^{2n}$ can be described by conditions of the type (3.12). Thus the set of self–adjoint extensions of the operator L^0 coincides with the set of singular interactions with finite or infinite coupling.

3.2.4   Second order differential operator in one dimension

We consider in this section finite rank perturbations of the second derivative operator in dimension one. It follows from the previous section that such perturbed operators are described by the following formal expression

$$L_T = -D_x^2 + t_{00}\delta(\cdot)\delta + t_{01}\delta^{(1)}(\cdot)\delta + t_{10}\delta(\cdot)\delta^{(1)} + t_{11}\delta^{(1)}(\cdot)\delta^{(1)}, \qquad (3.77)$$

where the coefficients $t_{00}, t_{01}, t_{10}, t_{11}$ form a Hermitian 2×2 matrix. The operator $-D_x^2 + t_{00}\delta(\cdot)\delta$ can be determined using the standard perturbation theory, since δ is an element from the Hilbert space $\mathcal{H}_{-1}(-D_x^2)$. The self–adjoint operator corresponding to this formal expression is described by Theorem 3.2.2. This operator was determined using the distribution theory with discontinuous test functions. This theory allows one to determine the self–adjoint operators corresponding to the following formal expressions:

• Schrödinger operator with singular potential

$$L_{X_1 X_2} = -D_x^2 + X_1\delta + X_2\delta^{(1)}; \qquad (3.78)$$

• regularized Schrödinger operator with singular gauge field

$$L_{X_3} = (iD_x + X_3\delta)^2 - (X_3\delta)^2; \qquad (3.79)$$

• Schrödinger operator with singular density

$$L_{X_4} = -D_x(1 + X_4\delta)D_x. \qquad (3.80)$$

The coefficients X_1, X_2, X_3, X_4 in these formulas are real. One can combine these formal expressions to obtain the four–parameter family of operators with point interactions

$$L_X = -D_x^2(1 + X_4\delta) + iD_x(2X_3\delta - iX_4\delta^{(1)}) + X_1\delta + (X_2 - iX_3)\delta^{(1)}, \quad (3.81)$$

where $X = (X_1, X_2, X_3, X_4) \in \mathbf{R}^4$. We are going to prove that the corresponding four–parameter family of self–adjoint operators describes all finite rank point perturbations of the second derivative operator. We denote by L^0 the second derivative operator restricted to the set of functions with support separated from the origin. Then every self–adjoint extension of this operator coincides with a certain operator with singular interaction if one considers coefficients which are elements from the projective space. Therefore all the self–adjoint extensions of the operator L^0 can be classified. To prove this we need a more detailed version of Theorem 3.2.1, which describes the self–adjoint extensions of the operator L^0.

Theorem 3.2.3 *Every self–adjoint extension of the operator L^0 coincides with the operator L^{0*}, restricted to the set of functions, satisfying the boundary conditions at the origin of one of the following types*

1.

$$\left(\begin{array}{c} \psi(+0) \\ \psi'(+0) \end{array} \right) = J \left(\begin{array}{c} \psi(-0) \\ \psi'(-0) \end{array} \right) \tag{3.82}$$

with the matrix J equal to

$$J = e^{i\varphi} \left(\begin{array}{cc} a & b \\ c & d \end{array} \right) \tag{3.83}$$

with the real parameters $\varphi \in [-\pi/2, \pi/2], a, b, c, d \in \mathbf{R}$ fulfilling the condition $ad - bc = 1$; or

2.

$$\left\{ \begin{array}{l} h_0^+ \psi'(+0) = h_1^+ \psi(+0) \\ h_0^- \psi'(-0) = h_1^- \psi(-0), \end{array} \right. \tag{3.84}$$

with the parameters $\mathbf{h}^{\pm} = (h_0^{\pm}, h_1^{\pm})$ from the projective space $\mathbf{P}^1$.

Proof The deficiency elements $g_{\pm}(\lambda)$ for the operator L^0 at any point $\lambda = k^2, \Im\lambda \neq 0, \Im k > 0$, are equal to

$$g_1(\lambda) = \Theta(x)e^{ikx}; \quad g_2(\lambda) = \Theta(-x)e^{-ikx},$$

where Θ is the Heaviside function. Every element from the domain of L^{0*} can be presented in the form

$$\psi = \tilde{\psi} + a_{+1}g_1(\lambda) + a_{+2}g_2(\lambda) + a_{-1}g_1(\bar{\lambda}) + a_{-2}g_2(\bar{\lambda}), \tag{3.85}$$

where $\tilde{\psi}$ belongs to the closure of $\mathrm{Dom}(L^0)$ in the operator norm. Then the family of the self–adjoint extensions of the operator L^0 can be described by the 2×2 unitary matrices $\mathbf{V} = \left(\begin{array}{cc} v_{11} & v_{12} \\ v_{21} & v_{22} \end{array} \right)$. The domain of the self–adjoint

144 CHAPTER 3   FINITE RANK PERTURBATIONS

extension coincides with the subset of functions f from $\mathrm{Dom}(L^{0*})$ satisfying the following condition

$$\mathbf{V}\begin{pmatrix} a_{+1}(\psi) \\ a_{+2}(\psi) \end{pmatrix} = \begin{pmatrix} a_{-1}(\psi) \\ a_{-2}(\psi) \end{pmatrix}. \tag{3.86}$$

The boundary values of the function f on the left and right hand sides of the origin are equal to

$$\begin{aligned}
\psi(+0) &= (1+v_{11})a_{+1} + v_{12}a_{+2}, \\
\psi'(+0) &= (ik - i\overline{k}v_{11})a_{+1} - i\overline{k}v_{12}a_{+2};
\end{aligned} \tag{3.87}$$

$$\begin{aligned}
\psi(-0) &= v_{21}a_{+1} + (1+v_{22})a_{+2}, \\
\psi'(-0) &= i\overline{k}v_{21}a_{+1} - (ik - i\overline{k}v_{22})a_{+2}.
\end{aligned} \tag{3.88}$$

Conditions (3.86) can be written in terms of the boundary values of the function only. There are two ways to write these boundary conditions. If the matrix $\mathbf{V}$ is diagonal, then formulas (3.87) can be simplified

$$\begin{aligned} \psi(+0) &= (1+v_{11})a_{+1} \\ \psi'(+0) &= (ik - i\overline{k}v_{11})a_{+1} \end{aligned} \Rightarrow (1+v_{11})\psi'(+0) = (ik - i\overline{k}v_{11})\psi(+0).$$

Multiplying the last equation by $(1 + \overline{v_{11}})$ we get a boundary condition of the same type with the real parameters

$$(1+\overline{v_{11}})(1+v_{11})\psi'(+0) = (i(k - \overline{k}) - i\overline{k}v_{11} + ik\overline{v_{11}})\psi(+0). \tag{3.89}$$

The second condition can be derived in the same way from the linear system (3.88)

$$(1+\overline{v_{22}})(1+v_{22})\psi'(-0) = (i(k - \overline{k}) - i\overline{k}v_{22} + ik\overline{v_{22}})\psi(-0). \tag{3.90}$$

Equations (3.89),(3.90) give us the boundary conditions of the second type (3.84).

If the matrix $\mathbf{V}$ is not diagonal, then the determinant of the linear system (3.88) is equal to

$$iv_{21}(k + \overline{k}) \neq 0.$$

The second item does not vanish because the real part of k is not trivial. It follows that equation (3.86) can be written in the form

$$\begin{pmatrix} \psi(+0) \\ \psi'(+0) \end{pmatrix} = J\begin{pmatrix} \psi(-0) \\ \psi'(-0) \end{pmatrix}.$$

Since the operator L^{0*} restricted to the domain of functions satisfying these boundary conditions should be self-adjoint, the coefficients of the matrix $J = \begin{pmatrix} j_{11} & j_{12} \\ j_{21} & j_{22} \end{pmatrix}$ should satisfy the following conditions

$$j_{11}\bar{j}_{22} - j_{21}\bar{j}_{12} = 1;$$

$$j_{11}\bar{j}_{21} \in \mathbf{R}; j_{12}\bar{j}_{22} \in \mathbf{R}.$$

Every such matrix J can be presented in the form (3.83) with the real parameters φ, a, b, c, d. The parameter φ can be chosen from the interval $[-\pi/2, \pi/2]$. The boundary form of the operator vanishes on the subset of functions defined by the boundary conditions of the first or second type in the above sense (3.82), respectively (3.84). The operator L^{0*} restricted to the linear set defined by these boundary conditions is self-adjoint and the theorem is proven.

$\square$

Similar results have been proven in [218, 860] earlier. We note, that the diagonal matrices V define the operators, which can be presented by the orthogonal sum of two second derivative operators with the domains of functions defined on the separated half lines and satisfying certain boundary conditions at the origin. Such self-adjoint perturbations will be called **separated**. Self-adjoint operators corresponding to the matrices J cannot be presented by such an orthogonal sum. Self-adjoint perturbations of this type will be called **connected**.

We are now going to study the four-parameter family of second derivative operators with singular interactions. The operator L_X is an operator acting in the spaces $W_2^2(\mathbf{R} \setminus \{0\}) \to K'$. Therefor we again compose the operator L_X with the map η similar to Section 3.2.3

Theorem 3.2.4 *The second order differential operator with the singular interaction at the origin*

$$\mathbf{L}_X = \eta \left(-D_x^2(1 + X_4\delta) + iD_x(2X_3\delta - iX_4\delta^{(1)}) + X_1\delta + (X_2 - iX_3)\delta^{(1)} \right),$$
(3.91)

$X = (X_1, X_2, X_3, X_4) \in \mathbf{R}^4$, *coincides with the second derivative operator* $-D_x^2$ *defined on the domain of functions* $\psi \in W_2^2(\mathbf{R} \setminus \{0\})$, *satisfying the following boundary conditions at the origin:*

1.

$$\begin{pmatrix} \psi(+0) \\ \frac{d}{dx}\psi(+0) \end{pmatrix}$$
$$= \begin{pmatrix} \frac{(2+X_2)^2 - X_1 X_4 + X_3^2}{(2-iX_3)^2 + X_1 X_4 - X_2^2} & \frac{-4X_4}{(2-iX_3)^2 + X_1 X_4 - X_2^2} \\ \frac{4X_1}{(2-iX_3)^2 + X_1 X_4 - X_2^2} & \frac{(2-X_2)^2 - X_1 X_4 + X_3^2}{(2-iX_3)^2 + X_1 X_4 - X_2^2} \end{pmatrix} \begin{pmatrix} \psi(-0) \\ \frac{d}{dx}\psi(-0) \end{pmatrix},$$

$$\text{(3.92)}$$

if $(2 - iX_3)^2 + X_1 X_4 - X_2^2 \neq 0$;

2.

$$\begin{pmatrix} \frac{d}{dx}\psi(+0) \\ \frac{d}{dx}\psi(-0) \end{pmatrix} = \frac{1}{X_4} \begin{pmatrix} X_2 - 2 & 0 \\ 0 & X_2 + 2 \end{pmatrix} \begin{pmatrix} \psi(+0) \\ \psi(-0) \end{pmatrix}$$

$$\text{(3.93)}$$

if $4 + X_1 X_4 - X_2^2 = 0, X_3 = 0, X_4 \neq 0$;

3.

$$\begin{cases} \frac{d}{dx}\psi(+0) = \frac{X_1}{4}\psi(+0) \\ \psi(-0) = 0 \end{cases}$$

$$\text{(3.94)}$$

if $X_2 = 2, X_3 = 0, X_4 = 0$;

4.

$$\begin{cases} \psi(+0) = 0 \\ \frac{d}{dx}\psi(-0) = -\frac{X_1}{4}\psi(-0) \end{cases}$$

$$\text{(3.95)}$$

if $X_2 = -2, X_3 = 0, X_4 = 0$.

Proof The domain of the operator $\mathbf{L}_X$ coincides with the set of functions $\psi \in L_2(\mathbf{R})$, which are solutions of the equation $\mathbf{L}_X \psi = f$ for some function $f \in L_2(\mathbf{R})$. We consider the last equation in the generalized sense with the set of test functions D. Considering this equation for the test functions with support separated from the origin we deduce that $\psi \in W_2^2(\mathbf{R} \setminus \{0\})$. The functions from this Sobolev space are continuous outside the origin and have a continuous bounded first derivative there. The differential expression (3.91) is defined on such functions. The distributions δ and β and their first derivatives are defined on the functions ψ from $W_2^2(\mathbf{R} \setminus \{0\})$ as follows:

$$\delta(\psi) = \frac{\psi(+0) + \psi(-0)}{2}; \quad \delta^{(1)}(\psi) = -\frac{\psi'(+0) + \psi'(-0)}{2};$$

$$\beta(\psi) = \frac{\psi(+0) - \psi(-0)}{2}; \quad \beta^{(1)}(\psi) = -\frac{\psi'(+0) - \psi'(-0)}{2}.$$

Formulas (3.70), (3.71) define the product of the delta function or its derivative and any function from $W_2^2(\mathbf{R} \setminus \{0\})$. The distribution $\mathbf{L}_X \psi \in K', \psi \in$

$W_2^2(\mathbf{R} \setminus \{0\})$ can have singular support only at the origin. The singular term is equal to the linear combination of the distributions β and δ and their first derivatives. The distributions β and $\beta^{(1)}$ are mapped to zero by η. Then the distribution $L_X \psi$ is equivalent to some function from $L_2(\mathbf{R})$ if and only if the coefficients in front of the delta function and its derivative are equal to zero. Using (3.69) and (3.70) we get the following linear system

$$\begin{aligned} \delta: \quad & 2\beta^{(1)}(\psi) + X_1\delta(\psi) + (X_2 - iX_3)\delta^{(1)}(\psi) = 0 \\ \delta^{(1)}: \quad & -2\beta(\psi) + (X_2 + iX_3)\delta(\psi) + X_4\delta^{(1)}(\psi) = 0 \end{aligned} \Rightarrow$$

$$\begin{pmatrix} X_1 & -2 - X_2 + iX_3 & X_1 & 2 - X_2 + iX_3 \\ -2 + X_2 + iX_3 & -X_4 & 2 + X_2 + iX_3 & -X_4 \end{pmatrix}$$

$$\times \begin{pmatrix} \psi(+0) \\ \frac{d}{dx}\psi(+0) \\ \psi(-0) \\ \frac{d}{dx}\psi(-0) \end{pmatrix} = 0.$$

$$(3.96)$$

The rank of the matrix in the last equation is equal to 2 and it defines a two dimensional subspace in the four dimensional space of the boundary values $\{(\psi(+0), \frac{d}{dx}\psi(+0), \psi(-0), \frac{d}{dx}\psi(-0))\}$. We write conditions (3.96) in the form:

$$\begin{pmatrix} -\frac{X_1}{2} & 1 + \frac{X_2 - iX_3}{2} \\ 1 - \frac{X_2 + iX_3}{2} & \frac{X_4}{2} \end{pmatrix} \begin{pmatrix} \psi(+0) \\ \psi'(+0) \end{pmatrix}$$

$$= \begin{pmatrix} \frac{X_1}{2} & 1 - \frac{X_2 - iX_3}{2} \\ 1 + \frac{X_2 + iX_3}{2} & -\frac{X_4}{2} \end{pmatrix} \begin{pmatrix} \psi(-0) \\ \psi'(-0) \end{pmatrix}.$$

$$(3.97)$$

The determinant of the matrix in the left hand side of the latter equation is equal to

$$\Delta = \frac{-1}{4}\left((2 - iX_3)^2 + X_1 X_4 - X_2^2\right).$$

If $\Delta \neq 0$, then the matrix is invertible and these boundary conditions can be written in the form (3.92).

Consider the case $\Delta = 0$. In this case the coefficient X_3 is equal to zero. The boundary conditions (3.96) can be written as

$$\begin{pmatrix} 1 + \frac{X_2}{2} & -1 + \frac{X_2}{2} \\ \frac{X_4}{2} & \frac{X_4}{2} \end{pmatrix} \begin{pmatrix} \frac{d}{dx}\psi(+0) \\ \frac{d}{dx}\psi(-0) \end{pmatrix}$$

$$= \begin{pmatrix} \frac{X_1}{2} & \frac{X_1}{2} \\ -1 + \frac{X_2}{2} & 1 + \frac{X_2}{2} \end{pmatrix} \begin{pmatrix} \psi(+0) \\ \psi(-0) \end{pmatrix}.$$

The determinant of the matrix in the left-hand side of the latter equation is equal to X_4. If $X_4 \neq 0$, then the inverse matrix can be calculated and the boundary conditions have the form (3.93).

Consider now the case $\Delta = 0, X_4 = 0$. In this case $X_2 = \pm 2$. The boundary conditions, defined by $X_2 = 2, X_4 = 0$, and $X_2 = -2, X_4 = 0$, are equal to (3.94) and (3.95) correspondingly. All possible values of the coefficients $X_1, X_2, X_3, X_4 \in \mathbf{R}$ have been considered.

Moreover, the image of every function $\psi \in W_2^2(\mathbf{R} \setminus \{0\})$ satisfying these boundary conditions is equivalent to a certain function from $L_2(\mathbf{R})$ on the set of test functions from D. This completes the proof of the theorem.

$\square$

The boundary conditions (3.92)–(3.95) can be considered for infinite values of the parameters X_1, X_2, X_3, X_4. A good parameterization for this case can be done by using the formalism of projective spaces. We are going to parameterize all singular interactions by $\mathbf{X} \in \mathbf{P}^4$. We get the boundary conditions for all elements from the projective space with nonzero component X_0 with the help of the standard embedding of the space $\mathbf{R}^4$ into the space $\mathbf{P}^4 : (X_1, X_2, X_3, X_4) \mapsto (1, X_1, X_2, X_3, X_4)$. The boundary conditions corresponding to the other elements from the projective space will be defined using the homogenized analogue of the linear system (3.96)

$$\begin{pmatrix} X_1 & -2X_0 - X_2 + iX_3 & X_1 & 2X_0 - X_2 + iX_3 \\ -2X_0 + X_2 + iX_3 & -X_4 & 2X_0 + X_2 + iX_3 & -X_4 \end{pmatrix} \times$$

$$\times \begin{pmatrix} \psi(+0) \\ \frac{d}{dx}\psi(+0) \\ \psi(-0) \\ \frac{d}{dx}\psi(-0) \end{pmatrix} = 0. \tag{3.98}$$

We shall use the following definition.

Definition 3.2.6 *The algebraic set* $\mathbf{W}$ *is the set of elements from the projective space* $\mathbf{P}^4$, *satisfying the following two algebraic equations simultaneously*

$$X_0 = 0; \tag{3.99}$$

$$(X_0 - iX_3)^2 + X_1 X_4 - X_2^2 = 0. \tag{3.100}$$

The rank of the linear system (3.98) is not equal to 2 if and only if $\mathbf{X}$ is an element from the set $\mathbf{W}$.

Theorem 3.2.5 *Every element* $\mathbf{X}$ *from the projective space* $\mathbf{P}^4$, *which does not belong to the algebraic set* $\mathbf{W}$, *determines a unique self-adjoint extension* $L_{\mathbf{X}}$ *of the operator* L^0, *described by the following boundary conditions*

1.

$$
\begin{pmatrix} \psi(+0) \\ \frac{d}{dx}\psi(+0) \end{pmatrix}
$$

$$
= \begin{pmatrix} \frac{(2X_0+X_2)^2-X_1X_4+X_3^2}{(2X_0-iX_3)^2+X_1X_4-X_2^2} & \frac{-4X_0X_4}{(2X_0-iX_3)^2+X_1X_4-X_2^2} \\ \frac{4X_0X_1}{(2X_0-iX_3)^2+X_1X_4-X_2^2} & \frac{(2X_0-X_2)^2-X_1X_4+X_3^2}{(2X_0-iX_3)^2+X_1X_4-X_2^2} \end{pmatrix} \begin{pmatrix} \psi(-0) \\ \frac{d}{dx}\psi(-0) \end{pmatrix}, \tag{3.101}
$$

if $\mathbf{X} \in \mathbf{G}_1 = \{(2X_0 - iX_3)^2 + X_1X_4 - X_2^2 \neq 0\}$;

2.

$$
\begin{pmatrix} \frac{d}{dx}\psi(+0) \\ \frac{d}{dx}\psi(-0) \end{pmatrix} = \frac{1}{X_4} \begin{pmatrix} X_2 - 2X_0 & 0 \\ 0 & X_2 + 2X_0 \end{pmatrix} \begin{pmatrix} \psi(+0) \\ \psi(-0) \end{pmatrix} \tag{3.102}
$$

if $\mathbf{X} \in \mathbf{G}_2 = \{(2X_0 - iX_3)^2 + X_1X_4 - X_2^2 = 0, X_0 \neq 0, X_3 = 0, X_4 \neq 0\}$;

3.

$$
\begin{cases} 4X_0\frac{d}{dx}\psi(+0) = X_1\psi(+0) \\ \psi(-0) = 0 \end{cases} \tag{3.103}
$$

if $\mathbf{X} \in \mathbf{G}_3 = \{(2X_0 - iX_3)^2 + X_1X_4 - X_2^2 = 0, X_2 = 2X_0, X_0 \neq 0, X_3 = 0, X_4 = 0\}$.

Proof The rank of the matrix in the linear system (3.98) is equal to 1 if and only if $X_0 = 0$ and $(X_0 - iX_3)^2 + X_1X_4 - X_2^2 = 0$, i.e. if $\mathbf{X} \in \mathbf{W}$. If the rank of the matrix is equal to 2, then the homogeneous linear system defines two different boundary conditions as was shown during the proof of Theorem 3.2.4. Corresponding boundary conditions are the homogenized analogues of the boundary conditions (3.92)–(3.95) and cover the cases 1-3 of the present theorem.

$\square$

Theorem 3.2.6 *Every element* $\mathbf{X} \in \mathbf{W}$ *determines families of self–adjoint extensions of the operator* L^0, *described by the following boundary conditions*

1. *if* $\mathbf{X} \in \mathbf{W}_1 = \{\mathbf{X} \in \mathbf{W} : X_3 \neq 0\}$, *then*

$$\left(\begin{array}{c} \psi(+0) \\ \frac{d}{dx}\psi(+0) \end{array} \right)$$

$$= e^{i\varphi} \left(\begin{array}{cc} \frac{X_2}{X_3}\sin\varphi - \cos\varphi & -\frac{X_4}{X_3}\sin\varphi \\ \frac{X_2^2 + X_3^2}{X_4 X_3}\sin\varphi & -\cos\varphi - \frac{X_2}{X_3}\sin\varphi \end{array} \right) \left(\begin{array}{c} \psi(-0) \\ \frac{d}{dx}\psi(-0) \end{array} \right), \tag{3.104}$$

$\varphi \in [-\pi/2, \pi/2]$;

2. *if* $\mathbf{X} \in \mathbf{W}_2 = \{\mathbf{X} \in \mathbf{W} : X_3 = 0, X_1 \neq 0, X_4 \neq 0\}$, *then*

$$\left(\begin{array}{c} \psi(+0) \\ \frac{d}{dx}\psi(+0) \end{array} \right) = \left(\begin{array}{cc} a & -\frac{X_2}{X_1}(a+1) \\ \frac{X_1}{X_2}(a+1) & -2-a \end{array} \right) \left(\begin{array}{c} \psi(-0) \\ \frac{d}{dx}\psi(-0) \end{array} \right),$$

$$a \in \mathbf{R}, \tag{3.105}$$

or

$$\left\{ \begin{array}{l} X_2 \frac{d}{dx}\psi(+0) = X_1\psi(+0) \\ X_2 \frac{d}{dx}\psi(-0) = X_1\psi(-0); \end{array} \right. \tag{3.106}$$

3. *if* $\mathbf{X} \in \mathbf{W}_3 = \{\mathbf{X} \in \mathbf{W} : X_3 = 0, X_1 = 0\}$, *then*

$$\left(\begin{array}{c} \psi(+0) \\ \frac{d}{dx}\psi(+0) \end{array} \right) = \left(\begin{array}{cc} -1 & b \\ 0 & -1 \end{array} \right) \left(\begin{array}{c} \psi(-0) \\ \frac{d}{dx}\psi(-0) \end{array} \right), b \in \mathbf{R}, \tag{3.107}$$

or

$$\left\{ \begin{array}{l} \frac{d}{dx}\psi(+0) = 0 \\ \frac{d}{dx}\psi(-0) = 0; \end{array} \right. \tag{3.108}$$

4. *if* $\mathbf{X} \in \mathbf{W}_4 = \{\mathbf{X} \in \mathbf{W} : X_3 = 0, X_4 = 0\}$, *then*

$$\left(\begin{array}{c} \psi(+0) \\ \frac{d}{dx}\psi(+0) \end{array} \right) = \left(\begin{array}{cc} -1 & 0 \\ c & -1 \end{array} \right) \left(\begin{array}{c} \psi(-0) \\ \frac{d}{dx}\psi(-0) \end{array} \right), c \in \mathbf{R}, \tag{3.109}$$

or

$$\left\{ \begin{array}{l} \psi(+0) = 0 \\ \psi(-0) = 0. \end{array} \right. \tag{3.110}$$

Proof If **X** belongs to the set **W** then the system (3.98) defines only one boundary condition. The restriction of the adjoint operator L^{0*} to the set Q of all functions satisfying this boundary condition is not symmteric. The corresponding self–adjoint operator is not defined uniquely. In general elements from **W** define one–parameter families of self–adjoint operators. We are going to consider the four different cases separately.

1. Let $\mathbf{X} \in \mathbf{W}_1$. It follows that X_4 is not equal to zero. Otherwise the real parameters X_3 and X_2 would satisfy the equation

$$-X_3^2 - X_2^2 = 0$$

which has only the trivial solution.

The unique boundary condition, defined by the system (3.98), is equal to

$$(X_2 + iX_3)(\psi(+0) + \psi(-0)) - X_4(\psi'(+0) + \psi'(-0)) = 0. \qquad (3.111)$$

Every separate self–adjoint perturbation in this case should be described by the Dirichlet boundary conditions. This is only possible if $X_4 = 0$. Thus, no separated self–adjoint perturbation corresponds to such an element **X**.

Consider the connected self–adjoint perturbations. Substitution of the boundary condition (3.82) into the equation (3.111) leads to the equation

$$
\begin{aligned}
(X_2 + iX_3)(a\psi(-0) + b\psi'(-0)) &- X_4(c\psi(-0) + d\psi'(-0)) \\
+(X_2 + iX_3)e^{-i\varphi}\psi(-0) &- X_4 e^{-i\varphi}\psi'(-0) = 0,
\end{aligned}
$$

which should be satisfied for all values of $\psi(-0)$ and $\psi'(-0)$. It follows that the real coefficients a, b, c, d are solutions to the following linear system with real coefficients

$$
\begin{pmatrix}
X_2 & 0 & -X_4 & 0 \\
X_3 & 0 & 0 & 0 \\
0 & X_2 & 0 & -X_4 \\
0 & X_3 & 0 & 0
\end{pmatrix}
\begin{pmatrix}
a \\ b \\ c \\ d
\end{pmatrix}
=
\begin{pmatrix}
-X_2 \cos \varphi - X_3 \sin \varphi \\
X_2 \sin \varphi - X_3 \cos \varphi \\
X_4 \cos \varphi \\
-X_4 \sin \varphi.
\end{pmatrix}
$$

The coefficients a, b, c, d can be calculated, since $X_3 \neq 0, X_4 \neq 0$:

$$
\begin{array}{ll}
a = \frac{X_2}{X_3} \sin \varphi - \cos \varphi; & b = -\frac{X_4}{X_3} \sin \varphi; \\
c = \frac{X_2^2 + X_3^2}{X_4 X_3} \sin \varphi; & d = -\cos \varphi - \frac{X_2}{X_3} \sin \varphi
\end{array}
\qquad (3.112)
$$

The coefficients a, b, c, d satisfy the condition $ad - bc = 1$. It follows that the family of self–adjoint operators corresponding to this element **X** is described by boundary conditions (3.104).

2. Suppose that $\mathbf{X} \in \mathbf{W}_2$. It follows that $X_2 \neq 0$. The linear system (3.98) defines the unique condition

$$X_1(\psi(+0) + \psi(-0)) - X_2 \left(\frac{d}{dx}\psi(+0) + \frac{d}{dx}\psi(-0) \right) = 0.$$

Every connected self-adjoint perturbation, corresponding to $\mathbf{X}$, is defined by the matrix $J = e^{i\varphi} \begin{pmatrix} a & b \\ c & d \end{pmatrix}$. This matrix should be real $(\varphi = 0)$ and the coefficients should satisfy the following linear system

$$\begin{cases} aX_1 - cX_2 = -X_1 \\ bX_1 - dX_2 = X_2 \end{cases} \Rightarrow \begin{cases} c = \frac{X_1}{X_2}(a+1) \\ b = \frac{X_2}{X_1}(d+1). \end{cases}$$

The condition $ad - bc = 1$ leads to the equation $a + d = -2$. Then the matrix J, corresponding to the element $\mathbf{X}$ should be of the form (3.105). Every such matrix defines a self-adjoint perturbation. Every separated self-adjoint perturbation corresponding to the element $\mathbf{X}$ is defined by the boundary conditions (3.106).

3. Suppose that $\mathbf{X} \in \mathbf{W}, X_1 = 0$. Then $X_2 = 0$. The unique boundary condition, defined by $\mathbf{X}$, is equal to

$$\frac{d}{dx}\psi(+0) + \frac{d}{dx}\psi(-0) = 0.$$

This boundary condition leads to the matrices J of the following type

$$J = \begin{pmatrix} -1 & b \\ 0 & -1 \end{pmatrix}, \quad b \in \mathbf{R}.$$

Corresponding separated self-adjoint perturbations are defined by the Neumann boundary conditions.

4. The case $\mathbf{X} \in \mathbf{W}, X_4 = 0$ can be considered in a similar way.

The theorem is proved.

□

Thus the elements from $\mathbf{W}$ do not determine the self-adjoint perturbation uniquely. Any operator from the corresponding family can be used to describe the singular interaction.

Theorems 3.2.5 and 3.2.6 cover all possible values of $\mathbf{X}$ from the projective space.

Lemma 3.2.5 *The sets* $\mathbf{G}_1, \mathbf{G}_2, \mathbf{G}_3, \mathbf{G}_4$ *and* $\mathbf{W}$ *cover the projective space* $\mathbf{P}^4$.

Proof Every point from $\mathbf{P}^4$, which does not belong to $\mathbf{G}_1$, is an element from the algebraic set $\mathbf{V}_1 = \{(2X_0 - iX_3)^2 + X_1X_4 - X_2^2 = 0\}$. Every element of $\mathbf{V}_1$, which is not an element of $\mathbf{W}_1$ belongs to the algebraic set $\mathbf{V}_2 = \{(2X_0 - iX_3)^2 + X_1X_4 - X_2^2 = 0, X_0 \neq 0, X_3 = 0\}$ or $\mathbf{W}$. Every element of $\mathbf{V}_2$, which is not an element of $\mathbf{G}_2$ belongs to the set $\mathbf{V}_3 = \{(2X_0 - iX_3)^2 + X_1X_4 - X_2^2 = 0, X_0 \neq 0, X_3 = 0, X_4 = 0\}$. If $(X_0, X_1, X_2, X_3, X_4) \in \mathbf{V}_3$, then $(2X_0)^2 - X_2^2 = 0$ and it follows that $2X_0 = X_2$ or $2X_0 = -X_2$. This means that the sets $\mathbf{G}_3$ and $\mathbf{G}_4$ cover together the set $\mathbf{V}_3$. This completes the proof of the lemma.

$\square$

We note that the element $\mathbf{X}$ cannot be uniquely defined by the domain of the operator. For example elements with $X_0 = 0, X_2^2 - X_1X_4 + X_3^2 \neq 0$ correspond to the same self–adjoint operator, defined by the boundary conditions $\begin{pmatrix} \psi(+0) \\ \psi'(+0) \end{pmatrix} = -\begin{pmatrix} \psi(-0) \\ \psi'(-0) \end{pmatrix}$.

We are ready now to prove our main result in this section.

Theorem 3.2.7 *The set of all self–adjoint point perturbations at the origin of the second derivative operator in* $L_2(\mathbf{R})$ *coincides with the family of operators with singular interactions at the origin* $\{L_\mathbf{X}, \mathbf{X} \in \mathbf{P}^4\}$.

Proof Every operator $L_\mathbf{X}$ is defined as the restriction of the second derivative operator in $W_2^2(\mathbf{R} \setminus \{0\})$ on a certain linear set. The boundary conditions defining the operators $L_\mathbf{X}$ are of the type (3.82) or (3.84). It follows that every operator $L_\mathbf{X}$ is a self–adjoint extension of the operator L^0. We have to prove only that every such extension can be described by a certain singular interaction.

Consider first the arbitrary connected perturbation, defined by a certain matrix J (3.82). We are going to use the homogenized analogue of the conditions (3.97). If the element $\mathbf{X}$ defines the boundary conditions (3.82), then the following equation is fulfilled :

$$\begin{pmatrix} -X_1 & 2X_0 + X_2 - iX_3 \\ 2X_0 - X_2 - iX_3 & X_4 \end{pmatrix} e^{i\varphi} \begin{pmatrix} a & b \\ c & d \end{pmatrix}$$

$$= \begin{pmatrix} X_1 & 2X_0 - X_2 + iX_3 \\ 2X_0 + X_2 + iX_3 & -X_4 \end{pmatrix}.$$

The latter equation can be written as a 4×5 homogeneous linear system

$$
\begin{pmatrix}
2e^{i\varphi}c & -e^{i\varphi}a - 1 & e^{i\varphi}c & -ie^{i\varphi}c & 0 \\
-2 + 2e^{i\varphi}d & -e^{i\varphi}b & e^{i\varphi}d + 1 & -ie^{i\varphi}d - i & 0 \\
-2 + 2e^{i\varphi}a & 0 & -e^{i\varphi}a - 1 & -ie^{i\varphi}a - i & e^{i\varphi}c \\
2e^{i\varphi}b & 0 & -e^{i\varphi}b & -ie^{i\varphi}b & e^{i\varphi}d + 1
\end{pmatrix}
$$

$$
\times
\begin{pmatrix}
X_0 \\
X_1 \\
X_2 \\
X_3 \\
X_4
\end{pmatrix}
= 0.
\qquad (3.113)
$$

Let us denote by $\Delta_j, j = 0, 1, 2, 3, 4$, the determinants of the 4×4 matrices obtained from the 4×5 matrix by erasing the j-th column. These determinants are equal to

$$
\begin{aligned}
\Delta_0 &= 2ie^{2i\varphi}(2\cos\varphi + a + d)^2; \\
\Delta_1 &= -8ie^{2i\varphi}c(2\cos\varphi + a + d); \\
\Delta_2 &= 4ie^{2i\varphi}(a - d)(2\cos\varphi + a + d); \\
\Delta_3 &= -8i\sin\varphi e^{2i\varphi}(2\cos\varphi + a + d); \\
\Delta_4 &= -8ie^{2i\varphi}b(2\cos\varphi + a + d).
\end{aligned}
$$

All the determinants are equal to the product of the phase factor $ie^{i\varphi}$ and a certain real factor.

If $\Delta_0 \neq 0$, then the rank of the 4×5 matrix is equal to 4. The solution of the system (3.113) is equal to

$$
(-ie^{-i\varphi}\Delta_0, ie^{-i\varphi}\Delta_1, ie^{-i\varphi}\Delta_2, ie^{-i\varphi}\Delta_3, ie^{-i\varphi}\Delta_4, ie^{-i\varphi}\Delta_5) \in \mathbf{P}^4.
$$

This element does not belong to $\mathbf{W}$ and it defines the self–adjoint perturbation uniquely. This perturbation necessarily coincides with the one defined by the matrix J.

If $\Delta_0 = 0$, then the element $(0, c, a + \cos\varphi, \sin\varphi, -b) \in \mathbf{P}^4$ is a solution of the linear system. If $\sin\varphi \neq 0$, then these boundary conditions are described by the elements $\mathbf{X}$ from the set $\mathbf{W}_1$. In fact if $a + d + 2\cos\varphi = 0$ and $ad - bc = 1$ then the coordinates X_2, X_3, X_4 of $\mathbf{X}$ can be obtained from the system of nonlinear equations (3.112)

$$
X_2 = \frac{a + \cos\varphi}{\sin\varphi} X_3, \quad X_4 = -\frac{b}{\sin\varphi} X_3.
$$

The coordinate X_1 can be calculated from the equation $-X_3^2 + X_1 X_4 - X_2^2 = 0$

$$
X_1 = \frac{c}{\sin\varphi} X_3.
$$

Therefore the vector

$$\left(0, \frac{c}{\sin\varphi}X_3, \frac{a+\cos\varphi}{\sin\varphi}X_3, X_3, -\frac{b}{\sin\varphi}X_3\right) \in \mathbf{P}^4$$

determines the family of boundary conditions which includes the condition under discussion.

Consider the remaining case $\varphi = 0, a + d = -2$. The set of all such matrices is covered by the families

$$\left\{\begin{pmatrix} a & b \\ -\frac{(a+1)^2}{b} & -2-a \end{pmatrix}, a, b \in \mathbf{R}, b \neq 0\right\}$$

or

$$\left\{\begin{pmatrix} -1 & 0 \\ c & -1 \end{pmatrix}, c \in \mathbf{R}\right\}.$$

These matrices can be described by the singular interactions $\mathbf{X}$ from the algebraic set $\mathbf{W}$. Both families are covered by the boundary conditions (3.105), (3.107), (3.109). Thus we have proved that every connected self–adjoint perturbation is defined by a certain singular interaction.

Consider now the separated perturbations defined by the boundary conditions (3.84). Suppose that both zero components of the elements $\mathbf{h}^{\pm}$ are not equal to zero: $h_0^+ \neq 0, h_0^- \neq 0$. The coordinates of the element $\mathbf{X}$ can be calculated

$$\begin{cases} 2X_2 = X_4\left(\frac{h_1^-}{h_0^-} + \frac{h_1^+}{h_0^+}\right) \\ 4X_0 = X_4\left(\frac{h_1^-}{h_0^-} - \frac{h_1^+}{h_0^+}\right). \end{cases}$$

If $\mathbf{h}^- \neq \mathbf{h}^+$ as elements of $\mathbf{P}^1$, then the element $\mathbf{X}$ from $\mathbf{G}_2$ will define the boundary conditions. The coordinate X_3 can be chosen equal to zero. The first coordinate should be calculated from the condition

$$4X_0^2 + X_1X_4 - X_2^2 = 0 \Rightarrow X_1 = \frac{-4X_0^2 + X_2^2}{X_4}.$$

The case $\mathbf{h}^- = \mathbf{h}^+$ is described by the elements from $\mathbf{W}$. The boundary conditions (3.106),(3.108),(3.110) cover all conditions of this type.

If $h_0^- = 0$, then the element $\mathbf{X} = (h_0^+, 4h_1^+, 2h_0^+, 0, 0)$ defines such separated boundary conditions. If $h_0^+ = 0$, then the boundary conditions are defined by the element $\mathbf{X} = (h_0^-, -4h_1^-, -2h_0^-, 0, 0)$. The theorem is proved.

$\square$

We are going to prove here that only this four–parameter family of singular interactions can be described by self–adjoint operators in the framework of the approach developed. Every second derivative operator with singular interaction having support at the origin has the following form:

$$L = -D_x^2 \left(1 + \sum_{n=0}^{N_2} a_2^n \delta^{(n)}\right) + iD_x \left(\sum_{n=0}^{N_1} a_1^n \delta^{(n)}\right) + \sum_{n=0}^{N_0} a_0^n \delta^{(n)}.$$

This differential expression is defined on the functions from $W_2^2(\mathbf{R} \setminus \{0\})$ only if the coefficients $a_2^n, a_1^n, a_0^n, n = 2, 3, 4, ...,$ are equal to zero. Consider the formal operator

$$L = -D_x^2 (1 + a_2^0 \delta + a_2^1 \delta^{(1)}) + iD_x(a_1^0 \delta + a_1^1 \delta^{(1)}) + a_0^0 \delta + a_0^1 \delta^{(1)}.$$

If $\psi \in W_2^2(\mathbf{R} \setminus \{0\})$, then the singular part of the distribution $\eta L \psi$ is equal to the linear combination of the distributions δ and its first three derivatives. This distribution is equivalent to a function from $L_2(\mathbf{R})$ only if the coefficients in front of the δ function and its derivatives are equal to zero. We get the following linear system:

$$\begin{pmatrix} -a_2^1 & 0 & 0 & 0 \\ -a_2^0 + ia_1^1 & -a_2^1 & 0 & 0 \\ ia_1^1 + a_0^1 & ia_1^1 & -1 & 0 \\ a_0^0 & a_0^1 & 0 & 1 \end{pmatrix} \begin{pmatrix} \delta(\psi) \\ \delta^{(1)}(\psi) \\ 2\beta(\psi) \\ 2\beta^{(1)}(\psi) \end{pmatrix} = 0.$$

This linear system defines a self–adjoint operator only if its rank is equal to 2. Thus the following conditions should be satisfied

$$a_2^1 = 0, \quad -a_2^0 + ia_1^1 = 0. \tag{3.114}$$

The boundary conditions defined by the linear system can be written as

$$\begin{pmatrix} 2\beta(\psi) \\ 2\beta^{(1)}(\psi) \end{pmatrix} = \begin{pmatrix} ia_1^0 + a_0^1 & ia_1^1 \\ -a_0^0 & -a_0^1 \end{pmatrix} \begin{pmatrix} \delta(\psi) \\ \delta^{(1)}(\psi) \end{pmatrix}.$$

These boundary conditions define a symmetric operator if and only if the coefficients a_n^k satisfy the following homogeneous linear system

$$\begin{cases} a_0^0 - \bar{a}_0^0 = 0, \\ a_0^1 + i\bar{a}_1^1 - \bar{a}_0^1 = 0, \\ i\bar{a}_1^1 + ia_1^1 = 0. \end{cases}$$

These equations together with equations (3.114) lead to the following conditions on the coefficients

$$a_0^0, a_2^0, a_1^0 \in \mathbf{R}; \quad a_2^1 = 0; \quad a_1^1 = -2\Im a_0^1.$$

These coefficients describe the four–parameter family of singular interactions considered here. The following theorem is thus proved.

Theorem 3.2.8 *The set of self–adjoint second derivative operators with singular interaction of finite strength coincides with the four–parameter family of operators* $\{L_X, X \in \mathbf{R}^4\}$.

We are going to discuss the interpretation of the parameters X_1, X_2, X_3, and X_4, defining the four–parameter family of singular interactions. Three different subfamilies of the operators, which appear in different problems of mathematical physics, will be considered.

Let us study first the two dimensional subfamily of Schrödinger operators with generalized potentials

$$L_{X_1 X_2} \psi = -D_x^2 \psi + (X_1 \delta + X_2 \delta^{(1)}) \psi. \tag{3.115}$$

Every such operator coincides with the second derivative operator defined on the domain of functions from $W_2^2(\mathbf{R} \backslash \{0\})$ satisfying the boundary conditions

$$\begin{pmatrix} \psi(+0) \\ \frac{d}{dx}\psi(+0) \end{pmatrix} = \begin{pmatrix} \frac{2+X_2}{2-X_2} & 0 \\ \frac{4X_1}{4-X_2^2} & \frac{2-X_2}{2+X_2} \end{pmatrix} \begin{pmatrix} \psi(-0) \\ \frac{d}{dx}\psi(-0) \end{pmatrix}. \tag{3.116}$$

The regularized Schrödinger operator with the singular gauge field $(iD_x + X_3\delta)^2 - (X_3\delta)^2$ is the operator

$$L_{X_3} = -D_x^2 + iX_3 \left(2D_x\delta - \delta^{(1)} \right). \tag{3.117}$$

It is defined by the boundary conditions

$$\begin{pmatrix} \psi(+0) \\ \frac{d}{dx}\psi(+0) \end{pmatrix} = \begin{pmatrix} \frac{2+iX_3}{2-iX_3} & 0 \\ 0 & \frac{2+iX_3}{2-iX_3} \end{pmatrix} \begin{pmatrix} \psi(-0) \\ \frac{d}{dx}\psi(-0) \end{pmatrix}. \tag{3.118}$$

The Schrödinger operator with the singular density $-D_x(1 + X_4\delta)D_x$ is the heuristic operator

$$L_{X_4} = -D_x^2(1 + X_4\delta) + X_4 D_x \delta^{(1)}. \tag{3.119}$$

It is equal to the second derivative operator with the domain of functions from $W_2^2(\mathbf{R} \backslash \{0\})$ satisfying the boundary conditions

$$\begin{pmatrix} \psi(+0) \\ \frac{d}{dx}\psi(+0) \end{pmatrix} = \begin{pmatrix} 1 & -X_4 \\ 0 & 1 \end{pmatrix} \begin{pmatrix} \psi(-0) \\ \frac{d}{dx}\psi(-0) \end{pmatrix}. \tag{3.120}$$

It follows that the coefficients X_1 and X_2 in the four–parameter family of singular interactions (3.91) can be interpreted as the coefficients in front of the δ and δ' potentials. The coefficient X_3 defines the strength of the gauge field with singularity at the origin. The coefficient X_4 corresponds to the singular density.

Chapter 4

Scattering theory for finite rank perturbations

This chapter is devoted to the construction of a scattering theory for the model operators described above. We have calculated the scattering matrix for the generalized point interaction in Section 2.2.5, where the asymptotics of the eigenfunctions were considered. We develop here systematically a rigorous mathematical scattering theory for finite rank perturbations and generalized perturbations with internal structure. First perturbations of rank one will be considered. The wave operators and the scattering operator will be calculated explicitly. These results will then be extended to arbitrary perturbations of finite rank. The scattering theory is developed for two self–adjoint operators, which are two self–adjoint extensions of one symmetric operator with finite deficiency indices. The formulas obtained are used to calculate the scattering matrix for finite rank perturbations and generalized perturbations with internal structure.

4.1 Scattering theory for rank one perturbations

This section is devoted to the scattering theory for two self–adjoint operators $\mathcal{A}$ and $\mathcal{A}^*$, which are two self–adjoint extensions of one symmetric operator with deficiency indices $(1,1)$. We suppose without loss of generality that the perturbed and unperturbed operators are different $\mathcal{A} \neq \mathcal{A}^*$. In particular we are going to study the case where the operator $\mathcal{A}^*$ is a rank one perturbation of the operator $\mathcal{A}$:

$$\mathcal{A}^* = \mathcal{A}_\alpha = \mathcal{A} + \alpha \langle \varphi, \cdot \rangle \varphi.$$

The stationary scattering theory for the operators $\mathcal{A}$ and $\mathcal{A}_\alpha$ will be constructed in this section. We prove first that the absolutely continuous components of the spectra of the operators $\mathcal{A}$ and $\mathcal{A}_\alpha$ coincide. The wave operators for rank one perturbations are then calculated explicitly.

The behaviour of the absolutely continuous spectrum under finite rank bounded perturbations was first studied by T.Kato [517, 518]. The wave operators for the trace class perturbations have been calculated. Similar results were obtained by M.Rosenblum [848]. This approach was extended by S.T.Kuroda [645, 646, 648], in particular relatively small perturbations has been studied and the scattering operator have been calculated.

The difference of the perturbed and unperturbed operators for finite rank singular perturbations is not an operator. But the difference of their resolvents given by Krein's formula is a finite rank operator. The case where the difference of a certain function of the perturbed and unperturbed operators is from the trace class has been studied by M.Birman and M.Krein [153, 144, 147, 146, 148]. In particular the invariance of the absolutely continuous spectrum and the existence of the wave operators for two self–adjoint operators operators has been proven when the difference of their resolvents is from the trace class. These results were extended by T.Kato and S.T.Kuroda [520, 524]. See [522] for an excellent reveiw of this theory.

4.1.1   Rank one perturbations and operators with a simple spectrum

In this section we show that the study of the problem of a rank one perturbation can be reduced to the case of operators with a simple spectrum. Consider first bounded rank one perturbations, i.e the case where $\varphi \in \mathcal{H}$. Then the operators $\mathcal{A}$ and $\mathcal{A}_\alpha = \mathcal{A} + \alpha \langle \varphi, \cdot \rangle \varphi$ have the same domain and the difference operator $\mathcal{B} = \mathcal{A}_\alpha - \mathcal{A} = \alpha \langle \varphi, \cdot \rangle \varphi$ is bounded. Consider the subspace $\mathcal{K}$ which is equal to the closure of the linear hull of the vectors $E(\lambda)\varphi$, where $E(\lambda)$ denotes the spectral projector for the operator $\mathcal{A}$

$$\mathcal{K} = \overline{\mathcal{L}\{E(\lambda)\varphi\}_{\lambda \in \mathbf{R}}}.$$

The subspace $\mathcal{K}$ reduces the operators $\mathcal{A}$ and $\mathcal{B}$. Therefore it reduces the operator $\mathcal{A}_\alpha$ also. The orthogonal complement $\mathcal{H} \ominus \mathcal{K}$ reduces both the operators $\mathcal{A}$ and $\mathcal{A}_\alpha$ since the following equality holds

$$\mathcal{A}|_{(\mathcal{H}\ominus\mathcal{K})\cap\mathrm{Dom}(\mathcal{A})} = \mathcal{A}_\alpha|_{(\mathcal{H}\ominus\mathcal{K})\cap\mathrm{Dom}(\mathcal{A})}.$$

Thus the rank one perturbation affects only the subspace $\mathcal{K}$ which is a separable Hilbert space. The restrictions of the operators $\mathcal{A}$ and $\mathcal{A}_\alpha$ to this subspace have simple spectra, since this subspace is generated by one element φ. A similar result holds also for rank one $\mathcal{H}_{-2}$-perturbations. We are going to prove the same result for two self-adjoint operators which are extensions of one symmetric operator with arbitrary finite deficiency indices.

Theorem 4.1.1 *Let $\mathcal{A}$ and $\mathcal{A}^*$ be two self-adjoint extensions of the symmetric operator $\mathcal{A}^0$ having finite deficiency indices and acting in the Hilbert space $\mathcal{H}$. Then there exists a separable subspace $\mathcal{K} \subset \mathcal{H}$ with the properties:*

1. *$\mathcal{K}$ reduces both operators $\mathcal{A}$ and $\mathcal{A}^*$;*

2. *the space $\mathcal{K}$ is generated by a finite dimensional subspace and the operator $\mathcal{A}$ (or the operator $\mathcal{A}^*$);*

3. *the operators $\mathcal{A}$ and $\mathcal{A}^*$ restricted to the orthogonal complement of $\mathcal{K}$ coincide: $\mathcal{A}|_{(\mathcal{H}\ominus\mathcal{K})\cap\mathrm{Dom}\,(\mathcal{A})} = \mathcal{A}^*|_{(\mathcal{H}\ominus\mathcal{K})\cap\mathrm{Dom}\,(\mathcal{A}^*)}$.*

Proof Let us denote by $N(-i) = \mathrm{Ker}\,(\mathcal{A}^{0*} + i)$ the finite dimensional deficiency subspace for the operator $\mathcal{A}^0$ at point $+i$. Let us denote by $\mathcal{K}$ and $\mathcal{K}^*$ the closures of the linear hulls of the vectors $E(\lambda)N(-i)$ and $E^*(\lambda)N(-i)$, respectively:

$$\mathcal{K} = \overline{\mathcal{L}\{E(\lambda)g\}_{g\in N(-i),\lambda\in\mathbf{R}}};$$

$$\mathcal{K}^* = \overline{\mathcal{L}\{E^*(\lambda)g\}_{g\in N(-i),\lambda\in\mathbf{R}}}.$$

(4.1)

Consider Krein's formula (3.47) for the resolvent of the operator $\mathcal{A}^*$ at the point $\lambda = i$

$$\frac{1}{\mathcal{A}^* - i} = \frac{1}{\mathcal{A} - i} + \frac{\mathcal{A} + i}{\mathcal{A} - i}\frac{1}{\Gamma - Q(i)}P_{N(-i)}.$$

(4.2)

The subspace $\mathcal{K}$ reduces the operator $\mathcal{A}$ and therefore the subspace $\mathcal{H} \ominus \mathcal{K}$ reduces the operator $\mathcal{A}$. The latter formula implies that the subspace $\mathcal{H} \ominus \mathcal{K}$ reduces also the operator $\mathcal{A}^*$. It follows that the subspace $\mathcal{K}$ reduces $\mathcal{A}^*$. It follows that $\mathcal{K}^* \subset \mathcal{K}$, since the subspace $\mathcal{K}^*$ is the minimal subspace that contains $N(-i) \subset \mathcal{K}$ and reduces the operator $\mathcal{A}^*$. Changing the roles of the operators $\mathcal{A}$ and $\mathcal{A}^*$ we prove that $\mathcal{K} \subset \mathcal{K}^*$. It follows that $\mathcal{K} = \mathcal{K}^*$ and reduces both operators $\mathcal{A}$ and $\mathcal{A}^*$. It is generated by the subspace $N(-i)$. The operators $\mathcal{A}$ and $\mathcal{A}^*$ restricted to the orthgonal complement to $\mathcal{K}$ coincide. The theorem is proven.

$\square$

Corollary *If the deficiency indices of the operator $\mathcal{A}^0$ are equal to $(1,1)$ then* the space $\mathcal{K}$ is generated by any deficiency element g. Therefore the spectra of the operators $\mathcal{A}$ and $\mathcal{A}^*$ restricted to $\mathcal{K}$ are simple. The subspace $\mathcal{K}$ is the smallest subspace that contains $N(-i)$ and reduces the operator $\mathcal{A}$ ($\mathcal{A}^*$).

The decomposition

$$\mathcal{H} = \mathcal{K} \oplus (\mathcal{H} \ominus \mathcal{K})$$

of the Hilbert space we constructed in the latter theorem will be used to study arbitrary rank one perturbations of self–adjoint operators. It follows that without loss of generality one can restrict the study of such perturbations to the case where $\mathcal{H} = \mathcal{K}$. We thus confine our considerations to this case in the following sections.

4.1.2　Invariance of the absolutely continuous spectrum

We are going to prove that the absolutely continuous component of the spectrum is invariant under rank one perturbations. Let us consider first the case of a bounded perturbation.

Lemma 4.1.1 *Let $\mathcal{A}$ be a self–adjoint operator acting in the Hilbert space $\mathcal{H}$. Let $\varphi \in \mathcal{H}$. Then the absolutely continuous parts of the operators $\mathcal{A}$ and $\mathcal{A}_\alpha = \mathcal{A} + \alpha\langle\varphi, \cdot\rangle\varphi$, $\alpha \in \mathbf{R}$, are unitary equivalent.*

Proof Theorem 4.1.1 implies that it is enough to consider the case where the element φ is a generating element for the operators $\mathcal{A}$, $\mathcal{A}_\alpha$ and the Hilbert space $\mathcal{H}$. The operators $\mathcal{A}$ and $\mathcal{A}_\alpha$ have simple spectra in this case. Consider then the spectral function associated with the element φ:

$$\rho_\varphi(\lambda) = \langle\varphi, E(\lambda)\varphi\rangle,$$

where $E(\lambda)$ is the spectral projector for the operator $\mathcal{A}$. The function $\lambda \mapsto \rho_\varphi(\lambda)$ is a nondecreasing function of finite variation on the real line. The absolutely continuous spectrum of the operator $\mathcal{A}$ coincides with the support of the absolutely continous part of the measure $d\rho_\varphi$, since the vector φ is generating. To calculate the support let us consider the following Cauchy–Stieltjes integral

$$F(z) = \int_{-\infty}^{\infty} \frac{d\rho_\varphi(\lambda)}{\lambda - z} = \left\langle\varphi, \frac{1}{\mathcal{A} - z}\varphi\right\rangle.$$

It follows from Privalov's theorem [840, 839] that almost everywhere on the real line the limit

$$\lim_{\epsilon \to 0} F(z + i\epsilon) = F(z + i0)$$

exists and is finite. Moreover almost everywhere on the real line the following equality holds:

$$\pi \frac{\partial \rho_\varphi(z)}{\partial z} = \Im\left(F(z+i0)\right), \ z \in \mathbf{R}.$$

Therefore the support of the absolutely continuous part of the measure $d\rho_\varphi$ and the set of points where the function $F(z+i0)$ is defined and has nontrivial imaginary part are almost equal (with respect to the Lebesgue measure).

Let us denote by $d\rho_{\varphi,\alpha}$ the spectral measure corresponding to the element φ and the operator $\mathcal{A}_\alpha$

$$d\rho_{\varphi\alpha}(\lambda) = d\langle \varphi, E_\alpha(\lambda)\varphi \rangle$$

where $E_\alpha(\lambda)$ is the spectral projector for $\mathcal{A}_\alpha$. Then the absolutely continuous part of the spectrum of the operator $\mathcal{A}_\alpha$ is almost equal to the set where the function

$$F_\alpha(z+i0) = \int_{-\infty}^{\infty} \frac{d\rho_{\varphi\alpha}(\lambda)}{\lambda - (z+i0)} = \left\langle \varphi, \frac{1}{\mathcal{A}_\alpha - (z+i0)}\varphi \right\rangle$$

is defined and has nonreal values.

The functions $F(z)$ and $F_\alpha(z)$ are related as follows (1.9):

$$(1 - \alpha F_\alpha(z))(1 + \alpha F(z)) = 1.$$

Then the imaginary parts of the functions F and F_α are related by the following equality

$$\Im F_\alpha(z) = \frac{\Im F(z)}{|1 + \alpha F(z)|^2}, \tag{4.3}$$

which is valid for real z if and only if the limits of the left and right hand sides exist and are finite. The set of points where $F(z)$ and $F_\alpha(z)$ do not have finite limits has measure zero. The set of points where $F(z) = -1/\alpha$ also has measure zero. It follows that the sets where the functions $F(z+i0)$ and $F_\alpha(z+i0)$ have nonreal limit values agree up to sets of measure zero. Therefore the supports of the absolutely continuous parts of the measures $d\rho_\varphi$ and $d\rho_{\varphi\alpha}$ are almost equal and therefore coincide. The operators $\mathcal{A}$ and $\mathcal{A}_\alpha$ have simple spectra in accordance with our assumption and it follows that the absolutely continuous parts of the operators $\mathcal{A}$ and $\mathcal{A}_\alpha$ are unitary equivalent.

$\square$

Similar arguments can be used to prove that the absolutely continuous spectrum is invariant under arbitrary finite rank $\mathcal{H}_{-2}$-perturbations. Moreover

the absolutely continuous spectra of two operators being self–adjoint extensions of one symmetric operator with deficiency indices $(1,1)$ are equal.

Theorem 4.1.2 *Let $\mathcal{A}$ and $\mathcal{A}^\gamma$ be two self–adjoint extensions of the operator $\mathcal{A}^0$ having deficiency indices $(1,1)$. Then the absolutely continuous parts of the operators $\mathcal{A}$ and $\mathcal{A}^\gamma$ are unitary equivalent.*

Proof Let us denote by g the unit deficiency element for the operator $\mathcal{A}^0$ corresponding to the point $-i$. Consider the spectral measures $\rho_g(\lambda)$ and $\rho_g^\gamma(\lambda)$ corresponding to the element g and the operators $\mathcal{A}$ and $\mathcal{A}^\gamma$ respectively. Consider two Nevanlinna functions given by

$$Q(z) \equiv \left\langle g, \frac{1+z\mathcal{A}}{\mathcal{A}-z}g \right\rangle = \int_{-\infty}^{\infty} \frac{1+z\lambda}{\lambda-z}d\rho_g(\lambda),$$

$$G^\gamma(z) \equiv \left\langle g, \frac{1+z\mathcal{A}^\gamma}{\mathcal{A}^\gamma-z}g \right\rangle = \int_{-\infty}^{\infty} \frac{1+z\lambda}{\lambda-z}d\rho_g^\gamma,$$

where
$$\int_{-\infty}^{\infty} d\rho_g(\lambda) = 1; \quad \int_{-\infty}^{\infty} d\rho_g^\gamma(\lambda) = 1.$$

The latter two equalities follow from the fact that g has unit norm in $\mathcal{H}$.

We suppose that the element $g \in \mathcal{L}\{g\} = N(-i)$ is generating for the operator $\mathcal{A}$ and the Hilbert space $\mathcal{H}$. The latter assumption is not restrictive due to Theorem 4.1.1. Then the absolutely continuous parts of the spectra of the operators $\mathcal{A}$ and $\mathcal{A}^\gamma$ coincide with the supports of the absolutely continuous parts of the measures $d\rho_g$ and $d\rho_g^\gamma$. The functions $Q(z)$ and $G^\gamma(z)$ are related to the corresponding Cauchy–Stieltjes integrals as follows

$$Q(z) = z + (1+z^2)\left\langle g, \frac{1}{\mathcal{A}-z}g \right\rangle,$$

$$G^\gamma(z) = z + (1+z^2)\left\langle g, \frac{1}{\mathcal{A}^\gamma-z}g \right\rangle.$$

One can apply Privalov's theorem [840, 839], since the measures $d\rho_g$ and $d\rho_g^\gamma$ have finite variation on each compact interval, and prove that the absolutely continuous parts of the spectrum are almost equal respectively to the sets of points where the functions $Q(z+i0)$ and $G^\gamma(z+i0)$ exist and have nontrivial imaginary parts.

The two functions $Q(z)$ and $G^\gamma(z)$ are related as follows

$$G^\gamma(z) = \frac{\gamma Q(z) + 1}{\gamma - Q(z)}$$

where γ is the real parameter describing the extension $\mathcal{A}^\gamma$ of $\mathcal{A}^0$. It follows that

$$\Im G^\gamma(z) = \frac{(\gamma^2 + 1)\Im Q(z)}{(\gamma - \Re Q(z))^2 + (\Im Q(z))^2}.$$

Repeating the arguments used during the proof of Lemma 4.1.1 we prove that the absolutely continuous parts of the spectrum of the operators $\mathcal{A}$ and $\mathcal{A}^\gamma$ coincide and therefore the absolutely continuous parts of these operators are unitary equivalent.

$\square$

Consider now generalized rank one perturbations.

Theorem 4.1.3 *Let $\mathcal{A}$ be a self-adjoint operator acting in the Hilbert space $\mathcal{H}$ and let $\varphi \in \mathcal{H}_{-2}(\mathcal{A})\backslash\mathcal{H}$. Let $\mathbf{A}^\Gamma$ be a generalized perturbation with internal structure of the operator $\mathcal{A}$ defined in the Hilbert space $\mathbf{H} = \mathcal{H}\oplus\mathcal{H}'$, $\dim\mathcal{H}' < \infty$, by the element $\varphi' \in \mathcal{H}'$, the Hermitian operator A' and the Hermitian boundary operator Γ. Then the absolutely continuous parts of the operators $\mathcal{A}$ and $\mathbf{A}^\Gamma$ are unitary equivalent.*

Proof Following Theorem 4.1.1 we restrict our consideration to the case where the spaces $\mathcal{H}$ and $\mathcal{H}'$ are generated by the vectors $g = \frac{1}{\mathcal{A}+i}\varphi$, $g' = \frac{1}{\mathcal{A}'+i}\varphi'$ and the operators $\mathcal{A}$ and $\mathcal{A}'$ respectively.

The vectors g and g' are generating for the operator $\mathbf{A}^\Gamma$ and the Hilbert space $\mathbf{H}$. Therefore the absolutely continuous spectrum of the operator $\mathbf{A}^\Gamma$ coincides with the union of the supports of the absolutely continuous parts of the spectral measures corresponding to the elements g and g'. To calculate these supports let us consider the following two functions

$$
\begin{aligned}
G_0^\Gamma(z) &= \left\langle g, \frac{1 + z\mathbf{A}^\Gamma}{\mathbf{A}^\Gamma - z}g \right\rangle_{\mathcal{H}} = \int_{-\infty}^{+\infty} \frac{1 + \lambda z}{\lambda - z}d\rho_g^\Gamma(\lambda); \\
G_1^\Gamma(z) &= \left\langle g', \frac{1 + z\mathbf{A}^\Gamma}{\mathbf{A}^\Gamma - z}g' \right\rangle_{\mathcal{H}'} = \int_{-\infty}^{+\infty} \frac{1 + \lambda z}{\lambda - z}d\rho_{g'}^\Gamma(\lambda),
\end{aligned}
\tag{4.4}
$$

where $d\rho_g^\Gamma(\lambda) = d\langle g, E^\Gamma(\lambda)g\rangle$, $d\rho_{g'}^\Gamma(\lambda) = d\langle g', E^\Gamma(\lambda)g'\rangle$ are the spectral measures corresponding to the vectors g, g' and the operator $\mathbf{A}^\Gamma$. ($E^\Gamma(\lambda)$ denotes here the spectral projector corresponding to the operator $\mathbf{A}^\Gamma$.) Consider the

Q-functions corresponding to the original operators $\mathcal{A}$ and $\mathcal{A}'$

$$Q_0(\lambda) = \left\langle g, \frac{1+z\mathcal{A}}{\mathcal{A}-z} g \right\rangle_{\mathcal{H}} = \int_{-\infty}^{+\infty} \frac{1+z\lambda}{\lambda-z} d\rho_g(\lambda);$$

$$Q_1(\lambda) = \left\langle g', \frac{1+z\mathcal{A}'}{\mathcal{A}'-z} g' \right\rangle_{\mathcal{H}'} = \int_{-\infty}^{+\infty} \frac{1+z\lambda}{\lambda-z} d\rho_{g'}(\lambda),$$

(4.5)

where $d\rho_g(\lambda)$ and $d\rho_{g'}(\lambda)$ are the spectral measures corresponding to the vectors g, g' and the operator $\mathbf{A} = \mathcal{A} \oplus \mathcal{A}'$. Then the formula for the restricted resolvent of the operator $\mathbf{A}^\Gamma$ implies that

$$P_\mathcal{H} \frac{1}{\mathbf{A}^\Gamma - \lambda} P_\mathcal{H} = \frac{1}{\mathcal{A}-\lambda} - \frac{1}{Q_0(\lambda)+Q_0^+(\lambda)} \left\langle \frac{\mathcal{A}+i}{\mathcal{A}-\bar{\lambda}}g, \cdot \right\rangle \frac{\mathcal{A}+i}{\mathcal{A}-\lambda}g, \quad (4.6)$$

where $Q_0^+ = \frac{|\gamma_{01}|^2}{\gamma_{11}-Q_1(\lambda)} - \gamma_{00}$. It follows that the Nevanlinna functions G_0^Γ, Q_0 and Q_0^+ are related as follows

$$Q_0^\Gamma(\lambda) = \frac{Q_0(\lambda)Q_0^+(\lambda)+1}{Q_0(\lambda)+Q_0^+(\lambda)}, \quad (4.7)$$

which implies that the imaginary parts of these functions are related by

$$\Im G_0^\Gamma(\lambda) = \frac{\Im Q_0(\lambda)\left(|Q_0^+(\lambda)|^2+1\right)+\Im Q_0^+(\lambda)\left(|Q_0(\lambda)|^2+1\right)}{|Q_0(\lambda)+Q_0^+(\lambda)|^2}. \quad (4.8)$$

The dimension of the space $\mathcal{H}'$ is finite and it follows that the function $Q_1(\lambda)$ is rational and has only a finite number of singularities on the real axis, real outside the singular points. The same is true for the function $Q_0^+(\lambda)$ which is a rational transformation of $Q_1(\lambda)$. We conclude that the sets of points where the imaginary parts of the functions $Q_0^+(\lambda+i0)$ and $Q_0(\lambda+i0)$ exist and are not equal to zero coincide up to a set of measure zero.

Consider the function $G_1^\Gamma(\lambda)$ which is related to the functions $Q_1(\lambda)$ and $Q_1^+(\lambda) = \frac{|\gamma_{01}|^2}{\gamma_{00}-Q_0(\lambda)} - \gamma_{11}$ as follows

$$G_1^\Gamma(\lambda) = \frac{Q_1(\lambda)Q_1^+(\lambda)+1}{Q_1(\lambda)+Q_1^+(\lambda)}. \quad (4.9)$$

The latter formula implies that

$$\Im G_1^\Gamma = \frac{\Im Q_1(\lambda)\left(|Q_1^+(\lambda)|^2+1\right)+\Im Q_1^+(\lambda)\left(|Q_1(\lambda)|^2+1\right)}{|Q_1(\lambda)+Q_1^+(\lambda)|^2}. \quad (4.10)$$

The imaginary part of the function $Q_1^+(\lambda + i0)$ is different from zero only if the imaginary part of $Q_0(\lambda + i0)$ is not trivial. Therefore the sets where the imaginary parts of the functions $G_1^\Gamma(\lambda + i0)$ and $Q_0(\lambda + i0)$ exist and are different from zero coincide up to a set of measure zero.

Using Privalov's theorem we conclude that the absolutely continuous part of the spectrum of the operator $\mathbf{A}^\Gamma$ coincides with the union of the points where the functions $G_0^\Gamma(\lambda + i0)$ and $G_1^\Gamma(\lambda + i0)$ have nontrivial imaginary parts. But these sets almost coincide with the set where the function $Q_0(\lambda + i0)$ has nontrivial imaginary part, i.e. with the absolutely continuous spectrum of the operator $\mathcal{A}$. Thus we have proven that the absolutely continuous spectra of the operators $\mathbf{A}^\Gamma$ and $\mathcal{A}$ almost coincide. Therefore these two open sets coincide. The theorem is proven.

$\square$

The result proven can be generalized for any finite rank perturbation. Let us study the behavior of the continuous component of the spectrum under singular finite rank perturbations.

Theorem 4.1.4 *Let $\mathcal{A}^0$ be a symmetric operator with finite deficiency indices acting in the Hilbert space $\mathcal{H}$. Let $\mathcal{A}$ and $\mathcal{A}^*$ be two self–adjoint extensions of the operator $\mathcal{A}^0$. Then the continuous parts of the spectra of the operators $\mathcal{A}$ and $\mathcal{A}^*$ coincide.*

Proof Let us denote the difference of the resolvents of the operators $\mathcal{A}$ and $\mathcal{A}^*$ as follows

$$\frac{1}{\mathcal{A} - i} - \frac{1}{\mathcal{A}^* - i} \equiv \mathcal{B},$$

which is a finite dimensional operator due to Krein's formula. In fact its rank is not larger than the deficiency indices of the operator $\mathcal{A}^0$. We denote by σ_c and σ_c^* the continuous components of the spectra of the operators $\mathcal{A}$ and $\mathcal{A}^*$ respectively. The continuous spectrum σ_c consists of nonisolated points of the spectrum. Suppose that $\lambda \notin \sigma_c$. Then there exists $\epsilon > 0$ such that

$$(\lambda, \lambda + \epsilon) \cap \sigma = \emptyset,$$

$$(\lambda, \lambda - \epsilon) \cap \sigma = \emptyset, \tag{4.11}$$

where σ denotes the spectrum of $\mathcal{A}$. Let us denote by $P(\lambda, \lambda + \epsilon)$ and $P(\lambda - \epsilon, \lambda)$ the spectral projectors for the operator $\mathcal{A}$ on the intervals $(\lambda, \lambda + \epsilon)$ and $(\lambda - \epsilon, \lambda)$ respectively. Similarly $P^*(\lambda, \lambda + \epsilon)$ and $P^*(\lambda - \epsilon, \lambda)$ denote the spectral projectors for the operator $\mathcal{A}^*$. Then (4.11) implies that $P(\lambda, \lambda + \epsilon) =$

$P(\lambda - \epsilon, \lambda) = 0$. Thus to prove that the point λ does not belong to σ_c^* it is enough to show that $\dim P^*(\lambda, \lambda + \epsilon)$ and $\dim P^*(\lambda - \epsilon, \lambda)$ are finite. We are going to prove that

$$\dim P^*(\lambda, \lambda + \epsilon) \leq \operatorname{rank} B.$$

Suppose that the latter inequality does not hold. It follows that there exists a nontrivial function $g \in P^*(\lambda, \lambda + \epsilon)\mathcal{H} \cap \operatorname{Ker} B$. Consider the function $f = \frac{1}{A - i}g = \frac{1}{A^* - i}g$. The function f belongs to the domains of both operators A and A^*. Moreover

$$f = \frac{1}{A^* - i}P^*(\lambda, \lambda + \epsilon)g = P^*(\lambda, \lambda + \epsilon)f \in P^*(\lambda, \lambda + \epsilon)\mathcal{H}.$$

Similarly $f \perp P(\lambda, \lambda + \epsilon)\mathcal{H}$. Then the following two inequalities hold

$$\| (A^* - \lambda - \epsilon/2)f \|^2 \;=\; \int_{t \in (\lambda, \lambda + \epsilon)} (t - \lambda - \epsilon/2)^2 d\langle f, E^*(\lambda)f \rangle$$

$$< \; \frac{\epsilon^2}{4} \| f \|^2;$$

$$\| (A - \lambda - \epsilon/2)f \|^2 \;=\; \int_{t \notin (\lambda, \lambda + \epsilon)} (t - \lambda - \epsilon/2)^2 d\langle f, E(\lambda)f \rangle$$

$$\geq \; \frac{\epsilon^2}{4} \| f \|^2 .$$

Taking into account the equality $Af = A^*f$ we get a contradiction which proves the statement for the interval $(\lambda, \lambda + \epsilon)$. The proof for the interval $(\lambda - \epsilon, \lambda)$ is similar. Changing the roles of the operators A and A^* we prove that if λ does not belong to the continuous spectrum of A^* then it is not in the continuous spectrum of A. It follows that the continuous spectra of the operators A and A^* coincide. The theorem is proven.

$\square$

The latter theorem can be used to get information about the relations between the discrete spectra of the operators A and A^* (see [154] for more details). We have proven that the absolutely continuous and continuous components of the spectrum are invariant under rank one perturbations. This does not imply that the singular continuous component of the spectrum is invariant. In fact it has been shown that the singular continuous components of the spectrum for an operator and its rank one perturbation can be different [239, 903, 909].

4.1.3 Wave operators and scattering operator for rank one perturbations

We are going to calculate the wave operators for rank one perturbations. Finite rank perturbations and perturbations with internal structure will be studied in the following section. The existence of wave operators for finite rank bounded perturbations was first proven by T.Kato [517, 518] and S.T.Kuroda [648]. Finite dimensional singular perturbation can be studied using the results of M.G.Krein and M.S.Birman [153, 144, 147, 146, 148]. The case of bounded rank one perturbations is described in detail in the book by N.I.Akhiezer and I.M. Glazman [11]. Therefore we begin with singular rank one perturbations. But we are going to use the following technical lemma from [11].

Lemma 4.1.2 *Let $f \in L_2(\mathbf{R})$. Then*

$$\lim_{t \to \infty} \int_{-\infty}^{\infty} | \int_{-\infty}^{\infty} \frac{e^{it(\mu-\lambda)} - 1}{\mu - \lambda} f(\lambda) d\lambda - \int_{-\infty}^{\infty} \frac{f(\lambda) d\lambda}{\lambda - (\mu + i0)} |^2 d\mu = 0.$$

Proof For any $f \in L_2(\mathbf{R})$ we have

$$\frac{1}{\pi} \int_{-\infty}^{+\infty} \frac{f(\lambda) d\lambda}{\lambda - (\mu + i0)} = i f(\mu) + \frac{1}{\pi} \int_{-\infty}^{+\infty} \frac{f(\lambda)}{\lambda - \mu} d\lambda.$$

Let $\hat{f}$ be the Fourier transform of f. Then the functions $f(\mu)$ and $\tilde{f}(\mu)$ are given by

$$f(\mu) = \lim_{A \to \infty} \frac{1}{\sqrt{2\pi}} \int_{-A}^{A} \hat{f}(p) e^{i\mu p} dp,$$

$$\tilde{f}(\mu) = \lim_{A \to \infty} \frac{i}{\sqrt{2\pi}} \int_{-A}^{A} \hat{f}(p) \operatorname{sign} p \, e^{i\mu p} dp,$$

and it follows that

$$\frac{1}{2\pi i} \int_{-\infty}^{+\infty} \frac{f(\lambda)}{\lambda - (\mu + i0)} = \lim_{A \to \infty} \frac{1}{\sqrt{2\pi}} \int_{0}^{A} \hat{f}(p) e^{i\mu p} dp.$$

The convolution theorem implies

$$\frac{1}{2\pi i} \int_{-\infty}^{\infty} \frac{e^{it(\mu-\lambda)} - 1}{\mu - \lambda} f(\lambda) d\lambda$$

$$= \frac{1}{2\pi i} \frac{1}{\sqrt{2\pi}} \int_{-\infty}^{+\infty} \hat{f}(p) e^{i\mu p} dp \int_{-\infty}^{+\infty} \frac{1 - e^{-ity}}{y} e^{iyp} dy$$

$$= \frac{1}{\sqrt{2\pi}} \int_{0}^{t} \hat{f}(p) e^{i\mu p} dp.$$

Comparison of the latter two relations proves the lemma.

$\square$

The wave operators will be calculated directly in order to obtain an explicit formula for the scattering matrix. The scattering matrix will be calculated in terms of the Q-function determined by the original and perturbed operators. Consider a self–adjoint operator A and its rank one perturbation $A_\alpha = A + \alpha\langle\varphi,\cdot\rangle\varphi$. We are going to write all formulas for $\varphi \in \mathcal{H}_{-1}(A)$ but using the regularization (1.53) the result can be generalized for any $\varphi \in \mathcal{H}_{-2}(A)$. We denote by P^a and P^a_α the spectral projectors to the absolutely continuous parts of the spectra of the operators A and A_α respectively and by $\mathcal{H}^a$ and $\mathcal{H}^a_\alpha$ the corresponding absolutely continuous subspaces

$$\mathcal{H}^a = P^a\mathcal{H}, \quad \mathcal{H}^a_\alpha = P^a_\alpha\mathcal{H}.$$

We are going to use the following notation

$$U_t = e^{iA_\alpha t}e^{-iAt}P^a.$$

Suppose without loss of generality that the vector $g = \frac{1}{A+i}\varphi, \parallel g \parallel = 1$ is generating for the space $\mathcal{H}$ and each of operators A and A_α. The same is true for the vector $g_\alpha = \frac{1}{A_\alpha+i}\varphi = \frac{1}{1+\alpha F(-i)}g$. Every element $f \in \mathcal{H}$ possesses two different spectral representations

$$f = \int_{\mathbf{R}} \hat{f}(\lambda)dE(\lambda)g,$$

$$f = \int_{\mathbf{R}} \hat{f}_\alpha(\lambda)dE_\alpha(\lambda)g_\alpha,$$

associated with the self–adjoint operators A and A_α respectively. In what follows we shall need the relationship between the two spectral representations $\hat{f}$ and $\hat{f}_\alpha$. The scalar product of two vectors f and q from the Hilbert space can be calculated as follows

$$\langle f,q\rangle = \int_{-\infty}^{\infty} \overline{\hat{f}(\lambda)}\hat{q}(\lambda)d\rho(\lambda) = \int_{-\infty}^{\infty} \overline{\hat{f}_\alpha(\lambda)}\hat{q}_\alpha(\lambda)d\rho_\alpha(\lambda), \qquad (4.12)$$

where $\rho(\lambda) = \langle g, E(\lambda)g\rangle$ and $\rho_\alpha(\lambda) = \langle g_\alpha, E_\alpha(\lambda)g_\alpha\rangle$ are nondecreasing functions of finite variation. Since $\varphi \in \mathcal{H}_{-1}(A)$, the integral $\int_{-\infty}^{\infty}|\lambda|d\rho(\lambda) < \infty$ converges absolutely. If $f \in \mathcal{H}$ then the following equalities hold

$$\left\langle\varphi,\frac{1}{A-z}f\right\rangle = \int_{-\infty}^{\infty}\frac{\lambda-i}{\lambda-z}\hat{f}(\lambda)d\rho(\lambda),$$

$$\left\langle\varphi,\frac{1}{A_\alpha-z}f\right\rangle = \int_{-\infty}^{\infty}\frac{\lambda-i}{\lambda-z}\hat{f}_\alpha(\lambda)d\rho_\alpha(\lambda).$$

These follow from the fact that the spectral representations are associated with the operators $\mathcal{A}$ and $\mathcal{A}_\alpha$ respectively. Krein's formula for the resolvents of the operators $\mathcal{A}$ and $\mathcal{A}_\alpha$ implies the following equality for every $f \in \mathcal{H}$

$$\left\langle \varphi, \frac{1}{\mathcal{A}_\alpha - z} f \right\rangle = \frac{1}{1 + \alpha F(z)} \left\langle \varphi, \frac{1}{\mathcal{A} - z} f \right\rangle.$$

The latter equation can be written in the spectral representation as

$$\int_\infty^\infty \frac{\lambda - i}{\lambda - z} \hat{f}_\alpha(\lambda) d\rho_\alpha(\lambda) = \frac{1}{1 + \alpha F(z)} \int_{-\infty}^\infty \frac{\lambda - i}{\lambda - z} \hat{f}(\lambda) d\rho(\lambda),$$

which implies for almost all real values of $\mu \in \mathbf{R}$

$$\pm \pi i (\mu - i) \hat{f}_\alpha(\mu) \rho_\alpha'(\mu) + \text{v.p.} \int_{-\infty}^\infty \frac{\lambda - i}{\lambda - \mu} \hat{f}_\alpha(\lambda) d\rho_\alpha(\lambda)$$

$$= \frac{1}{1 + \alpha F(\mu \pm i0)} \int_{-\infty}^\infty \frac{\lambda - i}{\lambda - (\mu \pm i0)} \hat{f}(\lambda) d\rho(\lambda).$$

Considering the difference of the latter two equalities we can carry out the following calculations

$$2\pi i (\mu - i) \hat{f}_\alpha(\mu) \rho_\alpha'(\mu)$$

$$= \frac{1}{1 + \alpha F(\mu + i0)} \int_{-\infty}^\infty \frac{\lambda - i}{\lambda - (\mu + i0)} \hat{f}(\lambda) d\rho(\lambda)$$

$$- \frac{1}{1 + \alpha F(\mu - i0)} \int_{-\infty}^\infty \frac{\lambda - i}{\lambda - (\mu - i0)} \hat{f}(\lambda) d\rho(\lambda)$$

$$= - \frac{2i\alpha \Im F(\mu + i0)}{|1 + \alpha F(\mu + i0)|^2} \int_{-\infty}^\infty \frac{\lambda - i}{\lambda - (\mu + i0)} \hat{f}(\lambda) d\rho(\lambda)$$

$$+ \frac{1}{1 + \alpha F(\mu - i0)} \times \qquad\qquad (4.13)$$

$$\left[\int_{-\infty}^\infty \frac{\lambda - i}{\lambda - (\mu + i0)} \hat{f}(\lambda) d\rho(\lambda) - \int_{-\infty}^\infty \frac{\lambda - i}{\lambda - (\mu - i0)} \hat{f}(\lambda) d\rho(\lambda) \right]$$

$$= - \frac{2i\pi \alpha (\mu^2 + 1) \rho'(\mu)}{|1 + \alpha F(\mu + i0)|^2} \int_{-\infty}^\infty \frac{\lambda - i}{\lambda - (\mu + i0)} \hat{f}(\lambda) d\rho(\lambda)$$

$$+ \frac{2\pi i}{1 + \alpha F(\mu - i0)} (\mu - i) \hat{f}(\mu) \rho'(\mu).$$

To derive the latter equality we used the fact that for almost all $\mu \in \mathbf{R}$

$$\pi\alpha(\mu^2 + 1)\rho'(\mu) = \Im F(\mu + i0),$$

which follows from the spectral representation for the function $F(z)$

$$F(z) = \left\langle \varphi, \frac{1}{A - z}\varphi \right\rangle = \int_{-\infty}^{\infty} \frac{\lambda^2 + 1}{\lambda - z}d\rho(\lambda)$$

($\varphi \in \mathcal{H}_{-1}(A)$) and Privalov's theorem.

To determine the relationship between the absolutely continuous parts of the measures $d\rho(\lambda)$ and $d\rho_\alpha(\lambda)$ we consider the following generalized Krein formula

$$\frac{1}{A_\alpha - z}\varphi = \frac{1}{1 + \alpha F(z)}\frac{1}{A - z}\varphi,$$

which implies

$$\left\langle \frac{1}{A_\alpha - z}\varphi, \frac{1}{A_\alpha - z}\varphi \right\rangle = \frac{1}{|1 + \alpha F(z)|^2}\left\langle \frac{1}{A - z}\varphi, \frac{1}{A - z}\varphi \right\rangle.$$

Using the spectral representations the latter equality can be written as

$$\int_{-\infty}^{\infty} \frac{\lambda^2 + 1}{(\lambda - z)(\lambda - \bar{z})}d\rho_\alpha(\lambda) = \frac{1}{|1 + \alpha F(z)|^2}\int_{-\infty}^{\infty} \frac{\lambda^2 + 1}{(\lambda - z)(\lambda - \bar{z})}d\rho(\lambda)$$

which implies that the following equality is fulfilled almost everywhere on the real axis

$$\rho_\alpha'(\mu) = \frac{1}{|1 + \alpha F(\mu + i0)|^2}\rho'(\mu). \tag{4.14}$$

The latter equation establishes a relationship between the absolutely continuous parts of the measures $d\rho$ and $d\rho_\alpha$.

Equations (4.13) and (4.14) imply therefore the following relation between the spectral representations $\hat{f}$ and $\hat{f}_\alpha$ of the function f

$$\hat{f}_\alpha(\mu) = -\alpha(\mu + i)\int_{-\infty}^{\infty} \frac{\lambda - i}{\lambda - (\mu + i0)}\hat{f}(\lambda)d\rho(\lambda)$$

$$+ (1 + \alpha F(\mu + i0))\hat{f}(\mu) \tag{4.15}$$

valid almost everywhere on the real axis.

Our main result in this section can be formulated as follows.

Theorem 4.1.5 *Let $\mathcal{A}$ be a self-adjoint operator acting in the Hilbert space $\mathcal{H}$ and let $\mathcal{A}_\alpha = \mathcal{A} + \alpha \langle \varphi, \cdot \rangle \varphi$ be its form bounded rank one perturbation, i.e. $\varphi \in \mathcal{H}_{-1}(\mathcal{A}), \alpha \in \mathbf{R}$. Then the wave operators*

$$W_\pm(\mathcal{A}_\alpha, \mathcal{A}) = \mathrm{s} - \lim_{t \to \pm\infty} U_t = \mathrm{s} - \lim_{t \to \pm\infty} e^{i\mathcal{A}_\alpha t} e^{-i\mathcal{A}t} P^a \qquad (4.16)$$

exist and are defined by the equality:

$$W_\pm f = \int_{-\infty}^{\infty} (1 + \alpha F(\lambda \pm i0)) \hat{f}^a(\lambda) dE_\alpha(\lambda) \frac{1}{\mathcal{A}_\alpha + i} \varphi, \qquad (4.17)$$

where

$$P^a f = \int_{-\infty}^{\infty} \hat{f}^a(\lambda) dE(\lambda) \frac{1}{\mathcal{A} + i} \varphi.$$

Proof We consider only $f \in \mathcal{H}^a$, since the operators U_t and $W_\pm$ are equal to zero on the orthogonal complement of $\mathcal{H}^a$. Then $\hat{f}^a = \hat{f}$. We prove the theorem for the wave operator W_+, since the proof for the operator W_- is almost identical.

We are going to prove first the existence of the weak wave operator, i.e. that the limit (4.16) holds in the weak sense. Consider two arbitrary functions f and q satisfying the following conditions

$$f \in \mathrm{Dom}(\mathcal{A}) \cap \mathcal{H}^a, \quad q \in \mathrm{Dom}(\mathcal{A}_\alpha) \cap \mathcal{H}^a_\alpha. \qquad (4.18)$$

Then the following equality can be obtained by differentiation

$$\langle q, U_t f \rangle - \langle q, f \rangle = i\alpha \int_0^t \langle e^{-i\tau \mathcal{A}_\alpha} q, \varphi \rangle \langle \varphi, e^{-i\tau \mathcal{A}} f \rangle d\tau,$$

since the domains of the operators $\mathcal{A}$ and $\mathcal{A}_\alpha$ and the subspaces $\mathcal{H}^a$ and $\mathcal{H}^a_\alpha$ are invariant under the action of the operators $e^{-i\mathcal{A}t}$ and $e^{-i\mathcal{A}_\alpha t}$ respectively. Using the spectral representation and changing the order of integration one obtains the following equality

$$\langle q, U_t f \rangle - \langle q, f \rangle$$

$$= i\alpha \int_0^t \left[\int_{-\infty}^{\infty} e^{i\tau\lambda} \overline{\hat{q}_\alpha(\lambda)} (\lambda + i) d\rho_\alpha(\lambda) \right] \left[\int_{-\infty}^{\infty} e^{-i\tau\mu} \hat{f}(\mu)(\mu - i) d\rho(\mu) \right] d\tau$$

$$= \alpha \int_{-\infty}^{\infty} \int_{-\infty}^{\infty} \frac{e^{it(\lambda - \mu)} - 1}{\lambda - \mu} \hat{f}(\mu)(\mu - i)\overline{\hat{q}_\alpha(\lambda)}(\lambda + i) d\rho_\alpha(\lambda) d\rho(\mu).$$

Suppose in addition that

$$\hat{f}\rho' \in L_2(\mathbf{R}), \quad \hat{q}_\alpha \rho'_\alpha \in L_2(\mathbf{R}). \qquad (4.19)$$

Then using Lemma 4.1.2 the following limit can be calculated:

$$\lim_{t \to \infty} \langle q, U_t f \rangle = \langle q, f \rangle$$

$$+ \alpha \int_{-\infty}^{\infty} \int_{-\infty}^{\infty} \frac{\hat{f}(\mu)(\mu - i)\overline{\hat{q}_\alpha(\lambda)}(\lambda + i)}{\mu - (\lambda + i0)} \rho'_\alpha(\lambda)\rho'(\mu) d\lambda d\mu.$$

The latter expression can be simplified using (4.15) and (4.12) as follows

$$\lim_{t \to \infty} \langle q, U_t f \rangle = \langle q, f \rangle + \int_{-\infty}^{\infty} (1 + \alpha F(\lambda + i0)) \hat{f}(\lambda)\overline{\hat{q}_\alpha(\lambda)} \rho'_\alpha(\lambda) d\lambda$$

$$- \int_{-\infty}^{\infty} \hat{f}_\alpha(\lambda)\overline{\hat{q}_\alpha(\lambda)} \rho'_\alpha(\lambda) d\lambda$$

$$= \langle q, W_+ f \rangle,$$

where the operator W_+ has been defined by equation (4.17). We have proven that the weak limit (4.17) holds for all f, q satisfying conditions (4.18) and (4.19). But such functions form a dense subsets in $\mathcal{H}^a$ and $\mathcal{H}^a_\alpha$ respectively. Therefore we have proven that the following limit holds in the weak sense

$$\lim_{t \to \infty} P^a_\alpha U_t = W_+. \tag{4.20}$$

To prove that the strong wave operator exists it is enough to show that the weak wave operator is isometric on $\mathcal{H}^a$ [969]. Let $f \in \mathcal{H}^a$. Then the latter fact follows from the definition (4.17) and relation (4.14)

$$\| W_+ f \|^2 = \int_{-\infty}^{\infty} |1 + \alpha F(\lambda + i0)|^2 |\hat{f}(\lambda)|^2 d\rho_\alpha(\lambda)$$

$$= \int_{-\infty}^{\infty} |1 + \alpha F(\lambda + i0)|^2 |\hat{f}(\lambda)|^2 \rho'_\alpha(\lambda) d\lambda$$

$$= \int_{-\infty}^{\infty} |\hat{f}(\lambda)|^2 \rho'(\lambda) d\lambda$$

$$= \| f \|^2.$$

Thus we have proven that $\| W_+ f \| = \| P^a f \|$. Therefore the wave operator exists in the strong sense and coincides with the operator W_+. The theorem is proven.

□

The wave operators we calculated are complete due to the isometricity and the fact that the function $1 + \alpha F(\lambda + i0)$ has a null set of measure zero on the real axis. To calculate the adjoint operator W_+^* we consider two arbitrary functions $f \in \mathcal{H}^a, q \in \mathcal{H}_\alpha^a$. We have

$$\langle q, W_+ f \rangle = \int_{-\infty}^{\infty} \overline{\hat{q}_\alpha(\lambda)}(1 + \alpha F(\lambda + i0))\hat{f}(\lambda)\rho'_\alpha(\lambda)d\lambda$$

$$= \int_{-\infty}^{\infty} \frac{\overline{\hat{q}_\alpha(\lambda)}}{1 + \alpha F(\lambda - i0)} \hat{f}(\lambda)\rho'(\lambda)d\lambda.$$

Thus the adjoint wave operator is defined by the following equality

$$W_+^* g = \int_{-\infty}^{\infty} \frac{1}{1 + \alpha F(\lambda + i0)} \hat{g}_\alpha^a(\lambda) dE \frac{1}{A+i}\varphi \qquad (4.21)$$

if

$$P_\alpha^a g = \int_{-\infty}^{\infty} \hat{g}_\alpha^a(\lambda) dE_\alpha \frac{1}{A_\alpha + i}\varphi \in \mathcal{H}_\alpha^a.$$

Then the scattering operator $S(A_\alpha, A) = W_+^* W_-$ in the spectral representation of the original operator A, the scattering matrix $S(A_\alpha, A, \lambda)$, is the operator of multiplication by the following function

$$S(A_\alpha, A, \lambda) = \frac{1 + \alpha F(\lambda - i0)}{1 + \alpha F(\lambda + i0)}. \qquad (4.22)$$

In the following section we are going to derive similar formulae for arbitrary finite rank perturbations.

4.2   Scattering theory for self–adjoint extensions

4.2.1   Wave operators for self–adjoint extensions

The wave operators for bounded finite rank perturbations were first calculated by T.Kato [517]. The scattering operator for such perturbations was obtained by S.T.Kuroda [648]. The scattering theory for the case where the difference of the resolvents of the perturbed and unperturbed operators is finite dimensional (or even from the trace class) has been developed by M.Birman and M.Krein [153], M.Birman [146, 147] and T.Kato [520]. We are going to follow here the approach suggested by V.Adamyan and B.Pavlov [5] for the case where the perturbed and unperturbed operators are two self–adjoint extensions of one symmetric operator with finite deficiency indices.

It will be used in the following section to calculate the scattering matrix for
singular and generalized finite rank perturbations.

Let $\mathcal{A}^1$ and $\mathcal{A}^2$ be two different self–adjoint extensions of one symmetric
operator A^0 acting in the Hilbert space $\mathcal{H}$. Suppose that the operator A^0
is densely defined and its deficiency indices are equal to (m, m). The case
where the operator A^0 is not densely defined can be considered in the same
way, since our approach uses only the Cayley transforms $\mathcal{U}_1$ and $\mathcal{U}_2$ of the
operators $\mathcal{A}^1$ and $\mathcal{A}^2$ respectively

$$\mathcal{U}_\beta = \frac{A^\beta + i}{A^\beta - i}, \; \beta = 1, 2.$$

Let us denote the deficiency subspace for the operator A^0 by $N(z)$. Then the
following equality holds

$$N(i) = \mathcal{U}_1 N(-i) = \mathcal{U}_2 N(-i).$$

The corresponding orthonormal projectors on the deficiency subspaces will
be denoted by P_z. We denote by P_z^* the embedding operators of $N(z)$ into
the Hilbert space. The operators $\mathcal{U}_1$ and $\mathcal{U}_2$ coincide on $\mathcal{H} \ominus N(-i)$ and the
difference of these operators is determined by the finite dimensional operator
$\mathcal{E} = \mathcal{U}_2^{-1}\mathcal{U}_1|_{N(-i)}$. Without loss of generality we suppose that unity is not
an eigenvalue of the operator $\mathcal{E}$, i.e. that the operator A^0 is the greatest
common symmetric restriction of the operators $\mathcal{A}^1$ and $\mathcal{A}^2$.

Lemma 4.2.1 *The wave operators $W_\pm(\mathcal{A}^2, \mathcal{A}^1)$ exist and are given by the
following formulae*

$$W_-(\mathcal{A}^2, \mathcal{A}^1) = \text{s} - \lim_{r \nearrow 1} \frac{1 - r^2}{2\pi r} \int_{-\pi}^{\pi} d\varphi \, e^{i\varphi}(I - re^{-i\varphi}\mathcal{U}_2)^{-1}(\mathcal{U}_1 - re^{i\varphi})^{-1}P_1^a,$$

$$(4.23)$$

$$W_+(\mathcal{A}^2, \mathcal{A}^1) = \text{s} - \lim_{r \nearrow 1} \frac{1 - r^2}{2\pi r} \int_{-\pi}^{\pi} d\varphi \, e^{-i\varphi}(\mathcal{U}_2 - re^{-i\varphi})^{-1}(I - re^{i\varphi}\mathcal{U}_1)^{-1}P_1^a,$$

$$(4.24)$$

where the limits are understood in the strong sense.

Proof The wave operators can be calculated using the formula

$$W_\pm(\mathcal{A}^2, \mathcal{A}^1) = \lim_{n \to \infty} \mathcal{U}_2^n \mathcal{U}_1^{-n} P_1^a.$$

To calculate the limit we use Abel's method and the discrete Fourier transform

$W_-(\mathcal{A}^2, \mathcal{A}^1)f$

$$
= P_1^a f + \lim_{r \nearrow 1} \sum_{k=0}^{\infty} r^{2k} \mathcal{U}_2^k (\mathcal{U}_2 \mathcal{U}_1^{-1} - I) \mathcal{U}_1^{-k} P_1^a f
$$

$$
= P_1^a f + \lim_{r \nearrow 1} \frac{1}{2\pi} \int_{-\pi}^{\pi} d\varphi \, \frac{1}{1 - re^{-i\varphi} \mathcal{U}_2} (\mathcal{U}_2 \mathcal{U}_1^{-1} - 1) \frac{1}{1 - re^{i\varphi} \mathcal{U}_1^{-1}} P_1^a f
$$

$$
= \frac{1}{2\pi} \lim_{r \nearrow 1} \int_{-\pi}^{\pi} d\varphi \, \frac{1}{1 - re^{-i\varphi} \mathcal{U}_2}
$$

$$
\left(-re^{-i\varphi} \mathcal{U}_2 - re^{i\varphi} \mathcal{U}_1^{-1} + r^2 \mathcal{U}_2 \mathcal{U}_1^{-1} + \mathcal{U}_2 \mathcal{U}_1^{-1} \right) \frac{1}{1 - re^{i\varphi} \mathcal{U}_1^{-1}} P_1^a f.
$$

The latter expression can be simplified using the following equalities, which follow from Cauchy's residue theorem taking into account that the operators $\mathcal{U}_1$ and $\mathcal{U}_2$ are unitary and $r < 1$

$$
\frac{1}{2\pi} \int_{-\pi}^{\pi} d\varphi \frac{e^{i\varphi}}{\mathcal{U}_1 - re^{i\varphi}} P_1^a f = 0; \qquad \frac{1}{2\pi} \int_{-\pi}^{\pi} d\varphi \frac{e^{-i\varphi} \mathcal{U}_2}{1 - re^{-i\varphi} \mathcal{U}_2} P_1^a = 0.
$$

Then the wave operator is given by

$$
W_-(\mathcal{A}^2, \mathcal{A}^1) P_1^a f
$$

$$
= \lim_{r \nearrow 1} \frac{1 - r^2}{2\pi r} \int_{-\pi}^{\pi} d\varphi \, \frac{1}{1 - re^{-i\varphi} \mathcal{U}_2} e^{i\varphi} \mathcal{U}_1^{-1} \frac{1}{1 - re^{i\varphi} \mathcal{U}_1^{-1}} P_1^a f,
$$

which implies (4.23). Formula (4.24) can be obtained using the same method.
□

A similar method will be applied in the following section to calculate the scattering operator.

4.2.2   Scattering operator for self–adjoint extensions

The following lemma will be used to obtain an explicit formula for the scattering matrix for self-adjoint extensions.

Lemma 4.2.2 *Let*

$$
G(z) = P_{-i} \frac{\mathcal{U}_1 + z}{\mathcal{U}_1 - z} P_{-i}^*,
$$

$$\Gamma = -i\frac{\mathcal{E}+1}{\mathcal{E}-1}, \quad \mathcal{E} = \mathcal{U}_2^{-1}\mathcal{U}_1|_{N(-i)}.$$

Then the scattering operator $S(\mathcal{U}_2, \mathcal{U}_1)$ is given by the formula

$$S(\mathcal{U}_2, \mathcal{U}_1)$$

$$= I - w - \lim_{r \nearrow 1} \frac{(1-r^2)^2}{\pi r} \int_{-\pi}^{\pi} d\varphi \, e^{i\varphi} P_1^a \frac{1}{\mathcal{U}_1^{-1} - re^{-i\varphi}}$$

$$\times \frac{1}{\mathcal{U}_1 - re^{i\varphi}} P_{-i}^* \frac{1}{G^*(re^{i\varphi}) + i\Gamma} P_{-i} \frac{1}{\mathcal{U}_1 - re^{i\varphi}} \frac{1}{\mathcal{U}_1^{-1} - re^{-i\varphi}} \mathcal{U}_1^{-1} P_1^a$$

(4.25)

Proof It follows from the previous lemma that the wave operators exist and therefore the scattering operator can be calculated using the limit procedure described above. Let us use the notation

$$S_{r,\rho}(\mathcal{U}_2, \mathcal{U}_1) = \frac{(1-r^2)(1-\rho^2)}{4\pi^2 r \rho} \int_{-\pi}^{\pi} d\varphi \int_{-\pi}^{\pi} d\psi \, P_1^{a*} \frac{1}{1-re^{-i\varphi}\mathcal{U}_1^*} \frac{e^{i\varphi}}{\mathcal{U}_2^* - re^{i\varphi}}$$

$$\times \frac{e^{i\psi}}{1-\rho e^{-i\psi}\mathcal{U}_2} \frac{1}{\mathcal{U}_1 - \rho e^{i\psi}} P_1^a.$$

Then the scattering operator is equal to the double limit

$$S(\mathcal{U}_2, \mathcal{U}_1) = \lim_{r \nearrow 1} \lim_{\rho \nearrow r} S_{r,\rho}(\mathcal{U}_2, \mathcal{U}_1),$$

which will be calculated step by step in what follows. Using the fact that $U_j^* = U_j^{-1}$, $j = 1, 2$, and using the Hilbert identity

$$\frac{1}{\mathcal{U}_2^{-1} - re^{i\varphi}} \frac{1}{1 - \rho e^{-i\psi}\mathcal{U}_2} = \frac{1}{re^{i\varphi} - \rho e^{-i\psi}} \left(\frac{1}{1 - re^{i\varphi}\mathcal{U}_2} - \frac{1}{1 - \rho e^{-i\psi}\mathcal{U}_2} \right)$$

$S_{r,\rho}(\mathcal{U}_2, \mathcal{U}_1)$ can be presented as a sum of two integrals as follows

$$S_{r,\rho}(\mathcal{U}_2, \mathcal{U}_1)$$

$$= \frac{(1-r^2)(1-\rho^2)}{4\pi^2 r \rho} \int_{-\pi}^{\pi} d\varphi \int_{-\pi}^{\pi} d\psi \, e^{i(\varphi+\psi)} P_1^a$$

$$\times \frac{1}{1-re^{-i\varphi}\mathcal{U}_1^{-1}} \frac{1}{re^{i\varphi} - \rho e^{-i\psi}} \frac{1}{1-re^{i\varphi}\mathcal{U}_2} \frac{1}{\mathcal{U}_1 - \rho e^{i\psi}} P_1^a$$

$$- \frac{(1-r^2)(1-\rho^2)}{4\pi^2 r \rho} \int_{-\pi}^{\pi} d\varphi \int_{-\pi}^{\pi} d\psi \, e^{i(\varphi+\psi)} P_1^a$$

$$\times \frac{1}{1-re^{-i\varphi}\mathcal{U}_1^{-1}} \frac{1}{re^{i\varphi} - \rho e^{-i\psi}} \frac{1}{1-\rho e^{-i\psi}\mathcal{U}_2} \frac{1}{\mathcal{U}_1 - \rho e^{i\psi}} P_1^a$$

$$\equiv J_1 - J_2.$$

Each of the two double integrals in the latter expression can be reduced to simple integrals if one integrates with respect to ψ in the first integral and with respect to φ in the second integral. The integrals can be calculated using the Cauchy residue theorem taking into account that $\rho < r$

$$
\begin{aligned}
J_1 &= \frac{(1-r^2)(1-\rho^2)}{4\pi^2 r\rho} \int_{-\pi}^{\pi} d\varphi \frac{1}{i} \int_{|\xi|=1} d\xi \, e^{i\varphi} P_1^a \\
&\quad \times \frac{1}{1-re^{-i\varphi}\mathcal{U}_1^{-1}} \frac{1}{re^{i\varphi}-\rho\xi^{-1}} \frac{1}{1-re^{i\varphi}\mathcal{U}_2} \frac{1}{\mathcal{U}_1-\rho\xi} P_1^a \\
&= \frac{(1-r^2)(1-\rho^2)}{2\pi} \int_{-\pi}^{\pi} \frac{e^{-i\varphi}}{r^3} P_1^a \frac{1}{1-re^{-i\varphi}\mathcal{U}_1^{-1}} \frac{1}{1-re^{i\varphi}\mathcal{U}_2} \\
&\quad \times \frac{1}{\mathcal{U}_1 - \frac{\rho^2}{r}e^{-i\varphi}} P_1^a d\varphi;
\end{aligned}
$$

$$
\begin{aligned}
J_2 &= \frac{(1-r^2)(1-\rho^2)}{4\pi^2 r\rho} \int_{-\pi}^{\pi} d\psi \frac{1}{i} \int_{|z|=1} dz \, e^{i\psi} P_1^a \\
&\quad \times \frac{1}{1-rz^{-1}\mathcal{U}_1^{-1}} \frac{1}{rz-\rho e^{-i\psi}} \frac{1}{1-\rho e^{-i\psi}\mathcal{U}_2} \frac{1}{\mathcal{U}_1-\rho e^{i\psi}} P_1^a \\
&= \frac{(1-r^2)(1-\rho^2)}{2\pi r\rho} \int_{-\pi}^{\pi} d\psi \frac{e^{i\psi}}{r} P_1^a \left\{ \frac{r\mathcal{U}_1^{-1}}{r\mathcal{U}_1^{-1}-\frac{\rho}{r}e^{-i\psi}} + \frac{\frac{\rho}{r}e^{-i\psi}}{\frac{\rho}{r}e^{-i\psi}-r\mathcal{U}_1^{-1}} \right\} \\
&\quad \times \frac{1}{1-\rho e^{-i\psi}\mathcal{U}_2} \frac{1}{\mathcal{U}_1-\rho e^{i\psi}} P_1^a \\
&= \frac{(1-r^2)(1-\rho^2)}{2\pi} \int_{-\pi}^{\pi} d\psi \frac{e^{i\psi}}{r^2\rho} P_1^a \frac{1}{1-\rho e^{-i\psi}\mathcal{U}_2} \frac{1}{\mathcal{U}_1-\rho e^{i\psi}} P_1^a.
\end{aligned}
$$

Taking the limit $\rho \to r$ and changing the integration variable in the second term $\psi \to \varphi = -\psi$ we get the formula

$S_{r,r}(\mathcal{U}_2,\mathcal{U}_1)$

$$= \frac{(1-r^2)^2}{2\pi r^3} \int_{-\pi}^{\pi} d\varphi \, e^{-i\varphi} P_1^a \frac{1}{1-re^{-i\varphi}\mathcal{U}_1^{-1}} \frac{1}{1-re^{i\varphi}\mathcal{U}_2} \frac{1}{\mathcal{U}_1-re^{-i\varphi}} P_1^a$$

$$-\frac{(1-r^2)^2}{2\pi r^3} \int_{-\pi}^{\pi} d\varphi \, e^{-i\varphi} P_1^a \frac{1}{1-re^{i\varphi}\mathcal{U}_2} \frac{1}{\mathcal{U}_1-re^{-i\varphi}} P_1^a$$

$$= \frac{(1-r^2)^2}{2\pi r^2} \int_{-\pi}^{\pi} d\varphi \, e^{-2i\varphi} P_1^a \frac{1}{\mathcal{U}_1-re^{-i\varphi}} \frac{1}{1-re^{i\varphi}\mathcal{U}_2} \frac{1}{\mathcal{U}_1-re^{-i\varphi}} P_1^a.$$

$$= \frac{(1-r^2)^2}{2\pi r^2} \int_{-\pi}^{\pi} d\varphi \, e^{2i\varphi} P_1^a \frac{1}{\mathcal{U}_1-re^{i\varphi}} \frac{1}{1-re^{-i\varphi}\mathcal{U}_2} \frac{1}{\mathcal{U}_1-re^{i\varphi}} P_1^a.$$

Similarly Cauchy's residue theorem implies that

$$\frac{(1-r^2)^2}{2\pi r^2} \int_{-\pi}^{\pi} d\varphi \, e^{2i\varphi} P_1^a \frac{1}{\mathcal{U}_1-re^{i\varphi}} \frac{1}{1-re^{-i\varphi}\mathcal{U}_1} \frac{1}{\mathcal{U}_1-re^{i\varphi}} P_1^a$$

$$= \frac{(1-r^2)^2}{2\pi r^2} \frac{1}{i} \int_{|z|=1} dz \, P_1^a \frac{1}{\mathcal{U}_1-rz} \frac{z^2}{z-r\mathcal{U}_1} \frac{1}{\mathcal{U}_1-rz} P_1^a = P_1^a,$$

since $r < 1$. Therefore we get the following formula for the scattering operator:

$$S(\mathcal{U}_2,\mathcal{U}_1) = \mathrm{w} - \lim_{r \nearrow 1} S_{r,r}(\mathcal{U}_2,\mathcal{U}_1)$$

$$= P_1^a + \mathrm{w} - \lim_{r \nearrow 1} \frac{(1-r^2)^2}{2\pi r^2} \int_{-\pi}^{\pi} d\varphi \, e^{2i\varphi} P_1^a \frac{1}{\mathcal{U}_1-re^{i\varphi}} \quad (4.26)$$

$$\times \left[\frac{1}{1-re^{-i\varphi}\mathcal{U}_2} - \frac{1}{1-re^{-i\varphi}\mathcal{U}_1}\right] \frac{1}{\mathcal{U}_1-re^{i\varphi}} P_1^a.$$

The expression in the square brackets can be simplified as follows

$$\frac{1}{1-re^{-i\varphi}\mathcal{U}_2} - \frac{1}{1-re^{-i\varphi}\mathcal{U}_1}$$

$$= \frac{re^{-i\varphi}\mathcal{U}_2}{1-re^{-i\varphi}\mathcal{U}_2} - \frac{re^{-i\varphi}\mathcal{U}_1}{1-re^{-i\varphi}\mathcal{U}_1} \quad (4.27)$$

$$= re^{-i\varphi} \left\{\frac{1}{\mathcal{U}_2-re^{i\varphi}} - \frac{1}{\mathcal{U}_1-re^{i\varphi}}\right\}^*.$$

The difference of the resolvents of the operators $\mathcal{U}_2$ and $\mathcal{U}_1$ can be calculated using Krein's formula for the resolvents of the unitary operators. This formula can be easily obtained from Krein's formula for the resolvents of two self-adjoint extensions of one symmetric operator. But here we are going to derive this formula using the operator $\mathcal{E}$. The operators $\mathcal{U}_2$ and $\mathcal{U}_1$ coincide on the orthogonal complement to the subspace $N(-i)$ by assumption. Therefore

$$\mathcal{U}_2 = \mathcal{U}_2(1 - P_{-i}) + \mathcal{U}_2 P_{-i}$$

$$= \mathcal{U}_1(1 - P_{-i}) + \mathcal{U}_1 \mathcal{E}^{-1} P_{-i}$$

$$= \mathcal{U}_1 \left(1 - (1 - \mathcal{E}^{-1}) P_{-i} \right).$$

Then the resolvent of the operator $\mathcal{U}_2$ can be calculated as follows. Let

$$\frac{1}{\mathcal{U}_2 - z} f = g. \tag{4.28}$$

Then

$$f = (\mathcal{U}_2 - z)g = (\mathcal{U}_1 - z)g - \mathcal{U}_1(1 - \mathcal{E}^{-1})P_{-i}g.$$

Applying the resolvent of the operator $\mathcal{U}_1$ to the latter equality we get

$$\frac{1}{\mathcal{U}_1 - z} f = g - \frac{\mathcal{U}_1}{\mathcal{U}_1 - z}(1 - \mathcal{E}^{-1})P_{-i}g. \tag{4.29}$$

Projection onto the subspace $N(-i)$ gives the equality

$$P_{-i} \frac{1}{\mathcal{U}_1 - z} f = P_{-i}g - P_{-i}\frac{\mathcal{U}_1}{\mathcal{U}_1 - z}(1 - \mathcal{E}^{-1})P_{-i}g$$

$$= P_{-i}\left(1 - \frac{\mathcal{U}_1}{\mathcal{U}_1 - z}(1 - \mathcal{E}^{-1}) \right) P_{-i}g.$$

Then the projection $P_{-i}g$ is given by

$$P_{-i}g = \left(1 - P_{-i}\frac{\mathcal{U}_1}{\mathcal{U}_1 - z}P^*_{-i}(1 - \mathcal{E}^{-1}) \right)^{-1} P_{-i}\frac{1}{\mathcal{U}_1 - z}f. \tag{4.30}$$

Then (4.28), (4.29) and (4.30) imply that

$$\frac{1}{\mathcal{U}_2 - z} f = \frac{1}{\mathcal{U}_1 - z} f$$

$$- \frac{2\mathcal{U}_1}{\mathcal{U}_1 - z} P^*_{-i}\left\{ P_{-i}\frac{\mathcal{U}_1 + z}{\mathcal{U}_1 - z}P^*_{-i} + \frac{1 + \mathcal{E}}{1 - \mathcal{E}} \right\}^{-1} P_{-i}\frac{1}{\mathcal{U}_1 - z}f. \tag{4.31}$$

Therefore the expression in the square brackets in formula (4.26) can be written as folllows using (4.27) and (4.31):

$$\frac{1}{1 - re^{-i\varphi}\mathcal{U}_2} - \frac{1}{1 - re^{-i\varphi}\mathcal{U}_1}$$

$$= -2re^{-i\varphi}\left[\frac{\mathcal{U}_1}{\mathcal{U}_1 - re^{i\varphi}}\, P_{-i}^*\left\{G(re^{i\varphi}) - i\Gamma\right\}^{-1} P_{-i}\frac{1}{\mathcal{U}_1 - re^{i\varphi}}\right]^*$$

$$= -2re^{-i\varphi}\frac{1}{\mathcal{U}_1^{-1} - re^{-i\varphi}}\, P_{-i}^*\left[G^*(re^{i\varphi}) + i\Gamma\right]^{-1} P_{-i}\frac{\mathcal{U}_1^{-1}}{\mathcal{U}_1^{-1} - re^{-i\varphi}},$$

since the operator $\Gamma = -i\frac{\mathcal{E}+1}{\mathcal{E}-1}$ is Hermitian. The latter equation together with (4.26) imply (4.25) if one takes into account that the operator P_1^a is equal to the unit operator on the abslolutely continuous subspace of the operator $\mathcal{U}_1$. The lemma is proven.

$\square$

The operator Γ is Hermitian and acts in the finite dimensional space $N(-i)$. The operator-function

$$iG(z) = iP_{-i}\frac{\mathcal{U}_1 + z}{\mathcal{U}_1 - z}P_{-i}^* = i\int_{-\pi}^{\pi}\frac{e^{i\varphi} + z}{e^{i\varphi} - z}d\mathcal{E}^1(\varphi)$$

has positive imaginary part in the unit disk, where $\mathcal{E}^1(\varphi)$ denotes the spectral projector associated with the operator $\mathcal{U}_1$.

To obtain an explicit formula for the scattering operator let us denote by $\{e_\nu\}_{\nu=1}^m$ an arbitrary orthonormal basis in the m–dimensional subspace $N(-i)$. Then the projector P_{-i} is given by

$$P_{-i} = \sum_{\nu=1}^m \langle e_\nu, \cdot\rangle e_\nu.$$

Theorem 4.2.1 *The limit values of the operator function* $[G^*(z) + i\Gamma]^{-1}$ *from inside the unit circle exist almost everywhere. If these limit values are bounded in some neighborhood of the absolutely continuous spectrum of the operator* $\mathcal{U}_1$, *then for every two functions* f_1, f_2 *from* $\mathcal{H}_1^a$ *satisfying the condition*

$$\int_{-\pi}^{\pi} \| P_{-i}\frac{d}{d\theta}\mathcal{E}^1(\theta)f_j \|^2 \, d\theta < \infty, \quad j = 1, 2$$

the following equality holds

$$\langle f_2, S(\mathcal{U}_2, \mathcal{U}_1)f_1 \rangle$$

$$= \quad \langle f_2, f_1 \rangle$$

$$-4\pi \sum_{\mu,\nu=1}^{m} \int_{-\pi}^{\pi} \left\langle \frac{d\mathcal{E}^1(\varphi)}{d\varphi} f_2, e_\mu \right\rangle \tag{4.32}$$

$$\left\langle e_\mu, \left[G^*(e^{i\varphi}) + i\Gamma \right]^{-1} e_\nu \right\rangle \left\langle e_\nu, \frac{d\mathcal{E}^1(\varphi)}{d\varphi} f_1 \right\rangle d\varphi.$$

Proof The first statement follows from Privalov's theorem [840, 839] applied to the operator valued function with positive imaginary part inside the unit disk. To prove the second statement we calculate the limits

$$\lim_{r \nearrow 1}(1 - r^2) \left\langle f_2, P_1^a \frac{1}{\mathcal{U}_1^{-1} - re^{-i\varphi}} \frac{1}{\mathcal{U}_1 - re^{i\varphi}} e_\mu \right\rangle$$

$$= \quad \lim_{r \nearrow 1}(1 - r^2) \int_{-\pi}^{\pi} \frac{\frac{d}{d\theta}\langle f_2, \mathcal{E}^1(\theta)e_\mu \rangle}{1 + r^2 - 2r\cos(\theta - \varphi)} d\theta$$

$$= \quad 2\pi \frac{d}{d\varphi}\langle f_2, \mathcal{E}^1(\varphi)e_\mu \rangle;$$

$$\lim_{r \nearrow 1}(1 - r^2)e^{i\varphi} \left\langle e_\nu, \frac{1}{\mathcal{U}_1 - re^{i\varphi}} \frac{1}{\mathcal{U}_1^{-1} - re^{-i\varphi}} \mathcal{U}_1^{-1} P_1^a f_1 \right\rangle$$

$$= \quad \lim_{r \nearrow 1}(1 - r^2) \int_{-\pi}^{\pi} \frac{e^{i(\varphi-\psi)}\frac{d}{d\psi}\langle e_\nu, \mathcal{E}^1(\psi)f_1 \rangle}{1 + r^2 - 2r\cos(\psi - \varphi)} d\psi \tag{4.33}$$

$$= \quad 2\pi \frac{d}{d\varphi}\langle e_\nu, \mathcal{E}^1(\varphi)f_1 \rangle.$$

Then (4.25) and (4.33) imply that

$$\langle f_2, S(\mathcal{U}_2, \mathcal{U}_1) f_1 \rangle$$

$$= \langle f_2, f_1 \rangle$$

$$- \lim_{r \nearrow 1} \frac{1}{\pi r} \int_{-\pi}^{\pi} d\varphi \sum_{\mu,\nu=1}^{m} (1 - r^2) \left\langle f_2, P_1^a \frac{1}{\mathcal{U}_1^{-1} - re^{-i\varphi}} \frac{1}{\mathcal{U}_1 - re^{i\varphi}} e_\mu \right\rangle$$

$$\langle e_\mu, \left[G^*(e^{i\varphi}) + i\Gamma \right]^{-1} e_\nu \rangle (1 - r^2) e^{i\varphi}$$

$$\left\langle e_\nu, \frac{1}{\mathcal{U}_1 - re^{i\varphi}} \frac{1}{\mathcal{U}_1^{-1} - re^{-i\varphi}} \mathcal{U}_1^{-1} P_1^a f_1 \right\rangle$$

$$= \langle f_2, f_1 \rangle$$

$$- 4\pi \sum_{\mu,\nu=1}^{m} \int_{-\pi}^{\pi} \left\langle \frac{d\mathcal{E}^1(\varphi)}{d\varphi} f_2, e_\mu \right\rangle \langle e_\mu, \left[G^*(e^{i\varphi}) + i\Gamma \right]^{-1} e_\nu \rangle \left\langle e_\nu, \frac{d\mathcal{E}_\varphi^1}{d\varphi} f_1 \right\rangle d\varphi.$$

The theorem is proven.

$\square$

We note that $G(e^{i\varphi})$ denotes in the latter theorem the boundary values of
the operator function $G(re^{i\varphi})$ from inside the unit disk. Equality (4.32) is
valid even if the boundary values of the operator function $[G^*(z) + i\Gamma]^{-1}$ are
not bounded in a neighborhood of the absolutely continuous spectrum of the
operator $\mathcal{U}_1$ but only for functions f_1, f_2 with supports separated from the
singular points of the operator function. Such points form a set of measure
zero on the unit circle and it follows that the set of admissible functions f is
dense in $\mathcal{H}_1^a$.

4.2.3   Scattering matrix for self–adjoint extensions

We are going to calculate the scattering matrix $\hat{S}(\mathcal{U}_2, \mathcal{U}_1, \varphi)$ determined by
the scattering operator $S(\mathcal{U}_2, \mathcal{U}_1)$. Consider the spectral decomposition of $\mathcal{H}_1^a$
associated with the operator $\mathcal{U}_1$

$$\mathcal{H}_1^a = \int_{-\pi}^{\pi} \oplus K(\varphi) d\varphi.$$

The spectral representation of an arbitrary element $f \in \mathcal{H}_1^a$ will be denoted
by $\hat{f}(\varphi)$. It is a function taking values in the space $K(\varphi)$. Without loss of

generality we can suppose that the dimension of the space $K(\varphi)$ is finite (dim $K(\varphi) \leq m$), since the symmetric operator $\mathcal{A}^0$ has deficiency indices (m, m) (see Theorem 4.1.1). Without loss of generality we can also suppose that the basis elements e_ν belong to the space $\mathcal{H}_1^a$. Then formula (4.32) for $f_1, f_2 \in \mathcal{H}_1^a$ can be written as follows in the spectral representation

$$\langle \hat{f}_2, \hat{S}(\mathcal{U}_2, \mathcal{U}_1)\hat{f}_1 \rangle$$

$$= \langle \hat{f}_2, \hat{f}_1 \rangle - 4\pi \sum_{\mu,\nu=1}^{m} \int_{-\pi}^{\pi} d\varphi \int_{-\pi}^{\pi} d\psi \, \langle \delta(\psi - \varphi)\hat{f}_2(\psi), \hat{e}_\mu(\psi)\rangle_{K(\psi)}$$

$$\times \langle e_\mu, \left[G^*(e^{i\varphi}) + i\Gamma\right]^{-1} e_\nu \rangle \times \int_{-\pi}^{\pi} d\theta \, \langle \hat{e}_\nu(\theta), \delta(\theta - \varphi)\hat{f}_1(\theta)\rangle_{K(\theta)}$$

$$= \langle \hat{f}_2, \hat{f}_1 \rangle$$

$$- 4\pi \sum_{\mu,\nu=1}^{m} \int_{-\pi}^{\pi} d\varphi \, \langle \hat{f}_2(\varphi), \hat{e}_\mu(\varphi)\rangle_{K(\varphi)} \langle e_\mu, \left[G^*(e^{i\varphi}) + i\Gamma\right]^{-1} e_\nu \rangle$$

$$\langle \hat{e}_\nu(\varphi), \hat{f}_1(\varphi)\rangle_{K(\varphi)}.$$

One can see that in the spectral representation the scattering operator is equal to multiplication by the matrix valued function

$$\hat{S}(\mathcal{U}_2, \mathcal{U}_1, \varphi) = 1 - 4\pi \sum_{\mu,\nu=1}^{m} \langle e_\mu, \left[G^*(e^{i\varphi}) + i\Gamma\right]^{-1} e_\nu \rangle \langle \hat{e}_\nu(\varphi), \cdot \rangle_{K(\varphi)} \, \hat{e}_\mu(\varphi).$$

The latter formula gives the scattering matrix written in the spectral representation of the unitary operators $\mathcal{U}_1$. We are going to calculate now the scattering matrix for the self-adjoint operators $\mathcal{A}^1$ and $\mathcal{A}^2$ in the spectral representation of the operator $\mathcal{A}^1$, which is the Cayley transform of $\mathcal{U}_1$. For this it is only necessary to change the spectral variables. The real spectral parameter λ is related to the variable φ as follows

$$\lambda = \cot \varphi/2, \quad -\infty < \lambda < \infty.$$

The modulus of the Jacobian of the transformation $\lambda = i\frac{e^{i\varphi}+1}{e^{i\varphi}-1}$ is equal to $\frac{\lambda^2+1}{2}$. Let us denote by $\tilde{f}(\lambda)$ the spectral representation of the function f associated with the self-adjoint operator $\mathcal{A}^1$. Taking into account the Jacobian of the linear-fractional transformation the spectral representation $\tilde{f}$ can be chosen to be equal to

$$\tilde{f}(\lambda) = \frac{\sqrt{2}}{\lambda - i}\hat{f}(2 \operatorname{arccot} \lambda).$$

Then the spectral representations of the basis elements e_ν are equal to

$$\tilde{e}_\nu(\lambda) = \frac{\sqrt{2}}{\lambda - i}\hat{e}_\nu(2\,\text{arccot}\,\lambda).$$

The operator function $G^*(z)$ can be expressed in terms of the self–adjoint operator $\mathcal{A}^1$ as

$$G^*(z) \quad = \quad P_{-i}\frac{\mathcal{U}_1^{-1} + \bar{z}}{\mathcal{U}_1^{-1} - \bar{z}}P_{-1}^*$$

$$= \quad -iP_{-i}\frac{1 + \mathcal{A}^1\bar{\lambda}}{\mathcal{A}^1 - \bar{\lambda}}P_{-i}^*$$

$$\equiv \quad -iQ^1(\bar{\lambda}).$$

The operator function

$$Q(\lambda) = P_{-i}\frac{\mathcal{A}^1\lambda + 1}{\mathcal{A}^1 - \lambda}P_{-i}^*$$

has positive imaginary part in the upper half plane of the spectral parameter λ. The limit $\lim_{r \nearrow 1}$ corresponds to the limit from the lower half plane for the parameter λ. Thus the following formula for the scattering matrix $\tilde{S}(\mathcal{A}^2, \mathcal{A}^1, \lambda)$ has been proven

$$\tilde{S}(\mathcal{A}^2, \mathcal{A}^1, \lambda) \quad = \quad 1 - 2\pi i(\lambda^2 + 1)\sum_{\mu,\nu=1}^{m}\langle e_\mu, [Q(\lambda + i0) - \Gamma]^{-1}e_\nu\rangle$$

$$\times\langle\tilde{e}_\nu(\lambda), \cdot\rangle_{H(\lambda)}\,\tilde{e}_\mu(\lambda), \tag{4.34}$$

where $H(\lambda) = K(2\,\text{arccot}\,\lambda)$. The latter formula gives us the scattering matrix determined by two different self–adjoint extensions of one symmetric operator with the deficiency indices (m, m). The difference between the self–adjoint extensions is parameterized by the $m \times m$ Hermitian matrix $\Gamma = -i\frac{\mathcal{E}+1}{\mathcal{E}-1}$.

We are now going to prove that formula (4.34) determines a unitary operator in $H(\lambda)$ for almost all λ. Let us consider the following integral representation for the matrix elements

$$Q_{\mu\nu}(\lambda) = \langle e_\mu, Q(\lambda)e_\nu\rangle = \int_{-\infty}^{\infty}\frac{1 + x\lambda}{x - \lambda}\langle\tilde{e}_\mu(x), \tilde{e}_\nu(x)\rangle_{H(\lambda)}d\lambda.$$

Therefore the imaginary part of $Q_{\mu\nu}(\lambda)$ on the real axis is given by

$$2i\Im Q_{\mu\nu}(\lambda + i0) = Q_{\mu\nu}(\lambda + i0) - Q_{\mu\nu}(\lambda - i0) = 2\pi i(\lambda^2 + 1)\langle\tilde{e}_\mu(\lambda), \tilde{e}_\nu(\lambda)\rangle$$

$$\Rightarrow Q(\lambda + i0) - Q(\lambda - i0) = 2\pi i(\lambda^2 + 1)I, \qquad (4.35)$$

since the unit operator has the matrix elements $I_{\mu\nu}(\lambda) = \langle \tilde{e}_\mu(\lambda), \tilde{e}_\nu(\lambda) \rangle_{H(\lambda)}$ and the vectors $\tilde{e}_\mu(\lambda)$ span $H(\lambda)$. The adjoint scattering matrix is given by

$$\tilde{S}^*(\mathcal{A}^2, \mathcal{A}^1, \lambda) = 1 + 2\pi i(\lambda^2 + 1) \sum_{\mu,\nu=1}^{m} \langle e_\mu, [Q(\lambda - i0) - \Gamma]^{-1} e_\nu \rangle \langle \tilde{e}_\mu(\lambda), \cdot \rangle_{H(\lambda)} \tilde{e}_\nu(\lambda).$$

Let us use the notation

$$W_{\mu\nu}(\lambda) \equiv \langle e_\mu, [Q(\lambda) - \Gamma]^{-1} e_\nu \rangle.$$

Then the following equality holds

$$\tilde{S}(\mathcal{A}^2, \mathcal{A}^1, \lambda) \tilde{S}^*(\mathcal{A}^2, \mathcal{A}^1, \lambda) - I$$

$$= 2\pi i(\lambda^2 + 1) \sum_{\mu,\nu=1}^{m} \left\{ - W_{\mu\nu}(\lambda + i0) + W_{\nu\mu}(\lambda - i0) \right.$$

$$\left. - 2\pi i(\lambda^2 + 1) \sum_{\mu',\nu'=1}^{m} W_{\mu\mu'}(\lambda + i0) \langle \tilde{e}_{\mu'}(\lambda), \tilde{e}_{\nu'}(\lambda) \rangle_{H(\lambda)} W_{\nu\nu'}(\lambda - i0) \right\}$$

$$\times \langle \tilde{e}_\nu(\lambda), \cdot \rangle_{H(\lambda)} \tilde{e}_\mu(\lambda).$$

To prove that the expression in braces is equal to zero one can use the fact that $\{e_\nu\}$ is an orthogonal basis in $N(-i)$, equality (4.35) and the following equality

$$\frac{1}{Q(\lambda+i0)-\Gamma} - \frac{1}{Q(\lambda-i0)-\Gamma}$$

$$= \frac{1}{Q(\lambda + i0) - \Gamma} (Q(\lambda - i0) - Q(\lambda + i0)) \frac{1}{Q(\lambda - i0) - \Gamma}.$$

It follows that the calculated scattering matrix is unitary. Note that we have carried out all calculations in the absolutely continuous subspace of the operator $\mathcal{A}^1$.

The operator function $Q(\lambda) - \Gamma \equiv R(\lambda)$ is analytic in the upper half plane and has a positive imaginary part there. This function is analogous to Wigner's R–function [303, 961, 962, 960, 957]. This function is called the R-function for two self-adjoint extensions. Formula (4.34) establishes the relationship between the scattering matrix and the R-function.

4.3   Scattering theory for finite rank perturbations

This section is devoted to scattering theory for finite rank perturbations. The wave operators for perturbations of rank one have been calculated in Section 4.1. We study here the case of finite rank singular and generalized perturbations. The scattering matrix for such problems can be calculated using the results of the previous section, since the perturbed and unperturbed operators in all these problems are two self–adjoint extensions of one and the same symmetric operator with finite deficiency indices.

4.3.1   Scattering theory for finite rank perturbations

Let $\mathcal{A}$ be a self–adjoint operator acting in the Hilbert space $\mathcal{H}$. Then all finite rank perturbations of the operator are defined by the formula

$$\mathcal{A}_T = \mathcal{A} + \sum_{i,j=1}^{n} t_{ij} \langle \varphi_j, \cdot \rangle \varphi_i,$$

where $\varphi_j \in \mathcal{H}_{-2}(\mathcal{A})$ and the matrix $\mathbf{T} = \{t_{ij}\}_{i,j=1}^{n}$ is Hermitian. We suppose also that the vectors φ_j form an orthonormal system as elements from $\mathcal{H}_{-2}(\mathcal{A})$, i.e. $\langle \frac{1}{A+i}\varphi_j, \frac{1}{A+i}\varphi_k \rangle = \delta_{jk}$. The operator $\mathcal{A}_T$ was defined in Section 3.1.3. We suppose that an admissible matrix $\mathbf{R} = \{R_{jk}\}_{j,k=1}^{n}$ (see Definition 3.1.2) has been determined. Then the operator $\mathcal{A}_T$ is defined by Theorem 3.1.2. We are going to use formula (4.34) to calculate the scattering matrix determined by the operators $\mathcal{A}$ and $\mathcal{A}_T$. Let us choose an orthonormal basis $\{e_\nu\}_{\nu=1}^{m}$ in the deficiency subspace $N(-i)$ as follows

$$e_\nu = \frac{1}{A+i}\varphi_\nu.$$

The matrix elements of the operator $Q(\lambda)$ in this basis can be easily calculated:

$$Q_{\mu,\nu}(\lambda) = \int_{-\infty}^{\infty} \frac{z\lambda+1}{z-\lambda}\frac{1}{z^2+1}d\rho_{\nu,\mu}(z),$$

where

$$\langle e_\nu, E(\lambda)e_\mu \rangle = \int_{-\infty}^{\lambda} \frac{1}{z^2+1}d\rho_{\nu,\mu}(z).$$

The matrix $\mathbf{E}$ of the operator $\mathcal{E} = \mathcal{U}_T^{-1}\mathcal{U}$, $\mathcal{U}_T = \frac{\mathcal{A}_T+i}{\mathcal{A}_T-i}$, $\mathcal{U} = \frac{\mathcal{A}+i}{\mathcal{A}-i}$ in the basis $\{e_\nu\}_{\nu=1}^{M}$ is then given by

$$\mathbf{E} = \mathbf{V}^{-1}.$$

The matrix $\mathbf{V}$ is exactly the matrix connecting the boundary values $\vec{a}_{\pm}(\Psi)$ which appeared in formula (3.25). Therefore the matrix Γ of the operator Γ in the basis $\{e_\nu\}_{\nu=1}^M$ is equal to

$$\Gamma = -i\frac{\mathbf{I} + \mathbf{V}}{\mathbf{I} - \mathbf{V}} = -\frac{\mathbf{I} + \mathbf{T}\Re\Phi}{\mathbf{T}\Im\Phi} = -\mathbf{T}^{-1} - \mathbf{R},$$

since $\Im\Phi = \mathbf{I}$. Then the scattering matrix $\tilde{S}(\mathcal{A}_T, \mathcal{A}, \lambda)$ is given by

$$\tilde{S}(\mathcal{A}_T, \mathcal{A}, \lambda)$$

$$= I$$

$$-2\pi i(\lambda^2 + 1)\sum_{\mu,\nu=1}^m \left(\left[\mathbf{Q}(\lambda + i0) + \mathbf{T}^{-1} + \mathbf{R}\right]^{-1}\right)_{\mu\nu} \langle \tilde{e}_\nu(\lambda), \cdot \rangle_{H(\lambda)} \tilde{e}_\mu(\lambda).$$

$$(4.36)$$

The latter formula gives the scattering matrix for finite rank perturbations. The dimension of the scattering matrix is equal to dim $H(\lambda)$, which coincides with the multiplicity of the absolutely continuous spectrum at every point λ.

If all vectors φ_ν are elements from the Hilbert space $\mathcal{H}_{-1}(\mathcal{A})$ then the formula for the scattering matrix can be simplified as follows. The matrix $\mathbf{R}$ is determined uniquely by the elements φ_ν in this case, and we have

$$\mathbf{R}_{\mu\nu} = \left\langle \varphi_\nu, \frac{\mathcal{A}}{\mathcal{A}^2 + 1}\varphi_\nu \right\rangle.$$

Therefore the following equality holds

$$\left(\mathbf{Q}(\lambda) + \mathbf{T}^{-1} + \mathbf{R}\right)_{\mu\nu}$$

$$= \left\langle \varphi_\mu, \frac{1 + \mathcal{A}\lambda}{\mathcal{A} - \lambda}\frac{1}{\mathcal{A}^2 + 1}\varphi_\nu \right\rangle + \left(\mathbf{T}^{-1}\right)_{\mu\nu} + \left\langle \varphi_\nu, \frac{\mathcal{A}}{\mathcal{A}^2 + 1}\varphi_\nu \right\rangle$$

$$= \left\langle \varphi_\mu, \frac{1}{\mathcal{A} - \lambda}\varphi_\nu \right\rangle + \left(\mathbf{T}^{-1}\right)_{\mu\nu}.$$

Let us introduce the matrix $\mathbf{F}_{\mu\nu}(\lambda) = \langle \varphi_\mu, \frac{1}{\mathcal{A}-\lambda}\varphi_\nu \rangle$, which defines an operator with positive imaginary part in the upper half plane. Then the scattering matrix is equal to

$$\tilde{S}(\mathcal{A}_T, \mathcal{A}, \lambda)$$

$$= I - 2\pi i(\lambda^2 + 1)\sum_{\mu,\nu=1}^m \left(\left[\mathbf{F}(\lambda + i0) + \mathbf{T}^{-1}\right]^{-1}\right)_{\mu\nu} \langle \tilde{e}_\nu(\lambda), \cdot \rangle_{H(\lambda)} \tilde{e}_\mu(\lambda).$$

$$(4.37)$$

The unitarity of the scattering matrix follows easily from the following equality which is quite analogous to (4.35)

$$\mathbf{F}(\lambda + i0) - \mathbf{F}(\lambda - i0) = 2\pi i(\lambda^2 + 1)\mathbf{I}.$$

Again we have calculated the scattering matrix in the absolutely continuous subspace $\mathcal{H}^a$. The singularities of the scattering matrix coincide with the singularities of the resolvent given by (3.29).

Consider now the case of rank one perturbations. The scattering matrix is then given by

$$\tilde{S}(A_\alpha, A, \lambda) = \frac{1 + \alpha F(\lambda - i0)}{1 + \alpha F(\lambda + i0)}$$

if $\lambda \in \sigma_{ac}(A)$ and $F(\lambda) = \langle \varphi, \frac{1}{A - \lambda}\varphi \rangle$. The latter formula coincides with (4.22).

4.3.2   Scattering matrix for rank two perturbations

Let us consider rank two perturbations in the case where the operator $\mathbf{Q}(\lambda)$ is diagonal. Suppose that subspaces H_1 and H_2 generating by the elements $\frac{1}{A+i}\varphi_1$ and $\frac{1}{A+i}\varphi_2$ respectively are orthogonal. Then the following decomposition of the Hilbert space $\mathcal{H} = H_1 \oplus H_2$ reduces the operator $\mathcal{A} = A^1 \oplus A^2$, where $A_\beta = \mathcal{A}|_{H_\beta}, \beta = 1, 2$. The matrix function $\mathbf{Q}(\lambda)$ is diagonal

$$\mathbf{Q}(\lambda) = \begin{pmatrix} Q_1(\lambda) & 0 \\ 0 & Q_2(\lambda) \end{pmatrix},$$

where

$$Q_\beta(\lambda) = \left\langle \varphi_\beta, \frac{A_\beta \lambda + 1}{A_\beta - \lambda} \frac{1}{A_\beta^2 + 1} \varphi_\beta \right\rangle_{H_\beta} , \quad \beta = 1, 2.$$

The self–adjoint extensions of the operator

$$\mathcal{A}^0 = \mathcal{A}|_{\text{Dom}(\mathcal{A}^0)}, \quad \text{Dom}(\mathcal{A}^0) = \{\psi \in \text{Dom}(\mathcal{A}) : \langle \psi, \varphi_\beta \rangle = 0, \beta = 1, 2\}$$

are described by the self–adjoint 2×2 matrix

$$\Gamma = \begin{pmatrix} \gamma_{11} & \gamma_{12} \\ \gamma_{21} & \gamma_{22} \end{pmatrix}.$$

The inverse matrix $[\mathbf{Q}(\lambda + i0) - \Gamma]^{-1}$ can easily be calculated

$$\begin{pmatrix} Q_1 - \gamma_{11} & -\gamma_{12} \\ -\gamma_{21} & Q_2 - \gamma_{22} \end{pmatrix}^{-1}$$

$$= \frac{1}{(Q_1 - \gamma_{11})(Q_2 - \gamma_{22}) - |\gamma_{12}|^2} \begin{pmatrix} Q_2 - \gamma_{22} & \gamma_{12} \\ \gamma_{21} & Q_1 - \gamma_{11} \end{pmatrix}.$$

The scattering matrix is given by the formula

$\tilde{S}(\mathcal{A}_T, \mathcal{A}, \lambda)$

$$= \mathbf{I} - \frac{2\pi i(\lambda^2 + 1)}{(Q_1(\lambda + i0) - \gamma_{11})(Q_2(\lambda + i0) - \gamma_{22}) - |\gamma_{12}|^2}$$

$$\times \{ (Q_2(\lambda + i0) - \gamma_{22})\langle \tilde{e}_1(\lambda), \cdot \rangle \tilde{e}_1(\lambda) + \gamma_{12}\langle \tilde{e}_2(\lambda), \cdot \rangle \tilde{e}_1(\lambda) \qquad (4.38)$$

$$+ \gamma_{21}\langle \tilde{e}_1(\lambda), \cdot \rangle \tilde{e}_2(\lambda) + (Q_1(\lambda + i0) - \gamma_{11})\langle \tilde{e}_1(\lambda), \cdot \rangle \tilde{e}_1(\lambda) \}.$$

Let us consider in more detail the case where $\lambda \in \sigma_{ac}(A^1)$ and $\lambda \notin \text{supp } \sigma_{ac}(A^2)$. For such a λ the function $Q_2(\lambda + i0) = Q_2(\lambda)$ is real and the scattering matrix is then equal to the unitary function

$$\tilde{S}(\mathcal{A}_T, \mathcal{A}, \lambda) = 1 - \frac{2\pi i(\lambda^2 + 1)|\tilde{e}_1(\lambda)|^2}{Q_1(\lambda + i0) - \gamma_{11} - \frac{|\gamma_{12}|^2}{Q_2(\lambda) - \gamma_{22}}}$$

$$= \frac{Q_1(\lambda - i0) - \gamma_{11} - \frac{|\gamma_{12}|^2}{Q_2(\lambda) - \gamma_{22}}}{Q_1(\lambda + i0) - \gamma_{11} - \frac{|\gamma_{12}|^2}{Q_2(\lambda) - \gamma_{22}}}.$$

This formula will be used in what follows to obtain the scattering matrix for generalized perturbations with internal structure.

4.3.3 Scattering matrix for generalized perturbations

We are going to calculate the scattering matrix for rank one perturbations with internal structure described in Chapter 2. Consider first the case of a densely defined restricted operator. Every self–adjoint operator $\mathbf{A}^\Gamma$ is an extension of the symmetric operator $\mathbf{A}^0$ having deficiency indices $(2, 2)$. Therefore formula (4.38) can be used to calculate the scattering matrix for generalized perturbations

$\tilde{S}(\mathbf{A}^\Gamma, \mathbf{A}, \lambda)$

$$= \mathbf{I} - \frac{2\pi i(\lambda^2 + 1)}{(Q_0(\lambda + i0) - \gamma_{00})(Q_1(\lambda + i0) - \gamma_{11}) - |\gamma_{01}|^2}$$

$$\times \{ (Q_1(\lambda + i0) - \gamma_{11})\langle \tilde{e}_0(\lambda), \cdot \rangle \tilde{e}_0(\lambda) + \gamma_{01}\langle \tilde{e}_1(\lambda), \cdot \rangle \tilde{e}_0(\lambda) \qquad (4.39)$$

$$+ \gamma_{10}\langle \tilde{e}_0(\lambda), \cdot \rangle \tilde{e}_1(\lambda) + (Q_0(\lambda + i0) - \gamma_{00})\langle \tilde{e}_0(\lambda), \cdot \rangle \tilde{e}_0(\lambda) \}.$$

The scattering matrix restricted to the original Hilbert space $\mathcal{H}$ is not necessarily unitary. In fact we have

$$\tilde{S}_{00}(\mathbf{A}^\Gamma, \mathbf{A}, \lambda) = 1 - \frac{2\pi i (\lambda^2 + 1)|\tilde{e}_0(\lambda)|^2}{Q_0(\lambda + i0) + Q_0^+(\lambda + i0)}$$

$$= \frac{Q_0(\lambda - i0) + Q_0^+(\lambda + i0)}{Q_0(\lambda + i0) + Q_0^+(\lambda + i0)},$$

since the boundary value on the real axis of the function

$$Q_0^+(\lambda) = \frac{1}{\gamma_{11} - Q_1(\lambda)}|\gamma_{01}|^2 - \gamma_{00}$$

can have nontrivial positive imaginary part. The absolute value of this component of the scattering matrix

$$|\tilde{S}_{00}(\mathbf{A}^\Gamma, \mathbf{A}, \lambda)|^2$$

$$= 1 - \frac{4\Im Q_0(\lambda + i0)\Im Q_0^+(\lambda + i0)}{(\Re Q_0(\lambda + i0) + \Re Q_0^+(\lambda + i0))^2 + (\Im Q_0(\lambda + i0) + \Im Q_0^+(\lambda + i0))^2}$$

can easily be estimated as folllows

$$|\tilde{S}_{00}(\mathbf{A}^\Gamma, \mathbf{A}, \lambda)|^2 \leq 1$$

if one takes into account that both functions Q_0 and Q_0^+ have positive imaginary parts in the upper half plane. This estimate shows that the zero component of the scattering matrix is unitary if and only if the imaginary part of the function $Q_0^+(\lambda + i0)$ is trivial. (If the imaginary part of the function $Q_0(\lambda + i0)$ is trivial then the point λ is not in the absolutely continuous spectrum of $\mathcal{A}$ or the corresponding density is equal to zero.)

Consider now the case of a finite dimensional internal space $\mathcal{H}^1$. The restricted operator $\mathbf{A}^0$ is not densely defined in this case. The absolutely continuous spectrum of the operator $\mathbf{A}^\Gamma$ coincides with the absolutely continuous spectrum of the operator $\mathcal{A}$ and has multiplicity one. It follows that the scattering matrix in this case is a unitary function. Calculating the scattering matrix for self–adjoint extensions we have used only the fact that the Cayley transforms of the perturbed and unperturbed operators act in a different way on a finite dimensional subspace only. This condition is satisfied for the operator with internal structure $\mathbf{A}^\Gamma$ and original operator $\mathbf{A}$. Therefore the scattering matrix is given by the formula

$$\tilde{S}(\mathbf{A}^\Gamma, \mathbf{A}, \lambda) = \frac{Q_0(\lambda - i0) + Q_0^+(\lambda)}{Q_0(\lambda + i0) + Q_0^+(\lambda)}. \qquad (4.40)$$

It is a unitary function on the absolutely continuous spectrum of the operator $\mathcal{A}$, since the function $Q_0^+(\lambda)$ is real almost everywhere on the real axis. The latter formula coincides with the formula for the zero component of the scattering matrix in the case where the restricted operator is densely defined.

It can easily be shown that formula (4.39) gives the scattering matrix for arbitrary generalized perturbations with internal structure. The singularities of the scattering matrix coincide with the energies of the bound states and resonances of the perturbed operator.

Consider now the generalized perturbation of the Laplace operator in $L_2(\mathbf{R}^3)$ constructed in Section 2.2.5. The scattering matrix has been calculated there from the asymptotics of the continuous spectrum eigenfunctions. The subspace of $L_2(\mathbf{R}^3)$ generated by the delta function and Laplace operator consists only of the functions which are invariant under rotations of the three dimensional space with respect to the origin. Therefore formula (4.40) gives only the symmetric component of the scattering matrix. The Q-function of the external operator has been calculated to be $Q_0(\lambda) = 1 + c^2 \frac{i\sqrt{\lambda}}{4\pi}$. Therefore the scattering matrix is given by

$$\tilde{S}(\mathcal{L}^{\Gamma}, \mathcal{L}, \lambda) = \frac{1 - c^2 \frac{i\sqrt{\lambda}}{4\pi} + Q_0^+(\lambda)}{1 + c^2 \frac{i\sqrt{\lambda}}{4\pi} + Q_0^+(\lambda)}, \tag{4.41}$$

where $c = 2\sqrt[4]{2}\sqrt{\pi}$ is the normalizing constant for the delta function as an element from $\mathcal{H}_{-2}(L)$. Formula (4.41) gives only the nontrivial component of the scattering matrix (2.73), since we have restricted our consideration to the subspace generated by the delta function.

The scattering matrix for an arbitrary generalized finite rank perturbation can be calculated using formula (4.34). But this formula cannot be used in the case where the restricted symmetric operator has infinite deficiency indices. Several examples of scattering operators for such model operators are presented in the following chapters.

Chapter 5

Krein's formula for infinite deficiency indices and two-body problems

5.1 Infinite rank perturbations

5.1.1 Krein's formula for infinite deficiency indices

Let us consider Krein's formula for the resolvents of two self–adjoint extensions of one symmetric operator with arbitrary deficiency indices. We are going to concentrate our attention on the case of infinite deficiency indices; the case of arbitrary finite indices has already been studied in Chapter 3. Krein's formula for the resolvents of two self–adjoint extensions of one symmetric operator with arbitrary equal deficiency indices was obtained by Sh.N.Saakjan [851]. We going to derive the same formula using the boundary operator connecting the boundary values of the functions from the domain of the extended operator.

Let A be a self–adjoint operator acting in the Hilbert space $\mathcal{H}$. Let the operator A^0 be its densely defined symmetric closed restriction. We are going to describe the resolvents of all self–adjoint extensions of the operator A^0. Let us denote by N_z the deficiency subspaces for the operator A^0

$$N_z = \text{Ker} \left(A^{0*} - z \right).$$

In particular we are going to use the subspace $M = N_{-i}$, which can have infinite dimension. The subspace M is closed and therefore it is itself a Hilbert space. Then every element ψ from the domain Dom (A^{0*}) of the adjoint operator $\mathcal{A}^{0*}$ possesses the representation

$$\psi = \hat{\psi} + \frac{\mathcal{A}}{\mathcal{A} - i}\xi_+(\psi) + \frac{1}{\mathcal{A} - i}\xi_-(\psi), \tag{5.1}$$

where $\hat{\psi} \in \mathrm{Dom}(\mathcal{A}^0), \xi_\pm(\psi) \in M$. The elements $\xi_\pm \in M$ will be called *boundary values* of the vector ψ. The map $l : \psi \mapsto (\xi_+(\psi), \xi_-(\psi))$ acting in the spaces $\mathrm{Dom}\,(\mathcal{A}^{0*}) \rightarrow M \oplus M$ is called *the boundary map*.

The adjoint operator acts as follows

$$\begin{aligned} \mathcal{A}^{0*}\psi &= \mathcal{A}^{0*}\left(\hat{\psi} + \frac{\mathcal{A}}{\mathcal{A} - i}\xi_+(\psi) + \frac{1}{\mathcal{A} - i}\xi_-(\psi)\right) \\[2mm] &= \mathcal{A}^0\hat{\psi} - \frac{1}{\mathcal{A} - i}\xi_+(\psi) + \frac{\mathcal{A}}{\mathcal{A} - i}\xi_-(\psi). \end{aligned} \tag{5.2}$$

The boundary form of the adjoint operator can be calculated using the method of Chapter 2.

Theorem 5.1.1 *The boundary form*

$$B_{\mathcal{A}}[u, v] = \langle u, \mathcal{A}^{0*}v \rangle - \langle \mathcal{A}^{0*}u, v \rangle \tag{5.3}$$

of the operator $\mathcal{A}^{0*}$ is given by

$$B_{\mathcal{A}}[u, v] = \langle \xi_+(u), \xi_-(v) \rangle_M - \langle \xi_-(u), \xi_+(v) \rangle_M. \tag{5.4}$$

Proof The proof follows the same lines as the proof of Theorem 2.2.1.

$\square$

In Chapter 3 we established a one to one correspondence between the set of self–adjoint extensions of the operator $\mathcal{A}^0$ and the subspaces of $M \oplus M$ which are Lagrangian with respect to the boundary form $B_{\mathcal{A}}$. This statement is true only in the case where the restricted operator $\mathcal{A}^0$ has finite deficiency indices, i.e. the space $M \oplus M$ has finite dimension. If the dimension of the space $M \oplus M$ is not finite, then not every maximal subspace on which the boundary form vanishes determines a self–adjoint operator. The reason is that among the extensions of the operator $\mathcal{A}^0$ there are symmetric operators with nonequal deficiency indices. Such operators have no self–adjoint extension. Therefore let us consider directly boundary conditions of the form

$$\xi_-(\psi) = \Gamma\xi_+(\psi). \tag{5.5}$$

We suppose that Γ is a self–adjoint operator acting in the Hilbert space M. Such boundary conditions do not describe all self–adjoint extensions of the operator $\mathcal{A}^0$ but we do not actually aim here to describe all such extensions (described by unitary operators acting between the deficiency subspaces using von Neumann formulas). In fact the self–adjoint extensions of the operator $\mathcal{A}^0$ which cannot be described by the latter boundary conditions have the following property: the operator $\mathcal{A}^0$ is not the maximal common symmetric restriction of the operator $\mathcal{A}$ and the self–adjoint extension (see [637]).

In the case of infinite deficiency indices, i.e. where the space M has infinite dimension the boundary condition (5.5) may contain one additional restriction which is fulfilled automatically in the finite dimensional case: $\xi_+(\psi)$ belongs to the domain of the self–adjoint operator Γ

$$\xi_+(\psi) \in \text{Dom } (\Gamma). \tag{5.6}$$

This condition is fulfilled automatically in the case where the operator Γ is bounded, but the case of unbounded operators Γ is interesting in applications. The following definition will be used.

Definition 5.1.1 *Let $\mathcal{A}^0$ be a symmetric operator in $\mathcal{H}$ and let Γ be a self–adjoint operator in $M = \mathcal{R}(\mathcal{A}^0 - i)^\perp$. Then the operator $\mathcal{A}^\Gamma$ – **the extension of the operator** $\mathcal{A}^0$ **determined by** Γ – is the restriction of the operator $\mathcal{A}^{0*}$ to the set of functions*

$$\psi = \hat{\psi} + \frac{\mathcal{A}}{\mathcal{A} - i}\xi_+(\psi) + \frac{1}{\mathcal{A} - i}\xi_-(\psi) \in \text{Dom } (\mathcal{A}^{0*})$$

with the boundary values satisfying conditions (5.5) and (5.6).

Every function ψ from the domain Dom $(\mathcal{A}^\Gamma)$ of the operator $\mathcal{A}^\Gamma$ possesses the representation

$$\psi = \hat{\psi} + \frac{1}{\mathcal{A} - i}(\mathcal{A} + \Gamma)\xi_+(\psi), \tag{5.7}$$

where $\hat{\psi} \in \text{Dom } (\mathcal{A}^0), \xi_+(\psi) \in \text{Dom } (\Gamma) \subset M$.

If the operator $\mathcal{A}^0$ is not closed, then the latter formula defines an essentially self–adjoint operator and therefore the unique self–adjoint extension of $\mathcal{A}^0$.

The resolvent of the operator $\mathcal{A}^\Gamma$ can be calculated using the following theorem.

Theorem 5.1.2 *Let the operator $\mathcal{A}$ be a self–adjoint operator in $\mathcal{H}$ and the operator $\mathcal{A}^0$ be its symmetric closed densely defined restriction. Let the operator Γ in $M = \mathcal{R}(\mathcal{A}^0 - i)^\perp$ be self–adjoint. Then the extension $\mathcal{A}^\Gamma$ of the operator $\mathcal{A}^0$ determined by Γ is self–adjoint and its resolvent is given by*

$$\frac{1}{\mathcal{A}_\Gamma - \lambda} = \frac{1}{\mathcal{A} - \lambda} + \frac{\mathcal{A} + i}{\mathcal{A} - \lambda}\frac{1}{\Gamma - Q(\lambda)}P_M\frac{\mathcal{A} - i}{\mathcal{A} - \lambda} \tag{5.8}$$

for any $\lambda, \Im\lambda \neq 0$. *The operator* $Q(\lambda)$ *is by definition the operator valued R-function defined on* M *by the formula*

$$Q(\lambda) = P_M \frac{1 + \lambda\mathcal{A}}{\mathcal{A} - \lambda} P_M. \qquad (5.9)$$

Proof The proof is similar to the proof of Theorem 2.2.3. The operator $\mathcal{A}^\Gamma$ is symmetric, since the boundary form (5.4) vanishes on any two elements satisfying the boundary conditions (5.5). The resolvent of the operator $\mathcal{A}^\Gamma$ can be calculated if we suppose that the operator $\Gamma - Q(\lambda)$ is invertible. This operator is invertible, since its imaginary part $-P_M \frac{\Im\lambda(\mathcal{A}^2+1)}{(\mathcal{A}-\Re\lambda)^2+(\Im\lambda)^2} P_M$ is separated from zero.

To prove that the operator $\mathcal{A}^\Gamma$ is self–adjoint it is enough to show that the range of $\mathcal{A}^\Gamma - i$ coincides with the whole Hilbert space, i.e. that the resolvent $\frac{1}{\mathcal{A}^\Gamma-i}$ is defined on $\mathcal{H}$. The operator $Q(i)$ is bounded and can easily be calculated as $Q(i) = iI_M$, where I_M denotes the unit operator in M. Thus the operator $\Gamma - Q(i) = \Gamma - iI_M$ is invertible and the inverse operator coincides with the resolvent of the self–adjoint operator Γ calculated at the point i. Therefore the operator $\frac{1}{\Gamma-Q(i)}$ is defined on the whole space M. We have proven that the operator $\mathcal{A}^\Gamma$ is symmetric and its resolvent is defined on the whole space $\mathcal{H}$. The kernel of the resolvent is trivial, since the restricted operator $\mathcal{A}^0$ is densely defined. The theorem is proven.

□

5.1.2 Generalized perturbations of infinite rank

We consider the model with internal structure in the case where the restricted symmetric operator has equal (but otherwise arbitrary) deficiency indices. We are again going to concentrate our attention on the case where the deficiency indices are infinite. This construction is quite similar to that considered in the previous section. The difference is that we are going to suppose that the original Hilbert space, the original operator and the deficiency subspace possess orthogonal decompositions corresponding to the external and internal channels. Similarly we are going to suppose that the boundary operator possesses matrix decomposition. The main difference is that we are not going to suppose that the total restricted operator is densely defined. We are going to use the assumption that only the external restricted operator is densely defined. Let us consider the Hilbert space **H** decomposed into the orthogonal sum of two Hilbert spaces as follows

$$\mathbf{H} = \mathcal{H} \oplus \mathcal{H}'.$$

Consider the orthogonal sum $\mathbf{A}$ of two self–adjoint operators $\mathcal{A}$ and $\mathcal{A}'$ acting in the Hilbert spaces $\mathcal{H}$ and $\mathcal{H}'$ respectively

$$\mathbf{A} = \mathcal{A} \oplus \mathcal{A}'.$$

Then the operator $\mathbf{A}$ is self–adjoint in the Hilbert space $\mathbf{H}$. We restrict the self–adjoint operators $\mathcal{A}$ and $\mathcal{A}'$ to certain symmetric closed operators $\mathcal{A}^0$ and $\mathcal{A}'^0$ respectively. The corresponding symmetric closed restriction of the operator $\mathbf{A}$ will be denoted by $\mathbf{A}^0$

$$\mathbf{A}^0 = \mathcal{A}^0 \oplus \mathcal{A}'^0.$$

We suppose first that the restricted operator $\mathbf{A}^0$ is densely defined. This assumption will be removed later on in this section. Let us denote by N_z, N_z', the deficiency subspaces for the operators $\mathcal{A}^0$ and $\mathcal{A}'^0$. Then every element $\Psi = (\psi_0, \psi_1)$ from the domain Dom $(\mathbf{A}^{0*})$ of the adjoint operator $\mathbf{A}^{0*}$ possesses the following representation

$$\Psi = \hat{\Psi} + \frac{\mathbf{A}}{\mathbf{A} - i} \Xi_+(\Psi) + \frac{1}{\mathbf{A} - i} \Xi_-(\Psi), \qquad (5.10)$$

$\hat{\Psi} \in \text{Dom}\,(\mathbf{A}^0), \Xi_{\pm} = (\xi_{\pm}, \xi_{\pm}') \in \mathbf{M}$. The subspace

$$\mathbf{M} = M \oplus M' = N_{-i} \oplus N_{-i}' \subset \mathbf{H} \qquad (5.11)$$

has arbitrary, perhaps infinite dimension. The subspace $\mathbf{M}$ is a Hilbert space itself, since it is closed. The norm concides with the norm of the original Hilbert space $\mathbf{H}$.

The adjoint operator acts as follows

$$
\begin{aligned}
\mathbf{A}^{0*}\Psi &= \mathbf{A}^{0*}\left(\hat{\Psi} + \frac{\mathbf{A}}{\mathbf{A} - i} \Xi_+(\Psi) + \frac{1}{\mathbf{A} - i} \Xi_-(\Psi) \right) \\
&= \mathbf{A}^0 \hat{\Psi} - \frac{1}{\mathbf{A} - i} \Xi_+(\Psi) + \frac{\mathbf{A}}{\mathbf{A} - i} \Xi_-(\Psi).
\end{aligned}
\qquad (5.12)
$$

The boundary form

$$B_{\mathbf{A}}[U, V] = \langle U, \mathbf{A}^{0*}V \rangle - \langle \mathbf{A}^{0*}U, V \rangle \qquad (5.13)$$

of the operator $\mathbf{A}^{0*}$ is given by

$$B_{\mathbf{A}}[U, V] = \langle \Xi_+(U), \Xi_-(V) \rangle_{\mathbf{M}} - \langle \Xi_-(U), \Xi_+(V) \rangle_{\mathbf{M}}. \tag{5.14}$$

The self–adjoint extensions of the operator $\mathbf{A}^0$ can be defined with the help of boundary conditions. Let Γ be a self–adjoint operator in $\mathbf{M}$. Then the corresponding self–adjoint extension $\mathbf{A}^\Gamma$ can be defined using Definition 5.1.1. The boundary values $\Xi_\pm(\Psi)$ of any element

$$\Psi = \hat{\Psi} + \frac{\mathbf{A}}{\mathbf{A} - i} \Xi_+(\Psi) + \frac{1}{\mathbf{A} - i} \Xi_-(\Psi) \in \mathrm{Dom}\,(\mathbf{A}^{0*})$$

from the domain of the operator satisfy the following conditions

$$\Xi_-(\Psi) = \Gamma \Xi_+(\Psi) \tag{5.15}$$

and

$$\Xi_+(\Psi) \in \mathrm{Dom}\,(\Gamma). \tag{5.16}$$

Then Theorem 5.1.2 implies that the operator $\mathbf{A}^\Gamma$ is self–adjoint and its resolvent is given by

$$\frac{1}{\mathbf{A}^\Gamma - \lambda} = \frac{1}{\mathbf{A} - \lambda} + \frac{\mathbf{A} + i}{\mathbf{A} - \lambda} \frac{1}{\Gamma - \mathcal{Q}(\lambda)} P_{\mathbf{M}} \frac{\mathbf{A} - i}{\mathbf{A} - \lambda}. \tag{5.17}$$

for any $\lambda, \Im \lambda \neq 0$, where

$$\mathcal{Q}(\lambda) = P_{\mathbf{M}} \frac{1 + \lambda \mathbf{A}}{\mathbf{A} - \lambda} P_{\mathbf{M}}. \tag{5.18}$$

The space $\mathbf{M}$ possesses the natural orthogonal decomposition (5.11). The operator Γ is not necessarily bounded. In what follows we are going to use the following assumption. Let Γ_{00} and Γ_{11} be two self–adjoint operators acting in the Hilbert spaces $M = N_{-i}$ and $M' = N'_{-i}$ respectively. Let the operators Γ_{01} and Γ_{10} be bounded operators acting in the spaces $M' \to M$ and $M \to M'$ respectively. Suppose that $\Gamma^*_{01} = \Gamma_{10}$. Then the operator

$$\Gamma = \Gamma_{00} \oplus \Gamma_{11} + \begin{pmatrix} 0 & \Gamma_{01} \\ \Gamma_{10} & 0 \end{pmatrix} = \begin{pmatrix} \Gamma_{00} & \Gamma_{01} \\ \Gamma_{10} & \Gamma_{11} \end{pmatrix}$$

is self–adjoint on the domain $\mathrm{Dom}\,(\Gamma) = \mathrm{Dom}\,(\Gamma_{00}) \oplus \mathrm{Dom}\,(\Gamma_{11})$. The operators Γ possessing the latter decomposition will be called **decomposable**. Only decomposable boundary operators will be used in what follows.

Let us study the case where only the external restricted operator $\mathcal{A}^0$ is densely defined. Let us introduce the space

$$\mathbf{M} = M \oplus M' = N_{-i} \oplus \mathcal{R}(\mathcal{A}'^0 - i)^\perp.$$

We denote by $\mathrm{Dom}\,(\mathbf{A}^{\Gamma})$ the set of functions Ψ possessing the following decomposition

$$\Psi = \hat{\Psi} + \frac{1}{\mathbf{A} - i}(\mathbf{A} + \Gamma)\Xi_+(\Psi), \qquad (5.19)$$

where $\hat{\Psi} \in \mathrm{Dom}\,(\mathbf{A}^0), \Xi_+ \in \mathrm{Dom}\,(\Gamma) \subset \mathbf{M}$.

Lemma 5.1.1 *Let the operator $\mathcal{A}^0$ be densely defined, the self–adjoint operator Γ in $\mathbf{M}$ be decomposable and*

$$\mathrm{Ker}\,\Gamma_{01} \cap M' = \{0\}. \qquad (5.20)$$

Then the representation (5.19) is unique for every element $\Psi \in \mathrm{Dom}\,(\mathbf{A}^{\Gamma})$.

Proof Suppose that the representation is not unique, i.e. there exist $\hat{\Psi}, \hat{\Psi}^* \in \mathrm{Dom}\,(\mathbf{A}^0)$ and $\Xi_+, \Xi_+^* \in \mathbf{M}$ such that

$$\hat{\Psi} - \hat{\Psi}^* + \frac{1}{\mathbf{A} - i}(\Gamma + i)(\Xi_+ - \Xi_+^*) = -(\Xi_+ - \Xi_+^*).$$

The left hand side of the latter equality belongs to the domain of the operator $\mathbf{A}$. The intersection $\mathrm{Dom}\,(\mathcal{A}) \cap N_{-i}$ is trivial, since the operator $\mathcal{A}^0$ is densely defined. Therefore we conclude that $\Xi_+ - \Xi_+^* \in M'$. Applying the operator $\mathbf{A} - i$ to the equality and projecting to $\mathbf{M}$ we get in particular

$$\Gamma_{01}(\Xi_+ - \Xi_+^*) = 0.$$

Then (5.20) implies the lemma.

$\square$

Now it is possible to modify Definition 5.1.1 as follows

Definition 5.1.2 The Γ-modified operator $\mathbf{A}^{\Gamma}$ *is defined on the domain* $\mathrm{Dom}\,(\mathbf{A}^{\Gamma})$ *by the formula*

$$
\begin{aligned}
\mathbf{A}^{\Gamma}\Psi &= \mathbf{A}^{\Gamma}\left(\hat{\Psi} + \frac{1}{\mathbf{A} - i}(\mathbf{A} + \Gamma)\Xi_+(\Psi)\right) \\
&= \mathbf{A}\hat{\Psi} + \frac{1}{\mathbf{A} - i}(-1 + \mathbf{A}\Gamma)\Xi_+(\Psi).
\end{aligned}
\qquad (5.21)
$$

Note that the formula for the resolvent of the operator $\mathbf{A}^{\Gamma}$ just coincides with the formula for the resolvent (5.17) derived for the case of a densely defined restricted operator.

One can use the fact that the operator Γ is decomposable and the operator $\mathcal{Q}(\lambda)$ is equal to the orthogonal sum of the operators $Q(\lambda) = P_M \frac{1+\lambda\mathcal{A}}{\mathcal{A}-\lambda} P_M$, $Q(\lambda) = P_{M'} \frac{1+\lambda\mathcal{A}'}{\mathcal{A}'-\lambda} P_{M'}$:

$$\mathcal{Q}(\lambda) = Q_0(\lambda) \oplus Q_1(\lambda)$$

to calculate the inverse operator

$$(\Gamma - \mathcal{Q}(\lambda))^{-1}$$

$$= \begin{pmatrix} [\Gamma_{00} - Q_0(\lambda) - \Gamma_{01}\frac{1}{\Gamma_{11}-Q_1(\lambda)}\Gamma_{10}]^{-1} & 0 \\ 0 & [\Gamma_{11} - Q_1(\lambda) - \Gamma_{10}\frac{1}{\Gamma_{00}-Q_0(\lambda)}\Gamma_{01}]^{-1} \end{pmatrix}$$

$$\times \begin{pmatrix} 1 & -\Gamma_{01}\frac{1}{\Gamma_{11}-Q_1(\lambda)} \\ -\Gamma_{10}\frac{1}{\Gamma_{00}-Q_0(\lambda)} & 1 \end{pmatrix}.$$

5.1.3   Resolvent formula for functionals

We have determined certain self–adjoint extensions of the symmetric operator $\mathcal{A}^0$ using the boundary conditions which connect the boundary values of the functions from the domain of the adjoint operator. These boundary values are elements from the deficiency subspace corresponding to the point $-i$, i.e. these values are elements from the Hilbert space. But the symmetric restriction of the operator are determined by elements from the space $\mathcal{H}_{-2}(\mathcal{A})$ containing all bounded linear functionals on the domain of the operator $\mathcal{A}$. Therefore it will be more convenient in the future to write the boundary conditions and the resolvent formula in terms of such functionals.

Let us consider a closed subspace $\Phi \subset \mathcal{H}_{-2}(\mathcal{A})$ and the corresponding restriction $\mathcal{A}^0$ of the operator $\mathcal{A}$ to the domain

$$\text{Dom} (\mathcal{A}^0) = \{\Psi \in \text{Dom} (\mathcal{A}) : (\Phi \in \Phi \Rightarrow \langle\Psi, \Phi\rangle = 0)\}.$$

Then every element Ψ from the domain of the adjoint operator $\mathcal{A}^{0*}$ possesses the representation

$$\Psi = \hat{\Psi} + \frac{\mathcal{A}}{\mathcal{A}^2 + 1}\Phi_+(\Psi) + \frac{1}{\mathcal{A}^2 + 1}\Phi_-(\Psi), \qquad (5.22)$$

where $\hat{\Psi} \in \text{Dom} (\mathcal{A}^0)$, $\Phi_\pm(\Psi) \in \Phi$. One can write another representation for every element Ψ from the domain of the adjoint operator

$$\Psi = \tilde{\Psi} + \frac{\mathcal{A}}{\mathcal{A}^2 + 1}\Phi_+(\Psi), \qquad (5.23)$$

where $\tilde{\Psi} = \hat{\Psi} + \frac{1}{\mathcal{A}^2+1}\Phi_-(\Psi) \in \text{Dom }(\mathcal{A})$. The vector $\Phi_-(\Psi)$ can be calculated from the vector $\hat{\Psi}$ using the projector $\tilde{P} = (\mathcal{A}+i)P_M(\mathcal{A}-i)$ acting in the space $\mathcal{H}_2(\mathcal{A})$

$$\Phi_-(\Psi) = \tilde{P}\hat{\Psi}. \tag{5.24}$$

The adjoint operator acts in accordance to the following formula

$$\begin{aligned}
\mathcal{A}^{0*}\Psi &= \mathcal{A}^{0*}\left(\hat{\Psi} + \frac{\mathcal{A}}{\mathcal{A}^2+1}\Phi_+(\Psi) + \frac{1}{\mathcal{A}^2+1}\Phi_-(\Psi)\right) \\
&= \mathcal{A}\hat{\Psi} - \frac{1}{\mathcal{A}^2+1}\Phi_+(\Psi) + \frac{\mathcal{A}}{\mathcal{A}^2+1}\Phi_-(\Psi),
\end{aligned}$$

or using the representation (5.23)

$$\begin{aligned}
\mathcal{A}^{0*}\Psi &= \mathcal{A}^{0*}\left(\tilde{\Psi} + \frac{\mathcal{A}}{\mathcal{A}^2+1}\Phi_+(\Psi)\right) \\
&= \mathcal{A}\tilde{\Psi} - \frac{1}{\mathcal{A}^2+1}\Phi_+(\Psi).
\end{aligned} \tag{5.25}$$

The boundary form of the adjoint operator is given by

$$\begin{aligned}
B_{\mathcal{A}^{0*}}[U,V] &= \left\langle \Phi_+(U), \frac{1}{\mathcal{A}^2+1}\Phi_-(V)\right\rangle - \left\langle \Phi_-(U), \frac{1}{\mathcal{A}^2+1}\Phi_+(V)\right\rangle \\
&= \left\langle \Phi_+(U), \frac{1}{\mathcal{A}^2+1}\tilde{P}_M\hat{V}\right\rangle - \left\langle \frac{1}{\mathcal{A}^2+1}\tilde{P}_M\hat{U}, \Phi_+(V)\right\rangle.
\end{aligned}$$

The boundary form can easily be written using the scalar product in the Hilbert space $\mathcal{H}_{-2}(\mathcal{A})$ related to the standard scalar product as follows:

$$\langle F, G\rangle_{\mathcal{H}_{-2}(\mathcal{A})} = \left\langle F, \frac{1}{\mathcal{A}^2+1}G\right\rangle.$$

The boundary conditions describing the extension $\mathcal{A}^\Gamma$ of the operator $\mathcal{A}^0$ can be written as

$$\Phi_-(U) = (\mathcal{A}+i)\Gamma\frac{1}{\mathcal{A}+i}\Phi_+(U)$$

$$\Rightarrow \tilde{P}\tilde{U} = (\mathcal{A}+i)\Gamma\frac{1}{\mathcal{A}+i}\Phi_+(U) \tag{5.26}$$

with the additional condition

$$\Phi_+(U) \in (\mathcal{A}+i)\text{Dom }(\Gamma).$$

We introduce the operator $\tilde{\Gamma} = (\mathcal{A} + i)\Gamma\dfrac{1}{\mathcal{A} + i}$ which is self–adjoint in the space Φ with the scalar product $\langle \cdot, \cdot \rangle_{\mathcal{H}_{-2}(\mathcal{A})}$. The resolvent of the operator $\mathcal{A}^{\Gamma}$ is given by (5.8). This formula can be modified as follows:

$$\frac{1}{\mathcal{A}^{\Gamma} - \lambda} = \frac{1}{\mathcal{A} - \lambda} + \frac{1}{\mathcal{A} - \lambda}\frac{1}{\tilde{\Gamma} - \tilde{Q}(\lambda)}\tilde{P}_{M}\frac{1}{\mathcal{A} - \lambda}. \qquad (5.27)$$

The latter formula will be used in the next section to obtain the two-body analogue of Krein formula. This formula is similar to the original Krein's formula (1.31) obtained for rank one perturbations.

The resolvent formula for functionals we derived above can be considered even in the case where the original operator is equal to the orthogonal sum of two self-adjoint operators $\mathbf{A} = \mathcal{A} \oplus \mathcal{A}'$ acting in the Hilbert spaces $\mathcal{H}$ and $\mathcal{H}'$ respectively. We suppose that the space Φ possesses a similar decomposition $\Phi = \Phi \oplus \Phi'$, $\Phi \subset \mathcal{H}_{-2}(\mathcal{A})$, $\Phi' \subset \mathcal{H}_{-2}(\mathcal{A}')$. Suppose that the operator Γ is decomposable. The operator $\tilde{\Gamma}$ is then given by

$$\tilde{\Gamma} = \begin{pmatrix} (\mathcal{A} + i)\Gamma_{00}\dfrac{1}{\mathcal{A} + i} & (\mathcal{A} + i)\Gamma_{01}\dfrac{1}{\mathcal{A}' + i} \\ (\mathcal{A}' + i)\Gamma_{10}\dfrac{1}{\mathcal{A} + i} & (\mathcal{A}' + i)\Gamma_{11}\dfrac{1}{\mathcal{A}' + i} \end{pmatrix} \equiv \begin{pmatrix} \tilde{\Gamma}_{00} & \tilde{\Gamma}_{01} \\ \tilde{\Gamma}_{10} & \tilde{\Gamma}_{11} \end{pmatrix}.$$

The inverse operator $(\tilde{\Gamma} - \tilde{Q}(\lambda))^{-1}$ can be calculated:

$$(\tilde{\Gamma} - \tilde{Q}(\lambda))^{-1}$$

$$= \begin{pmatrix} \dfrac{1}{\tilde{\Gamma}_{00} - \tilde{Q}_{0}(\lambda) - \tilde{\Gamma}_{01}\frac{1}{\tilde{\Gamma}_{11} - \tilde{Q}_{1}(\lambda)}\tilde{\Gamma}_{10}} & 0 \\ 0 & \dfrac{1}{\tilde{\Gamma}_{11} - \tilde{Q}_{1}(\lambda) - \tilde{\Gamma}_{10}\frac{1}{\tilde{\Gamma}_{00} - \tilde{Q}_{0}(\lambda)}\tilde{\Gamma}_{01}} \end{pmatrix}$$

$$\times \begin{pmatrix} 1 & -\tilde{\Gamma}_{01}\frac{1}{\tilde{\Gamma}_{11} - \tilde{Q}_{1}(\lambda)} \\ -\tilde{\Gamma}_{10}\frac{1}{\tilde{\Gamma}_{00} - \tilde{Q}_{0}(\lambda)} & 1 \end{pmatrix}.$$

The resolvent of the operator $\mathbf{A}^{\Gamma}$ restricted to the Hilbert space $\mathcal{H}$ can easily be calculated:

$$P_{\mathcal{H}}\frac{1}{\mathbf{A}^{\Gamma} - \lambda}P_{\mathcal{H}} = \frac{1}{\mathcal{A} - \lambda}$$
$$+ \frac{1}{\mathcal{A} - \lambda}\left[\tilde{\Gamma}_{00} - \tilde{Q}(\lambda) - \tilde{\Gamma}_{01}\frac{1}{\tilde{\Gamma}_{11} - \tilde{Q}_{1}(\lambda)}\tilde{\Gamma}_{10} \right]^{-1}\tilde{P}_{M}\frac{1}{\mathcal{A} - \lambda},$$
$$(5.28)$$

where $\tilde{P}_M = (\mathcal{A} + i)P_M(\mathcal{A} - i)$.

5.2 Two-body problems

5.2.1 Two-body operator with interaction of rank one

Let us consider the case where the Hilbert space $\mathcal{H}$ possesses the following tensor decomposition

$$\mathcal{H} = K \otimes h, \tag{5.29}$$

where K and h are two Hilbert spaces. The sign $\otimes$ denotes the complete tensor product. Let B and a be two self–adjoint operators acting in the Hilbert spaces K and h respectively. Then the operator

$$\mathcal{A} = B \otimes I_h + I_K \otimes a \tag{5.30}$$

is well defined on the algebraic tensor product $\mathcal{L}$ of the domains $\mathrm{Dom}(B)$ and $\mathrm{Dom}(a)$. The standard way to define the corresponding self–adjoint operator A acting in $\mathcal{H} = K \otimes h$ has been described by Yu.Berezansky [132, 130, 129, 697, 696, 699]. One has to introduce the standard graph norm of the operator $\mathcal{A}$ on $\mathcal{L}$ given by

$$\| \psi \|^2_{\mathcal{H}_2(A)} = \| \mathcal{A}\psi \|^2_{\mathcal{H}} + \| \psi \|^2_{\mathcal{H}}. \tag{5.31}$$

Then the domain of the self–adjoint operator $\mathcal{A}$ is equal to the closure of $\mathcal{L}$ in the latter norm, i.e. the self–adjoint operator $\mathcal{A}$ is essentially self–adjoint on the algebraic tensor product of the domains of the operators B and a. The completion of the algebraic tensor product of $\mathrm{Dom}(B)$ and $\mathrm{Dom}(a)$ will be denoted as follows

$$\mathrm{Dom}(\mathcal{A}) = \mathrm{Dom}(B)\bar{\otimes}\mathrm{Dom}(a). \tag{5.32}$$

We are not going to distinguish between the operator $\mathcal{A}$ given by (5.30) on the algebraic tensor product of the domains $\mathrm{Dom}(B)$ and $\mathrm{Dom}(a)$ and its closure defined on $\mathrm{Dom}(\mathcal{A})$ given by (5.32).

Another norm on $\mathcal{L}$ can be introduced using the graph norms of the operators $B \otimes I_h$ and $I_K \otimes a$ as follows

$$\| \psi \|^{t2}_{\mathcal{H}_2(A)} = \left(\| (B \otimes I_h)\psi \|^2_{\mathcal{H}} + \| \psi \|^2_{\mathcal{H}} \right) + \left(\| (I_K \otimes a)\psi \|^2_{\mathcal{H}} + \| \psi \|^2_{\mathcal{H}} \right).$$

The latter norm is associated with the tensor decomposition.

In what follows we are going to use:

Assumption 5.2.1 *The operator $\mathcal{A}$ acting in the Hilbert space $\mathcal{H} = K \otimes h$ possesses the decomposition*

$$\mathcal{A} = B \otimes I_h + I_K \otimes a$$

and at least one of the following conditions is satisfied:

1. *the operator $B \otimes a$ is positive;*

2. *the operator a is finite dimensional.*

Condition 1 in the latter assumption implies that the operator $\mathcal{A}$ is semi-bounded. Condition 2 means that the Hilbert space h is in fact finite dimensional. The following lemma proves that the two norms we introduced are equivalent if at least one of the latter conditions is satisfied.

Lemma 5.2.1 *Let Assumption 5.2.1 be satisfied. Then the norms $\| \cdot \|_{\mathcal{H}_2(A)}$ and $\| \cdot \|_{\mathcal{H}_2(A)}^t$ are equivalent on the algebraic tensor product $\mathcal{L}$ of the domains* Dom (B) *and* Dom (a).

Proof Let $\psi \in \mathcal{L}$. Then the following inequalities hold

$$\| \psi \|_{\mathcal{H}_2(A)}^2$$

$$= \| (B \otimes I_h + I_K \otimes a)\psi \|_{\mathcal{H}}^2 + \| \psi \|_{\mathcal{H}}^2$$

$$= \langle \psi, (B^2 \otimes I_h)\psi \rangle + \langle \psi, (I_K \otimes a^2)\psi \rangle + 2\langle \psi, (B \otimes a)\psi \rangle + \| \psi \|_{\mathcal{H}}^2$$

$$\leq 2 \| (B \otimes I_h)\psi \|_{\mathcal{H}}^2 + 2 \| (I_K \otimes a)\psi \|_{\mathcal{H}}^2 + \| \psi \|_{\mathcal{H}}^2$$

$$\leq 2 \| \psi \|_{\mathcal{H}_2(A)}^{t2}$$

Suppose that the operator $B \otimes a$ is positive. Then the following inequalities hold

$$\| \psi \|_{\mathcal{H}_2(A)}^2$$

$$= \| (B \otimes I_h + I_K \otimes a)\psi \|_{\mathcal{H}}^2 + \| \psi \|_{\mathcal{H}}^2$$

$$= \langle \psi, (B^2 \otimes I_h)\psi \rangle + \langle \psi, (I_K \otimes a^2)\psi \rangle + 2\langle \psi, (B \otimes a)\psi \rangle + \| \psi \|_{\mathcal{H}}^2$$

$$\geq \langle \psi, (B^2 \otimes I_h)\psi \rangle + \langle \psi, (I_K \otimes a^2\psi) \rangle + \| \psi \|_{\mathcal{H}}^2$$

$$\geq \frac{1}{2} \| \psi \|_{\mathcal{H}_2(A)}^{t2}$$

Therefore the two norms are equivalent in the case where the operator $B \otimes a$ is positive.

Suppose now that the operator a is finite dimensional. Without loss of generality we can suppose in addition that the space h is finite dimensional, i.e. $h = \mathbf{C}^n, n \in \mathbf{N}$, and the operator a is diagonal, i.e. it is given by a diagonal self–adjoint matrix with eigenvalues $\lambda_l, l = 1, 2, \ldots . n$. Since the operator a is bounded, the norm $\| \psi \|^t_{\mathcal{H}_2(A)}$ is equivalent to the graph norm of the operator B

$$ \| B\psi \|^2_{\mathcal{H}} + \| \psi \|^2_{\mathcal{H}} . $$

Then the following inequalities prove the lemma for $\psi \in \mathcal{L}$

$$ \| \psi \|^2_{\mathcal{H}_2(A)} = \sum_{l=1}^{n} \left\{ \| (B + \lambda_l)\psi^l \|^2_K + \| \psi^l \|^2_K \right\} $$

$$ = \sum_{l=1}^{n} \left\{ \| B\psi^l \|^2_K + 2\langle \psi^l, \lambda_l B\psi^l \rangle_K + (\lambda_l^2 + 1) \| \psi^l \|^2_K \right\} $$

$$ \geq \sum_{l=1}^{n} \left\{ \frac{1}{2\lambda_l^2 + 1} \| B\psi^l \|^2_K + \frac{1}{2} \| \psi^l \|^2_K \right\} $$

$$ \geq \min \left\{ \frac{1}{2\lambda_l^2 + 1}, \frac{1}{2} \right\} \left(\| B\psi \|^2_{\mathcal{H}} + \| \psi \|^2_{\mathcal{H}} \right), $$

where ψ^l denotes the component of the function ψ in the space of vector functions $K \otimes h = K \otimes \mathbf{C}^n$. This finishes the proof of the lemma for both cases.

□

In the course of the proof of the previous lemma we have shown that the following estimate holds:

$$ \| \psi \|^{t2}_{\mathcal{H}_2(A)} \leq 2 \| \psi \|^2_{\mathcal{H}_2(A)} \tag{5.33} $$

for every $\psi \in \mathcal{L}$ providing that Assumption 5.2.1 is satisfied.

The previous lemma implies that the operator A is self–adjoint on the completion of $\mathcal{L}$ in the second tensor operator $\| \cdot \|^t$ norm if the operators B and a satisfy Assumption 5.2.1. One can prove Lemma 5.2.1 even in the case where the operator a is equal to the orthogonal sum of two operators, each of them satisfying assumption 5.2.1. Therefore we are also going to use the following:

Assumption 5.2.2 *The operator $\mathcal{A}$ acting in the Hilbert space $\mathcal{H} = K \otimes H$ is given by*

$$\mathcal{A} = B \otimes I_H + I_K \otimes A$$

where B and A are self-adjoint operators in K and H respectively. The operator A is equal to the orthogonal sum of two self-adjoint operators $A = a \oplus a'$, where a' is a finite dimensional operator and the operator $B \otimes a$ is positive.

The following lemma is analogous to the embedding theorems for Sobolev spaces [912, 913]

Lemma 5.2.2 *Let Assumption 5.2.1 be satisfied. Let ψ be an element from the domain of the operator $\mathcal{A} = B \otimes I_h + I_K \otimes a$ in $K \otimes h$: $\psi \in \mathrm{Dom}\,(\mathcal{A})$. Let f be a certain element from the Hilbert space $\mathcal{H}$. Then the following inclusions hold*

$$\langle f, \sqrt{|a| + 1}\psi\rangle_h \in \mathcal{H}_1(B); \tag{5.34}$$

$$\langle f, (a + i)\psi\rangle_h \in K = \mathcal{H}_0(B). \tag{5.35}$$

Proof To prove that the vector $\langle f, \sqrt{|a| + 1}\psi\rangle_h$ is an element from $\mathcal{H}_1(B)$ it is enough to show that the following inequality holds for some constant C

$$\langle \psi, (|a| + 1) \otimes (|B| + 1)\psi\rangle < C\langle \psi, (\mathcal{A}^2 + 1)\psi\rangle = C \parallel \psi \parallel^2_{\mathcal{H}_2(\mathcal{A})} \tag{5.36}$$

and for every $\psi \in \mathrm{Dom}\,(\mathcal{A})$. The following chain of inequalities proves (5.36):

$$\langle \psi, (|a| + 1) \otimes (|B| + 1)\psi\rangle$$

$$= \langle \psi, (|a| \otimes |B|)\psi\rangle + \langle \psi, |a|\psi\rangle + \langle \psi, |B|\psi\rangle + \langle \psi, \psi\rangle$$

$$\leq \frac{\langle \psi, a^2\psi\rangle + \langle \psi, b^2\psi\rangle}{2} + \frac{\langle \psi, (a^2 + 1)\psi\rangle + \langle \psi, (b^2 + 1)\psi\rangle}{2} + \langle \psi, \psi\rangle$$

$$= \parallel \psi \parallel^{t2}_{\mathcal{H}_2(\mathcal{A})}.$$

In the course of the proof of Lemma 5.2.1 we have shown that the norm $\parallel \cdot \parallel^t_{\mathcal{H}_2(\mathcal{A})}$ can be estimated from above by the norm $\parallel \cdot \parallel_{\mathcal{H}_2(\mathcal{A})}$ and the desired inequality is proven.

The second inclusion follows from the following inequality

$$\parallel (a + i)\psi \parallel^2 = \langle \psi, (a^2 + 1)\psi\rangle \leq 2 \parallel \psi \parallel_{\mathcal{H}_2(\mathcal{A})},$$

which is also a corollary of Lemma 5.2.1 (see (5.33)). The lemma is proven.

□

We would like to construct a self–adjoint perturbation of the operator $\mathcal{A}$ possessing the tensor decomposition (5.30). One can obtain such an operator by first restricting and then extending the self–adjoint operator a in h. Let us first consider the operator

$$\mathcal{A}_\alpha = \mathcal{A} + \alpha\langle \varphi, \cdot\rangle_h \varphi, \tag{5.37}$$

where $\varphi \in \mathcal{H}_{-2}(a), \alpha \in \mathbf{R}$. The latter operator can be presented formally as follows

$$\mathcal{A}_\alpha = B \otimes I_h + I_k \otimes \left(a + \alpha\langle \varphi, \cdot\rangle_h \varphi\right).$$

Therefore to define the operator $\mathcal{A}_\alpha$ it is enough to determine the operator

$$a_\alpha = a + \alpha\langle \varphi, \cdot\rangle_h \varphi \tag{5.38}$$

acting in the Hilbert space h. This operator has already been studied in Chapter 1. If $\varphi \in h$, then the perturbation $\alpha\langle \varphi, \cdot\rangle_h \varphi$ is a bounded operator and the perturbed operator a_α is self–adjoint on the domain Dom (a). Let us study in detail the case $\varphi \notin h$. It has been proven that the operator corresponding to the formal expression (5.38) is a certain self–adjoint extension of the operator a^0 being the restriction of the operator a to the domain

$$\text{Dom } (a^0) = \{\psi \in \text{Dom } (a) : \langle \varphi, \psi\rangle_h = 0\}.$$

The operator a^0 is densely defined, since $\varphi \notin h$. Then every element ψ from the domain Dom (a^{0*}) of the adjoint operator a^{0*} possesses the following decomposition

$$\psi = \tilde{\psi} + \xi_+(\psi)\frac{a}{a^2+1}\varphi,$$

where $\tilde{\psi} \in \text{Dom}(a), \xi_+(\psi) \in \mathbf{C}$. The adjoint operator a^{0*} acts as follows

$$a^{0*}\psi = a^{0*}\left(\tilde{\psi} + \xi_+(\psi)\frac{a}{a^2+1}\varphi\right) = a\tilde{\psi} - \xi_+(\psi)\frac{1}{a^2+1}\varphi.$$

We define the self–adjoint operator $a^\gamma, \gamma \in \mathbf{R}$, as the restriction of the operator a^{0*} to the domain

$$\text{Dom } (a^\gamma) = \{\psi \in \text{Dom } (a^{0*}) : \langle \varphi, \tilde{\psi}\rangle_h = \gamma\xi_+(\psi)\}.$$

The real parameter γ is related to the interaction parameter α as follows (see 1.54)

$$\gamma = -\frac{1}{\alpha} - c. \tag{5.39}$$

The constant c in the latter formula is uniquely defined when $\varphi \in \mathcal{H}_{-1}(a)$:

$$c = \left\langle \varphi, \frac{a}{a^2 + 1}\varphi \right\rangle_h.$$

If $\varphi \in \mathcal{H}_{-2}(a) \setminus \mathcal{H}_{-1}(a)$ then the real parameter c determines the extension of the functional φ to the element $\frac{a}{a^2+1}\varphi$ (see Section 1.3). The resolvent of the operator a^γ is given by Krein's formula

$$\frac{1}{a^\gamma - \lambda} = \frac{1}{a - \lambda} + \frac{1}{\gamma - q(\lambda)}\left\langle \frac{1}{a - \bar{\lambda}}\varphi, \cdot \right\rangle_h \frac{1}{a - \lambda}\varphi, \qquad (5.40)$$

where $q(\lambda) = \langle \varphi, \frac{1+\lambda a}{a-\lambda}\frac{1}{a^2+1}\varphi \rangle_h$.

The operator $\mathcal{A}^\gamma = B \otimes I_h + I_K \otimes a^\gamma$ is self-adjoint on the domain Dom $(B) \bar\otimes$ Dom (a_γ). Let the operator a satisfy Assumption 5.2.1. Then the operator a^γ satisfies Assumption 5.2.2, since it is a rank one perturbation of a. The resolvent of the operator A_γ can be calculated using the spectral representation for the operator B.

Let us prove first some facts concerning the Nevanlinna functions from the **Stieltjes class**, i.e. possessing the representation

$$F(z) = \int_A^\infty \frac{1 + z\lambda}{\lambda - z}d\rho(\lambda), \qquad (5.41)$$

where the real measure $d\rho(\lambda)$ is finite $\int_A^\infty d\rho(\lambda) < \infty$.

Lemma 5.2.3 *Let $F(z)$ be a Stieltjes function. Then for any real y and any positive $\epsilon > 0$ there exists a certain $b = b(y, \epsilon) > 0$ such that the following estimate holds*

$$|F(x + iy)| < \epsilon|x| + b \qquad (5.42)$$

for all $x < A$.

Proof Consider the real and imaginary parts of the function

$$F(x + iy) = \int_A^\infty \frac{1 + (x + iy)\lambda}{\lambda - x - iy}d\rho(\lambda).$$

The imaginary part

$$\Im F(x + iy) = y\int_A^\infty \frac{\lambda^2 + 1}{(\lambda - x)^2 + y^2}d\rho(\lambda)$$

is uniformly bounded for all $x < A$. The real part is given by the sum of two integrals

$$\Re F(x + iy) = \int_A^\infty \frac{\lambda(1 - y^2) - x}{(\lambda - x)^2 + y^2}d\rho(\lambda) + \int_A^\infty \frac{x\lambda(\lambda - x)}{(\lambda - x)^2 + y^2}d\rho(\lambda).$$

The first integral is uniformly bounded. The second integral can be estimated as follows

$$\left| \int_A^\infty \frac{x\lambda(\lambda - x)}{(\lambda - x)^2 + y^2} d\rho(\lambda) \right| \leq \int_A^\infty \frac{|x||\lambda|}{\lambda - x} d\rho(\lambda)$$

for all $x < A - 1$. To estimate the latter integral we choose $C > A$ such that $\int_C^\infty d\rho(\lambda) < \epsilon/2$ and we get

$$\int_A^\infty \frac{|x||\lambda|}{\lambda - x} d\rho(\lambda) \leq |x| \frac{C}{C - x} \int_A^\infty d\rho(\lambda) + |x| \int_C^\infty d\rho(\lambda)$$

$$\leq \left(\frac{C}{C - x} \int_A^\infty d\rho(\lambda) + \epsilon/2 \right) |x|.$$

For all $x \leq x_0 = C(1 - \frac{1}{\epsilon} \int_A^\infty d\rho(\lambda))$ the latter expression is estimated by $\epsilon|x|$. The function $F(x + iy)$ is continuous on the interval $x_0 \leq x \leq A$ and is therefore uniformly bounded on this interval (If $x_0 > A$ then this interval is empty). The lemma is proven.

□

The following lemma describes rational transformations of Stieltjes functions.

Lemma 5.2.4 *Let $F(z)$ be a Stieltjes function. Let a, b, c, d be real numbers such that*

$$ad - bc = 1. \tag{5.43}$$

Then there exists a real number A_1, such that for any real y and any positive $\epsilon > 0$ there exists $b_1 = b_1(y, \epsilon)$ such that the function

$$G(z) = \frac{aF(z) + b}{cF(z) + d} \tag{5.44}$$

possesses the representation

$$G(z) = \beta z + g(z), \quad \beta > 0, \tag{5.45}$$

where

$$|g(x + iy)| < \epsilon|x| + b_1 \tag{5.46}$$

for all $x < A_1$.

Proof Condition (5.43) guarantees that the function $G(z)$ is a Nevanlinna function and possesses the representation

$$G(z) = \alpha + \beta z + \int_{-\infty}^\infty \frac{1 + \lambda z}{\lambda - z} d\rho_1(\lambda),$$

where $\int_{-\infty}^{\infty} d\rho_1(\lambda) < \infty$, $\alpha, \beta \in \mathbf{R}$, $\beta > 0$. The support of the measure $d\rho_1(\lambda)$ coincides with the set of real points z where the boundary values $G(z + i0)$ are not real or do not exist. The function $G(z)$ is real on the interval $(-\infty, A)$ outside the points where $cF(z) + d = 0$. The derivative

$$\frac{dF(z)}{dz} = \int_A^{\infty} \frac{\lambda^2 + 1}{(\lambda - z)^2} d\rho_1(\lambda)$$

is positive for all $z < A$. It follows that there exists at most one point where $F(z) = -\frac{d}{c}$. Therefore the support of the measure $d\rho_1(\lambda)$ is bounded from below. Now Lemma 5.2.3 implies estimate (5.46). The lemma is proven.

$\square$

Note that the constant β appearing in (5.46) is different from zero only if the function $F(z)$ has a finite limit at infinity and $F(\infty) = -\frac{d}{c}$.

To calculate the resolvent of the operator $\mathcal{A}^\gamma$ we will need the following corollary of the two previous Lemmas.

But first we need the following technical lemma.

Lemma 5.2.5 *Let the self-adjoint operator a in h be positive. Let y be a certain positive real number. Consider the Nevanlinna function*

$$G(\lambda) = \frac{1}{\gamma - q(\lambda)} = \frac{1}{\gamma - \langle \varphi, \frac{1+\lambda a}{a-\lambda} \frac{1}{a^2+1} \varphi \rangle_h}.$$

If $\varphi \in \mathcal{H}_{-1}(a)$, then the function satisfies the estimate

$$|G(x + iy)| \leq C_1(y)(1 + |x|) \tag{5.47}$$

for all negative $x < 0$ and a certain $C_1(y) > 0$. If $\varphi \in \mathcal{H}_{-2}(a) \setminus \mathcal{H}_{-1}(a)$, then the function can be estimated as

$$|G(x + iy)| \leq C_2(y) \tag{5.48}$$

for all negative $x < 0$ and a certain positive $C_2(y) > 0$.

Proof The function function $q(\lambda) = \langle \varphi, \frac{1+a\lambda}{a-\lambda} \frac{1}{a^2+1} \varphi \rangle_h$ is a Stieltjes function, since the operator a is bounded from below. Lemma 5.2.3 implies that the estimate (5.47) holds for all $\varphi \in \mathcal{H}_{-2}(a)$.

Consider now the case $\varphi \in \mathcal{H}_{-2}(a) \setminus \mathcal{H}_{-1}(a)$. We are going to prove that the real part of $q(x + iy)$ tends to minus infinity when $x \to -\infty$. In fact the function $q(\lambda)$ can be presented by the following integral

$$q(\lambda) = \int_0^{\infty} \frac{1 + \mu\lambda}{\mu - \lambda} d\rho(\lambda),$$

where the masure $d\rho(\mu)$ is finite $\int_0^\infty d\rho(\mu) < \infty$, but the integral $\int_0^\infty \mu d\rho(\mu) = \infty$ diverges.

The real part of $q(x + iy)$ is given by

$$\Re q(x + iy) = \int_0^\infty \frac{(\mu - x) - \mu y^2}{(\mu - x)^2 + y^2} d\rho(\mu) + \int_0^\infty \frac{x\mu(\mu - x)}{(\mu - x)^2 + y^2} d\rho(\mu). \quad (5.49)$$

The first integral in the latter formula is bounded for negative values of x :

$$\left| \int_0^\infty \frac{(\mu - x) - \mu y^2}{(\mu - x)^2 + y^2} d\rho(\mu) \right|$$

$$\leq \int_0^\infty \frac{\mu - x}{(\mu - x)^2 + y^2} d\rho(\mu) + \int_0^\infty \frac{\mu y^2}{\mu^2 + y^2} d\rho(\mu)$$

$$\leq \int_0^\infty \frac{1}{2y} d\rho(\mu) + \int_0^\infty \frac{y}{2} d\rho(\mu).$$

The second integral in (5.49) is negative and can be estimated as

$$\left| \int_0^\infty \frac{x\mu(\mu - x)}{(\mu - x)^2 + y^2} d\rho(\mu) \right|$$

$$\geq \frac{x^2}{x^2 + y^2} \int_0^\infty \frac{|x|\mu}{\mu - x} d\rho(\mu)$$

$$\geq \frac{x^2}{x^2 + y^2} \frac{1}{2} \int_0^{|x|} \mu d\rho(\mu).$$

The latter integral tends to infinity when $x \to -\infty$. It follows that

$$\lim_{x \to -\infty} G(x + iy) = \lim_{x \to -\infty} \frac{1}{\gamma - q(x + iy)} = 0.$$

Therefore the continuous function $G(x + iy)$ is uniformly bounded on the interval $x < 0$, i.e. the estimate (5.48) holds. The lemma is proven.

□

Note that if $y = 0$ then the estimates (5.47) and (5.48) hold for $x \in (-\infty, A_1)$, where A_1 is a certain real constant.

Theorem 5.2.1 *Let the operator* $\mathcal{A} = B \otimes I_h + I_K \otimes a$ *satisfy Assumption 5.2.1. Then the resolvent of the operator* $\mathcal{A}^\gamma = B \otimes I_h + I_K \otimes a^\gamma$ *at a certain point* $\lambda, \Im\lambda \neq 0$, *is given by the formula*

$$\frac{1}{\mathcal{A}^\gamma - \lambda} = \frac{1}{\mathcal{A} - \lambda} + \frac{1}{\mathcal{A} - \lambda}\left(\frac{1}{\gamma - q(\lambda - B)}\langle\frac{1}{\mathcal{A} - \bar\lambda}\varphi, \cdot\rangle_h \otimes \varphi\right) \quad (5.50)$$

where $q(\lambda - B) = \langle\varphi, \dfrac{1 + (\lambda - B) \otimes a}{\mathcal{A} - \lambda}\dfrac{1}{a^2 + 1}\varphi\rangle_h.$

Comment Let us discuss formula (5.50) first. If the operator a is finite dimensional then this formula obviously defines an operator in the Hilbert space $\mathcal{H}$. Consider the case where the operator $B \otimes a$ is positive. It is enough to study the case where both operators B and a are positive. The case of negative operators is similar.

Let $\varphi \in \mathcal{H}_{-1}(a)$. Then Lemma 5.2.2 implies that for any $f \in \mathcal{H}$ the following inclusion holds

$$\langle\frac{1}{\mathcal{A} - \bar\lambda}\varphi, f\rangle_h = \langle\frac{1}{\sqrt{|a| + 1}}\varphi, \frac{\sqrt{|a| + 1}}{\mathcal{A} - \bar\lambda}f\rangle_h \in \mathcal{H}_1(B).$$

The function $\frac{1}{\gamma - q(x+iy)}$ satisfies the estimate (5.47) and it follows that the operator

$$\frac{1}{\gamma - q(\lambda - B)}$$

maps $\mathcal{H}_1(B)$ onto $\mathcal{H}_{-1}(B)$. Taking into account Lemma 5.2.2 this implies that

$$\frac{1}{\gamma - q(\lambda - B)}\left\langle\frac{1}{\mathcal{A} - \bar\lambda}\varphi, f\right\rangle_h \otimes \varphi \in \mathcal{H}_{-2}(\mathcal{A}). \quad (5.51)$$

This means that formula (5.50) defines an operator in the Hilbert space $\mathcal{H}$ for $\varphi \in \mathcal{H}_{-1}(a)$.

Consider now the case $\varphi \in \mathcal{H}_{-2}(a) \setminus \mathcal{H}_{-1}(a)$. Lemma 5.2.1 implies that for any $f \in \mathcal{H}$ the vector $\langle\frac{1}{\mathcal{A} - \bar\lambda}\varphi, f\rangle_h = \langle\frac{1}{a+i}\varphi, (a+i)\frac{1}{\mathcal{A} - \bar\lambda}f\rangle_h$ belongs to the space K. The function $\frac{1}{\gamma - q(x+iy)}$ is bounded for negative x (see (5.48)) and the operator $\frac{1}{\gamma - q(\lambda - B)}$ is bounded in K. This implies that condition (5.51) holds. Therefore formula (5.50) defines an operator acting in the Hilbert space for any $\varphi \in \mathcal{H}_{-2}(a)$.

Proof of Theorem 5.2.1 Let us denote by $\mathcal{F}_B$ the operator of spectral transformation for B – the linear operator which maps the operator B into the operator of multiplication by the independent variable x. In the above formula $k(x)$ is a certain Hilbert space. Then the resolvent on a dense set can easily be calculated as follows

$$\frac{1}{\mathcal{A}^\gamma - \lambda} f = \psi$$

$$\Rightarrow f \ = \ (\mathcal{A}^\gamma - \lambda)\psi$$

$$\Rightarrow (\mathcal{F}_B f)(x) \ = \ (x + a^\gamma - \lambda)(\mathcal{F}_B \psi)(x)$$

$$\Rightarrow (\mathcal{F}_B \psi)(x) \ = \ \frac{1}{a^\gamma - (\lambda - x)}(\mathcal{F}_B f)(x)$$

$$= \ \frac{1}{a - (\lambda - x)}(\mathcal{F}_B f)(x)$$
$$+ \left(\frac{1}{\gamma - q(\lambda - x)} \left\langle \frac{1}{a - (\bar{\lambda} - x)} \varphi, (\mathcal{F}_B f)(x) \right\rangle_h \right)$$
$$\otimes \frac{1}{a - (\lambda - x)} \varphi$$

$$\Rightarrow \psi \ = \ \frac{1}{A - \lambda} f + \frac{1}{A - \lambda} \left(\left\{ \frac{1}{\gamma - q(\lambda - B)} \langle \frac{1}{A - \bar{\lambda}} \varphi, f \rangle_h \right\} \otimes \varphi \right).$$

We have supposed that $\psi \in \mathcal{L}^\gamma$, where $\mathcal{L}^\gamma$ is an algebraic tensor product of $\text{Dom}\,(B)$ and $\text{Dom}\,(a^\gamma)$. The operator $\mathcal{A}^\gamma$ is essentially self–adjoint on this domain and this completes the proof of the theorem.

$\square$

The resolvent of the operator $\mathcal{A}^\gamma$ has been calculated using the tensor decomposition. The operator $\mathcal{A}^\gamma$ is a self–adjoint extension of the symmetric operator $\mathcal{A}^0 = B \otimes I_h + I_K \otimes a^0$ with infinite deficiency indices. Consider the annulating set Φ_{reg} of regular functionals for the operator $\mathcal{A}^0$ defined as follows:

• if $\varphi \in \mathcal{H}_{-1}(a)$ then $\Phi_{\text{reg}} = \{\Phi : \Phi = \rho(\Phi) \otimes \varphi, \ \rho(\Phi) \in \mathcal{H}_{-1}(B)\}$,

• if $\varphi \in \mathcal{H}_{-2}(a) \setminus \mathcal{H}_{-1}(a)$ then $\Phi_{\text{reg}} = \{\Phi : \Phi = \rho(\Phi) \otimes \varphi, \ \rho(\Phi) \in K = \mathcal{H}_0(B)\}$.

Lemma 5.2.2 implies $\Phi_{\text{reg}} \subset \mathcal{H}_{-2}(A)$. The annulating space of functionals for the operator $\mathcal{A}^0$ is equal to the closure of the set of regular functionals in the norm of the space $\mathcal{H}_{-2}(A)$.

Let us consider the corresponding subspace of regular elements from the domain of the adjoint operator $\mathcal{A}^{0*}$:

- if $\varphi \in \mathcal{H}_{-1}(a)$ then $\text{Dom}_{\text{reg}}(\mathcal{A}^{0*}) = \{\psi : \psi = \tilde{\psi} + \frac{\mathcal{A}}{\mathcal{A}^2+1}\rho(\psi) \otimes \varphi, \tilde{\psi} \in \text{Dom}(\mathcal{A}), \rho(\psi) \in \mathcal{H}_{-1}(B)\}$;

- if $\varphi \in \mathcal{H}_{-2}(a) \backslash \mathcal{H}_{-1}(a)$ then $\text{Dom}_{\text{reg}}(\mathcal{A}^{0*}) = \{\psi : \psi = \tilde{\psi} + \frac{\mathcal{A}}{\mathcal{A}^2+1}\rho(\psi) \otimes \varphi, \tilde{\psi} \in \text{Dom}(\mathcal{A}), \rho(\psi) \in K = \mathcal{H}_0(B)\}$.

The boundary form of the adjoint operator calculated on the regular elements is given by

$$U, V \in \text{Dom}_{\text{reg}}(\mathcal{A}^{0*}) \Rightarrow$$

$$
\begin{aligned}
\langle U, \mathcal{A}^{0*}V \rangle - \langle \mathcal{A}^{0*}U, V \rangle &= \langle \rho(U) \otimes \varphi, \tilde{V} \rangle - \langle \tilde{U}, \rho(V) \otimes \varphi \rangle \\
&= \left\langle \rho(U), \langle \varphi, \tilde{V} \rangle_h \right\rangle_K - \left\langle \langle \varphi, \tilde{U} \rangle_h, \rho(V) \right\rangle_K.
\end{aligned}
$$
(5.52)

A symmetric extension of the operator $\mathcal{A}^0$ can be defined in terms of any symmetric operator Γ by restricting the operator $\mathcal{A}^{0*}$ to the domain of functions from $\text{Dom}_{\text{reg}}(\mathcal{A}^{0*})$ satisfying the boundary condition

$$\langle \varphi, \hat{U} \rangle = \Gamma \rho(U).$$
(5.53)

In order to obtain the perturbed operator possessing the tensor decomposition (5.30) let us consider the symmetric operator $\mathcal{A}^\Gamma$ determined by the following boundary operator

$$\Gamma = \gamma + B \left\langle \varphi, \frac{a\mathcal{A} - 1}{(\mathcal{A}^2 + 1)(a^2 + 1)} \varphi \right\rangle_h.$$
(5.54)

The operator $\langle \varphi, \frac{a\mathcal{A}-1}{(\mathcal{A}^2+1)(a^2+1)}\varphi \rangle_h$ is a bounded self–adjoint operator in K commuting with the operator B. The norm of this operator is less then or equal to 1. Therefore the operator Γ is essentially self–adjoint on the domain $\text{Dom}(B)$ of the operator B. We are going to keep the same notation Γ for the corresponding self–adjoint operator.

Let us calculate the resolvent of the operator $\mathcal{A}^\Gamma$. Consider an arbitrary $f \in \mathcal{H}$ and suppose that $\psi = \tilde{\psi} + \frac{\mathcal{A}}{\mathcal{A}^2+1}\rho(\psi) \otimes \varphi = \frac{1}{\mathcal{A}^\Gamma - \lambda}f$. Then the function ψ satisfies the following equation

$$(\mathcal{A} - \lambda)\tilde{\psi} - \frac{1 + \lambda\mathcal{A}}{\mathcal{A}^2 + 1}(\rho(\psi) \otimes \varphi) = f$$

and the boundary conditions (5.53). Applying the resolvent of the original operator $\mathcal{A}$ to the previous equation we get

$$\tilde{\psi} - \frac{1 + \lambda\mathcal{A}}{\mathcal{A} - \lambda}\frac{1}{\mathcal{A}^2 + 1}\rho(\psi) \otimes \varphi = \frac{1}{\mathcal{A} - \lambda}f$$

$$\Rightarrow \left(\Gamma - \langle\varphi, \frac{1+\lambda\mathcal{A}}{\mathcal{A}-\lambda}\frac{1}{\mathcal{A}^2+1}\varphi\rangle_h\right)\rho(\psi) = \langle\varphi, \frac{1}{\mathcal{A}-\lambda}f\rangle_h. \qquad (5.55)$$

This equation can be solved and the function $\rho(\psi)$ can be calculated if the operator $\Gamma - \langle\varphi, \frac{1+\lambda\mathcal{A}}{\mathcal{A}-\lambda}\frac{1}{\mathcal{A}^2+1}\varphi\rangle_h$ is invertible. The operator can be simplified as follows taking into account equality (5.54)

$$\Gamma - \langle\varphi, \frac{1+\lambda\mathcal{A}}{\mathcal{A}-\lambda}\frac{1}{\mathcal{A}^2+1}\varphi\rangle_h$$

$$= \gamma + B\left\langle\varphi, \frac{a\mathcal{A}-1}{(\mathcal{A}^2+1)(a^2+1)}\varphi\right\rangle_h - \left\langle\varphi, \frac{1+\lambda\mathcal{A}}{\mathcal{A}-\lambda}\frac{1}{\mathcal{A}^2+1}\varphi\right\rangle_h \qquad (5.56)$$

$$= \gamma - \left\langle\varphi, \frac{1+(\lambda-B)\otimes a}{\mathcal{A}-\lambda}\frac{1}{a^2+1}\varphi\right\rangle_h.$$

Lemma 5.2.2 implies that $\langle\varphi, \frac{1}{\mathcal{A}-\lambda}f\rangle_h \in K$. The comment after Theorem 5.2.1 shows that the operator $\Gamma - \langle\varphi, \frac{1+\lambda\mathcal{A}}{\mathcal{A}-\lambda}\frac{1}{\mathcal{A}^2+1}\varphi\rangle_h$ is invertible in K. Therefore there exists $\rho(\psi) \in \text{Dom}\,(\Gamma) \subset K$ which satisfies the equation (5.55).

The component $\tilde{\psi}$ of the function ψ can be calculated using the formula

$$\tilde{\psi} = \frac{1}{\mathcal{A}-\lambda}f + \frac{1}{\mathcal{A}-\lambda}\frac{1+\lambda\mathcal{A}}{\mathcal{A}^2+1}\rho(\psi)\otimes\varphi.$$

Thus the function ψ is given by

$$\psi = \frac{1}{\mathcal{A}-\lambda}f + \frac{1}{\mathcal{A}-\lambda}\left(\left\{\frac{1}{\Gamma - \langle\varphi, \frac{1+\lambda\mathcal{A}}{\mathcal{A}-\lambda}\frac{1}{\mathcal{A}^2+1}\varphi\rangle_h}\langle\varphi, \frac{1}{\mathcal{A}-\lambda}f\rangle_h\right\}\otimes\varphi\right).$$

The resolvent of the operator $\mathcal{A}^\Gamma$ coincides with the resolvent of the self-adjoint operator $\mathcal{A}^\gamma$. This implies that the operator $\mathcal{A}^\Gamma$ is in fact self-adjoint even if it has been defined only on the regular elements.

Thus the following theorem has been proven.

Theorem 5.2.2 *The operator $\mathcal{A}^\Gamma$ which is the restriction of the operator $\mathcal{A}^{0*}$ to the set of regular elements*

$$\psi = \tilde{\psi} + \frac{\mathcal{A}}{\mathcal{A}^2+1}\left(\rho(\psi)\otimes\varphi\right) \in \text{Dom}_{\text{reg}}(\mathcal{A}^{0*}) \qquad (5.57)$$

satisfying the boundary condition

$$\langle\varphi, \tilde{\psi}\rangle_h = \Gamma\rho(\psi) \qquad (5.58)$$

is self-adjoint and its resolvent is given by

$$\frac{1}{\mathcal{A}^\Gamma - \lambda} = \frac{1}{\mathcal{A} - \lambda} + \frac{1}{\mathcal{A} - \lambda} \left(\left\{ \frac{1}{\gamma - \langle \varphi, \frac{1+(\lambda-B)\otimes a}{\mathcal{A}-\lambda} \frac{1}{a^2+1}\varphi\rangle_h} \langle \varphi, \frac{1}{\mathcal{A} - \lambda} \cdot \rangle_h \right\} \otimes \varphi \right)$$

(5.59)

for any $\lambda; \Im\lambda \neq 0$.

Comment In the course of the proof of the previous theorem we have shown that the density $\rho(\psi)$ is an element from the domain of the operator Γ. It is possible to prove that the restriction of the operator $\mathcal{A}^\Gamma$ to the domain of functions possessing the representation (5.57), boundary conditions (5.58) and having $\rho(\psi) \in \text{Dom}(B)$ is essentially self–adjoint.

Suppose that $\varphi \in \mathcal{H}_{-1}(a)$. Then the boundary conditions (5.58) can be simplified as follows. Consider the scalar product $\langle \varphi, \psi \rangle_h$, where ψ is any function from the domain of the operator $\mathcal{A}^\Gamma$. Then the following equalities hold

$$
\begin{aligned}
\langle \varphi, \psi \rangle_h &= \left\langle \varphi, \tilde{\psi} \right\rangle_h + \left\langle \varphi, \frac{A}{A^2+1}\rho \otimes \varphi \right\rangle_h \\
&= \Gamma\rho(\psi) + \left\langle \varphi, \frac{A}{A^2+1}\rho \otimes \varphi \right\rangle_h \\
&= \gamma\rho(\psi) + \left\langle \varphi, \left(\frac{B(Aa-1)}{(A^2+1)(a^2+1)} + \frac{A}{A^2+1} \right)\rho \otimes \varphi \right\rangle_h \quad (5.60) \\
&= \gamma\rho(\psi) + \left\langle \varphi, \frac{a}{a^2+1}\varphi \right\rangle_h \rho(\psi) \\
&= (\gamma + c)\,\rho(\psi),
\end{aligned}
$$

where we have used the fact that the function ψ satisfies boundary condition (5.58). Taking into account (5.39) the latter condition can be written as

$$-\alpha\langle \varphi, \psi \rangle_h = \rho(\psi).$$

One can define the operator $\mathcal{A}^\Gamma$ using this boundary condition, but this condition cannot be generalized to the case of $\mathcal{H}_{-2}$ interactions, since the scalar product

$$\left\langle \varphi, \frac{A}{A^2+1}\rho \otimes \varphi \right\rangle_h$$

does not necessarily define a function from K in this case. Therefore in what follows we are going to use the definition given by Theorem 5.2.2.

5.2.2 Two-body operator with generalized interaction of rank one

Let us consider the model operators with internal structure in the special case where the external and internal spaces can be presented by tensor products of a Hilbert space K and two other Hilbert spaces h and h':

$$\mathcal{H} = K \otimes h,$$

$$\mathcal{H}' = K \otimes h'.$$

We suppose that the operators $\mathcal{A}$ and $\mathcal{A}'$ acting in $\mathcal{H}$ and $\mathcal{H}'$ respectively possess similar representations

$$\mathcal{A} = B \otimes I_h + I_K \otimes a;$$

$$\mathcal{A}' = B \otimes I_{h'} + I_K \otimes a',$$

where B is a self–adjoint operator acting in K and a and a' are two self-adjoint operators in h and h' respectively. Then the total unperturbed operator can be written in the form

$$
\begin{aligned}
\mathbf{A} &= (B \otimes I_h + I_K \otimes a) \oplus (B \otimes I_{h'} + I_K \otimes a') \\[4pt]
&= B \otimes (I_h \oplus I_{h'}) + I_K \otimes (a \oplus a') \qquad (5.61) \\[4pt]
&= B \otimes I_{h \oplus h'} + I_K \otimes A,
\end{aligned}
$$

where $A = a \oplus a'$. It acts in the space $\mathbf{H} = \mathcal{H} \oplus \mathcal{H}' = K \otimes (h \oplus h')$.

The domain Dom $(\mathbf{A})$ of the operator $\mathbf{A}$ is equal to the closure of the algebraic tensor product $\mathbf{L}$ of Dom (B) and Dom $(A) = $ Dom $(a) \oplus$ Dom (a') with respect to the standard graph norm

$$\| \Psi \|^2_{\mathcal{H}_2(\mathbf{A})} = \| \mathbf{A}\Psi \|^2_{\mathbf{H}} + \| \Psi \|^2_{\mathbf{H}} = \| \psi \|^2_{\mathcal{H}_2(\mathcal{A})} + \| \psi_1 \|^2_{\mathcal{H}_2(\mathcal{A}')},$$

where $\Psi = (\psi, \psi')$ and

$$\| \psi \|^2_{\mathcal{H}_2(\mathcal{A})} = \| \mathcal{A}\psi \|^2_{\mathcal{H}} + \| \psi \|^2_{\mathcal{H}};$$

$$\| \psi' \|^2_{\mathcal{H}_2(\mathcal{A}')} = \| \mathcal{A}'\psi' \|^2_{\mathcal{H}'} + \| \psi' \|^2_{\mathcal{H}'}.$$

The norm on $\mathbf{L}$ associated with the tensor decomposition is defined as follows:

$$\| \Psi \|^{t2}_{\mathcal{H}_2(\mathbf{A})} = \| \Psi_0 \|^{t2}_{\mathcal{H}_2(A_0)} + \| \Psi_1 \|^{t2}_{\mathcal{H}_2(A_1)}.$$

We suppose that the operators $\mathcal{A}$ and $\mathcal{A}'$ satisfy Assumption 5.2.1 from the previous section. The operators $\mathcal{A}, \mathcal{A}'$ are self-adjoint on the domains $\mathrm{Dom}\,(B) \bar{\otimes} \mathrm{Dom}\,(a)$ and $\mathrm{Dom}\,(B) \bar{\otimes} \mathrm{Dom}\,(a')$ respectively. It follows that the operator $\mathbf{A} = \mathcal{A} \oplus \mathcal{A}'$ is self-adjoint on the domain

$$\mathrm{Dom}\,(\mathbf{A}) = \mathrm{Dom}\,(B) \bar{\otimes} \mathrm{Dom}\,(A)$$

We are going to construct the perturbed operator possessing a decomposition similar to (5.61). Therefore the restricted operator will be defined by restricting the self-adjoint operators a and a'. Let us consider two vectors φ and φ' which are normalized elements of the spaces $\mathcal{H}_{-2}(a)$ and $\mathcal{H}_{-2}(a')$ respectively. Consider the operators a^0 and a'^0 being the restrictions of the operators a and a' to the domains

$$\mathrm{Dom}\,(a^0) \;=\; \{\psi \in \mathrm{Dom}\,(a) : \langle \psi, \varphi \rangle_h = 0\};$$

$$\mathrm{Dom}\,(a'^0) \;=\; \{\psi' \in \mathrm{Dom}\,(a') : \langle \psi', \varphi \rangle_{h'} = 0\}.$$

The orthogonal sum of the operators a^0 and a'^0

$$A^0 = a^0 \oplus a'^0,$$

is a symmetric restriction of the operator A. Then the total restricted operator is defined as follows

$$\mathbf{A}^0 = B \otimes I_{h \oplus h'} + I_K \otimes A^0.$$

In what follows we are going to consider the case where the restricted operator A^0 is not densely defined. We suppose as in Section 2.2.3 that $\varphi \in \mathcal{H}_{-2}(a) \backslash h$ and $\varphi' \in h'$. The case where the operator A^0 is densely defined can be studied using the same method. In order to define a self-adjoint extension of the operator A^0 in $H = h \oplus h'$ we consider the two dimensional subspace m of H spanned by the vectors $\frac{1}{a+i}\varphi$ and $\frac{1}{a'+i}\varphi'$. Let γ be a Hermitian 2×2 matrix. Such a matrix determines a self-adjoint operator in m using the standard orthonormal basis $\frac{1}{a+i}\varphi, \frac{1}{a'+i}\varphi'$. The operator in m will also be denoted by γ. Then the self-adjoint operator A^γ is defined on the domain $\mathrm{Dom}(A^\gamma)$ of functions from $\mathrm{Dom}(a^{0*}) \oplus \mathrm{Dom}(a') \subset H$ possessing the representation

$$\Psi = (\psi, \psi'),$$

$$\psi \;=\; \hat{\psi} + \frac{a}{a^2+1}\varphi \xi_+(\Psi) + \frac{1}{a^2+1}\varphi \xi_-(\Psi),$$

$$\psi' \;=\; \hat{\psi}' + \frac{a'}{a'^2+1}\varphi' \xi'_+(\Psi) + \frac{1}{a'^2+1}\varphi' \xi'_-(\Psi),$$

$$(5.62)$$

where $\hat{\psi} \in \text{Dom}\,(a^0), \hat{\psi}' \in \text{Dom}\,(a'^0), \xi_{\pm}(\Psi) \in \mathbf{C}, \xi'_{\pm}(\Psi) \in \mathbf{C}$ and

$$
\begin{pmatrix} \xi_-(\Psi) \\ \xi'_-(\Psi) \end{pmatrix} = \begin{pmatrix} \gamma_{00} & \gamma_{01} \\ \gamma_{10} & \gamma_{11} \end{pmatrix} \begin{pmatrix} \xi_+(\Psi) \\ \xi'_+(\Psi) \end{pmatrix}.
$$

The operator A^γ acts as follows

$$
\begin{aligned}
A^\gamma \psi &= A^\gamma \left(\hat{\psi} + \hat{\psi}' + \left(\frac{1}{a^2+1}(a+\gamma_{00})\varphi + \frac{\gamma_{10}}{a'^2+1}\varphi' \right) \xi_+(\Psi) \right. \\
&\quad \left. + \left(\frac{1}{a'^2+1}(a'+\gamma_{11})\varphi' + \frac{\gamma_{01}}{a^2+1}\varphi \right) \xi'_+(\Psi) \right) \\[2mm]
&= a\hat{\psi} + a'\hat{\psi}' + \left(\frac{1}{a^2+1}(-1+\gamma_{00}a)\varphi + \frac{\gamma_{10}a'}{a'^2+1}\varphi' \right) \xi(\Psi) \\
&\quad + \left(\frac{1}{a'^2+1}(-1+\gamma_{11}a')\varphi' + \frac{\gamma_{01}a}{a^2+1}\varphi \right) \xi'(\Psi).
\end{aligned}
\tag{5.63}
$$

Lemma 2.2.1 implies that the representation (5.62) is unique if $\gamma_{01} \neq 0$. We are going to use this assumption in what follows. Using Theorem 2.2.4 we conclude that the operator A^γ is self–adjoint in H. The resolvent of the operator A^γ is given by

$$
\frac{1}{A^\gamma - \lambda} = \frac{1}{A - \lambda} + \frac{A+i}{A-\lambda} \frac{1}{\gamma - q(\lambda)} P_m \frac{A-i}{A-\lambda},
\tag{5.64}
$$

where $q(\lambda) = \begin{pmatrix} q_0(\lambda) & 0 \\ 0 & q_1(\lambda) \end{pmatrix}$;

$$
q_0(\lambda) = \langle \varphi, \frac{1+\lambda a}{a-\lambda} \frac{1}{a^2+1}\varphi \rangle_h,
$$

$$
q_1(\lambda) = \langle \varphi', \frac{1+\lambda a'}{a'-\lambda} \frac{1}{a'^2+1}\varphi' \rangle_{h'}.
$$

The latter formula can be written using the functionals φ, φ' for $\Psi =$

(ψ, ψ') as follows:

$$
\frac{1}{A^\gamma - \lambda}\Psi = \frac{1}{a - \lambda}\psi + \frac{1}{a' - \lambda}\psi'
$$

$$
+\frac{1}{a - \lambda}\varphi \frac{1}{\gamma_{00} - q_0(\lambda) - \frac{|\gamma_{01}|^2}{\gamma_{11} - q_1(\lambda)}}
$$

$$
\times \left(\left\langle \varphi, \frac{1}{a - \lambda}\psi \right\rangle_h - \frac{\gamma_{01}}{\gamma_{11} - q_1(\lambda)}\left\langle \varphi', \frac{1}{a' - \lambda}\psi' \right\rangle_{h'} \right)
$$

$$
+\frac{1}{a' - \lambda}\varphi' \frac{1}{\gamma_{11} - q_1(\lambda) - \frac{|\gamma_{01}|^2}{\gamma_{00} - q_0(\lambda)}}
$$

$$
\times \left(-\frac{\gamma_{10}}{\gamma_{00} - q_0(\lambda)}\left\langle \varphi, \frac{1}{a - \lambda}\psi \right\rangle_h + \left\langle \varphi', \frac{1}{a' - \lambda}\psi' \right\rangle_{h'} \right).
$$

$$(5.65)$$

The resolvent restricted to the spaces h and h' can easily be calculated:

$$
P_h \frac{1}{A^\gamma - \lambda} P_h = \frac{1}{a - \lambda} - \frac{1}{a - \lambda}\varphi \left(\frac{1}{q_0(\lambda) + q_0^+(\lambda)}\langle \varphi, \frac{1}{a - \lambda}\cdot\rangle_h \right),
$$

$$
P_{h'} \frac{1}{A^\gamma - \lambda} P_{h'} = \frac{1}{a' - \lambda} - \frac{1}{a' - \lambda}\varphi' \left(\frac{1}{q_1(\lambda) + q_1^+(\lambda)}\langle \varphi, \frac{1}{a' - \lambda}\cdot\rangle_{h'} \right),
$$

$$(5.66)$$

where $q_0^+(\lambda) = \frac{|\gamma_{01}|^2}{\gamma_{11} - q_1(\lambda)} - \gamma_{00}$; $q_1^+(\lambda) = \frac{|\gamma_{10}|^2}{\gamma_{00} - q_0(\lambda)} - \gamma_{11}$.

We define the perturbed operator A^γ acting in the Hilbert space $\mathbf{H} = K \otimes (H)$ using the tensor decomposition

$$
\mathbf{A}^\gamma = B \otimes I_H + I_K \otimes A^\gamma. \qquad (5.67)
$$

Theorem 5.2.3 *Let the operators A and A' satisfy Assumption 5.2.1. Then* *the resolvent of the operator* $\mathbf{A}^\gamma$ *at a certain point* $\lambda, \Im\lambda \neq 0$, *is given by*

$$\frac{1}{\mathbf{A}^\gamma - \lambda}\Psi$$

$$= \frac{1}{A - \lambda}\Psi - \frac{1}{A - \lambda}\left[\left\{\frac{1}{q_0(\lambda - B) + q_0^+(\lambda - B)}\right.\right.$$

$$\left.\left(\langle\varphi, \frac{1}{A - \lambda}\psi\rangle_h - \frac{\gamma_{01}}{\gamma_{11} - q_1(\lambda - B)}\langle\varphi', \frac{1}{A' - \lambda}\psi'\rangle_{h'}\right)\right\} \otimes \varphi\right]$$

$$-\frac{1}{A' - \lambda}\left[\left\{\frac{1}{q_1(\lambda - B) + q_1^+(\lambda - B)}\right.\right.$$

$$\left.\left(-\frac{\gamma_{10}}{\gamma_{00} - q_0(\lambda - B)}\langle\varphi_0, \frac{1}{A - \lambda}\psi\rangle_h + \langle\varphi', \frac{1}{A' - \lambda}\psi'\rangle_{h'}\right)\right\} \otimes \varphi'\right].$$

$$(5.68)$$

Comment Let the operators a and a' be finite dimensional. Then formula (5.68) defines an operator in $\mathbf{H}$, since the functions q_0, q_1, q_0^+, q_1^+ are rational. Therefore let us discuss the case where the operator $B \otimes a$ is positive and a' is finite dimensional. We are going to use this case in the following chapter. It is enough to study the case where both operators B and a are positive.

The vector φ' is an element from the Hilbert space. Therefore the function $q_1(\lambda) = \langle\varphi', \frac{1+\lambda a'}{a'-\lambda}\frac{1}{a'^2+1}\varphi'\rangle_{h'}$ has a finite limit at infinity: $\lim_{x\to-\infty} q_1(x+iy) = -\langle\varphi', \frac{a'}{a'^2+1}\varphi'\rangle_{h'}$. Following the comment after Theorem 5.2.1 one can consider four possible cases

1. $\varphi \in \mathcal{H}_{-1}(a_0) \setminus H^0$, $\gamma_{11} \neq -\langle\varphi', \frac{a'}{a'^2+1}\varphi'\rangle_{h'}$;

2. $\varphi \in \mathcal{H}_{-1}(a_0) \setminus H^0$, $\gamma_{11} = -\langle\varphi', \frac{a'}{a'^2+1}\varphi'\rangle_{h'}$;

3. $\varphi \in \mathcal{H}_{-2}(a_0) \setminus \mathcal{H}_{-1}(a_0)$, $\gamma_{11} \neq -\langle\varphi', \frac{a'}{a'^2+1}\varphi'\rangle_{h'}$;

4. $\varphi \in \mathcal{H}_{-2}(a_0) \setminus \mathcal{H}_{-1}(a_0)$, $\gamma_{11} = -\langle\varphi', \frac{a'}{a'^2+1}\varphi'\rangle_{h'}$.

Let us study the first possible case in detail. The other three possibilities can be studied similarly.

Lemma 5.2.2 implies that $\langle\varphi', \frac{1}{A'-\lambda}\psi'\rangle_{h'} \in \mathcal{H}_2(B)$ and $\langle\varphi, \frac{1}{A-\lambda}\psi\rangle_h \in \mathcal{H}_1(B)$. This implies in the first case that

$$\left\langle\varphi, \frac{1}{A - \lambda}\psi\right\rangle_h - \frac{\gamma_{01}}{\gamma_{11} - q_1(\lambda - B)}\left\langle\varphi', \frac{1}{A' - \lambda}\psi'\right\rangle_{h'} \in \mathcal{H}_1(B),$$

since the function $\frac{\gamma_{01}}{\gamma_{11}-q_1(x+iy)}$ is uniformly bounded for $x < 0$. The function q_0^+ is rational, therefore the following estimate holds

$$\frac{1}{|q_0(x+iy) + q_0^+(x+iy)|} \leq C(y)(1+|x|)$$

for all negative x. The latter estimate is similar to that used for the proof of Lemma 5.2.5. It implies that

$$\frac{1}{\mathcal{A} - \lambda}\left[\left\{\frac{1}{q_0(\lambda - B) + q_0^+(\lambda - B)}\right.\right.$$

$$\left.\left.\left(\left\langle \varphi, \frac{1}{\mathcal{A} - \lambda}\psi\right\rangle_h - \frac{\gamma_{01}}{\gamma_{11} - q_1(\lambda - B)}\langle \varphi', \frac{1}{\mathcal{A}' - \lambda}\psi'\rangle_{h'}\right)\right\} \otimes \varphi\right] \in \mathbf{H}. \tag{5.69}$$

If $\gamma_{00} \neq q_0(\infty)$ then the function $\frac{\gamma_{10}}{\gamma_{00}-q_0(x+iy)}$ is uniformly bounded on every interval $y = \mathrm{const} \neq 0, x < 0$. This implies that

$$-\frac{\gamma_{10}}{\gamma_{00} - q_0(\lambda - B)}\left\langle \varphi, \frac{1}{\mathcal{A} - \lambda}\psi\right\rangle_h + \left\langle \varphi', \frac{1}{\mathcal{A} - \lambda}\psi'\right\rangle_{h'} \in \mathcal{H}_1(B)$$

and therefore

$$-\frac{1}{\mathcal{A}' - \lambda}\left[\left\{\frac{1}{q_1(\lambda - B) + q_1^+(\lambda - B)}\right.\right.$$

$$\left.\left.\left(-\frac{\gamma_{10}}{\gamma_{00}-q_0(\lambda-B)}\langle \varphi, \frac{1}{\mathcal{A}-\lambda}\psi\rangle_h + \langle \varphi', \frac{1}{\mathcal{A}'-\lambda}\psi'\rangle_{h'}\right)\right\} \otimes \varphi'\right] \in \mathbf{H}. \tag{5.70}$$

If $\gamma_{00} = q_0(\infty)$ then the function

$$\frac{\gamma_{10}}{(\gamma_{11} - q_1(x+iy))(\gamma_{00} - q_0(x+iy)) - |\gamma_{10}|^2}$$

is uniformly bounded for $x < 0$ and therefore (5.70) holds.

The other three cases can be studied similarly.

Proof of Lemma 5.2.3 The proof just coincides with the proof of Lemma 5.2.1.

$\square$

The operator $\mathbf{A}^\gamma$ is a self–adjoint extension of the operator $\mathbf{A}^0$. As in the previous section we are going to consider the annulating set of regular functionals for the operator $\mathbf{A}^0$:

- if $\varphi \in \mathcal{H}_{-1}(a)$ then

$$\Phi_{\text{reg}} = \{\Phi : \Phi = \rho(\Phi) \otimes \varphi + \rho'(\Phi) \otimes \varphi', \rho(\Phi) \in \mathcal{H}_{-1}(B), \rho'(\Phi) \in \mathcal{H}_{-2}(B)\};$$

- if $\varphi \in \mathcal{H}_{-2}(a) \setminus \mathcal{H}_{-1}(a)$ then

$$\Phi_{\text{reg}} = \{\Phi : \Phi = \rho(\Phi) \otimes \varphi + \rho'(\Phi) \otimes \varphi', \rho(\Phi) \in K, \rho'(\Phi) \in \mathcal{H}_{-2}(B)\}.$$

The corresponding subspace of regular elements is defined as follows

- if $\varphi \in \mathcal{H}_{-1}(a_0)$ then

$$
\begin{aligned}
&\text{Dom}_{\text{reg}} \\
&= \Big\{\Psi : \Psi = \tilde{\Psi} + \frac{\mathcal{A}}{\mathcal{A}^2 + 1}\left(\rho(\Psi) \otimes \varphi\right) + \frac{\mathcal{A}'}{\mathcal{A}'^2 + 1}\left(\rho'(\Psi) \otimes \varphi'\right), \\
&\qquad \tilde{\Psi} \in \text{Dom}(\mathbf{A}), \rho(\Psi) \in \mathcal{H}_{-1}(B), \rho'(\Psi) \in \mathcal{H}_{-2}(B)\Big\};
\end{aligned}
$$

- if $\varphi \in \mathcal{H}_{-2}(a) \setminus \mathcal{H}_{-1}(a)$ then

$$
\begin{aligned}
&\text{Dom}_{\text{reg}} \\
&= \Big\{\Psi : \Psi = \tilde{\Psi} + \frac{\mathcal{A}}{\mathcal{A}^2 + 1}\left(\rho(\Psi) \otimes \varphi\right) + \frac{\mathcal{A}'}{\mathcal{A}'^2 + 1}\left(\rho'(\Psi) \otimes \varphi'\right), \quad . \\
&\qquad \tilde{\Psi} \in \text{Dom}(\mathbf{A}), \rho(\Psi) \in K, \rho'(\Psi) \in \mathcal{H}_{-2}(B)\Big\}
\end{aligned}
$$

Consider the set $\text{Dom}\,(\mathbf{A}^\Gamma)$ of regular elements satisfying the boundary conditions

$$\left(\begin{array}{c} \langle\varphi, \tilde{U}\rangle_h \\ \langle\varphi', \tilde{U}'\rangle_{h'} \end{array}\right) = \Gamma \left(\begin{array}{c} \rho(U) \\ \rho'(U) \end{array}\right), \tag{5.71}$$

where Γ is a symmetric operator in $K \oplus K$. Then the modified operator $\mathbf{A}^\Gamma$ is defend on $\text{Dom}\,(\mathbf{A}^\Gamma)$ by the formula

$$
\begin{aligned}
\mathbf{A}^\Gamma &\left(\tilde{\Psi} + \frac{\mathcal{A}}{\mathcal{A}^2 + 1}(\rho \otimes \varphi) + \frac{\mathcal{A}'}{\mathcal{A}'^2 + 1}(\rho' \otimes \varphi')\right) \\
&= \mathbf{A}\tilde{\Psi} - \frac{1}{\mathcal{A}^2 + 1}(\rho \otimes \varphi) - \frac{1}{\mathcal{A}'^2 + 1}(\rho' \otimes \varphi').
\end{aligned}
\tag{5.72}
$$

Let us study the following question: what boundary operator Γ determines the modified operator equal to the operator $\mathbf{A}^\gamma$. Consider the boundary operator

$$\Gamma = \begin{pmatrix} \gamma_{00} & \gamma_{01} \\ \gamma_{10} & \gamma_{11} \end{pmatrix} + B \begin{pmatrix} \langle \varphi, \frac{aA-1}{(A^2+1)(a^2+1)}\varphi \rangle_h & 0 \\ 0 & \langle \varphi', \frac{a'A'-1}{(A'^2+1)(a'^2+1)}\varphi' \rangle_{h'} \end{pmatrix}.$$

(5.73)

The operator Γ is essentially self-adjoint on the domain $\mathrm{Dom}\,(B) \oplus \mathrm{Dom}\,(B)$. In what follows we are going to use the notation Γ for the corresponding self-adjoint operator.

Theorem 5.2.4 *The operator $\mathbf{A}^\Gamma$ is self-adjoint and coincides with the operator $\mathbf{A}^\gamma = B \otimes I_H + I_K \otimes A^\gamma$.*

Proof The proof is similar to the proof of Theorem 5.2.2.

$\square$

The densities $(\rho(\Psi), \rho'(\Psi))$ for the elements from the domain $\mathrm{Dom}(\mathbf{A}^\Gamma)$ belong to the domain of the operator Γ. One can prove that the restriction of the operator constructed to the set of functions having $\rho(\Psi), \rho'(\Psi)$ from the domain $\mathrm{Dom}\,(B)$ of the operator B is essentially self-adjoint.

Chapter 6

Few-body problems

6.1 Few-body formula I: self–adjoint extensions

6.1.1 Tensor structure of the few-body Hilbert space

Let us consider an operator $\mathcal{A}$ describing several noninteracting quantum mechanical particles. Every such operator admits several tensor decompositions

$$\mathcal{A} = B_n \otimes I_{h_n} + I_{K_n} \otimes a_n, \quad n = 1, 2, ..., N \tag{6.1}$$

in the tensor products of Hilbert spaces $\mathcal{H} = K_n \otimes h_n$. The operators B_n and a_n are self–adjoint in the Hilbert spaces K_n and h_n respectively. Each tensor decomposition corresponds to a certain cluster decomposition of the few-body system. We suppose in what follows that Assumption 5.2.1 from the previous chapter is satisfied for each pair of the operators $(B_n, a_n), n = 1, 2, ..., N$.

The formal few-body operator with rank one two-body interactions can be written in the form

$$\mathcal{A}_\alpha = \mathcal{A} + \sum_{n=1}^{N} \alpha_n \langle \varphi_n, \cdot \rangle_{h_n} \varphi_n, \tag{6.2}$$

where $\varphi_n \in \mathcal{H}_{-2}(a_n), n = 1, 2, ..., N$ are arbitrary elements characterizing the interaction. The coefficients α_n form a real vector of coupling constants $\alpha = (\alpha_1, \alpha_2, ..., \alpha_N) \in \mathbf{R}^N$. Our aim in this section is to define the self–adjoint operator in $\mathcal{H}$ corresponding to the formal expression (6.2).

To determine the perturbed operator $\mathcal{A}_\alpha$ we consider first the operator $\mathcal{A}^0$ which is equal to the restriction of the self-adjoint operator $\mathcal{A}$ to the following domain

$$\text{Dom}(\mathcal{A}^0) = \{\Psi \in \text{Dom}(\mathcal{A}) : \langle \varphi_n, \Psi \rangle_{h_n} = 0, \ n = 1, 2, \dots, N\}. \qquad (6.3)$$

We are going to assume in what follows that the vectors φ_n are not elements from the corresponding Hilbert spaces. This assumption is not very restrictive, since the operator $\langle \varphi_n, \cdot \rangle_{h_n} \varphi_n$ is bounded if the element φ_n belongs to the Hilbert space. Therefore the corresponding operator A_α is self–adjoint on the domain of the original operator Dom $(\mathcal{A})$ if all $\varphi_n \in h_n$; $n = 1, 2, ..., N$. We are going to use the assumption that $\varphi_n \notin h_n$, $n = 1, 2, ..., N$, everywhere in this section. The adjoint operator $\mathcal{A}^{0*}$ can be defined if the restricted operator $\mathcal{A}^0$ is densely defined. If the operator $\mathcal{A}^0$ is not densely defined then there exist self–adjoint extensions that are operator relations. In order to avoid the discussion of self–adjoint operator relations we confine our consideration to the case of densely defined restricted operators.

Suppose that the tensor decompositions associated with the vectors

$$\varphi_{n_1}, \varphi_{n_2}, \dots, \varphi_{n_m}$$

coincide $h_{n_1} = h_{n_2} = \dots = h_{n_m} = h_n$ and their linear combination $\sum_{j=1}^{m} a_j \varphi_{n_j}$ belongs to the Hilbert space h_n

$$\sum_{j=1}^{m} a_j \varphi_{n_j} \in h_n.$$

Then the operator $\mathcal{A}^0$ is not densely defined even if all vectors φ_{n_j} are singular. The following lemma gives us necessary and sufficient conditions for the restricted operator to be densely defined.

Lemma 6.1.1 *Let $\{\varphi_n\}_{n=1}^{N}$ be a set of singular vectors $\varphi \in \mathcal{H}_{-2}(a_n) \setminus h_n$. Consider the set $\Phi \in \mathcal{H}_{-2}(\mathcal{A})$ of functionals Φ possessing the representation*

$$\Phi = \sum_{n=1}^{N} \phi_n \otimes \varphi_n, \quad \phi_n \in K_n.$$

Let us denote by $\overline{\Phi}$ the closure of the set Φ in the norm of the Hilbert space $\mathcal{H}_{-2}(\mathcal{A})$. Then the restricted operator $\mathcal{A}^0$ is densely defined if and only if the set $\overline{\Phi}$ contains no nontrivial vectors from the Hilbert space, i.e.

$$\overline{\Phi} \cap \mathcal{H} = \{0\}.$$

Proof Suppose that the set $\overline{\Phi}$ contains a certain nonzero vector Φ from the Hilbert space. Then there exists a sequence Φ_n of vectors from Φ converging to Φ in $\mathcal{H}_{-2}(\mathcal{A})$. Consider an arbitrary vector Ψ from the domain of the restricted operator $\Psi \in \mathrm{Dom}\,(\mathcal{A}^0) \subset \mathcal{H}_2(\mathcal{A})$. Then for all n the equality $\langle \Psi, \Phi_n \rangle = 0$ holds and therefore $\langle \Psi, \Phi \rangle = 0$. The latter scalar product makes sense because $\psi \in \mathcal{H}_2(\mathcal{A})$ and $\Phi \in \mathcal{H}_{-2}(\mathcal{A})$. In other words we have proven that $\mathrm{Dom}\,(\mathcal{A}^0) \perp \Phi$, i.e. the operator $\mathcal{A}^0$ is not densely defined.

Suppose that the restricted operator $\mathcal{A}^0$ is not densely defined. Then there exists a vector Ψ from the Hilbert space which is orthogonal to the domain of the restricted operator. But $\mathrm{Dom}\,(\mathcal{A}^0)^{\perp} = \overline{\Phi}$ and it follows that the subspace $\overline{\Phi}$ contains an element from the Hilbert space. The lemma is proven.

$\square$

Definition 6.1.1 *The set of vectors $\varphi_n \in \mathcal{H}_{-2}(a_n) \setminus h_n$ is called* **singular** *if the set*

$$\overline{\Phi} = \overline{\{\sum_{n=1}^{N} \phi_n \otimes \varphi_n, \phi_n \in K_n\}}^{\mathcal{H}_{-2}(\mathcal{A})}$$

contains no nontrivial element from the Hilbert space $\mathcal{H}$.

We are also going to use the following generalization of Definition 3.1.1

Definition 6.1.2 *The set of singular vectors $\varphi_n \in \mathcal{H}_{-2}(a_n) \setminus h_n$ is called $\mathcal{H}$-* **independent** *if and only if for any set $f_n \in K_n$ the inclusion $\sum_{n=1}^{N} f_n \otimes \varphi_n \in \mathcal{H}$ implies that all elements f_n are equal to zero, i.e. $f_n = 0$.*

The tensor decompositions corresponding to different vectors φ_n are not necessarily different. Independent vectors φ_n corresponding to the same tensor decomposition of the original Hilbert space are linearly independent. No linear combination of these vectors belongs to the Hilbert space appearing in the tensor decomposition of the original space $\mathcal{H}$. We are going to consider in what follows only independent singular systems of vectors φ_n.

Let us define the **set of regular elements** from the domain of the adjoint operator as follows

$$\mathrm{Dom}_{\mathrm{reg}}(\mathcal{A}^{0*}) = \Big\{ \Psi : \Psi = \tilde{\Psi} + \sum_{n=1}^{N} \frac{\mathcal{A}}{\mathcal{A}^2 + 1} \left(\rho_n(\Psi) \otimes \varphi_n \right);$$

$$\rho_n(\Psi) \in K_n = \mathcal{H}_0(B_n) \Big\}. \tag{6.4}$$

Comment In the previous chapter considering the set of regular elements from the domain of the adjoint operator we have distinguished the following two cases

- $\varphi \in \mathcal{H}_{-1}(a_n)$;

- $\varphi \in \mathcal{H}_{-2}(a_n) \setminus \mathcal{H}_{-1}(a_n)$.

We are no longer going to consider these two different cases, since it has
been shown that the densities $\rho(\Psi)$ for the elements from the domain of the
perturbed operator belong to the domain of the operator Γ and are elements
from the Hilbert space K_n.

It is possible to describe the set of regular functions from the domain of the

adjoint operator with the help of boundary elements $\xi_{\pm}^n(\Psi)$ in a similar way
as we have done in previous chapters. We prefer to use the representation
(6.4) which is convenient for the second part of this chapter, where cluster
generalized interactions will be considered. The boundary form of the adjoint
operator can be calculated using the densities $\rho_n(\Psi)$ and $\langle \varphi_n, \tilde{\Psi} \rangle_{h_n}$.

Lemma 6.1.2 *The boundary form of the adjoint operator $\mathcal{A}^{0*}$ calculated on
the regular elements $U, V \in \mathrm{Dom}_{\mathrm{reg}}(\mathcal{A}^{0*})$ is given by*

$$\langle U, \mathcal{A}^{0*} V \rangle_{\mathcal{H}} - \langle \mathcal{A}^{0*} U, V \rangle_{\mathcal{H}}$$

$$= \sum_{n=1}^{N} \left(\left\langle \rho_n(U) \otimes \varphi_n, \tilde{V} \right\rangle_{\mathcal{H}} - \left\langle \tilde{U}, \rho_n(V) \otimes \varphi_n \right\rangle_{\mathcal{H}} \right)$$

$$\tag{6.5}$$

$$= \sum_{n=1}^{N} \left(\left\langle \rho_n(U), \langle \varphi_n, \tilde{V} \rangle_{h_n} \right\rangle_{K_n} - \left\langle \langle \varphi_n, \tilde{U} \rangle_{h_n}, \rho_n(V) \right\rangle_{K_n} \right).$$

Proof Let $U, V \in \mathrm{Dom}_{\mathrm{reg}}(\mathcal{A}^{0*})$, then the following calculations prove the
lemma:

$$\left\langle U, \mathcal{A}^{0*}V \right\rangle_{\mathcal{H}} - \left\langle \mathcal{A}^{0*}U, V \right\rangle_{\mathcal{H}}$$

$$= \left\langle \tilde{U} + \sum_{n=1}^{N} \frac{\mathcal{A}}{\mathcal{A}^2+1}(\rho_n(U) \otimes \varphi_n), \right.$$
$$\left. \mathcal{A}^{0*}\left(\tilde{V} + \sum_{m=1}^{N} \frac{\mathcal{A}}{\mathcal{A}^2+1}(\rho_m(V) \otimes \varphi_m) \right) \right\rangle_{\mathcal{H}}$$
$$- \left\langle \mathcal{A}^{0*}\left(\tilde{U} + \sum_{n=1}^{N} \frac{\mathcal{A}}{\mathcal{A}^2+1}(\rho_n(U) \otimes \varphi_n) \right), \right.$$
$$\left. \tilde{V} + \sum_{m=1}^{N} \frac{\mathcal{A}}{\mathcal{A}^2+1}(\rho_m(V) \otimes \varphi_m) \right\rangle_{\mathcal{H}}$$

$$= \left\langle \tilde{U} + \sum_{n=1}^{N} \frac{\mathcal{A}}{\mathcal{A}^2+1}(\rho_n(U) \otimes \varphi_n), \right.$$
$$\left. \mathcal{A}\tilde{V} - \sum_{m=1}^{N} \frac{1}{\mathcal{A}^2+1}(\rho_m(V) \otimes \varphi_m) \right\rangle_{\mathcal{H}}$$
$$- \left\langle \mathcal{A}\tilde{U} - \sum_{n=1}^{N} \frac{1}{\mathcal{A}^2+1}(\rho_n(U) \otimes \varphi_n), \right.$$
$$\left. \tilde{V} + \sum_{m=1}^{N} \frac{\mathcal{A}}{\mathcal{A}^2+1}(\rho_m(V) \otimes \varphi_m) \right\rangle_{\mathcal{H}}$$

$$= \sum_{n=1}^{N} \left\langle \rho_n(U) \otimes \varphi_n, \tilde{V} \right\rangle_{\mathcal{H}} - \sum_{m=1}^{N} \left\langle \tilde{U}, \rho_m(V) \otimes \varphi_m \right\rangle_{\mathcal{H}}$$
$$+ \left\langle \tilde{U}, \mathcal{A}\tilde{V} \right\rangle_{\mathcal{H}} - \left\langle \mathcal{A}\tilde{U}, \tilde{V} \right\rangle_{\mathcal{H}}$$
$$+ \sum_{n,m=1}^{N} \left\{ -\left\langle \frac{\mathcal{A}}{\mathcal{A}^2+1}(\rho_n(U) \otimes \varphi_n), \frac{1}{\mathcal{A}^2+1}(\rho_m(V) \otimes \varphi_m) \right\rangle_{\mathcal{H}} \right.$$
$$\left. + \left\langle \frac{1}{\mathcal{A}^2+1}(\rho_n(U) \otimes \varphi_n, \frac{\mathcal{A}}{\mathcal{A}^2+1}(\rho_m(V) \otimes \varphi_n) \right\rangle_{\mathcal{H}} \right\}$$

$$= \sum_{n=1}^{N} \left\langle \rho_n(U) \otimes \varphi_n, \tilde{V} \right\rangle_{\mathcal{H}} - \sum_{m=1}^{N} \left\langle \tilde{U}, \rho_m(V) \otimes \varphi_m \right\rangle_{\mathcal{H}}.$$

The expression in braces vanishes because, due to our assumptions, the functions $\frac{1}{\mathcal{A}^2+1}(\rho_n(U) \otimes \varphi_n), n = 1, 2, ..., N$, belong to the domain of the operator $\mathcal{A}$:

$$\frac{1}{\mathcal{A}^2 + 1} \left(\rho_n(U) \otimes \varphi_n \right) \in \mathcal{H}_2(A), n = 1, 2, ..., N$$

and the original operator $\mathcal{A}$ is self–adjoint. The lemma is proven.

$\square$

Symmetric extensions of the operator $\mathcal{A}^0$ can be defined by introducing the boundary conditions connecting the functions $\rho_n(U)$ and $\langle \varphi_n, \tilde{U} \rangle_{h_n}$. Such boundary conditions can be chosen in different ways. We are interested in the boundary conditions corresponding to the formal operator $\mathcal{A}_\alpha$ given by (6.2). These boundary conditions are uniquely defined in the case of $\mathcal{H}_{-1}$-interactions, i.e. when $\varphi_n \in \mathcal{H}_{-1}(a_n)$, $n = 1, 2, ..., N$. This case is studied in the following section. The few-body operator $\mathcal{A}_\alpha$ with arbitrary $\mathcal{H}_{-2}$-interaction will be defined using the regularization procedure considered in Chapter 1. This case will be discussed in Section 6.1.3.

6.1.2   Few-body operator with $\mathcal{H}_{-1}$ interaction

Let us consider first the case where all elements φ_n belong to the spaces $\mathcal{H}_{-1}(a_n)$ respectively: $\varphi_n \in \mathcal{H}_{-1}(a_n)$, $n = 1, 2, ..., N$. Moreover we suppose that the operator $\mathcal{A}$ satisfies Assumption 5.2.1. In this case the perturbation $\sum_{n=1}^{N} \alpha_n \langle \varphi_n, \cdot \rangle_{h_n} \varphi_n$ is infinitesimally form bounded with respect to the operator $\mathcal{A}$. The operator $\mathcal{A}_\alpha$ can easily be defined using the form perturbation approach. This operator is semibounded. In this section we are going to describe this operator using the extension theory for symmetric operators. The following theorem describes the domain of the operator $\mathcal{A}_\alpha$ restricted to the set of regular elements.

Theorem 6.1.1 *Let the self–adjoint operator $\mathcal{A}$ satisfy Assumption 5.2.1 and let the set of vectors $\varphi_n \in \mathcal{H}_{-1}(a_n)$ be singular and $\mathcal{H}$- independent. Consider the linear operator*

$$\mathcal{A}_\alpha = \mathcal{A} + \sum_{n=1}^{N} \alpha_n \langle \varphi_n, \cdot \rangle_{h_n} \varphi_n$$

defined on the functions from the regular domain

$$\text{Dom}_{\text{reg}}(\mathcal{A}^{0*}) = \{ \Psi = \tilde{\Psi} + \sum_{n=1}^{N} \frac{\mathcal{A}}{\mathcal{A}^2 + 1} (\rho_n(\Psi) \otimes \varphi_n), \rho_n(\Psi) \in K_n \};$$

$$\mathcal{A}_\alpha : \text{Dom}_{\text{reg}}(\mathcal{A}^{0*}) \to \mathcal{H}_{-2}(A).$$

Then the corresponding Hilbert space operator is defined on the domain of functions from $\mathrm{Dom}_{\mathrm{reg}}(\mathcal{A}^{0*})$ *satisfying the boundary conditions*

$$\langle \varphi_n, \Psi^n \rangle_{h_n} = \Gamma_n \rho_n(\Psi), \tag{6.6}$$

where

$$\Gamma_n = \gamma_n + B_n \langle \varphi_n, \frac{a_n \mathcal{A} - 1}{(\mathcal{A}^2 + 1)(a_n^2 + 1)} \varphi_n \rangle_{h_n}, \tag{6.7}$$

$$\Psi^n = \Psi - \frac{\mathcal{A}}{\mathcal{A}^2 + 1} (\rho_n(\Psi) \otimes \varphi_n). \tag{6.8}$$

The operator acts in accordance with the following formula

$$\begin{aligned}
\mathcal{A}_\alpha \Psi &= \mathcal{A}_\alpha \left(\tilde{\Psi} + \sum_{n=1}^{N} \frac{\mathcal{A}}{\mathcal{A}^2 + 1} (\rho_n(\Psi) \otimes \varphi_n) \right) \\
&= \mathcal{A}\tilde{\Psi} - \sum_{m=1}^{N} \frac{1}{\mathcal{A}^2 + 1} (\rho_n(\Psi) \otimes \varphi_n).
\end{aligned}$$

Comment The operator Γ_n defined by (6.7) is bounded. It is exactly the boundary operator corresponding to the perturbed operator

$$\mathcal{A}_n = \mathcal{A} + \alpha_n \langle \varphi_n, \cdot \rangle_{h_n} \varphi_n = B_n \otimes I_{h_n} + I_{K_n} \otimes (a_n + \alpha_n \langle \varphi_n, \cdot \rangle_{h_n} \varphi_n)$$

acting in $\mathcal{H} = K_n \otimes h_n$. (Note that the operator $\mathcal{A}_n$ defined by this formula contains only one nontrivial interaction.) Parameters γ_n are related to α_n as follows

$$\gamma = -\frac{1}{\alpha_n} - \left\langle \varphi_n, \frac{a_n}{a_n^2 + 1} \varphi_n \right\rangle_{h_n}.$$

The function Ψ^n defined by (6.8) possessing also the following representation

$$\Psi^n = \tilde{\Psi} + \sum_{m=1, m\neq n}^{N} \frac{\mathcal{A}}{\mathcal{A}^2 + 1} (\rho_m(\Psi) \otimes \varphi_m)$$

will be called the *n*-th **nonsingular component** of the function Ψ. Using such components the boundary conditions describing the few-body operator can be written in the form (6.6), i.e. using these components the boundary operator can be "diagonalized". These components will be used intensively in the following section to construct the few-body operator with $\mathcal{H}_{-2}$ singular interaction.

Proof The linear operator $\mathcal{A}_\alpha$ acts on the set of regular elements from the domain of the adjoint operator $\mathrm{Dom}_{\mathrm{reg}}(\mathcal{A}^{0*})$ as follows

$$
\begin{aligned}
\mathcal{A}_\alpha \Psi &= \left(A + \sum_{n=1}^{N} \alpha_n \langle \varphi_n, \cdot \rangle_{h_n} \otimes \varphi_n \right) \left(\tilde{\Psi} + \sum_{m=1}^{N} \frac{\mathcal{A}}{\mathcal{A}^2 + 1} (\rho_m(\Psi) \otimes \varphi_m) \right) \\
&= A\tilde{\Psi} + \sum_{n=1}^{N} \alpha_n \left\langle \varphi_n, \tilde{\Psi} \right\rangle_{h_n} \otimes \varphi_n + \sum_{m=1}^{N} \frac{\mathcal{A}^2}{\mathcal{A}^2 + 1} (\rho_m(\Psi) \otimes \varphi_m) \\
&\quad + \sum_{n,m=1}^{N} \alpha_n \langle \varphi_n, \frac{\mathcal{A}}{\mathcal{A}^2 + 1} (\rho_m(\Psi) \otimes \varphi_m) \rangle_{h_n} \otimes \varphi_n \\
&= A\tilde{\Psi} - \sum_{m=1}^{N} \frac{1}{\mathcal{A}^2 + 1} (\rho_m(\Psi) \otimes \varphi_m) \\
&\quad + \sum_{n=1}^{N} \left\{ \alpha_n \langle \varphi_n, \tilde{\Psi} \rangle_{h_n} + \rho_n(\Psi) \right. \\
&\quad + \left. \sum_{m=1}^{N} \alpha_n \left\langle \varphi_n, \frac{\mathcal{A}}{\mathcal{A}^2 + 1} (\rho_m(\Psi) \otimes \varphi_m) \right\rangle_{h_n} \right\} \otimes \varphi_n.
\end{aligned}
$$

$$(6.9)$$

The operator $\mathcal{A}_\alpha$ considered as an operator in the Hilbert space is defined on the following domain

$$\mathrm{Dom}\,(\mathcal{A}_\alpha) = \{ \Psi \in \mathrm{Dom}_{\mathrm{reg}}(\mathcal{A}^{0*}) : \mathcal{A}_\alpha \Psi \in \mathcal{H} \}.$$

The element $\mathcal{A}_\alpha \Psi$ belongs to the Hilbert space if and only if each expression in braces in formula (6.9) is identically equal to zero, since the set of singular vectors φ_n is independent. Therefore every function from the domain of the operator $\mathcal{A}_\alpha$ acting in the Hilbert space satisfies the conditions:

$$\alpha_n \langle \varphi_n, \tilde{\Psi} \rangle_{h_n} + \rho_n(\Psi) + \alpha_n \sum_{m=1}^{N} \langle \varphi_n, \frac{\mathcal{A}}{\mathcal{A}^2 + 1} (\rho_m(\Psi) \otimes \varphi_m) \rangle_{h_n} = 0,$$

$$n = 1, 2, ..., N. \tag{6.10}$$

The latter condition can be simplified as follows

$$\langle \varphi_n, \Psi^n \rangle_{H_n} = \Gamma_n \rho_n(\Psi), \tag{6.11}$$

if one considers the nonsingular components Ψ^n, $n = 1, 2, ..., N$ of the wave function Ψ defined by (6.8). The theorem is proven.

□

We are going to prove that the operator $\mathcal{A}_\alpha$ defined above is self–adjoint. The following lemma proves that the operator $\mathcal{A}_\alpha$ defined on the described domain is symmetric. This lemma follows from the fact that the operator we defined is a restriction of the self–adjoint operator determined by the form perturbation approach. We are going to further analyze the proof of this lemma in the following section.

Lemma 6.1.3 *The operator $\mathcal{A}_\alpha$ in $\mathcal{H}$ determined by Theorem 6.1.1 is symmetric.*

Proof The boundary form of the adjoint operator $\mathcal{A}^{0*}$ considered on the regular elements is given by (6.5). The functions $\rho_n \otimes \varphi_n$ belong to $\mathcal{H}_{-1}(\mathcal{A})$, since $\rho_n \in K_n$ and $\varphi_n \in \mathcal{H}_{-1}(a_n)$ and the operator $\mathcal{A}$ satisfies Assumption 5.2.1. Therefore the following sesquilinear form

$$Q_{nm}(\rho_n, \rho_m) \equiv \left\langle \frac{\mathcal{A}}{\mathcal{A}^2 + 1}(\rho_n \otimes \varphi_n), (\rho_m \otimes \varphi_m) \right\rangle_{\mathcal{H}} \tag{6.12}$$

is well defined for $\rho_n \in K_n$, $\rho_m \in K_m$. Moreover the following equality holds

$$\left\langle \frac{\mathcal{A}}{\mathcal{A}^2 + 1}(\rho_n \otimes \varphi_n), (\rho_m \otimes \varphi_m) \right\rangle_{\mathcal{H}} = \left\langle (\rho_n \otimes \varphi_n), \frac{\mathcal{A}}{\mathcal{A}^2 + 1}(\rho_m \otimes \varphi_m) \right\rangle_{\mathcal{H}},$$

since the operator $\mathcal{A}$ is self–adjoint. Let U, V be two arbitrary elements from $\mathrm{Dom}_{\mathrm{reg}}(\mathcal{A}^{0*})$ satisfying the boundary conditions (6.11). Then the latter equality implies that

$$\langle U, \mathcal{A}^{0*}V \rangle - \langle \mathcal{A}^{0*}U, V \rangle$$

$$= \sum_{n=1}^{N} \left\langle \rho_n(U) \otimes \varphi_n, \tilde{V} \right\rangle_{\mathcal{H}} - \sum_{m=1}^{N} \left\langle \tilde{U}, \rho_m(V) \otimes \varphi_m \right\rangle_{\mathcal{H}}$$

$$= \sum_{n=1}^{N} \left\langle \rho_n(U) \otimes \varphi_n, V^n \right\rangle_{\mathcal{H}} - \sum_{m=1}^{N} \langle U^m, \rho_m(V) \otimes \varphi_m \rangle_{\mathcal{H}}$$

$$- \sum_{\substack{n,m=1 \\ n \neq m}}^{N} \left\langle (\rho_n(U) \otimes \varphi_n), \frac{\mathcal{A}}{\mathcal{A}^2 + 1}(\rho_m(V) \otimes \varphi_m) \right\rangle_{\mathcal{H}}$$

$$+ \sum_{\substack{n,m=1 \\ n \neq m}}^{N} \left\langle \frac{\mathcal{A}}{\mathcal{A}^2 + 1}(\rho_n(U) \otimes \varphi_n), (\rho_m(V) \otimes \varphi_m) \right\rangle_{\mathcal{H}}$$

$$= \sum_{n=1}^{N} \langle \rho_n(U) \otimes \varphi_n, V^n \rangle_{\mathcal{H}} - \sum_{m=1}^{N} \langle U^m, \rho_m(V) \otimes \varphi_m \rangle_{\mathcal{H}}$$

$$= \sum_{n=1}^{N} \left(\langle \rho_n(U), \Gamma_n \rho_n(V) \rangle_{K_n} - \langle \Gamma_n \rho_n(U), \rho_n(V) \rangle_{K_n} \right)$$

$$= 0.$$

We have taken into account that the operators Γ_n acting in K_n are symmetric. The lemma is proven.

□

Comment In the course of the proof of the lemma we have rearranged the boundary form to separate the channels of interaction. The boundary form has been expressed in terms of the nonsingular components U^n, V^n and densities $\rho_n(U), \rho_n(V)$ as follows

$$\begin{aligned} &\langle U, \mathcal{A}^{0*} V \rangle_{\mathcal{H}} - \langle \mathcal{A}^{0*} U, V \rangle_{\mathcal{H}} \\ &= \sum_{n=1}^{N} \langle \rho_n(U) \otimes \varphi_n, V^n \rangle_{\mathcal{H}} - \sum_{m=1}^{N} \langle U^m, \rho_m(V) \otimes \varphi_m \rangle_{\mathcal{H}}. \end{aligned} \tag{6.13}$$

The sesquilinear form $\left\langle \dfrac{\mathcal{A}}{\mathcal{A}^2+1}(\rho_n \otimes \varphi_n), (\rho_m \otimes \varphi_m) \right\rangle_{\mathcal{H}}$ cannot always be defined if $\varphi_n \in \mathcal{H}_{-2}(a_n) \setminus \mathcal{H}_{-1}(a_n)$, $n = 1, 2, ..., N$. Therefore the boundary form of $\mathcal{A}^{0*}$ is not always given by (6.13) for such φ_n. We are going to discuss the possibility of representing the boundary form by (6.13) for $\varphi_n \in \mathcal{H}_{-2}(a_n)$ in the following section.

In fact the operator $\mathcal{A}_\alpha$ we defined is self-adjoint and therefore coincides with the self-adjoint operator defined using the form perturbation approach.

Theorem 6.1.2 *The operator $\mathcal{A}_\alpha$ in $\mathcal{H}$ determined by Theorem 6.1.1 is self-adjoint.*

Proof To prove the theorem we are going to calculate the resolvent $(\mathcal{A}-\lambda)^{-1}$ for negative λ with large absolute value. The resolvent equation

$$\frac{1}{\mathcal{A}-\lambda} F = \Psi$$

can be written using (6.4) as

$$F = (\mathcal{A}_\alpha - \lambda)\left(\tilde{\Psi} + \frac{\mathcal{A}}{\mathcal{A}^2 + 1}\sum_{n=1}^{N}\rho_n(\Psi) \otimes \varphi_n\right)$$

$$= (\mathcal{A} - \lambda)\tilde{\Psi} - \frac{1 + \lambda\mathcal{A}}{\mathcal{A}^2 + 1}\sum_{n=1}^{N}\rho_n \otimes \varphi_n.$$

Applying the resolvent of the original operator and projecting on the vectors φ_n we get

$$\frac{1}{\mathcal{A} - \lambda}F = \tilde{\Psi} - \frac{1 + \lambda\mathcal{A}}{\mathcal{A} - \lambda}\frac{1}{\mathcal{A}^2 + 1}\sum_{n=1}^{N}\rho_n(\Psi) \otimes \varphi_n; \qquad (6.14)$$

$$\left\langle \varphi_n, \frac{1}{\mathcal{A} - \lambda}F \right\rangle_{h_n} = \langle \varphi_n, \Psi^n \rangle_{h_n} - \sum_{m=1, m\neq n}^{N}\left\langle \varphi_n, \frac{1}{\mathcal{A} - \lambda}(\rho_m(\Psi) \otimes \varphi_m) \right\rangle_{h_n}.$$
$$(6.15)$$

The boundary conditions (6.6) imply

$$\left\langle \varphi_n, \frac{1}{\mathcal{A} - \lambda}F \right\rangle = -\frac{1}{\alpha_n}\rho_n(\Psi) - \sum_{m=1}^{N}\left\langle \varphi_n, \frac{1}{\mathcal{A} - \lambda}(\rho_m(\Psi) \otimes \varphi_m) \right\rangle_{h_n}. \quad (6.16)$$

We are going to show that this system of N linear equations is solvable for all λ with sufficiently large absolute value.

Let us consider the Hilbert space

$$K = K_1 \oplus K_2 \oplus \ldots \oplus K_N, \qquad (6.17)$$

and introduce the bounded operator $G(\lambda)$ given by its matrix elements in the orthogonal decomposition (6.17)

$$G_{nm}(\lambda) = \left\langle \varphi_n, \frac{1}{\mathcal{A} - \lambda}(\cdot \otimes \varphi_m) \right\rangle_{h_n}, \quad n, m = 1, 2, \cdots, N. \qquad (6.18)$$

The quadratic form of this operator can be estimated as

$$|\langle \rho, G(\lambda)\rho \rangle|_{\mathcal{H}}$$

$$\leq N \sum_{n=1}^{N} \left\langle \rho_n \otimes \varphi_n, \frac{1}{A - \lambda}(\rho_n \otimes \varphi_n) \right\rangle_{\mathcal{H}}$$

$$\leq N \sum_{n=1}^{N} \left\langle \rho_n \otimes \varphi_n, \frac{1}{a_n - \lambda}(\rho_n \otimes \varphi_n) \right\rangle_{\mathcal{H}}$$

$$= N \sum_{n=1}^{N} \| \rho_n \|_{K_n}^2 \left\langle \varphi_n, \frac{1}{a_n - \lambda}\varphi_n \right\rangle_{h_n}$$

$$\leq \left(N \min_{n} \left\langle \varphi_n, \frac{1}{a_n - \lambda}\varphi_n \right\rangle_{h_n} \right) \| \rho \|_K^2 .$$

The norm of the operator $G(\lambda)$ tends to zero as $\lambda \to -\infty$, since

$$\lim_{\lambda \to -\infty} \left\langle \varphi_n, \frac{1}{a_n - \lambda}\varphi_n \right\rangle_{h_n} = 0, \quad n = 1, 2, \ldots, N.$$

The linear system (6.16) can be written as

$$f(\lambda) = \left(-\frac{1}{\alpha} - G(\lambda) \right) \rho, \tag{6.19}$$

where

$$f_n(\lambda) = \left\langle \varphi_n, \frac{1}{A - \lambda}F \right\rangle_{h_n} \in K_n, \tag{6.20}$$

$$\frac{1}{\alpha} = \frac{1}{\alpha_1} \oplus \frac{1}{\alpha_2} \oplus \ldots \oplus \frac{1}{\alpha_N}.$$

This equation is solvable, since the spectrum of the operator $1/\alpha$ is separated from the origin and the operator $G(\lambda)$ tends to zero as $\lambda \to -\infty$. The component $\tilde{\Psi}$ can be calculated using (6.14).

Thus we have proven that the range of the resolvent of the symmetric operator $\mathcal{A}_\alpha$ is equal to the Hilbert space $\mathcal{H}$. Hence the operator $\mathcal{A}_\alpha$ is self-adjoint on the domain described by Theorem 6.1.1. The theorem is proven.

□

6.1.3   Few-body operator with $\mathcal{H}_{-2}$ interaction

In this section we are going to study the operator $\mathcal{A}_\alpha$ given formally by (6.2) in the case where the elements φ_n do not belong to the spaces $\mathcal{H}_{-1}(a_n)$, but rather to the spaces $\mathcal{H}_{-2}(a_n)$:

$$\varphi_n \in \mathcal{H}_{-2}(a_n) \setminus \mathcal{H}_{-1}(a_n), \quad n = 1, 2, ..., N.$$

The self–adjoint operator corresponding to the formal expression (6.2) should coincide with one of the extensions of the symmetric operator $\mathcal{A}^0$ which is the restriction of the operator $\mathcal{A}$ to the domain Dom $(\mathcal{A}^0)$ given by (6.3). We supposed in the previous section abd still suppose here that the set φ_n is singular and $\mathcal{H}$-independent. In particular this implies that the operator $\mathcal{A}^0$ is densely defined. The analysis carried out in the previous section shows that the boundary operator which determines the few-body operator $\mathcal{A}_\alpha$ is diagonal if one considers the boundary values $(\langle \varphi_n, \Psi^n \rangle_{H_n}, \rho_n), n = 1, 2, ..., N$, corresponding to separated channels of interaction. Then the boundary form is given by (6.13). The comment after Lemma 6.1.3 implies that the boundary form cannot always be written in this form. Therefore we are going to use the following two definitions.

Definition 6.1.3 *Let the self–adjoint operator $\mathcal{A}$ acting in $\mathcal{H}$ possess several tensor decompositions $\mathcal{A} = B_n \otimes I_{H_n} + I_{K_n} \otimes a_n$; $\mathcal{H} = K_n \otimes h_n$, $n = 1, 2, ..., N$. Then the set of vectors $\varphi_n \in \mathcal{H}_{-2}(a_n), n = 1, 2, ..., N$ is called* **separable** *if and only if the sesquilinear forms*

$$Q_{nm}(\rho_n, \rho_m) \equiv \left\langle (\rho_n \otimes \varphi_n), \frac{\mathcal{A}}{\mathcal{A}^2 + 1} (\rho_m \otimes \varphi_m) \right\rangle_{\mathcal{H}}, \quad n \neq m \qquad (6.21)$$

are bounded in the graph norms of the operators B_n and B_m, i.e. if there exist positive constants $C_{nm} \in \mathbf{R}_+$ such that the following inequalities hold

$$|Q_{nm}(\rho_n, \rho_m)| \leq C_{nm} \| \rho_n \|_{\mathcal{H}_2(B_n)} \| \rho_m \|_{\mathcal{H}_2(B_m)}, \quad n \neq m \qquad (6.22)$$

for arbitrary $\rho_n \in$ Dom (B_n), $\rho_m \in$ Dom (B_m). The set of singular vectors $\{\varphi_n\}_{n=1}^N$ is called **strongly separable** *if there exist positive constants D_{nm} such that the following inequalities hold*

$$|Q_{nm}(\rho_n, \rho_m)| \leq D_{nm} \| \rho_n \|_{K_n} \| \rho_m \|_{K_m}, \quad n \neq m. \qquad (6.23)$$

In Section 6.2.4 we are also going to define infinitesimally separable interactions.

If the set of singular vectors is strongly separable, then the sesquilinear forms Q_{nm} can be extended by continuity to sesquilinear bounded forms on $K_n \times K_m$. The same notation will be used for the extended forms.

If all singular vectors φ_n are elements from the spaces $\mathcal{H}_{-1}(a_n)$ and the operator $\mathcal{A}$ satisfies Assumption 5.2.1, then the system $\{\varphi_n\}$ is strongly separable. (This follows from the proof of Lemma 6.1.3.) But not every system of vectors $\varphi_n \in \mathcal{H}_{-2}(a_n)$ is even separable. Important examples of few-body operators with separable and nonseparable interactions are discussed in the following section. We are going to consider only few-body operators with separable interactions in this section.

Lemma 6.1.4 *Let the singular $\mathcal{H}$-independent set of vectors $\{\varphi_n\}_{n=1}^{N}$ be strongly separable. Let $\mathcal{A}^0$ be the restriction of the self-adjoint operator $\mathcal{A}$ to the domain*

$$\mathrm{Dom}\,(\mathcal{A}^0) = \{\Psi \in \mathrm{Dom}\,(\mathcal{A}) : \forall n, \langle \varphi_n, \Psi \rangle_{h_n} = 0\}. \qquad (6.24)$$

Then the boundary form of the adjoint operator $\mathcal{A}^{0*}$ calculated on the regular elements $U, V \in \mathrm{Dom}_{\mathrm{reg}}(\mathcal{A}^{0*})$ possessing the representation (6.4) is equal to

$$\begin{aligned}
&\left\langle U, \mathcal{A}^{0*}V \right\rangle_{\mathcal{H}} - \left\langle \mathcal{A}^{0*}U, V \right\rangle_{\mathcal{H}} \\
&= \sum_{n=1}^{N} \left(\langle \rho_n(U) \otimes \varphi_n, V^n \rangle_{\mathcal{H}} - \langle U^n, \rho_n(V) \otimes \varphi_n \rangle_{\mathcal{H}} \right).
\end{aligned} \qquad (6.25)$$

Proof Lemma 6.1.2 implies that the boundary form of the adjoint operator on the regular elements is given by (6.5). The proof of the present lemma follows using the main lines of the proof of Lemma 6.1.3, taking into account that the few-body interactions are strongly separable.

$\square$

Comment One can generalize the previous lemma to include separable, but not strongly separable interactions. In this case one has to restrict the adjoint operator to the set of regular elements with the densities ρ_n from $\mathrm{Dom}\,(B_n)$. Then the boundary form is given also by (6.25). We are going to use this to construct the self-adjoint operator with nonseparable generalized interactions.

The operators $\mathcal{A}_n = \mathcal{A} + \alpha_n \langle \varphi_n, \cdot \rangle_{h_n} \otimes \varphi_n$ with a single few-body interaction can be determined following the main lines of Section 5.2.1 for arbitrary φ_n being elements from the class $\mathcal{H}_{-2}(a_n)$. Every such operator can be defined by closing the symmetric operator originally defined on the domain of functions possessing the representation

$$\Psi = \tilde{\Psi} + \frac{\mathcal{A}}{\mathcal{A}^2 + 1} \left(\rho_n(\Psi) \otimes \varphi_n \right)$$

where

$$\tilde{\Psi} \in \mathrm{Dom}\,(\mathcal{A}), \ \rho_n(\Psi) \in \mathrm{Dom}\,(B_n) \subset \mathrm{Dom}\,(\Gamma_n) \subset K_n$$

and satisfying the boundary conditions

$$\left\langle \varphi_n, \tilde{\Psi} \right\rangle_{H_n} = \Gamma_n \rho_n(\Psi). \tag{6.26}$$

The boundary operator

$$\Gamma_n = \gamma_n + B_n \left\langle \varphi_n, \frac{a_n \mathcal{A} - 1}{(a^2 + 1)(\mathcal{A}^2 + 1)} (\cdot \otimes \varphi_n) \right\rangle_{h_n} \tag{6.27}$$

is essentially self–adjoint on the domain Dom (B_n), since $\varphi_n \in \mathcal{H}_{-2}(a_n)$. We are going to use the same notation for the corresponding self–adjoint operator. The constant γ_n is related to the interaction constant α_n and parameter $c_n \equiv \langle \varphi_n, \frac{a_n}{a_n^2+1}\varphi_n \rangle_{h_n}$ (introduced in Chapter 1 to parametrize $\mathcal{H}_{-2}$ rank one perturbations) as follows

$$\gamma_n = -\frac{1}{\alpha_n} - c_n.$$

The few-body operator with separable interactions can be determined using the following theorem.

Theorem 6.1.3 *Let the operator $\mathcal{A}$ satisfy Assumption 5.2.1 and let the singular $\mathcal{H}$-independent set of vectors $\varphi_n \in \mathcal{H}_{-2}(a_n) \setminus \mathcal{H}_{-1}(a_n)$ be strongly separable. Consider the linear operator*

$$\mathcal{A}_\alpha = \mathcal{A} + \sum_{n=1}^{N} \alpha_n \langle \varphi_n, \cdot \rangle_{h_n} \otimes \varphi_n$$

with the regularized cluster interactions defined on the domain of functions from

$$\mathrm{Dom}_{\mathrm{reg}}(\mathcal{A}^{0*}) = \{ \Psi = \tilde{\Psi} + \sum_{n=1}^{N} \frac{\mathcal{A}}{\mathcal{A}^2 + 1}(\rho_n(\Psi) \otimes \varphi_n), \rho_n(\Psi) \in K_n \},$$

$$\mathcal{A}_\alpha : \mathrm{Dom}_{\mathrm{reg}}(\mathcal{A}^{0*}) \to \mathcal{H}_{-2}(\mathcal{A}).$$

Then the corresponding Hilbert space operator is defined on the domain of functions from $\mathrm{Dom}_{\mathrm{reg}}(\mathcal{A}^{0*})$ satisfying the boundary conditions

$$\langle \varphi_n, \Psi^n \rangle_{h_n} = \Gamma_n \rho_n(\Psi);$$

$$\rho_n(\Psi) \in \mathrm{Dom}\,(\Gamma_n), \tag{6.28}$$

where Γ_n is the self–adjoint operator defined by (6.27). The operator acts in accordance with the following formula

$$
\begin{aligned}
\mathcal{A}_\alpha \Psi &= \mathcal{A}_\alpha \left(\tilde{\Psi} + \sum_{n=1}^{N} \frac{\mathcal{A}}{\mathcal{A}^2 + 1} \left(\rho_n(\Psi) \otimes \varphi_n \right) \right) \\
&= \mathcal{A}\tilde{\Psi} - \sum_{m=1}^{N} \frac{1}{\mathcal{A}^2 + 1} \left(\rho_n(\Psi) \otimes \varphi_n \right).
\end{aligned}
$$

Proof The proof of the theorem follows the same lines as that of Theorem 6.1.1. It is necessary to take into account that in the case $\varphi_n \in \mathcal{H}_{-2}(a_n) \setminus \mathcal{H}_{-1}(a_n)$ the following regularization should be used

$$
\begin{aligned}
&\left\langle \varphi_n, \frac{\mathcal{A}}{\mathcal{A}^2 + 1} (\rho_n \otimes \varphi_n) \right\rangle_{h_n} \\
&= \left\langle \varphi_n, \frac{a_n}{a_n^2 + 1} (\rho_n \otimes \varphi_n) \right\rangle_{h_n} \\
&\quad - B \left\langle \varphi_n, \frac{a_n \mathcal{A} - 1}{(\mathcal{A}^2 + 1)(a_n^2 + 1)} (\rho_n(\Psi) \otimes \varphi_n) \right\rangle_{h_n} \quad\quad (6.29) \\
&= c_n \rho_n(\Psi) - B \left\langle \varphi_n, \frac{a_n \mathcal{A} - 1}{(\mathcal{A}^2 + 1)(a_n^2 + 1)} (\rho_n(\Psi) \otimes \varphi_n) \right\rangle_{h_n}.
\end{aligned}
$$

The theorem is proven.

$\square$

Comment It is important to take into account that the definition of the few-body operator considered in the previous theorem contains several real parameters c_n corresponding to each $\varphi_n \notin \mathcal{H}_{-1}(a_n)$. One can carry out similar calculations to define few-body operators with arbitrary separable interactions. In this case one has to restrict the consideration to the set of regular elements with densities $\rho_n \in \mathrm{Dom}\,(B_n)$.

The regularization procedure described by (6.29) was used to define the few–body operator. This regularization appears natural, since the corresponding cluster interactions possess the same scaling properties as those in the heuristic formula (6.2).

One can easily prove that the calculated operator $\mathcal{A}_\alpha$ is symmetric following the main lines of that of Lemma 6.1.3.

Lemma 6.1.5 *The operator $\mathcal{A}_\alpha$ in $\mathcal{H}$ determined by Theorem 6.1.3 is symmetric.*

It is possible to use the boundary conditions (6.28) to define the few–body operator with separable but not strongly separable interactions. In addition one has to restrict the set of regular elements allowing only elements with the densities from Dom (B_n). Following the main lines of the previous Lemma one proves that the operator so defined is symmetric, but it can have nontrivial deficiency indices, since the boundary conditions (6.28) connect the densities $\rho_n(\Psi)$ for different n. Therefore the operator Γ_n appearing in (6.28) can be symmetric (not self–adjoint) on the set of allowed densities. The deficiency indices of these operators can be nontrivial. To illustrate this let us rewrite the boundary conditions (6.28) in the form

$$\left\langle \varphi_n, \tilde\Psi \right\rangle_{h_n} + \sum_{m\neq n} \left\langle \varphi_n, \frac{A}{A^2+1}\left(\rho_n(\Psi)\otimes\varphi_m\right)\right\rangle_{h_n} = \Gamma_n\rho_n(\Psi), \ n=1,2,\dots,N$$

$$\Rightarrow \Gamma_n\rho_n(\Psi) - \sum_{m\neq n}\left\langle \varphi_n, \frac{A}{A^2+1}\left(\rho_n(\Psi)\otimes\varphi_m\right)\right\rangle_{h_n} = \left\langle \varphi_n, \tilde\Psi\right\rangle_{h_n}. \quad (6.30)$$

This set of equations can be written in the operator form in the Hilbert space K. Let us introduce the following operators acting in the Hilbert space K:

• the self–adjoint operator Γ is equal to the orthogonal sum of the self–adjoint operators Γ_n

$$\Gamma = \Gamma_1 \oplus \Gamma_2 \oplus \dots \oplus \Gamma_N; \quad (6.31)$$

• the bounded operator R is given by its matrix elements in the orthogonal decomposition (6.17)

$$R_{nm} = \begin{cases} 0, & m=n, \\[2mm] \left\langle \varphi_n, \dfrac{A}{A^2+1}(\cdot\otimes\varphi_m)\right\rangle_{h_n}, & n\neq m. \end{cases} \quad (6.32)$$

Equation (6.30) is equivalent to the following equation

$$(\Gamma - R)\,\rho = P\tilde\Psi, \quad (6.33)$$

where

$$\rho = (\rho_1, \rho_2, \dots, \rho_N) \in K,$$

$$P\tilde\Psi = (\langle\varphi_1, \tilde\Psi\rangle_{h_1}, \langle\varphi_2, \tilde\Psi\rangle_{h_2}, \dots, \langle\varphi_N, \tilde\Psi\rangle_{h_N}) \in K.$$

If the vectors φ_n are not strongly separable, then the operator R is not bounded and the domain of the operator sum $\Gamma - R$ can be different from the domain of the operator Γ

$$\mathrm{Dom}\,(\Gamma) = \mathrm{Dom}\,(\Gamma_1) \oplus \mathrm{Dom}\,(\Gamma_2) \oplus \ldots \oplus \mathrm{Dom}\,(\Gamma_N).$$

In what follows we are going to define the few–body operator by the boundary conditions (6.33) whenever the operator sum $\Gamma - R$ can be defined as a self–adjoint operator on a certain domain (which can be different from the domain of the operator Γ). If the operator sum $\Gamma - R$ is not a self–adjoint operator, then the few body operator can be defined using densities from the common domain of the operators Γ and R. But the operator so defined is not necessarily essentially self–adjoint. We are going to discuss different possibilities to construct self–adjoint extensions of this operator in what follows.

Definition 6.1.4 *Let the operator $\Gamma - R$ be a self–adjoint operator in K. Then the* **few-body operator with rank one two-body interactions** $\mathcal{A}_\alpha$ *is defined on the set of regular elements from* $\mathrm{Dom}_{\mathrm{reg}}(\mathcal{A}^{0*})$ *satisfying the boundary conditions (6.33).*

If the vectors φ_n are strongly separable then the operator R is bounded and therefore the operator $\Gamma - R$ is self–adjoint on the domain of the operator Γ. Hence Definition 6.1.4 can be used to define the few-body operator with non $\mathcal{H}_{-1}$ cluster interactions. If the cluster interactions are from $\mathcal{H}_{-1}$ then Theorem 6.1.1 implies that just the same boundary conditions define the few-body operator. Therefore the latter definition can be used for all few-body operators with strongly separable cluster interactions.

6.1.4   Self–adjointness of the few-body operator with strongly separable interactions

In this section we intend to study the few-body operator $\mathcal{A}_\alpha$ with strongly separable interactions. We are going to concentrate our attention on the interactions defined by vectors from $\mathcal{H}_{-2}(a_n) \setminus \mathcal{H}_{-1}(a_n)$, since the few–body operators with $\mathcal{H}_{-1}(a_n)$ interactions were described in Section 6.1.2. Moreover without loss of generality we suppose that the operator $\mathcal{A}$ is positive.

Theorem 6.1.4 *Let the singular $\mathcal{H}$-independent set of vectors $\{\varphi_n\}_{n=1}^N$, $\varphi_n \in \mathcal{H}_{-2}(a_n) \setminus \mathcal{H}_{-1}(a_n)$ be strongly separable and let the operator $\mathcal{A}$ satisfy Assumption 5.2.1. Let the following estimate hold for all $\lambda < \lambda_0$, where λ_0 is a certain negative number*

$$\left\langle \varphi_n, \frac{1}{\mathcal{A} - \lambda}\,(\rho_m \otimes \varphi_m) \right\rangle_{h_n} \leq b \,\| \rho_m \|_{K_m}, \quad m = 1, \ldots, N; \; m \neq n. \quad (6.34)$$

Then the operator $\mathcal{A}_\alpha$ determined by Definition 6.1.4 is self-adjoint.

Proof The proof is similar to that of Theorem 6.1.2. Consider the resolvent equation

$$F = (\mathcal{A}_\alpha - \lambda)U \qquad (6.35)$$

for negative λ.

The equation can be written as follows

$$
\begin{aligned}
F &= (\mathcal{A}_\alpha - \lambda)\left(\tilde{U} + \sum_{m=1}^{N} \frac{\mathcal{A}}{\mathcal{A}^2 + 1}(\rho_m(U) \otimes \varphi_m)\right) \\
&= \mathcal{A}\tilde{U} - \lambda\tilde{U} - \lambda\sum_{m=1}^{N} \frac{\mathcal{A}}{\mathcal{A}^2 + 1}(\rho_m(U) \otimes \varphi_m) \\
&\quad - \sum_{m=1}^{N} \frac{1}{\mathcal{A}^2 + 1}(\rho_m(U) \otimes \varphi_m).
\end{aligned}
$$

To derive the latter equality we have taken into account that the operator $\mathcal{A}_\alpha$ is a restriction of the operator $\mathcal{A}^{0*}$. Applying the resolvent of the original operator $\mathcal{A}$ to the latter equation we get the following chain of equalities

$$
\begin{aligned}
\frac{1}{\mathcal{A} - \lambda}F &= \tilde{U} - \frac{1}{\mathcal{A} - \lambda}\sum_{m=1}^{N} \frac{1 + \lambda\mathcal{A}}{\mathcal{A}^2 + 1}(\rho_m(U) \otimes \varphi_m) \\
&= U^n - \frac{1 + \lambda\mathcal{A}}{\mathcal{A} - \lambda}\frac{1}{\mathcal{A}^2 + 1}(\rho_n(u) \otimes \varphi_n) \qquad (6.36) \\
&\quad - \sum_{\substack{m=1 \\ m\neq n}}^{N} \frac{1}{\mathcal{A} - \lambda}(\rho_m(U) \otimes \varphi_m).
\end{aligned}
$$

Projecting on the element φ_n and taking into account the boundary conditions (6.33) we get

$$
\begin{aligned}
\left\langle \varphi_n, \frac{1}{\mathcal{A} - \lambda}F \right\rangle_{h_n} &= \langle \varphi_n, U^n \rangle_{h_n} - \left\langle \varphi_n, \frac{1 + \lambda\mathcal{A}}{\mathcal{A} - \lambda}\frac{1}{\mathcal{A}^2 + 1}(\rho_n(U) \otimes \varphi_n) \right\rangle_{h_n} \\
&\quad - \sum_{\substack{m=1 \\ m\neq n}}^{N} \left\langle \varphi_n, \frac{1}{\mathcal{A} - \lambda}(\rho_m(U) \otimes \varphi_m) \right\rangle_{h_n} \\
&= \Gamma_n \rho_n(U) - \left\langle \varphi_n, \frac{1 + \lambda\mathcal{A}}{\mathcal{A} - \lambda}\frac{1}{\mathcal{A}^2 + 1}(\rho_n(U) \otimes \varphi_n) \right\rangle_{h_n} \\
&\quad - \sum_{\substack{m=1 \\ m\neq n}}^{N} \left\langle \varphi_n, \frac{1}{\mathcal{A} - \lambda}(\rho_m(U) \otimes \varphi_m) \right\rangle_{h_n}.
\end{aligned}
$$

$$(6.37)$$

Let us study the system of N linear equations that we have obtained in the Hilbert space K, introducing the following operators acting in K :

• the self–adjoint operator $D(\lambda)$ is equal to the orthogonal sum of the self-adjoint operators

$$D_n(\lambda) = \Gamma_n - \left\langle \varphi_n, \frac{1+\lambda A}{A-\lambda}\frac{1}{A^2+1}(\cdot \otimes \varphi_n)\right\rangle_{h_n} ;$$

• the bounded self–adjoint operator $R(\lambda)$ given by its matrix elements in the orthogonal decomposition (6.17)

$$R_{nm}(\lambda) = \begin{cases} 0, & n = m \\ \left\langle \varphi_n, \dfrac{1}{A-\lambda}(\cdot \otimes \varphi_m)\right\rangle_{h_n}, & n \neq m. \end{cases}$$

The system of linear equations (6.37) can be writtes as

$$(D(\lambda) - R(\lambda))\,\rho = f(\lambda), \qquad\qquad (6.38)$$

where $f(\lambda)$ is given by (6.20).

The operators $D_n(\lambda)$ are essentially self-adjoint on the corresponding domains Dom (B_n). For arbitrary $\rho_n \in$ Dom (B_n) the following estimate holds

$$\langle \rho_n, \Gamma_n\rho_n\rangle_{K_n} - \left\langle \rho_n \otimes \varphi_n, \frac{1+\lambda A}{A-\lambda}\frac{1}{A^2+1}(\rho_n \otimes \varphi_n)\right\rangle_{\mathcal{H}}$$

$$= \gamma_n\,\| \rho_n \|^2_{K_n} - \left\langle \rho_n \otimes \varphi_n, \frac{1+(\lambda - B_n)a_n}{a_n-(\lambda - B_n)}\frac{1}{a_n^2+1}(\rho_n \otimes \varphi_n)\right\rangle_{\mathcal{H}}$$

$$\geq (\gamma_n - q_n(\lambda))\,\| \rho_n \|^2_{K_n},$$

since the operators B_n are positive and the functions

$$q_n(\lambda) = \left\langle \varphi_n, \frac{1+\lambda a_n}{a_n-\lambda}\frac{1}{a_n^2+1}\varphi_n\right\rangle_{h_n}$$

are increasing on the interval $(-\infty, -1)$. The vectors $\varphi_n \in \mathcal{H}_{-2}(a_n)$ do not belong to $\mathcal{H}_{-1}(a_n)$. Therefore the functions $q_n(\lambda)$ tend to $-\infty$ as $\lambda \to -\infty$. The same estimate holds for all $\rho_n \in$ Dom $(D_n(\lambda))$, since the operators Γ_n and $D_n(\lambda)$ are essentially selfadjoint on Dom (B_n).

The operator $R(\lambda)$ is uniformly bounded for $\lambda \in (-\infty, -1)$ due to estimate (6.34).

Hence for sufficiently large negative λ the operator $D(\lambda) + R(\lambda)$ is invertible. Therefore the linear system (6.37) is solvable for arbitrary $f \in K$ and the densities ρ_n can be calculated. The function $\tilde{U}$ can be calculated using equation (6.36). Therefore the range of the operator $\mathcal{A}_\alpha - \lambda$ coincides with the Hilbert space $\mathcal{H}$ for all sufficiently large negative λ. Hence the operator $\mathcal{A}_\alpha$ is self–adjoint and bounded from below. The theorem is proven.

$\square$

In the course of the proof of Theorem 6.1.4 we have already calculated the resolvent of the operator $\mathcal{A}_\alpha$ and by this the following theorem has already been proven.

Theorem 6.1.5 *Let conditions of Theorem 6.1.4 be satisfied. Then the resolvent $(\mathcal{A}_\alpha - \lambda)^{-1}$ of the self–adjoint operator $\mathcal{A}_\alpha$ at an arbitrary point $\lambda, \Im \lambda \neq 0$, calculated on any $F \in \mathcal{H}$ is given by*

$$\frac{1}{\mathcal{A}_\alpha - \lambda} F = \frac{1}{\mathcal{A} - \lambda} F + \frac{1}{\mathcal{A} - \lambda} \sum_{m=1}^{N} (\rho_m \otimes \varphi_m), \qquad (6.39)$$

where $\rho_m, m = 1, 2, ..., N$ are the unique solutions to the system of linear equations (6.37).

6.1.5   Few-body operator with delta interaction.

Let us consider an important example which will be studied in more detail (in the one dimensional case) in the following chapter. We consider the operator M^3 describing the three-body system in $\mathbf{R}^3$ with two-body and three-body interactions given by delta functions. The operator can be considered as a perturbation of the Laplace operator $-\Delta$ acting in the Hilbert space $L_2(\mathbf{R}^9)$. The formal operator for the particles with equal masses is then given by

$$\begin{aligned}
M^3 = \ & -\Delta \\
& + \alpha_3 \delta(x_1 - x_2) + \alpha_1 \delta(x_2 - x_3) + \alpha_2 \delta(x_3 - x_1) \qquad (6.40) \\
& + \alpha_{123} \delta^0(x_1, x_2, x_3),
\end{aligned}$$

where $x_1, x_2, x_3 \in \mathbf{R}^3$ denote the coordinates of the particles and $\alpha_{12}, \alpha_{23}, \alpha_{31}, \alpha_{123}$ are certain real parameters. We denote the three dimensional delta functions describing the pairwise interactions by $\delta^j(x_i - x_k)$ and the delta functions with support at the origin in $\mathbf{R}^9$ by $\delta^0(x_1, x_2, x_3)$. The interactions in formula (6.40) are not separable, since the elements $\frac{-\Delta}{\Delta^2+1}(\rho_{12} \otimes \delta^3)$ are not bounded at the origin if $\rho_{12}(0) \neq 0$ and therefore do not belong to the domain of the delta function δ^0. (The vector δ^0 does not belong to $\mathcal{H}_{-2}(-\Delta)$ in $L_2(\mathbf{R}^9)$.) Consider a similar operator L^3 defined by (6.40) with vanishing three-body interaction, i.e.

$$
\begin{aligned}
L^3 \; = \; &-\Delta \\
&+\alpha_3\delta(x_1 - x_2) + \alpha_1\delta(x_2 - x_3) + \alpha_2\delta(x_3 - x_1).
\end{aligned} \tag{6.41}
$$

The interaction terms in this operator are separable but are not strongly separable. The operator sum $\Gamma - R$ is not a self–adjoint operator and the three-body operator can be defined only using densities from the common domain of the operators Γ and R. The operator defined in this way has nontrivial deficiency indices. The self–adjoint extensions of this operator have been studied by L.D. Faddeev and R.A. Minlos [730, 731]. (See Section A.4 for historical remarks.)

Similar operators can be used to describe the system of three particles in dimensions two and one, $x_1, x_2, x_3 \in \mathbf{R}^2, \mathbf{R}$. In the two dimensional case the two-body interactions are not from $\mathcal{H}_{-1}$ but are strongly separable. The corresponding three-body operator can be defined using Definition 6.1.4 and is bounded from below.

In the one dimensional case the two-body interactions are from $\mathcal{H}_{-1}$. Therefore the set of two-body interactions is strongly separable. The three-body interaction is from the class $\mathcal{H}_{-2}$. But the set of three-body and two-body interactions is separable in this case too. These operators will be studied in more detail in Chapter 7.

6.2   Few-body formula II: generalized self–adjoint extensions

6.2.1   Generalized unperturbed operator

The few-body operator describing a system of several particles with generalized interactions will be constructed in this section. This operator is equal to a generalized self–adjoint perturbation of the operator describing the system of noninteracting particles. The main problem in the construction of such operators is that the set of generalized perturbations includes operators having no physical sense. For example few-body operators with two-body interactions depending on the total energy of the system can be obtained. In order to avoid such problems we confine our consideration to the set of generalized interactions with internal structure. We consider first an arbitrary unperturbed self–adjoint operator $\mathcal{A}$ describing a certain system of noninteracting particles. Let us denote the corresponding Hilbert space by $\mathcal{H}$. The operator $\mathcal{A}$ in $\mathcal{H}$ is similar to the original operator defined in the previous section. In the second step we construct a generalized unperturbed operator $\mathbf{A}$ acting in a certain extended Hilbert space $\mathbf{H} \supset \mathcal{H}$. The operator $\mathbf{A}$ will be chosen in such a way that its restriction to the original Hilbert space $\mathcal{H}$ coincides with the operator $\mathcal{A}$. Then the generalized few-body operator $\mathbf{A}^\Gamma$ will be defined as a self–adjoint perturbation of the generalized unperturbed operator $\mathbf{A}$. Obviously such an operator is a generalized self–adjoint perturbation of the original operator $\mathcal{A}$. The operator $\mathbf{A}$ and its self–adjoint perturbations will be constructed in a special way in order to guarantee locality and translational invariance of the cluster interactions. The generalized unperturbed operator $\mathbf{A}$ is described in this section.

Let $\mathcal{A}$ be a self–adjoint operator acting in the Hilbert space $\mathcal{H}$ describing the system of several noninteracting particles. Let the operator possess tensor decompositions corresponding to any of the possible cluster decompositions of the system

$$\mathcal{A} = B_n \otimes I_{h_n} + I_{K_n} \otimes a_n, \quad n = 1, 2, ..., N. \qquad (6.42)$$
$$\mathcal{H} = K_n \otimes h_n,$$

The operators B_n and a_n are positive self–adjoint operators acting in the Hilbert spaces K_n and h_n, respectively. The operator $\mathcal{A}$ defined by the above formula is essentially self–adjoint on the algebraic tensor product of $\text{Dom}(B_n)$ and $\text{Dom}(a_n)$, but we are going to use the same notation for the corresponding self–adjoint operator. This operator $\mathcal{A}$ satisfies Assumption 5.2.1. We consider the set of extension Hilbert spaces $\{\mathcal{H}'_n\}_{n=1}^N$, each possessing at least one of the tensor decompositions described above

$$\mathcal{H}'_n = K_n \otimes h'_n \qquad (6.43)$$

and consider N self–adjoint operators $\mathcal{A}'_n$ each acting in the corresponding Hilbert space $\mathcal{H}'_n$. We suppose that each operator $\mathcal{A}'_n$ possesses a tensor decomposition similar to the corresponding decomposition of the operator $\mathcal{A}$

$$\mathcal{A}'_n = B_n \otimes I_{h'_n} + I_{K_n} \otimes a'_n. \qquad (6.44)$$

We suppose that the operator a'_n is finite dimensional and/or positive. It follows that the operators $\mathcal{A}'_n$ satisfy Assumption 5.2.1. We note that in the general situation the Hilbert spaces h_n and h'_n and the self–adjoint operators a_n and a'_n are different.

The generalized unperturbed few-body operator acts in the extended Hilbert space

$$\mathbf{H} = \mathcal{H} \oplus \left(\oplus \sum_{n=1}^{N} \mathcal{H}'_n \right) \tag{6.45}$$

in accordance with the formula

$$\mathbf{A} = \mathcal{A} \oplus \left(\oplus \sum_{n=1}^{N} \mathcal{A}'_n \right). \tag{6.46}$$

The self–adjoint perturbations of the generalized unperturbed operator will be studied in the following sections.

6.2.2   Inner-cluster generalized interaction

The self–adjoint perturbations of the generalized unperturbed operator can be described using certain boundary conditions. We have seen earlier (in Section 6.1) that the standard few-body operator can be defined using the cluster boundary operators. Therefore in the first step we construct a self-adjoint operator with the generalized interaction inside one cluster only. Such an operator describes a system where the particles can be separated into two classes:

• free particles which do not interact with the other particles and between themselves;

• cluster particles which do not interact with the free particles but interact with the other cluster particles and the interaction depends on the total energy of the particles in the cluster.

If one considers a cluster consisting of two particles then such an operator describes a few-body system with a unique nontrivial two-body interaction between the two particles forming the cluster. If one considers a cluster of n particles then the operator describes only n-body interactions between the cluster particles. This operator will be used in the following section to construct generalized few-body operators.

We are now going to calculate the boundary operator corresponding to the cluster m. For simplicity we consider the unperturbed operator

$$\mathbf{A}_m = \mathcal{A} \oplus \mathcal{A}'_m \tag{6.47}$$

acting in the Hilbert space $\mathbf{H}_m = \mathcal{H} \oplus \mathcal{H}'_m = K_m \otimes (h_m \oplus h'_m)$. The operator $\mathbf{A}_m$ possesses the following tensor decomposition

$$\mathbf{A}_m = B_m \otimes I_{h_m \oplus h'_m} + I_{K_m} \otimes A_m,$$

where the self-adjoint operator

$$A_m = a_m \oplus a'_m \tag{6.48}$$

acts in the Hilbert space $H_m = h_m \oplus h'_m$. We are going to construct self-adjoint perturbations of A_m possessing similar tensor decompositions. Such perturbations can be obtained by considering the self-adjoint perturbations of the operator A_m. We are going to modify the approach developed and describe self-adjoint perturbations of the operator A_m in terms of densities.

To introduce the interaction between the channels we restrict the operators a_m and a'_m to certain symmetric densely defined operators a^0_m and a'^0_m respectively. As in Chapter 2, we choose two finite sets of independent vectors $\{\varphi_{mj}\}_{j=1}^{d_m} \subset \mathcal{H}_{-2}(a_m) \setminus h_m$ and $\{\varphi'_{mk}\}_{k=1}^{d'_m} \subset \mathcal{H}_{-2}(a'_m) \setminus h'_m$ and restrict the operators a_m and a'_m to the following domains:

$$
\begin{aligned}
\mathrm{Dom}\,(a^0_m) &= \{\psi \in \mathrm{Dom}\,(a_m) : \langle \varphi_{mj}, \psi \rangle_{h_m} = 0, j = 1, 2, ..., d_m\}; \\
\mathrm{Dom}\,(a'^0_m) &= \{\psi' \in \mathrm{Dom}\,(a'_m) : \langle \varphi'_{mj}, \psi' \rangle_{h'_m} = 0, j = 1, 2, ..., d'_m\}.
\end{aligned}
$$

Then the deficiency subspaces for the operators a^0_m and a'^0_m are given by

$$
\begin{aligned}
N_m(-i) &= \frac{1}{a_m + i}\mathcal{L}\{\varphi_{mj}\}_{j=1}^{d_m}, \\
N'_m(-i) &= \frac{1}{a'_m + i}\mathcal{L}\{\varphi'_{mk}\}_{k=1}^{d'_m},
\end{aligned}
$$

where $\mathcal{L}$ denotes the linear hull. We can suppose without loss of generality that the systems $\{\varphi_{mj}\}_{j=1}^{d_m}$ and $\{\varphi'_{mk}\}_{k=1}^{d'_m}$ are orthonormal in the norms

$$\| \cdot \|_{\mathcal{H}_{-2}(a_m)} \quad \text{and} \quad \| \cdot \|_{\mathcal{H}_{-2}(a'_m)}$$

respectively. These elements will be used in the following section to define the few-body operator with several cluster interactions. We suppose that the operators a^0_m and a'^0_m are densely defined. The case where the operators a'^0_m are not densely defined can be studied using the main ideas of Section 5.2.2.

Every element $\psi = (\psi, \psi')$ from the domain $\mathrm{Dom}\,(A^{0*}_m)$ of the adjoint operator A^{0*}_m possesses the representation

$$
\begin{aligned}
\Psi = \hat{\Psi} &+ \frac{A_m}{A_m^2 + 1}\left(\sum_{j=1}^{d_m} \rho_{mj}(\psi) \otimes \varphi_{mj} + \sum_{j=1}^{d'_m} \rho'_{mj}(\psi) \otimes \varphi'_{mj}\right) \\
&+ \frac{1}{A_m^2 + 1}\left(\sum_{j=1}^{d_m} \sigma_{mj}(\psi) \otimes \varphi_{mj} + \sum_{j=1}^{d'_m} \sigma'_{mj}(\psi) \otimes \varphi'_{mj}\right),
\end{aligned} \tag{6.49}
$$

where $\hat{\psi}_m = (\hat{\psi}_m, \hat{\psi}'_m) \in \mathrm{Dom}\,(A_m^0)$, and

$$\rho_m(\psi) = \left(\rho_{m1}(\psi), ..., \rho_{md_m}(\psi); \rho'_{m1}(\psi), ..., \rho'_{md'_m}(\psi)\right) \in \mathbf{C}^{d_m + d'_m};$$

$$\sigma_m(\psi) = \left(\sigma_{m1}(\psi), ..., \sigma_{md_m}(\psi); \sigma'_{m1}(\psi), ..., \sigma'_{md'_m}(\psi)\right) \in \mathbf{C}^{d_m + d'_m}.$$

Let us introduce the map η_m defined by

$$\begin{aligned}\eta_m: \quad \mathbf{C}^{d_m + d'_m} &\to \mathcal{H}_{-2}(A_m)\\ \rho_m &\mapsto \sum_{j=1}^{d_m} \rho_{mj} \otimes \varphi_{mj} + \sum_{j'=1}^{d'_m} \rho'_{mj'} \otimes \varphi'_{mj'}.\end{aligned} \tag{6.50}$$

Then the following representation holds for every element ψ from the domain of the adjoint operator:

$$\psi = \tilde{\psi} + \frac{A_m}{A_m^2 + 1}\,(\eta_m \rho_m(\psi)), \tag{6.51}$$

where $\tilde{\psi} = \hat{\psi} + \dfrac{1}{A_m^2 + 1}\,(\eta_m \sigma_m(\psi)) \in \mathcal{H}_2(A_m)$. The vector $\sigma_m \in \mathbf{C}^{d_m + d'_m}$ can be calculated from the function $\tilde{\psi}$ using the map

$$\begin{aligned}P_m: \quad \mathcal{H}_2(A_m) &\to \mathbf{C}^{d_m + d'_m}\\ \tilde{\psi} &\mapsto \sigma_m(\psi);\end{aligned}$$

$$\begin{aligned}\sigma_{mj}(\psi) &= \langle \varphi_{mj}, \tilde{\psi} \rangle_{h_m};\\ \sigma'_{mj}(\psi) &= \langle \varphi'_{mj}, \tilde{\psi} \rangle_{h'_m}.\end{aligned} \tag{6.52}$$

The adjoint operator A_m^{0*} is equal to the orthogonal sum of the operators a_m^{0*} and $a_m^{0'*}$ and acts as follows:

$$\begin{aligned}A_m^{0*}\Psi &= A_m^{0*}\left(\tilde{\psi} + \frac{A_m}{A_m^2 + 1}\,(\eta_m \rho_m(\psi))\right)\\ &= A_m \tilde{\psi} - \frac{1}{A_m^2 + 1}\,(\eta_m \rho_m(\psi)).\end{aligned} \tag{6.53}$$

The boundary form of the adjoint operator can be calculated using the boundary values ρ_m, σ_m

$$\begin{aligned}B_{A_m^{0*}}[U, V] &= \langle U, A_m^{0*}V \rangle_{H_m} - \langle A_m^{0*}U, V \rangle_{H_m}\\ &= \langle \rho_m(U), \sigma_m(V) \rangle_{\mathbf{C}^{d_m + d'_m}} - \langle \sigma_m(U), \rho_m(V) \rangle_{\mathbf{C}^{d_m + d'_m}}.\end{aligned} \tag{6.54}$$

To define the inner-cluster generalized interaction one has to construct a self-adjoint extension of the operator A_m^0 different from the operator A_m. Such a self-adjoint extension can be determined by a cluster boundary operator γ_m which is a Hermitian (self-adjoint) $(d_m + d'_m) \times (d_m + d'_m)$ matrix.

Definition 6.2.1 *Let γ_m be a certain Hermitian (self–adjoint) $(d_m + d'_m) \times (d_m + d'_m)$ matrix. Then the operator A_{γ_m} - **the extension of the operator** A^0_m **determined by the boundary operator** γ_m - is the restriction of the operator A^{0*}_m to the set of functions*

$$\psi = \tilde{\psi} + \frac{A_m}{A^2_m + 1}\left(\eta_m \rho_m(\psi)\right) \in \text{Dom}\,(A^{0*}_m) \qquad (6.55)$$

with boundary values satisfying the following conditions

$$\sigma_m(\psi) = P_m \tilde{\psi} = \gamma_m \rho_m(\psi). \qquad (6.56)$$

The operator A_{γ_m} describes the inner cluster interaction in the cluster m with the separated centre of mass motion. The corresponding perturbation of the operator $\mathbf{A}_m$ can be defined using the formula

$$\mathbf{A}_{\gamma_m} = B_m \otimes I_{H_m} + I_{K_m} \otimes A_{\gamma_m}. \qquad (6.57)$$

This operator can be calculated using the spectral decomposition of the operator B_m. We are going to show now how to define the same operator in the framework of our approach by certain boundary conditions. The operator $\mathbf{A}_{\gamma_m}$ is a self–adjoint extension of the restricted symmetric operator

$$\mathbf{A}^0_m = \mathcal{A}^0_m \oplus \mathcal{A}'^0_m = B_m \otimes I_{H_m} + I_{K_m} \otimes A^0_m \qquad (6.58)$$

equal to the orthogonal sum of the symmetric operators

$$\begin{aligned} \mathcal{A}^0_m &= B_m \otimes I_{h_m} + I_{K_m} \otimes a^0_m; \\ \mathcal{A}'^0_m &= B_m \otimes I_{h'_m} + I_{K_m} \otimes a'^0_m, \end{aligned}$$

which are the restrictions of the operators $\mathcal{A}$ and $\mathcal{A}'_m$ respectively.

Let us introduce the set $\text{Dom}_{\text{reg}}(\mathbf{A}^{0*}_m)$ of regular elements from the domain $\text{Dom}\,(\mathbf{A}^{0*}_m)$ of the adjoint operator $\mathbf{A}^{0*}_m$. Let us use the following notation in what follows

$$\mathbf{K}_m = K_m \otimes \mathbf{C}^{d_m + d'_m}. \qquad (6.59)$$

Then the regular domain $\text{Dom}_{\text{reg}}(\mathbf{A}^{0*}_m)$ is given by

$$\begin{aligned} \text{Dom}_{\text{reg}}(\mathbf{A}^{0*}_m) = \Big\{ &\Psi = \tilde{\Psi} + \frac{\mathbf{A}_m}{\mathbf{A}^2_m + 1}\eta_m \rho_m(\Psi), \\ &\tilde{\Psi} \in \text{Dom}\,(\mathbf{A}_m), \rho_m(\Psi) \in \mathbf{K}_m \Big\}, \end{aligned} \qquad (6.60)$$

where the map $\eta : K_m \otimes \mathbf{C}^{d_m + d'_m} \to \mathcal{H}_{-2}(\mathbf{A}_m)$ is the extension of the map used earlier and is defined by the same formula (6.50). The adjoint operator acts as follows on every element Ψ from the regular domain

$$
\begin{aligned}
\mathbf{A}_m^{0*}\Psi &= \mathbf{A}_m^{0*}\left(\tilde{\Psi} + \frac{\mathbf{A}_m}{\mathbf{A}_m^2+1}\eta_m\rho_m(\Psi)\right)\\
&= \mathbf{A}_m\tilde{\Psi} - \frac{1}{\mathbf{A}_m^2+1}\eta_m\rho_m(\Psi).
\end{aligned}
\tag{6.61}
$$

The operator P_m can be extended to an operator acting from $\mathcal{H}_2(\mathbf{A}_m)$ to $\mathbf{K}_m$ using the same formula (6.52). We are going to keep the same notation for the extended operator. Then the boundary values of an element Ψ from the regular domain $\mathrm{Dom}_{\mathrm{reg}}(\mathbf{A}_m^{0*})$ can be chosen equal to $\rho_m(\Psi)$, $\sigma_m(\Psi)$, where

$$
\sigma_m(\Psi) = P_m\tilde{\Psi}. \tag{6.62}
$$

The boundary form of the adjoint operator $\mathbf{A}_m^{0*}$ restricted to the regular domain is given by

$$
\begin{aligned}
B_{\mathbf{A}_m^{0*}}[U,V] &= \langle U, \mathbf{A}_m^{0*}V\rangle_{\mathbf{H}} - \langle \mathbf{A}_m^{0*}U, V\rangle_{\mathbf{H}}\\
&= \langle \rho_m(U), \sigma_m(V)\rangle_{\mathbf{K}_m} - \langle \sigma_m(U), \rho_m(V)\rangle_{\mathbf{K}_m},
\end{aligned}
\tag{6.63}
$$

where $U, V \in \mathrm{Dom}_{\mathrm{reg}}(\mathbf{A}_m^{0*})$. This formula follows from formula (6.54) and representation (6.58). Self–adjoint extensions of the operator $\mathbf{A}_m^0$ can be defined by boundary conditions of the form

$$
\sigma_m(\Psi) = \Gamma_m\rho_m(\Psi)
$$

connecting the boundary values of any function from the domain of the extended operator. In other words the self–adjoint extension of the operator $\mathbf{A}_m^0$ defined by the boundary operator Γ_m is the restriction of the operator $\mathbf{A}_m^{0*}$ to the domain of regular elements satisfying the latter condition. The following theorem shows the relation between the operators Γ_m and γ_m corresponding to the same self–adjoint operator in $\mathbf{H}_m$.

Theorem 6.2.1 *The operator $\mathbf{A}_{\gamma_m}$ coincides with the closure of the adjoint operator $\mathbf{A}_m^{0*}$ restricted to the set of regular elements*

$$
\Psi = \tilde{\Psi} + \frac{\mathbf{A}_m}{\mathbf{A}_m^2+1}\eta_m\rho_m(\Psi) \in \mathrm{Dom}_{\mathrm{reg}}(\mathbf{A}_m^{0*}) \tag{6.64}
$$

satisfying the boundary conditions

$$
\sigma_m(\Psi) \equiv P_m\tilde{\Psi} = \Gamma_m\rho_m(\Psi), \tag{6.65}
$$

where Γ_m is the closure of the operator $\gamma_m + B_m P_m \dfrac{\mathbf{A}_m\mathbf{A}_m - 1}{\mathbf{A}_m^2+1}\dfrac{1}{\mathbf{A}_m^2+1}\eta_m$ defined on $\mathrm{Dom}\,(B_m) \times \mathbf{C}^{d_m+d'_m}$.

Comment The operator $P_m \dfrac{A_m A_m - 1}{A_m^2 + 1} \dfrac{1}{A_m^2 + 1} \eta_m$ is a bounded self-adjoint operator in $\mathbf{K}_m$, commuting with the operator B_m; moreover

$$P_m \frac{A_m A_m - 1}{A_m^2 + 1} \frac{1}{A_m^2 + 1} \eta_m \leq 1.$$

Therefore the operator

$$\Gamma_m = \gamma_m + B_m P_m \frac{A_m A_m - 1}{A_m^2 + 1} \frac{1}{A_m^2 + 1} \eta_m \tag{6.66}$$

is essentially self–adjoint on the domain of the operator B_m, since the operator γ_m is bounded and self–adjoint in $\mathbf{K}_m$. We are going to keep the same notation Γ_m for the corresponding self–adjoint operator. The operator Γ_m defined by the latter formula is the cluster boundary operator for the system with nonseparated centre of mass motion. The boundary condition (6.65) contains one additional condition $\rho_m(\Psi) \in \text{Dom}\,(\Gamma_m)$.

Proof Let us denote by $\mathbf{A}^{\Gamma_m}$ the closure of the operator $\mathbf{A}_m^{0*}$ restricted to the set of regular functions satisfying the boundary conditions (6.65). We are going to prove that the resolvents of the operators $\mathbf{A}^{\Gamma_m}$ and $\mathbf{A}_{\gamma_m}$ coincide on a certain dense domain. This will imply that the operator $\mathbf{A}^{\Gamma_m}$ is self–adjoint and coincides with $\mathbf{A}_{\gamma_m}$.

The resolvent of the operator $\mathbf{A}_{\gamma_m}$ can be calculated using the spectral representation for the operator B_m. Consider the operator

$$P_m \frac{1 + (\lambda - B_m)A_m}{A_m - \lambda} \frac{1}{A_m^2 + 1} \eta_m$$

which is defined on the domain $\text{Dom}\,(B_m) \times \mathbf{C}^{d_m + d'_m} \subset \mathbf{K}_m$ of the operator $B_m \otimes I_{\mathbf{C}^{d_m + d'_m}}$. Let us denote the closure of this operator by $Q_m(\lambda - B_m)$. The domain of the latter operator coincides with the domain of the operator Γ_m due to the following equality

$$\Gamma_m - P_m \frac{1 + \lambda A_m}{A_m - \lambda} \frac{1}{A_m^2 + 1} \eta_m = \gamma_m - Q_m(\lambda - B_m), \tag{6.67}$$

since the operators γ_m and $P_m \dfrac{1 + \lambda A_m}{A_m - \lambda} \dfrac{1}{A_m^2 + 1} \eta_m$ are bounded. The resolvent of the operator $\mathbf{A}_{\gamma_m}$ is given by

$$\frac{1}{\mathbf{A}_{\gamma_m} - \lambda} = \frac{1}{\mathbf{A}_m - \lambda} + \frac{1}{\mathbf{A}_m - \lambda} \eta_m \frac{1}{\gamma_m - Q_m(\lambda - B_m)} P_m \frac{1}{\mathbf{A}_m - \lambda} \tag{6.68}$$

on the range of the algebraic tensor product of Dom (B_m) and $h_m \oplus h'_m$. The considered domain is dense in the Hilbert space $\mathbf{H}$.

Let us calculate the resolvent of the operator $\mathbf{A}^{\Gamma_m}$ restricted to the same domain. Consider an arbitrary $\mathbf{F}$ from the described domain. To calculate the resolvent of the operator $\mathbf{A}^{\Gamma_m}$ one has to solve the following equation

$$(\mathbf{A}^{\Gamma_m} - \lambda)\Psi = \mathbf{F}.$$

The function Ψ which appears in this equation can be obtained from the set of regular elements satisfying the boundary conditions (6.65). Therefore the latter equation can be written as

$$
\begin{aligned}
\mathbf{F} &= (\mathbf{A}^{\Gamma_m} - \lambda)\left(\tilde{\Psi} + \frac{\mathbf{A}_m}{\mathbf{A}_m^2 + 1}\eta_m\rho_m(\Psi)\right) \\
&= (\mathbf{A}_m - \lambda)\tilde{\Psi} - \frac{1 + \lambda\mathbf{A}_m}{\mathbf{A}_m^2 + 1}\eta_m\rho_m(\Psi).
\end{aligned}
$$

Applying the resolvent of the original operator $\mathbf{A}_m$ and then the operator P_m we obtain

$$
\begin{aligned}
P_m\frac{1}{\mathbf{A}_m - \lambda}\mathbf{F} &= P_m\tilde{\Psi} - P_m\frac{1 + \lambda\mathbf{A}_m}{\mathbf{A}_m - \lambda}\frac{1}{\mathbf{A}_m^2 + 1}\eta_m\rho_m(\Psi) \\
&= \Gamma_m\rho_m(\Psi) - P_m\frac{1 + \lambda\mathbf{A}_m}{\mathbf{A}_m - \lambda}\frac{1}{\mathbf{A}_m^2 + 1}\eta_m\rho_m(\Psi),
\end{aligned}
$$

taking into account that the function Ψ satisfies boundary conditions (6.65). The latter equation can be written as follows using (6.67)

$$P_m\frac{1}{\mathbf{A}_m - \lambda}\mathbf{F} = (\gamma_m - Q_m(\lambda - B_m))\rho_m(\Psi).$$

This implies that the resolvent of the operator $\mathbf{A}^{\Gamma_m}$ is given by

$$\frac{1}{\mathbf{A}^{\Gamma_m} - \lambda} = \frac{1}{\mathbf{A}_m - \lambda} + \frac{1}{\mathbf{A}_m - \lambda}\eta_m\frac{1}{\gamma_m - Q_m(\lambda - B_m)}P_m\frac{1}{\mathbf{A}_m - \lambda} \quad (6.69)$$

on the described dense domain. Therefore the operators $\mathbf{A}_{\gamma_m}$ and $\mathbf{A}^{\Gamma_m}$ coincide, since both operators are closed. In particular this implies that the operator $\mathbf{A}^{\Gamma_m}$ is self-adjoint. The theorem is proven.

□

6.2.3 Few-body operator with generalized interaction

This section is devoted to the construction of the few-body operator with generalized interaction. We present here the definition of such an operator.

The self-adjointness of this operator will be proven in the following two sections under certain additional assumptions on the singular vectors.

Let us consider self-adjoint perturbations of the operator **A**. The interaction between the channels will be introduced using boundary conditions connecting the components Ψ and Ψ_m of the total wavefunction

$$\boldsymbol{\Psi} = (\Psi, \Psi_1', ..., \Psi_N') \in \mathbf{H} = \mathcal{H} \oplus \mathcal{H}_1' \oplus \mathcal{H}_2' \oplus ... \oplus \mathcal{H}_N'.$$

In what follows the component Ψ of $\boldsymbol{\Psi}$ will be called **external**, the components Ψ_n' will be called **internal**.

First we describe the restriction $\mathcal{A}^0$ of the operator $\mathcal{A}$. We have supposed that the operator $\mathcal{A}$ possesses N independent tensor decompositions (6.42). The symmetric operator $\mathcal{A}^0$ is defined on the domain Dom $(\mathcal{A}^0)$ equal to the intersection of the domains of the symmetric operators $\mathcal{A}_m^0$ considered in the previous section. More precisely

$$\text{Dom}\,(\mathcal{A}^0) = \Big\{ \Psi \in \text{Dom}\,(\mathcal{A}) : \langle \Psi, \varphi_{mj} \rangle_{h_m} = 0, \\ j = 1, 2, ..., d_m, \ m = 1, 2, ..., N \Big\}. \tag{6.70}$$

This definition makes sense, since $\varphi_{mj} \in \mathcal{H}_{-2}(a_m)$ and $\Psi \in \mathcal{H}_2(\mathcal{A}) =$ Dom $(\mathcal{A})$ imply that $\langle \Psi, \varphi_{mj} \rangle_{h_m} \in K_m$. Let the set of vectors $\{\varphi_{nj}\}$ be singular and $\mathcal{H}$-independent. Then the operator $\mathcal{A}^0$ is densely defined.

The restrictions $\mathcal{A}_m'^0$ of the operators $\mathcal{A}_m'$ have been defined in the previous section. We are going to identify each vector $\Psi_m' \in \mathcal{H}_m'$ with the vector

$$(0, 0, ..., 0, \Psi_m', 0, ..., 0) \in \mathcal{H} \oplus \mathcal{H}_1' \oplus ... \oplus \mathcal{H}_{m-1}' \oplus \mathcal{H}_m' \oplus \mathcal{H}_{m+1}' \oplus ... \oplus \mathcal{H}_N' = \mathbf{H}.$$

A similar identification will be used for the operators. We suppose that the sets $\{\varphi_{nj}'\}_{j=1}^{d_n'}$ are $\mathcal{H}_n'$-independent, $n = 1, 2, ..., N$.

To define the few-body Hamiltonian with singular $\mathcal{H}_{-2}$ interactions we have used the hypotheses that the set of vectors $\{\varphi_m\}_{m=1}^N$ is separable. In this section we suppose that every two vectors φ_{nj} and φ_{ml} are separable for $n \neq m$. We say that the set $\{\varphi_{ml}\}_{l=1, m=1}^{d_m, N}$ is **separable with respect to the cluster decomposition** if the latter condition is fulfilled. **Sets that are strongly separable with respect to the tensor decomposition** are defined similarly. The corresponding assumption for the vectors φ_{nj}' is fulfilled automatically. For simplicity we suppose that the vectors φ_{nj} and $\varphi_{nj'}'$ form orthonormal sets in the spaces $\mathcal{H}_{-2}(a_n)$ and $\mathcal{H}_{-2}(a_n')$ respectively.

The total restricted operator

$$\mathbf{A}^0 = \mathcal{A}^0 \oplus \Big(\oplus \sum_{m=1}^N \mathcal{A}_m'^0 \Big) \tag{6.71}$$

is densely defined, since all symmetric operators $\mathcal{A}^0$ and $\mathcal{A}'^0_n$ are densely defined. The deficiency subspace for the operator $\mathbf{A}^0$ at the point $-i$ will be denoted by $\mathbf{M}$.

We are not going to describe all self–adjoint extensions of the symmetric operator $\mathbf{A}^0$. All such extensions can be described by von Neumann formulas for infinite deficiency indices. Our aim in this section is to construct the self–adjoint extensions which have only translational invariant cluster interactions and do not have any interaction between the clusters. Therefore only the so-called regular elements from the domain of the adjoint operator will be considered.

To define the set of regular element from the domain of the adjoint operator we are going to use the space $\mathbf{K}_m$ of boundary values corresponding to the cluster decomposition m and the map η_m introduced in the previous section. Then the set of regular elements from the domain of the adjoint operator $\mathbf{A}^{0*}$ is defined as follows:

$$
\mathrm{Dom}_{\mathrm{reg}}(\mathbf{A}^{0*}) \;=\; \Big\{ \Psi = \tilde{\Psi} + \frac{\mathbf{A}}{\mathbf{A}^2+1} \sum_{m=1}^{N} \eta_m \rho_m, \tag{6.72}
$$
$$
\tilde{\Psi} \in \mathrm{Dom}\,(\mathbf{A}), \rho_m \in \mathbf{K}_m \Big\}.
$$

The set of regular elements from the domain of the adjoint operator does not coincide in general with the domain of the adjoint operator. In what follows we are also going to use the set

$$
\mathrm{Dom}_{\mathrm{reg}}(\mathcal{A}^{0*}) \;=\; P_{\mathcal{H}}\mathrm{Dom}_{\mathrm{reg}}(\mathbf{A}^{0*})
$$
$$
= \Big\{ \Psi = \tilde{\Psi} + \frac{\mathcal{A}}{\mathcal{A}_0^2+1} \sum_{m=1}^{N} \Big(\sum_{j=1}^{d_m} \rho_{mj}(\Psi) \otimes \varphi_{mj} \Big), \tag{6.73}
$$
$$
\tilde{\Psi} \in \mathrm{Dom}(\mathcal{A}), \rho_{mj}(\Psi) \in K_m \Big\}.
$$

We also define the map η acting from the space $\mathbf{K} = \mathbf{K}_1 \oplus \mathbf{K}_2 \oplus ... \oplus \mathbf{K}_N$ to the set of regular elements from the domain of the adjoint operator $\mathrm{Dom}_{\mathrm{reg}}(\mathbf{A}^{0*})$ as follows

$$
\eta\,(\rho_1, \rho_2, ..., \rho_N) \;=\; \eta_1\rho_1 + \eta_2\rho_2 + ... + \eta_N\rho_N
$$
$$
= \textstyle\sum_{m=1}^{N} \Big(\sum_{j=1}^{d_m} \rho_{mj} \otimes \varphi_{mj} + \sum_{j=1}^{d'_m} \rho'_{mj} \otimes \varphi'_{mj} \Big). \tag{6.74}
$$

The following lemma proves that the map η is injective.

Lemma 6.2.1 *Let the set of singular vectors* $\varphi_{nj} \in \mathcal{H}_{-2}(a_n) \setminus h_n$, $j = 1, 2, ..., d_j$, $n = 1, 2, ..., N$ *be* $\mathcal{H}-$ *independent. Then the map* η *is injective.*

Proof To prove that the map η is injective it is enough to show that the kernel of the map is trivial, since the map η is linear. Suppose that some element $\rho = (\rho_1, \rho_2, ..., \rho_N) \in \mathbf{K}$ is mapped to zero:

$$\eta\rho = 0. \tag{6.75}$$

This equality is equivalent to

$$\sum_{m=1}^{N} \left(\sum_{j=1}^{d_m} \rho_{mj} \otimes \varphi_{mj} + \sum_{j=1}^{d'_m} \rho'_{mj} \otimes \varphi'_{mj} \right) = 0,$$

which implies that

$$\sum_{j=1}^{d_m} \rho_{mj}\varphi_{mj} = 0;$$
$$\rho'_{mj} = 0, \ m = 1, 2, ..., N, \ j = 1, 2, ..., d'_m,$$

since the vectors φ'_{mj} have been chosen orthonormal in $\mathcal{H}_{-2}(a'_m)$. The vectors φ_{mj} are independent, therefore

$$\rho_{mj} = 0, \ m = 1, 2, ..., N, \ j = 1, 2, ..., d_m.$$

The lemma is proven.

□

Lemma 6.2.2 *The boundary form of the adjoint operator $\mathbf{A}^{0*}$ calculated on the domain of regular elements is given by*

$$\left\langle \mathbf{U}, \mathbf{A}^{0*}\mathbf{V} \right\rangle_{\mathbf{H}} - \left\langle \mathbf{A}^{0*}\mathbf{U}, \mathbf{V} \right\rangle_{\mathbf{H}}$$

$$= \sum_{n=1}^{N} \left\{ \left\langle \eta_n \rho_n(\mathbf{U}), \tilde{\mathbf{V}} \right\rangle_{\mathbf{H}} - \left\langle \tilde{\mathbf{U}}, \eta_n \rho_n(\mathbf{V}) \right\rangle_{\mathbf{H}} \right\}, \tag{6.76}$$

where $\mathbf{U}, \mathbf{V} \in \mathrm{Dom}_{\mathrm{reg}}(\mathbf{A}^{0*})$.

Proof The adjoint operator $\mathbf{A}^{0*}$ acts as follows on every element $\mathbf{U} \in \mathrm{Dom}_{\mathrm{reg}}(\mathbf{A}^{0*})$:

$$\mathbf{A}^{0*}\mathbf{U} = \mathbf{A}^{0*}\left(\tilde{\mathbf{U}} + \frac{\mathbf{A}}{\mathbf{A}^2 + 1} \sum_{n=1}^{N} \eta_n \rho_n(\mathbf{U}) \right)$$

$$= \mathbf{A}\tilde{\mathbf{U}} - \frac{1}{\mathbf{A}^2 + 1} \sum_{n=1}^{N} \eta_n \rho_n(\mathbf{U}).$$

Then the boundary form can be calculated as follows

$$
\left\langle \mathbf{U}, \mathbf{A}^{0*}\mathbf{V} \right\rangle_{\mathbf{H}} - \left\langle \mathbf{A}^{0*}\mathbf{U}, \mathbf{V} \right\rangle_{\mathbf{H}}
$$

$$
= \left\langle \tilde{\mathbf{U}} + \frac{\mathbf{A}}{\mathbf{A}^2+1}\sum_{m=1}^{N}\eta_m\rho_m(\mathbf{U}), \mathbf{A}\tilde{\mathbf{V}} - \frac{1}{\mathbf{A}^2+1}\sum_{m=1}^{N}\eta_m\rho_m(\mathbf{V}) \right\rangle_{\mathbf{H}}
$$
$$
- \left\langle \mathbf{A}\tilde{\mathbf{U}} - \frac{1}{\mathbf{A}^2+1}\sum_{m=1}^{N}\eta_m\rho_m(\mathbf{U}), \tilde{\mathbf{V}} + \frac{\mathbf{A}}{\mathbf{A}^2+1}\sum_{m=1}^{N}\eta_m\rho_m(\mathbf{V}) \right\rangle_{\mathbf{H}}
$$

$$
= \left\langle \tilde{\mathbf{U}}, \mathbf{A}\tilde{\mathbf{V}} \right\rangle_{\mathbf{H}} + \sum_{m=1}^{N}\left\langle \frac{\mathbf{A}}{\mathbf{A}^2+1}\eta_m\rho_m(\mathbf{U}), \mathbf{A}\tilde{\mathbf{V}} \right\rangle_{\mathbf{H}}
$$
$$
- \sum_{m=1}^{N}\left\langle \tilde{\mathbf{U}}, \frac{1}{\mathbf{A}^2+1}\eta_m\rho_m(\mathbf{V}) \right\rangle_{\mathbf{H}}
$$
$$
- \left\langle \frac{\mathbf{A}}{\mathbf{A}^2+1}\sum_{m=1}^{N}\eta_m\rho_m(\mathbf{U}), \frac{1}{\mathbf{A}^2+1}\sum_{m=1}^{N}\eta_m\rho_m(\mathbf{V}) \right\rangle_{\mathbf{H}}
$$
$$
- \left\langle \mathbf{A}\tilde{\mathbf{U}}, \tilde{\mathbf{V}} \right\rangle_{\mathbf{H}} + \sum_{m=1}^{N}\left\langle \frac{1}{\mathbf{A}^2+1}\eta_m\rho_m(\mathbf{U}), \tilde{\mathbf{V}} \right\rangle_{\mathbf{H}}
$$
$$
- \sum_{m=1}^{N}\left\langle \mathbf{A}\tilde{\mathbf{U}}, \frac{\mathbf{A}}{\mathbf{A}^2+1}\eta_m\rho_m(\mathbf{V}) \right\rangle_{\mathbf{H}}
$$
$$
+ \left\langle \frac{1}{\mathbf{A}^2+1}\sum_{m=1}^{N}\eta_m\rho_m(\mathbf{U}), \frac{\mathbf{A}}{\mathbf{A}^2+1}\sum_{m=1}^{N}\eta_m\rho_m(\mathbf{V}) \right\rangle_{\mathbf{H}}
$$

$$
= \sum_{m=1}^{N}\left\{ \left\langle \eta_m\rho_m(\mathbf{U}), \tilde{\mathbf{V}} \right\rangle_{\mathbf{H}} - \left\langle \tilde{\mathbf{U}}, \eta_m\rho_m(\mathbf{V}) \right\rangle_{\mathbf{H}} \right\}.
$$

The lemma is proven.

$$\square$$

Comment Lemma 6.2.2 implies that the boundary form of the adjoint operator can be written as

$$
\left\langle \mathbf{U}, \mathbf{A}^{0*}\mathbf{V} \right\rangle_{\mathbf{H}} - \left\langle \mathbf{A}^{0*}\mathbf{U}, \mathbf{V} \right\rangle_{\mathbf{H}}
$$

$$
= \sum_{n=1}^{N}\left\{ \sum_{j=1}^{d_n}\left(\left\langle \rho_{nj}(\mathbf{U}), \langle\varphi_{nj}, \tilde{V}\rangle_{h_n} \right\rangle_{K_n} - \left\langle \langle\varphi_{nj}, \tilde{U}\rangle_{h_n}, \rho_{nj}(\mathbf{V}) \right\rangle_{K_n} \right) \right.
$$
$$
\left. + \sum_{j'=1}^{d'_n}\left(\left\langle \rho'_{nj'}(\mathbf{U}), \langle\varphi'_{nj'}, \tilde{V}'_n\rangle_{h'_n} \right\rangle_{K_n} - \left\langle \langle\varphi'_{nj'}, \tilde{U}'_n\rangle_{h'_n}, \rho'_{nj'}(\mathbf{V}) \right\rangle_{K_n} \right) \right\},
$$
$$
(6.77)
$$

since $\langle \varphi'_{nj'}, \tilde{V}'_m \rangle = 0$ if $m \neq n$.

To define a translational invariant cluster interaction we have to rearrange the boundary form of the adjoint operator. This rearrangement affects only the component of the wave function in the original Hilbert space $\mathcal{H}$. As in the previous section we introduce **the nonsingular components**

$$
\begin{aligned}
V^n &= \tilde{V} + \frac{\mathcal{A}}{\mathcal{A}^2 + 1} \sum_{m=1, m\neq n}^{N} \sum_{j=1}^{d_m} \rho_{mj}(V)\varphi_{mj} \\
&= V - \sum_{j=1}^{d_n} \rho_{nj}(V)\varphi_{nj} \in \mathcal{H}.
\end{aligned}
\tag{6.78}
$$

Lemma 6.2.3 *Let the set of vectors $\{\varphi_{nj}\}_{j=1,n=1}^{d_n \, N}$ be strongly separable with respect to the cluster decompositions. Then the boundary form of the operator adjoint to the total restricted operator is given on the domain of regular elements by the following expression*

$$
\langle U, A^{0*}V \rangle_{\mathbf{H}} - \langle A^{0*}U, V \rangle_{\mathbf{H}}
$$

$$
\begin{aligned}
= \sum_{n=1}^{N} \Bigg\{ &\sum_{j=1}^{d_n} \left(\langle \rho_{nj}(U), \langle \varphi_{nj}, V^n \rangle_{h_n} \rangle_{K_n} - \langle \langle \varphi_{nj}, U^n \rangle_{h_n}, \rho_{nj}(V) \rangle_{K_n} \right) \\
&+ \sum_{j'=1}^{d'_n} \left(\langle \rho'_{nj'}(U), \langle \varphi'_{nj'}, \tilde{V}'_n \rangle_{h'_n} \rangle_{K_n} - \langle \langle \varphi'_{nj'}, \tilde{U}'_n \rangle_{h'_n}, \rho'_{nj'}(V) \rangle_{K_n} \right) \Bigg\},
\end{aligned}
\tag{6.79}
$$

where $U, V \in \mathrm{Dom}_{\mathrm{reg}}(A^{0*})$.

Proof The comment after Lemma 6.2.2 implies that the boundary form of the adjoint operator on the regular elements is given by (6.77). To prove the lemma one needs only to rearrange the terms in the first sum corresponding to the external operator $\mathcal{A}$. This is possible due to the assumption that the set of vectors $\{\varphi_{nj}\}_{j=1,n=1}^{d_n \, N}$ is separable. The proof follows the same lines as that of Lemma 6.1.4.

□

Lemma 6.2.3 shows that it is useful to introduce the following notation

$$
\begin{aligned}
\sigma_{nj}(U) &= \langle \varphi_{nj}, U^n \rangle_{h_n}, \quad j = 1, 2, ..., d_n \\
\sigma'_{nj'}(U) &= \langle \varphi'_{nj'}, \tilde{U}'_n \rangle_{h'_n}, \quad j' = 1, 2, ..., d'_n.
\end{aligned}
\tag{6.80}
$$

In order to define the boundary values of the element U we have used the nonsingular components of the function U, but we have not modified the internal components $\tilde{U}'_n$.

We are also going to use the following vector functions

$$\sigma_n(\mathbf{U}) = \big(\sigma_{n1}(\mathbf{U}), \sigma_{n2}(\mathbf{U}), ..., \sigma_{nd_n}(\mathbf{U}), \sigma'_{n1}(\mathbf{U}), \sigma'_{n2}(\mathbf{U}), ..., \sigma'_{nd'_n}(\mathbf{U})\big) \quad (6.81)$$

and

$$\sigma(\mathbf{U}) = \big(\sigma_1(\mathbf{U}), \sigma_2(\mathbf{U}), ..., \sigma_N(\mathbf{U})\big). \quad (6.82)$$

A similar result holds for separable, but not strongly separable interactions, if one considers densities $\rho_n \in \mathbf{C}^{d_n + d'_n} \times \mathrm{Dom}\,(B_n)$. In particular one can prove the following lemma.

Lemma 6.2.4 *Let the set of vectors $\{\varphi_{nj}\}_{j=1,n=1}^{d_n \quad N}$ be separable with respect to the cluster decompositions and let $\mathbf{U}$ be an arbitrary element from the regular set $\mathrm{Dom}_{\mathrm{reg}}(\mathbf{A}^{0*})$ such that $\rho_{nj}(\mathbf{U}) \in \mathcal{H}_2(B_n) = \mathrm{Dom}\,(B_n)$. Then the function $\sigma_n(\mathbf{U})$ belongs to the space $\mathbf{C}^{d_n + d'_n} \otimes \mathcal{H}_{-2}(B_n)$ for all n. The boundary form of the operator adjoint to the total restricted operator is given by*

$$\big\langle \mathbf{U}, \mathbf{A}^{0*}\mathbf{V} \big\rangle_{\mathbf{H}} - \big\langle \mathbf{A}^{0*}\mathbf{U}, \mathbf{V} \big\rangle_{\mathbf{H}}$$

$$= \sum_{n=1}^{N} \big(\langle \rho_n(\mathbf{U}), \sigma_n(\mathbf{V})\rangle_{K_n} - \langle \sigma_n(\mathbf{U}), \rho_n(\mathbf{V})\rangle_{K_n}\big). \quad (6.83)$$

Proof To prove the lemma it is enough to show that the functions $\sigma_{nj}(\mathbf{U})$ and $\sigma'_{nj}(\mathbf{U})$ belong to the space $\mathcal{H}_{-2}(B_n)$.

Consider an arbitrary function $g \in \mathcal{H}_2(B_n)$. The following equalities

$$\langle g, \sigma_{nj} \rangle_{K_n}$$

$$= \big\langle g \otimes \varphi_{nj}, U_0^n \big\rangle_{\mathbf{H}}$$

$$= \Big\langle g \otimes \varphi_{nj}, \tilde{U}_0 + \frac{\mathcal{A}}{\mathcal{A}^2 + 1} \sum_{m=1, m\neq n}^{N} \sum_{k=1}^{d_m} \rho_{mk} \otimes \varphi_{mk} \Big\rangle_{\mathbf{H}}$$

$$= \big\langle g \otimes \varphi_{nj}, \tilde{U}_0 \big\rangle_{\mathbf{H}} + \sum_{m=1, m\neq n}^{N} \sum_{k=1}^{d_m} \Big\langle g \otimes \varphi_{nj}, \frac{\mathcal{A}}{\mathcal{A}_0^2 + 1} \rho_{mk} \otimes \varphi_{mk} \Big\rangle_{\mathbf{H}}$$

imply that

$$|\langle g, \sigma_{nj} \rangle_{K_n}| \leq \|g\|_{\mathcal{H}_2(B_n)} \|\varphi_{nj}\|_{\mathcal{H}_{-2}(a_n)} \|\tilde{U}_0\|_{\mathcal{H}_2(\mathcal{A})}$$

$$+ \sum_{m=1, m\neq n}^{N} \sum_{k=1}^{d_n} C_{mn} \|g\|_{\mathcal{H}_2(B_n)} \|\rho_{mk}\|_{\mathcal{H}_2(B_m)},$$

since the set of vectors $\{\varphi_{nj}\}_{j=1,n=1}^{d_n\ N}$ is separable with respect to the tensor decomposition. The latter inequality implies that $\sigma_{nj} \in \mathcal{H}_{-2}(B_n)$.

Similarly one can prove that $\sigma'_{nj'} \in \mathcal{H}_{-2}(B_n)$. The boundary form given by Lemma 6.2.2 can be rearranged to get (6.83), since the interactions are separable and the densities for **U** and **V** are from Dom (B_n). The lemma is proven.

$\square$

Lemma 6.2.4 implies that symmetric extensions of the operator $\mathbf{A}^0$ can be described by the boundary conditions connecting the boundary vectors $\rho(\mathbf{\Psi})$ and $\sigma(\mathbf{\Psi})$ of each element $\mathbf{\Psi}$ from the domain of the extension. In order to construct a few-body operator with translational invariant cluster interactions and without interactions between the clusters we choose the same boundary conditions as in the previous section, where inner-cluster generalized interactions were considered. Using an analogy with the first part of the current chapter we can define the few-body operator with the generalized interactions $\mathbf{A}^\Gamma$ on the set of regular elements $\mathbf{\Psi}$

$$\mathbf{\Psi} = \tilde{\mathbf{\Psi}} + \frac{\mathbf{A}}{\mathbf{A}^2 + 1} \sum_{m=1}^{N} \eta_m \rho_m(\mathbf{\Psi})$$

satisfying the following boundary conditions

$$\sigma_m(\mathbf{\Psi}) = \Gamma_m \rho_m(\mathbf{\Psi}), \tag{6.84}$$

where Γ_m are the self–adjoint cluster boundary operators defined by (6.66). The boundary conditions (6.84) can contain extra conditions on the densities $\rho_m(\mathbf{\Psi})$ restricting the operators Γ_m to certain symmetric operators. Let us rewrite the latter boundary conditions in the form

$$\begin{pmatrix} \langle \varphi_{n1}, \tilde{\mathbf{\Psi}} \rangle_{h_n} \\ \cdots \\ \langle \varphi_{nd_n}, \tilde{\mathbf{\Psi}} \rangle_{h_n} \\ \langle \varphi'_{n1}, \tilde{\mathbf{\Psi}}'_n \rangle_{h'_n} \\ \cdots \\ \langle \varphi'_{nd'_n}, \tilde{\mathbf{\Psi}}'_n \rangle_{h'_n} \end{pmatrix} + \begin{pmatrix} \sum_{m \neq n} \sum_{j=1}^{d_m} \left\langle \varphi_{n1}, \frac{\mathbf{A}}{\mathbf{A}^2 + 1} (\rho_{mj}(\mathbf{\Psi}) \otimes \varphi_{mj}) \right\rangle_{h_n} \\ \cdots \\ \sum_{m \neq n} \sum_{j=1}^{d_m} \left\langle \varphi_{nd_n}, \frac{\mathbf{A}}{\mathbf{A}^2 + 1} (\rho_{mj}(\mathbf{\Psi}) \otimes \varphi_{mj}) \right\rangle_{h_n} \\ 0 \\ \cdots \\ 0 \end{pmatrix}$$

$$= \Gamma_n \begin{pmatrix} \rho_{n1}(\Psi) \\ \cdots \\ \rho_{nd_n}(\Psi) \\ \rho'_{n1}(\Psi) \\ \cdots \\ \rho'_{nd'_n}(\Psi) \end{pmatrix}, \quad n = 1, 2, \ldots, N.$$

This system of equations can be written in operator form in the space

$$\mathbf{K}_n = \mathbf{C}^{d_n + d'_n} \otimes K_n. \tag{6.85}$$

The following operators acting in the Hilbert space $\mathbf{K}$ will be used:

• the self–adjoint operator Γ is equal to the orthogonal sum of the self–adjoint cluster boundary operators Γ_n

$$\Gamma = \oplus \sum_{n=1}^{N} \Gamma_n; \tag{6.86}$$

• the bounded self–adjoint operator $\mathbf{R}$ is defined by its matrix elements

$$R_{nk,mj} = \left\langle \varphi_{nk}, \frac{A}{A^2 + 1} (\cdot \otimes \varphi_{mj}) \right\rangle_{h_n}, \quad 1 \leq k \leq d_n, 1 \leq j \leq d_m$$
$$n \neq m; \tag{6.87}$$

$$R_{nk,mj} = 0; \qquad \text{otherwise.}$$

Let us consider also the functions $P_n \tilde{\Psi}$ defined by (6.52) and the corresponding element from $\mathbf{K}$

$$P\tilde{\Psi} = (P_1 \tilde{\Psi}, \ldots, P_N \tilde{\Psi}).$$

Then the system of equations can be written as

$$P\tilde{\Psi} = (\Gamma - \mathbf{R})\rho. \tag{6.88}$$

The few-body operator can be defined only if the operator sum $\Gamma - \mathbf{R}$ is defined as a self–adjoint operator.

We are going to use the following two definitions

Definition 6.2.2 *Let the cluster interactions be strongly separable. Then the* **few-body operator with the generalized interactions** $\mathbf{A}^\Gamma$ *is the closure of the operator* $\mathbf{A}^{0*}$ *restricted to the set of regular elements from* $\mathrm{Dom}_{\mathrm{reg}}(\mathbf{A}^{0*})$ *satisfying the boundary conditions (6.88).*

Definition 6.2.3 *Let the cluster interactions be separable. Then the* **regularized few-body operator with the generalized interactions** A_{reg}^{Γ} *is the closure of the operator* A^{0*} *restricted to the set of regular elements from* $\text{Dom}_{reg}(A^{0*})$ *with the densities*

$$\rho_{nj} \in \text{Dom}(B_n)$$

satisfying the boundary conditions (6.88).

Definition 6.2.2 can be used whenever the operator $\Gamma - R$ is self–adjoint. If the interactions are not strongly separable then the operator $\Gamma - R$ is not well defined. Its domain contains functions from $\text{Dom}(B_n)$ and Definition 6.2.3 can be used to define a symmetric few-body operator. This operator can have nontrivial deficiency indices. In Section 6.2.4 we are going to show that the regularized operator can be self–adjoint under additional assumptions on the interactions.

6.2.4 Self–adjointness of the few-body operator with infinitesimally separable generalized interactions

We have proven that the few-body operator with the singular interactions which we have constructed is self–adjoint in the case where the singular vectors $\{\varphi_{nj}\}$ form a strongly separable system. In this section we are going to show that the few-body operator with generalized interactions can be self–adjoint even if the singular vectors $\{\varphi_{nj}\}$ are not strongly separable. We are going to study the case where the dimensions d_n and d'_n are trivial $d_n = d'_n = 1$. This assumption is not essential but it helps to simplify the calculations. Let us introduce the Hilbert space K equal to the orthogonal sum of the Hilbert spaces K_n

$$K = \oplus \sum_{n=1}^{N} K_n.$$

We consider two operators acting in this Hilbert space:

• the operator $R(\lambda)$ is defined by its matrix elements

$$R_{nm}(\lambda) = \begin{cases} 0, & n = m \\ \left\langle \varphi_n, \dfrac{1}{A - \lambda}(\cdot \otimes \varphi_m) \right\rangle_{h_n}, & n \neq m; \end{cases}$$

- the operator B is equal to the orthogonal sum of the operators B_n

$$B = \oplus \sum_{n=1}^{N} B_n.$$

The operator B is self–adjoint on the domain $\mathrm{Dom}\,(B) = \oplus \sum_{n=1}^{N} \mathrm{Dom}\,(B_n)$. B is positive, since all operators B_n are positive. If the singular vectors φ_n are separable then the operator $R(\lambda)$ is defined via its quadratic form on the domain $\mathrm{Dom}\,(B) \times \mathrm{Dom}\,(B)$.

Let us introduce the following definition.

Definition 6.2.4 *Let the self–adjoint operator A acting in $\mathcal{H}$ possess the tensor decompositions $A = B_n \otimes I_{h_n} + I_{K_n} \otimes a_n$; $\mathcal{H} = K_n \otimes h_n$, $n = 1, 2, \ldots, N$. Then the set of singular vectors $\varphi_n \in \mathcal{H}_{-2}(a_n)$ is called* **infinitesimally separable** *if and only if for any positive $\epsilon > 0$ there exists $b_R = b_R(\epsilon)$ such that*

$$\| R(\lambda)\rho \|_K \leq \epsilon \| (B - \lambda)\rho \|_K + b_R(\epsilon) \| \rho \|_K \qquad (6.89)$$

for arbitrary $\rho \in \mathrm{Dom}\,(B)$ and $\lambda < -1$.

Every strongly separable system of singular vectors satisfying (6.34) is infinitesimally separable, but not every separable system is infinitesimally separable. In fact (6.89) implies that

$$\left| \left\langle \rho_n \otimes \varphi_n, \frac{A}{A^2 + 1} (\rho_m \otimes \varphi_m) \right\rangle_{\mathcal{H}} \right|$$

$$\leq \left| \left\langle \rho_n \otimes \varphi_n, \frac{1}{A - \lambda} (\rho_m \otimes \varphi_m) \right\rangle_{\mathcal{H}} \right|$$

$$+ \left| \left\langle \rho_n \otimes \varphi_n, \frac{1 + \lambda A}{A - \lambda} \frac{1}{A^2 + 1} (\rho_m \otimes \varphi_m) \right\rangle_{\mathcal{H}} \right|$$

$$\leq \left(\epsilon \| B_m \rho_m \|_{K_m} + c(\epsilon, \lambda) \| \rho_m \|_{K_m} \right) \| \rho_n \|_{K_n},$$

where $c(\epsilon, \lambda)$ is a certain positive constant. The assumption that the system is infinitesimally separable is sufficient for what follows but not necessary. We are going to discuss at the end of this section how it is possible to modify this assumption to get even stronger results.

We are going to prove in this section that the regularized few-body operator with infinitesimally separable interactions is self–adjoint under some additional assumptions on the parameters. Without these assumptions one can show that this operator is symmetric, but it can have nontrivial deficiency indices (like the few-body operator with $\mathcal{H}_{-2}$ interactions studied in the first part of this chapter). In what follows we are going to use the following simplified notation $\mathbf{A}^{\Gamma} \equiv \mathbf{A}^{\Gamma}_{\mathrm{reg}}$ for the regularized few-body operator.

Theorem 6.2.2 *Let $\mathbf{A}^{\Gamma}$ be the regularized few-body operator with generalized interactions described in Definition 6.2.3. Let the following conditions be satisfied:*

- *the system of vectors $\{\varphi_n\}$ is independent and infinitesimally separable;*

- *the functions* $r^n(\lambda) = \left(\gamma^n_{11} - \left\langle \varphi'_n, \dfrac{1 + a'_n \lambda}{a'_n - \lambda} \dfrac{1}{a'^2_n + 1} \varphi'_n \right\rangle_{h'_n} \right)^{-1}$ *have nontrivial linear term in the standard representation (1.7), i.e.*

$$\lim_{y \to \infty} \frac{r^n(iy)}{y} = b^n > 0. \tag{6.90}$$

Then the operator $\mathbf{A}^{\Gamma}$ is self–adjoint.

Proof Let us note first that the function $q^n(\lambda)$ is from the Stieltjes class of Nevanlinna functions and the function $r^n(\lambda)$ is a rational transformation of the Stieltjes function $q'(\lambda) = \langle \varphi'_n, \dfrac{1 + a'_n \lambda}{a'_n - \lambda} \dfrac{1}{a'^2_n + 1} \varphi'_n \rangle_{h'_n}$. Therefore these functions satisfy the estimates

$$|q^n(\lambda)| \leq \epsilon |\lambda| + b^n_q(\epsilon) \tag{6.91}$$

for all $\lambda \leq -1$ and

$$|r^n(\lambda) - b^n \lambda| \leq \epsilon |\lambda| + b^n_r(\epsilon) \tag{6.92}$$

for all $\lambda \leq \lambda_0$, where λ_0 is a certain negative constant depending on the cluster interaction. The latter two estimates hold for arbitrary positive ϵ and certain constants b^n_q and b^n_r.

To prove that the operator is self–adjoint it is enough to show that the resolvent equation

$$\frac{1}{\mathbf{A}^{\Gamma} - \lambda} \mathbf{F} = \mathbf{\Psi}. \tag{6.93}$$

is solvable for any $\mathbf{F}$ from a certain dense subset of $\mathbf{H}$. Consider the following dense subset of $\mathbf{H}$

$$\mathbf{H}' = \{ \mathbf{\Psi} \in \mathbf{H} : \mathbf{\Psi} \in \mathcal{H}, \mathbf{\Psi}'_n \in \mathcal{H}_2(B_n) \otimes h'_n \}. \tag{6.94}$$

Since the operator $\mathbf{H}^{\Gamma}$ is symmetric in $\mathbf{H}$, it is enough to prove that the range of the resolvent is dense in $\mathbf{H}$. We are going to show that the resolvent equation is solvable for arbitrary $\mathbf{F} \in \mathbf{H}'$ and for all negative λ having sufficiently large absolute value. The resolvent equation can be written as follows using the densities

$$
\mathbf{F} = (\mathbf{A}^{\Gamma} - \lambda) \left(\tilde{\boldsymbol{\Psi}} + \frac{\mathbf{A}}{\mathbf{A}^2 + 1} \sum_{n=1}^{N} \eta_n \rho_n(\boldsymbol{\Psi}) \right)
$$

$$
= (\mathbf{A} - \lambda)\tilde{\boldsymbol{\Psi}} - \frac{1 + \lambda\mathbf{A}}{\mathbf{A}^2 + 1} \sum_{n=1}^{N} \eta_n \rho_n(\boldsymbol{\Psi}).
$$

Applying the resolvents $\dfrac{1}{\mathcal{A} - \lambda}$ and $\dfrac{1}{\mathcal{A}'_n - \lambda}$ to the corresponding external and internal components of the latter equation we get the equations

$$
\frac{1}{\mathcal{A} - \lambda} F = \Psi^n - \sum_{m=1, m\neq n}^{N} \frac{1}{\mathcal{A} - \lambda}(\rho_m(\boldsymbol{\Psi}) \otimes \varphi_m)
$$

$$
- \frac{1 + \lambda\mathcal{A}}{\mathcal{A} - \lambda} \frac{1}{\mathcal{A}^2 + 1}(\rho_n(\boldsymbol{\Psi}) \otimes \varphi_n) \tag{6.95}
$$

$$
\frac{1}{\mathcal{A}'_n - \lambda} F'_n = \tilde{\Psi}'_n - \frac{1 + \lambda\mathcal{A}'_n}{\mathcal{A}'_n - \lambda} \frac{1}{\mathcal{A}'^2_n + 1}(\rho'_n(\boldsymbol{\Psi}) \otimes \varphi'_n).
$$

Projecting on the vectors φ_n and φ'_n we get

$$
\left\langle \varphi_n, \frac{1}{\mathcal{A} - \lambda} F \right\rangle_{h_n} = \langle \varphi_n, \Psi^n \rangle_{h_n} - \sum_{m=1, m\neq n}^{N} \left\langle \varphi_n, \frac{1}{\mathcal{A} - \lambda}(\rho_m(\boldsymbol{\Psi}) \otimes \varphi_m) \right\rangle_{h_n}
$$

$$
- \left\langle \varphi_n, \frac{1 + \lambda\mathcal{A}}{\mathcal{A} - \lambda} \frac{1}{\mathcal{A}^2 + 1}(\rho_n(\boldsymbol{\Psi}) \otimes \varphi_n) \right\rangle_{h_n}
$$

$$
\left\langle \varphi'_n, \frac{1}{\mathcal{A}'_n - \lambda} F'_n \right\rangle_{h'_n} = \langle \varphi'_n, \tilde{\Psi}'_n \rangle_{h'_n}
$$

$$
- \left\langle \varphi'_n, \frac{1 + \lambda\mathcal{A}'_n}{\mathcal{A}'_n - \lambda} \frac{1}{\mathcal{A}'^2_n + 1}(\rho'_n(\boldsymbol{\Psi}) \otimes \varphi'_n) \right\rangle_{h'_n} .
$$

$$
\tag{6.96}
$$

The functions $\langle \varphi_n, \Psi^n \rangle_{h_n}$, $\langle \varphi'_n, \tilde{\Psi}'_n \rangle_{h'_n}$ are related to the densities $\rho_n(\boldsymbol{\Psi})$, $\rho'_n(\boldsymbol{\Psi})$ via the boundary conditions. We cannot use the boundary conditions directly, since the boundary conditions (6.88) can be written using the operators Γ_n given by (6.66) only if $\rho_n \in \mathrm{Dom}\,(B_n)$. Therefore let us suppose that the densities ρ_n are from the domains of the operators B_n. Then the boundary conditions can be written as

$$
\begin{aligned}
\langle \varphi_n, \Psi^n \rangle_{h_n} &= \gamma_{00}^n \rho_n(\Psi) + \gamma_{01}^n \rho_n'(\Psi) \\
&\quad + B_n \left\langle \varphi_n, \frac{Aa_n - 1}{A^2 + 1} \frac{1}{a_n^2 + 1} (\rho_n(\Psi) \otimes \varphi_n) \right\rangle_{h_n} ; \\
\langle \varphi_n', \tilde{\Psi}_n' \rangle_{h_n'} &= \gamma_{10}^n \rho_n(\Psi) + \gamma_{11}^n \rho_n'(\Psi) \\
&\quad + B_n \left\langle \varphi_n', \frac{A_n' a_n' - 1}{A_n'^2 + 1} \frac{1}{a_n'^2 + 1} (\rho_n'(\Psi) \otimes \varphi_n') \right\rangle_{h_n'} .
\end{aligned}
$$

(6.97)

We obtain the following system of linear equations

$$
\gamma_{00}^n \rho_n(\Psi) + \gamma_{01}^n \rho_n'(\Psi) - \left\langle \varphi_n, \frac{1 + a_n(\lambda - B_n)}{a_n - (\lambda - B_n)} \frac{1}{a_n^2 + 1} (\rho_n(\Psi) \otimes \varphi_n) \right\rangle_{h_n}
$$
$$
- \sum_{m=1, m \neq n}^N \left\langle \varphi_n, \frac{1}{A - \lambda} (\rho_m(\Psi) \otimes \varphi_m) \right\rangle_{h_n}
$$
$$
= \left\langle \varphi_n, \frac{1}{A - \lambda} \right\rangle_{h_n}
$$

$$
\gamma_{10}^n \rho_n(\Psi) + \gamma_{11}^n \rho_n'(\Psi) - \left\langle \varphi_n', \frac{1 + a_n'(\lambda - B_n)}{a_n' - (\lambda - B_n)} \frac{1}{a_n'^2 + 1} (\rho_n'(\Psi) \otimes \varphi_n') \right\rangle_{h_n'}
$$
$$
= \left\langle \varphi_n', \frac{1}{A_n' - \lambda} F_n' \right\rangle_{h_n'} .
$$

(6.98)

The function

$$
\gamma_{11}^n - q_n'(\lambda) = \gamma_{11}^n - \left\langle \varphi_n', \frac{1 + a_n' \lambda}{a_n' - \lambda} \frac{1}{a_n'^2 + 1} \varphi_n' \right\rangle_{h_n'}
$$

(6.99)

is not equal to zero for negative λ having sufficiently large absolute value, since the function r^n satisfies (6.92). Therefore the second equation (6.98) can be resolved as follows

$$
\rho_n'(\Psi) = (\gamma_{11}^n - q_n'(\lambda - B_n))^{-1} \left(-\gamma_{10}^n \rho_n(\Psi) + P_n' \frac{1}{A_n' - \lambda} F_n' \right).
$$

(6.100)

Excluding the densities $\rho_n'(\Psi)$ from the linear system (6.98) we get the following linear system of equations for the densities $\rho^n(\Psi)$

$$\left(\gamma_{00}^n - |\gamma_{01}^n|^2 r^n(\lambda - B_n) - q^n(\lambda - B_n)\right)\rho^n(\Psi)$$

$$- \sum_{m=1,m\neq n}^{N} \left\langle \varphi_n, \frac{1}{\mathcal{A}_n - \lambda}(\rho_m(\Psi) \otimes \varphi_m) \right\rangle_{h_n} \tag{6.101}$$

$$= -\gamma_{01}^n r^n(\lambda - B_n)\left\langle \varphi_n', \frac{1}{\mathcal{A}_n' - \lambda}F_n' \right\rangle_{h_n'} + \left\langle \varphi_n, \frac{1}{\mathcal{A} - \lambda}F \right\rangle_{h_n}.$$

One has to prove that the latter equation is solvable for all negative λ having sufficiently large absolute value. We consider this equation in the space K introduced earlier.

Let us introduce the vector $g(\lambda) \in K$ with the components

$$g^n(\lambda) = -\gamma_{01}^n r^n(\lambda - B_n)\left\langle \varphi_n', \frac{1}{\mathcal{A}_n' - \lambda}F_n' \right\rangle_{h_n'} + \left\langle \varphi_n, \frac{1}{\mathcal{A} - \lambda}F \right\rangle_{h_n}.$$

The functions $g^n(\lambda)$ belong to the spaces K_n since $\mathbf{F} \in \mathbf{H}'$.

We consider the operator $T(\lambda)$ defined by its matrix elements in the orthogonal decomposition of K :

$$T_{nm}(\lambda) = \begin{cases} \left(\gamma_{00}^n - |\gamma_{01}^n|^2 r^n(\lambda - B_n) - q^n(\lambda - B_n)\right), & n = m; \\[2mm] -\left\langle \varphi_n, \frac{1}{\mathcal{A} - \lambda}(\cdot \otimes \varphi_m) \right\rangle_{h_n}, & n \neq m. \end{cases} \tag{6.102}$$

The system of equations (6.101) can be written as

$$T(\lambda)\rho(\mathbf{psi}) = g(\lambda).$$

The operators γ_{00}^n are bounded. The operators $q^n(\lambda - B_n)$ are infinitesimally small with respect to the operator $-\lambda + B_n$ (estimate (6.91)). Every operator $r^n(\lambda - B_n)$ is equal to the sum of the operator $b^n(\lambda - B_n)$, $b^n > 0$ and a certain operator which is infinitesimally small with respect to the operator $-\lambda + B_n$ (estimate (6.92)). The nondiagonal components are also infinitesimally small with respect to the operator B, since the interactions are infinitesimally separable.

Thus the operator $T(\lambda)$ can be considered as an infinitesimally small perturbation of the operator

$$\oplus \sum_{n=1}^{N} |\gamma_{01}^n|^2 b^n(-\lambda + B_n).$$

Therefore the operator $T(\lambda)$ is essentially selfadjoint on Dom (B). (Without loss of generality we can suppose that all parameters γ_{01}^n are different from zero.) Let us prove that there exists a certain $\lambda_0 < 0$ such that the operator $T(\lambda)$ is positive definite:

$$T(\lambda) \geq c > 0$$

for all $\lambda < \lambda_0$. We consider the operator

$$T(\lambda) + \left(\oplus \sum_{n=1}^{N} |\gamma_{01}^n|^2 b^n \right) \lambda$$

as a perturbation of the positive operator

$$\oplus \sum_{n=1}^{N} b^n |\gamma_{01}^n|^2 B^n.$$

The perturbation term is infinitesimally small with respect to the operator B

$$\left\| \left(T(\lambda) + \oplus \sum_{n=1}^{N} b^n |\gamma_{01}^n|^2 (\lambda - B_n) \right) \rho \right\|_K \leq \epsilon \parallel B\rho \parallel_K + \epsilon |\lambda| \parallel \rho \parallel_K + b(\epsilon) \parallel \rho \parallel_F$$

for any $\epsilon > 0$ and where $b(\epsilon)$ is a certain positive constant. It follows that the perturbed operator is bounded from below by $-\epsilon |\lambda| - b(\epsilon)$. Hence choosing $\epsilon < \min_n \{ b^n |\gamma_{01}^n|^2 \}$, we prove that there exists a certain $\lambda_0 < 0$ such that the operator $T(\lambda)$ has a bounded inverse for all $\lambda < \lambda_0$. The system of equations (6.101) can be solved for all negative λ with sufficiently large absolute value. Thus the operator $\mathbf{A}^{\Gamma}$ is self–adjoint and bounded from below. The theorem is proven.

$\square$

We have mentioned that the conditions of the theorem are not optimal. For example it is enough to suppose that the estimate (6.89) is satisfied for a certain $\epsilon > 0$. Then one can prove that the operator $\mathbf{A}^{\Gamma}$ is self-adjoint for sufficiently large coupling constants $|\gamma_{01}^n|, n = 1, 2, \ldots, N$. One can show that in special cases it is enough to suppose that only some of the functions $q^n(\lambda)$ are growing linearly when $\lambda \to \infty$. For example for the three body system in $\mathbf{R}^3$ with pair interactions given by delta functions the operator is self–adjoint if two of the three cluster functions $q^n(\lambda)$ contain nontrivial linear term [669, 673, 671, 674, 786]. This result can be generalized to the case of an arbitrary number of particles [643]. (See Section A.4 for detailed discussion.)

Chapter 7

Three-body models in one dimension

7.1 Schrödinger operators constructed using self–adjoint extensions

7.1.1 Tensor structure of the Hilbert space

The Schrödinger operator $\mathcal{L}_\alpha$ describing a system of three one dimensional particles with pairwise interactions given by delta functions is studied in this section. We confine our consideration to the case where the particles have equal masses. The corresponding operator was introduced in Chapter 6 and is formally given by the differential expression

$$\mathcal{L}_\alpha = -\Delta + \alpha_1 \delta(r_2 - r_3) + \alpha_2 \delta(r_3 - r_1) + \alpha_3 \delta(r_1 - r_2). \qquad (7.1)$$

The coordinates of three particles are denoted by r_1, r_2, r_3. The real parameters α_γ characterize the two-body interactions in the system. Thus the interaction is determined by the three dimensional real vector $\alpha = (\alpha_1, \alpha_2, \alpha_3)$. This operator can be restricted to the plane of zero total momentum $\Lambda = \{(r_1, r_2, r_3) \in \mathbf{R}^3 : r_1 + r_2 + r_3 = 0\}$. Three different systems of Jacobi coordinates associated with different cluster decompositions of three particles can be introduced:

$$s_\gamma \equiv x_{\alpha,\beta} = r_\alpha - r_\beta,$$

$$t_\gamma \equiv x_{\alpha\beta,\gamma} = \sqrt{\frac{4}{3}} \left(\frac{1}{2}(r_\alpha + r_\beta) - r_\gamma \right). \qquad (7.2)$$

273

(Where $\{\alpha, \beta, \gamma\}$ is a permutation of $\{1, 2, 3\}$.) Then the Laplace operator
$$\mathcal{L} = -\Delta = -\frac{1}{2}\left(\frac{\partial^2}{\partial r_1^2} + \frac{\partial^2}{\partial r_2^2} + \frac{\partial^2}{\partial r_3^2}\right) \text{ considered in the Hilbert space } L_2(\Lambda)$$
possesses three tensor decompositions

$$-\Delta = -\left(\frac{\partial^2}{\partial t_\gamma^2} + \frac{\partial^2}{\partial s_\gamma^2}\right), \quad \gamma = 1, 2, 3. \tag{7.3}$$

This operator is self–adjoint on the domain $\text{Dom}(\mathcal{L}) = W_2^2(\Lambda)$. The corresponding tensor decompositions of the Hilbert space $L_2(\Lambda)$ are

$$L_2(\Lambda) = L_2(\mathbf{R}_{t_\gamma}) \otimes L_2(\mathbf{R}_{s_\gamma}), \quad \gamma = 1, 2, 3. \tag{7.4}$$

The delta functions $\delta(s_\gamma)$ are elements from the corresponding spaces

$$\mathcal{H}_{-1}(-\frac{d^2}{ds_\gamma^2})$$

and it follows that the two-body interactions are strongly separable. Moreover the set of delta functions $\delta^1(s_1), \delta^2(s_2), \delta^3(s_3)$ is independent. Thus the self–adjoint operator corresponding to the formal expression (7.1) can be defined using the general approach described in Chapter 6 for few-body operators with $\mathcal{H}_{-1}$ interactions. In this section we are going to show how to define the self–adjoint operator $\mathcal{L}_\alpha$ explicitly. The spectrum and eigenfunctions of the operator $\mathcal{L}_\alpha$ will be calculated.

Let us introduce several notations first. We denote the lines on Λ where the coordinates of two of the three particles coincide by $l_\gamma = \{(t_\gamma, s_\gamma) \in \Lambda : s_\gamma = 0\}$. These lines intersect at the origin where the coordinates of all three particles are equal to zero. The lines divide the plane Λ onto six equal sectors. By φ^γ we denote the normalized delta function $\varphi^\gamma = c\delta^\gamma(s_\gamma)$, $c = 2^{3/4}$, having support on the corresponding line l_γ. Each delta function is normalized as an element from the space $\mathcal{H}_{-2}(-\frac{\partial^2}{\partial s_\gamma^2})$ (see Section 2.4.2).

7.1.2   Definition of the Schrödinger operator as a self–adjoint operator.

Consider the restriction $\mathcal{L}^0$ of the operator $\mathcal{L}$ to the domain of functions from $W_2^2(\Lambda)$ vanishing on the lines $s_\gamma = 0$:

$$\text{Dom}(\mathcal{L}^0) = \{\Psi \in W_2^2(\Lambda) : \Psi(t_\gamma, s_\gamma)|_{s_\gamma=0} = 0, \gamma = 1, 2, 3\}. \tag{7.5}$$

The operator $\mathcal{L}^0$ is symmetric and densely defined. The set $\text{Dom}_{\text{reg}}(\mathcal{L}^{0*})$ of regular elements from the domain of the adjoint operator can be introduced as follows:

$$\text{Dom}_{\text{reg}}(\mathcal{L}^{0*}) = \left\{ \Psi : \Psi = \hat{\Psi} + \sum_{\gamma=1}^{3} \frac{\mathcal{L}}{\mathcal{L}^2 + 1} \left(\rho_\gamma(\Psi) \otimes \varphi^\gamma \right), \right.$$
$$\left. \rho_\gamma(\Psi) \in W_2^2(\mathbf{R}_{t_\gamma}), \ \hat{\Psi} \in \text{Dom}(\mathcal{L}^0) \right\}. \tag{7.6}$$

This is just the set of regular elements having densities from the class $W_2^2(\mathbf{R}_{t_\gamma})$. This set is contained in the set of regular elements introduced in Chapter 6. It was proven there that the densities for the functions from the domain of the self–adjoint operator $\mathcal{L}_\alpha$ are from the class $\mathcal{H}_{-2}(B_n)$ if the interactions are from the class $\mathcal{H}_{-1}$. It follows that to define the self–adjoint few-body operator $\mathcal{L}_\alpha$ it is enough to consider the densities ρ_γ appearing in (7.6).

The functions from the regular domain can be characterized by the following lemma.

Lemma 7.1.1 *Let us denote by* $\text{Dom}_{\text{reg}}^{\infty}(\mathcal{L}^{0*})$ *the set of functions* $\Psi = \hat{\Psi} +$ $\sum_{\gamma=1}^{3} \frac{\mathcal{L}}{\mathcal{L}^2 + 1} \left(\rho_\gamma(\Psi) \otimes \varphi^\gamma \right)$ *from the regular domain* $\text{Dom}_{\text{reg}}(\mathcal{L}^{0*})$, *such that* $\hat{\Psi} \in$ $C_0^{\infty}(\Lambda) \cap \text{Dom}(\mathcal{L}^0)$. *Then every function from* $\text{Dom}_{\text{reg}}^{\infty}(\mathcal{L}^{0*})$ *belongs to the space* $W_2^2(\Lambda \setminus \{l_\gamma\}_{\gamma=1}^{3}) \cap C(\Lambda)$. *The normal derivative* $\partial\Psi/\partial s_\gamma$ *has a jump discontinuity on each line* l_γ *such that*

$$[\partial\Psi/\partial s_\gamma]|_{s_\gamma=0} \in L_2(\mathbf{R}_{t_\gamma}). \tag{7.7}$$

Proof Each function $\Psi_\gamma^{\pm} = \dfrac{1}{\mathcal{L} \mp i} \left(\rho_\gamma(\Psi) \otimes \varphi^\gamma \right) = \dfrac{c}{\mathcal{L} \mp i} \left(\rho_\gamma(\Psi) \otimes \delta^\gamma \right)$ is an element from the space $W_2^2(\Lambda \setminus \{l_\gamma\}_{\gamma=1}^{3})$ and continuous on Λ. This function is equal to a simple layer potential with the density from the class $W_2^2(\mathbf{R}_{t_\gamma})$. Therefore the normal derivative of $\Psi_\gamma^{\pm}$ has a jump discontinuity on the line l_γ related to the density as follows:

$$[\partial\Psi_\gamma^{\pm}/ds_\gamma]|_{s_\gamma=0} = -c\rho_\gamma(\Psi). \tag{7.8}$$

Every function Ψ from $\text{Dom}_{\text{reg}}^{\infty}(\mathcal{L}^{0*})$ is equal to the sum of the $C_0^{\infty}(\Lambda)$ function $\hat{\Psi}$ and six simple layer potentials $\Psi_\gamma^{\pm}$. It follows that every such function Ψ is continuous on Λ and has jump discontinuities of the first normal derivatives on each line l_γ. The lemma is proven.

$\square$

The domain $\text{Dom}(\mathcal{L}_\alpha)$ of the self–adjoint operator $\mathcal{L}_\alpha$ can be introduced following Chapter 6. We are going to present here the explicit definition of this domain. We mention first that the domain $\text{Dom}^\infty(\mathcal{L}_\alpha) = \text{Dom}^\infty_{\text{reg}}(\mathcal{L}^{0*}) \cap \text{Dom}\,(\mathcal{L}_\alpha)$ is dense in the domain $\text{Dom}\,(\mathcal{L}_\alpha)$ in the graph norm of the operator $\mathcal{L}_\alpha$. This follows from the fact that the domain $\text{Dom}\,(\mathcal{L}^0) \cap C_0^\infty(\Lambda)$ is dense in the domain $\text{Dom}\,(\mathcal{L}^0)$ with respect to the graph norms of the operators $\mathcal{L}$ and $\mathcal{L}_\alpha$, since the restrictions of the operators $\mathcal{L}$ and $\mathcal{L}_\alpha$ to $\text{Dom}\,(\mathcal{L}^0)$ coincide. Therefore the operator $\mathcal{L}_\alpha$ is essentially self–adjoint on the domain $\text{Dom}^\infty(\mathcal{L}_\alpha)$. This domain is described by

Lemma 7.1.2 *Every function Ψ from the domain $\text{Dom}^\infty(\mathcal{L}_\alpha)$ is continuous on Λ and satisfies the boundary conditions on the lines l_γ*

$$\begin{cases} [\partial\Psi/\partial s_\gamma]|_{s_\gamma=0} = \alpha_\gamma\Psi|_{s_\gamma=0}, \\[2mm] [\Psi]|_{s_\gamma=0} = 0, \end{cases} \qquad \gamma=1,2,3. \qquad (7.9)$$

Note Every function $\Psi \in \text{Dom}^\infty(\mathcal{L}_\alpha)$ is bounded on Λ.

Proof Every function $\Psi \in \text{Dom}^\infty(\mathcal{L}_\alpha)$ is mapped by the linear operator $\mathcal{L}_\alpha$ to the following functional

$$\begin{aligned} \mathcal{L}_\alpha\Psi &= \left(\mathcal{L} + \sum_{\gamma=1}^3 \alpha_\gamma\delta^\gamma\right)\left(\hat{\Psi} + c\sum_{\beta=1}^3 \frac{\mathcal{L}}{\mathcal{L}^2+1}(\rho_\beta(\Psi)\otimes\delta^\beta)\right) \\[2mm] &= \mathcal{L}\hat{\Psi} - c\sum_{\gamma=1}^3 \frac{1}{\mathcal{L}^2+1}(\rho_\gamma(\Psi)\otimes\delta^\gamma) \\[2mm] &\quad + \sum_{\gamma=1}^3 \left\{c\rho_\gamma(\Psi) + \alpha_\gamma\Psi|_{s_\gamma=0}\right\}\otimes\delta^\gamma. \end{aligned} \qquad (7.10)$$

This expression belongs to the Hilbert space $L_2(\Lambda)$ if and only if each expression in braces is equal to zero, i.e. the following conditions hold (see (7.8))

$$-[\partial\Psi/\partial s_\gamma]|_{s_\gamma=0} + \alpha_\gamma\Psi|_{s_\gamma=0} = 0.$$

These conditions together with the continuity of the function Ψ on Λ imply (7.9). The lemma is proven.

$\square$

The functions from the domain $\text{Dom}^\infty(\mathcal{L}_\alpha)$ do not have compact support and are not necessarily differentiable in the classical sense outside the lines l_γ. Let us introduce the set $\text{Dom}^\infty_0(\mathcal{L}_\alpha)$ of all functions Ψ that

• are infinitely many times differentiable outside the lines l_γ:

$$\Psi \in C^\infty(\Lambda \setminus \{l_\gamma\}_{\gamma=1}^3);$$

• have compact support;

• satisfy boundary conditions (7.9).

One can prove that the operator $\mathcal{L}_\alpha$ is essentially self–adjoint on $\mathrm{Dom}_0^\infty(\mathcal{L}_\alpha)$.

We have just proven that the operator $\mathcal{L}_\alpha$ is equal to the Laplace operator on Λ defined on the set of functions satisfying the boundary conditions (7.9). These boundary conditions replace the boundary conditions (6.28) derived for the general operator defined in Chapter 6. It appears that if one uses the boundary values $\Psi|_{s_\gamma=0}$ and $[\partial\Psi/\partial s_\gamma]|_{s_\gamma=0}$ then the boundary operator coincides with the operator of multiplication by a constant (in the standard basis considered in Chapter 6 the boundary operator is a differential operator).

The spectrum and eigenfunctions of the operator defined here will be studied in the following section.

7.1.3 Spectrum, eigenfunctions and Bethe Ansatz

The spectrum of the operator $\mathcal{L}_\alpha$ has been studied intensively. It has been proven that the discrete spectrum is empty. The continuous spectrum consists of the branch $[0, \infty)$ of infinite multiplicity (corresponds to the case where the three particles are free) and perhaps two-body branches

$$[\lambda_j, \infty), \quad j = 1, 2, 3,$$

of multiplicity 2 (corresponding to the scattering processes where two of the three particles are in the bound state). The two-body branches of the continuous spectrum are present if and only if the interaction parameters α_γ are negative. Then the energy of the two-body bound state λ_γ is equal to

$$\lambda_\gamma = -\frac{\alpha_\gamma^2}{4}.$$

The eigenfunctions corresponding to the continuous spectrum can be calculated using the following analogy with geometric optics. The eigenfunction is a solution to the equation

$$-\Delta\Psi = \lambda\Psi \qquad (7.11)$$

inside each sector on the plane Λ. The three lines l_γ can be considered as screens with transition coefficient T_γ and reflection coefficient R_γ. To construct the eigenfunctions one considers an arbitrary initial plane wave and its reflections on the described system of screens. As a result one obtains a system of plane waves satisfying the boundary conditions on the screens. This way of constructing the eigenfunctions is called the Bethe Ansatz. Unfortunately it cannot be used to construct eigenfunctions for a system of arbitrary particles. Even if the particles have equal masses but the two-body interactions are characterized by delta functions of different strength α_γ, the Bethe Ansatz defines a system of plane waves which is discontinuous inside the sectors. It follows that the set of plane waves constructed using the two-body reflection and transition coefficients and laws of geometric optics is not a solution to equation (7.11) inside some of the sectors. Therefore the eigenfunctions should contain additional Fresnel waves. We are going to confine our consideration to the case where the eigenfunctions are given by the Bethe Ansatz. In fact we shall deduce a necessary and sufficient condition for this to happen.

Let us introduce some additional notation. Let (α, β, γ) be an arbitrary permutation of the numbers $(1, 2, 3)$. We denote by $\Lambda_{\alpha\beta\gamma}$ the sector on the plane Λ where the coordinates of the particles satisfy the estimate

$$r_\alpha > r_\beta > r_\gamma.$$

We introduce polar coordinates (r, φ) on the plane Λ with the center at the origin $r_1 = r_2 = r_3 = 0$. The angle φ, $0 \le \varphi \le 2\pi$, is defined in such a way that $\varphi = 0, \pi$ for all points on the line l_1 and sector Λ_{123} consists of all points with the angle satisfying the estimate $0 \le \varphi \le \pi/3$, i.e.

$$\Lambda_{123} = \{(r, \varphi), 0 \le r < \infty, 0 \le \varphi \le \pi/3\}.$$

Then the other five sectors are given by

$$\Lambda_{213} = \{(r, \varphi), 0 \le r < \infty, \pi/3 \le \varphi \le 2\pi/3\},$$

$$\Lambda_{231} = \{(r, \varphi), 0 \le r < \infty, 2\pi/3 \le \varphi \le \pi\},$$

$$\Lambda_{321} = \{(r, \varphi), 0 \le r < \infty, \pi \le \varphi \le 4\pi/3\},$$

$$\Lambda_{312} = \{(r, \varphi), 0 \le r < \infty, 4\pi/3 \le \varphi \le 5\pi/3\},$$

$$\Lambda_{132} = \{(r, \varphi), 0 \le r < \infty, 5\pi/3 \le \varphi \le 2\pi\}.$$

Let $\Psi_0 = \exp(ipr\cos(\varphi - \varphi_0))$, $p^2 = \lambda > 0$, be an inducing plane wave coming from the sector Λ_{123}. Then the initial angle φ_0 should satisfy the estimate

$$\pi \leq \varphi_0 \leq 4\pi/3.$$

Let us introduce the absolute values of the projections of the wave vector to the normals to the screens l_γ :

$$p_1 \;=\; p\cos(\varphi_0 - 3\pi/2) \;=\; p\sin(\varphi_0 - \pi),$$

$$p_2 \;=\; p\cos(\varphi_0 - 7\pi/6) \;=\; p\sin(\varphi_0 - 2\pi/3), \tag{7.12}$$

$$p_3 \;=\; p\cos(\varphi_0 - 5\pi/6) \;=\; p\sin(\varphi_0 - \pi/3).$$

These projections satisfy the equality

$$p_3 + p_1 = p_2. \tag{7.13}$$

An arbitrary incoming plane wave $\Psi_0 = \exp(ipr\cos(\varphi - \varphi_0))$ induces 18 plane waves in all six sectors of Λ. The plane wave coming from the sector Λ_{123} can induce plane waves with discontinuous amplitudes in sectors Λ_{123}, Λ_{213} and Λ_{132}. Even the reflected wave $\exp(ipr\cos(\varphi + \varphi_0 + 2\pi/3))$ in the sector Λ_{123} has a discontinuity $T_3(p_1)T_3(p_3) - T_1(p_1)T_1(p_3)$. The amplitudes of all plane waves in all sectors are continuous if and only if the following conditions are satisfied

$$T_3(p_1)T_3(p_3) = T_1(p_1)T_1(p_3); \tag{7.14}$$

$$R_3(p_1)R_2(p_2)T_3(p_3) + T_3(p_1)R_1(p_2)R_3(p_3) = R_2(p_3)T_3(p_2)R_1(p_1); \tag{7.15}$$

$$R_1(p_3)R_2(p_2)T_1(p_1) + T_1(p_3)R_3(p_2)R_1(p_1) = R_2(p_1)T_1(p_2)R_3(p_3). \tag{7.16}$$

The eigenfunctions are given by the Bethe Ansatz if and only if these three equalities are satisfied for arbitrary p_1, p_2, p_3 subject to condition (7.13). The two-body reflection and transition coefficients for the two-body interaction given by delta function have already been calculated:

$$R_\gamma(k) \;=\; \frac{\alpha_\gamma}{2ik - \alpha_\gamma};$$

$$T_\gamma(k) \;=\; \frac{2ik}{2ik - \alpha_\gamma}. \tag{7.17}$$

These two-body scattering matrices satisfy equation (7.14) for arbitrary p_1 and p_3 if and only if α_1 and α_3 are equal, i.e. if the particles are identical. (We do not consider the degenerate cases where one or several of the coefficients α_γ are equal to 0 or ∞.) Equalities (7.15), (7.16) imply that

$$\alpha_1\alpha_2 p_3 + \alpha_1^2 p_1 \;=\; \alpha_1\alpha_2 p_2,$$

$$\alpha_1\alpha_2 p_1 + \alpha_1^2 p_3 \;=\; \alpha_1\alpha_2 p_2,$$

and it follows that the following equality is satisfied for arbitrary p_1, p_3 :

$$\alpha_1(\alpha_2 - \alpha_1)p_1 + \alpha_1(\alpha_1 - \alpha_2)p_3 = 0.$$

This implies that all coefficients α_γ are equal: $\alpha_1 = \alpha_2 = \alpha_3$. [1] Only in this case does the set of plane waves gives a complete description of the continuous spectrum eigenfunctions for the operator $\mathcal{L}_\alpha$. If the particles are not identical then the solution of the problem can be calculated using the Sommerfeld integral (see [701, 702, 703, 705, 706]). We confine our consideration in what follows to the case of identical particles.

Let all coefficients α_γ be equal. Then the set Ψ^p of plane waves constructed using the Bethe Anstaz forms an eigenfunction for the operator $\mathcal{L}_\alpha$. We denote by $\Psi^p_{\alpha\beta\gamma}$ the restriction of the function Ψ^p to the sector $\Lambda_{\alpha\beta\gamma}$. Then the set of plane waves is given by

$$\Psi^p_{123} \;=\; e^{ipr\cos(\varphi-\varphi_0)}$$

$$+ R(p_3)e^{ipr\cos(\varphi+\varphi_0-2\pi/3)} + R(p_2)R(p_3)e^{ipr\cos(\varphi-\varphi_0+2\pi/3)}$$

$$+ R(p_1)e^{ipr\cos(\varphi+\varphi_0)} + R(p_2)R(p_1)e^{ipr\cos(\varphi-\varphi_0-2\pi/3)}$$

$$+ (T(p_1)R(p_2)T(p_3) + R(p_1)R(p_2)R(p_3))e^{ipr\cos(\varphi+\varphi_0+2\pi/3)};$$

$$\Psi^p_{213} \;=\; T(p_3)e^{ipr\cos(\varphi-\varphi_0)} + R(p_2)T(p_3)e^{ipr\cos(\varphi+\varphi_0+2\pi/3)}$$

$$+ T(p_2)R(p_1)e^{ipr\cos(\varphi+\varphi_0)} + R(p_3)T(p_2)R(p_1)e^{ipr\cos(\varphi-\varphi_0+2\pi/3)};$$

$$\Psi^p_{231} \;=\; T(p_2)T(p_3)e^{ipr\cos(\varphi-\varphi_0)} + T(p_3)T(p_2)R(p_1)e^{ipr\cos(\varphi+\varphi_0)};$$

$$\Psi^p_{321} \;=\; T(p_1)T(p_2)T(p_3)e^{ipr\cos(\varphi-\varphi_0)};$$

---

[1] For general boundary conditions given by four parameter family discussed in [24, 25] this conclusion is no longer true.

$$\Psi^p_{312} = T(p_2)T(p_1)e^{ipr\cos(\varphi-\varphi_0)} + R(p_3)T(p_2)T(p_1)e^{ipr\cos(\varphi+\varphi_0-2\pi/3)};$$

$$\Psi^p_{132} = T(p_1)e^{ipr\cos(\varphi-\varphi_0)} + R(p_2)T(p_1)e^{ipr\cos(\varphi+\varphi_0+2\pi/3)}$$

$$+T(p_2)R(p_3)e^{ipr\cos(\varphi+\varphi_0-2\pi/3)} \tag{7.18}$$

$$+R(p_1)T(p_2)R(p_3)e^{ipr\cos(\varphi-\varphi_0-2\pi/3)}.$$

If the parameters $\alpha_1 = \alpha_2 = \alpha_3$ are negative then the set of two–body waves can be constructed using a similar method. We are going to present corresponding formulas in the following section for the few-body operator with generalized two-body interactions.

7.2 Operators with generalized delta interactions

7.2.1 Two-body generalized delta interactions

We consider first the operator describing the three one dimensional particles with only one two-body generalized delta interaction. Suppose that we would like to introduce a generalized delta interaction between the particles α and β. We choose first the extension space $\mathcal{H}'_\gamma$ as follows

$$\mathcal{H}'_\gamma = L_2(\mathbf{R}_{t_\beta}) \otimes h'_\gamma, \tag{7.19}$$

where $h'_\gamma = \mathbf{C}^{n_\gamma}$ is a finite dimensional Hilbert space. Then the total unperturbed operator is defined in the Hilbert space

$$\mathbf{H}_\gamma = L_2(\Lambda) \oplus \mathcal{H}'_\gamma \tag{7.20}$$

by the formula

$$\mathbf{A}_\gamma = -\left(\frac{\partial^2}{\partial t_\gamma^2} + \frac{\partial^2}{\partial s_\gamma^2}\right) \oplus \left(-\frac{\partial^2}{\partial t_\gamma^2} + a'_\gamma\right), \tag{7.21}$$

where a'_γ is an arbitrary Hermitian matrix acting in h'_γ. The operator $\mathbf{A}_\gamma$ possesses the tensor decomposition

$$\mathbf{A}_\gamma = -\frac{\partial^2}{\partial t_\gamma^2} \otimes \left(-\frac{\partial^2}{\partial s_\gamma^2} \oplus a'_\gamma\right). \tag{7.22}$$

The perturbed operator $\mathbf{A}_\gamma^B$ will be constructed so as to possess a similar decomposition $\mathbf{A}_\gamma^B = -\frac{\partial^2}{\partial t_\gamma^2} \otimes A_\gamma^B$. The perturbed operator A_γ^B is a generalized delta perturbation of the second derivative in $L_2(\mathbf{R}_{s_\gamma})$. This operator has already been considered in Chapter 2 (see Section 2.4). To construct the perturbed operator we choose an arbitrary element $\varphi' \in h'_\gamma$ and an arbitrary

2×2 real matrix $B = \begin{pmatrix} \beta_{11} & \beta_{12} \\ \beta_{21} & \beta_{22} \end{pmatrix}$ with unit determinant $\beta_{11}\beta_{22} - \beta_{21}\beta_{12} = 1$.

Then the operator A_γ^B is defined by the formula

$$A_\gamma^B \begin{pmatrix} \Psi \\ \Psi'_\gamma \end{pmatrix} = \begin{pmatrix} \left(-\dfrac{\partial^2}{\partial s_\gamma^2}\right)^{0*} \Psi \\ a'_\gamma \Psi'_\gamma + \left(\beta_{11}\Psi|_{s_\gamma=0} + \beta_{12}[\partial\Psi/\partial s_\gamma]|_{s_\gamma=0}\right)\varphi' \end{pmatrix} \qquad (7.23)$$

on the functions from the set Dom (A_γ^B) of functions from

$$\left\{\Psi \in W_2^2(\mathbf{R}_{s_\gamma} \setminus \{0\}) : \Psi(-0) = \Psi(+0)\right\} \oplus h'_\gamma \qquad (7.24)$$

satisfying the boundary conditions

$$\langle\varphi', \Psi'\rangle_{h'_\gamma} = \beta_{21}\Psi|_{s_\gamma=0} + \beta_{22}[\partial\Psi/\partial s_\gamma]|_{s_\gamma=0}. \qquad (7.25)$$

Here $\left(-\dfrac{\partial^2}{\partial s_\gamma^2}\right)^0$ denotes the second derivative operator in $L_2(\mathbf{R}_{s_\gamma})$ restricted

to the set of functions equal to zero at the origin $s_\gamma = 0$; and $\left(-\dfrac{\partial^2}{\partial s_\gamma^2}\right)^{0*}$ is

its adjoint.

Let us denote by $\mathrm{Dom}_0^\infty(\mathbf{A}_\gamma^B)$ the set of functions from

$$C_0^\infty(\mathbf{R}_{t_\gamma}) \otimes \mathrm{Dom}\,(A_\gamma^B),$$

i.e. all functions from

$$C_0^\infty(\mathbf{R}_{t_\gamma}) \otimes \left(W_2^2(\mathbf{R}_{s_\gamma} \setminus \{0\}) \oplus h'_\gamma\right) \subset W_2^2(\Lambda \setminus l_\gamma) \oplus \left(C_0^\infty(\mathbf{R}_{t_\gamma}) \otimes h'_\gamma\right)$$

satisfying the boundary conditions

$$\begin{cases} \Psi(t_\gamma, s_\gamma)|_{s_\gamma=-0} = \Psi(t_\gamma, s_\gamma)|_{s_\gamma=+0}, \\ \langle\varphi', \Psi'(t_\gamma)\rangle_{h'_\gamma} = \beta_{21}\Psi(t_\gamma, s_\gamma)|_{s_\gamma=0} + \beta_{22}[\partial\Psi(t_\gamma, s_\gamma)/\partial s_\gamma]|_{s_\gamma=0}. \end{cases} \qquad (7.26)$$

Then the perturbed operator $\mathbf{A}_\gamma^B$ restricted to the domain $\mathrm{Dom}_0^\infty(\mathbf{A}_\gamma^B)$ is defined by the formula

$$\mathbf{A}_\gamma^B \begin{pmatrix} \Psi \\ \Psi'_\gamma \end{pmatrix} = \begin{pmatrix} (-\Delta)_\gamma^{0*}\Psi \\ a'_\gamma\Psi'_\gamma + \left(\beta_{11}\Psi|_{s_\gamma=0} + \beta_{12}[\partial\Psi/\partial s_\gamma]|_{s_\gamma=0}\right)\varphi' \end{pmatrix}, \qquad (7.27)$$

where $(-\Delta)^0_\gamma$ denotes the Laplace operator $-\frac{\partial^2}{\partial t^2_\gamma} - \frac{\partial^2}{\partial s^2_\gamma}$ restricted to the set of functions vanishing on the line $l_\gamma = \{(t_\gamma, s_\gamma) : s_\gamma = 0\}$; and $(-\Delta)^{0*}_\gamma$ is its adjoint. The perturbed operator $\mathbf{A}^B_\gamma$ is essentially self–adjoint on the domain $\mathrm{Dom}^\infty_0(\mathbf{A}^B_\gamma)$, since the operator $-\frac{\partial^2}{\partial t^2_\gamma}$ is essentially self–adjoint on the domain $C^\infty_0(\mathbf{R}_{t_\gamma})$.

7.2.2 Three-body Schrödinger operator with two-body generalized delta interactions

We are going to define a three-body operator with a generalized two-body delta interaction describing a system of equal particles. This operator can be defined using the general approach described in Chapter 6. We have seen in the previous section that the operator with generalized two-body delta interaction can be defined using the boundary values $[\partial\Psi/\partial s_\gamma]|_{s_\gamma=0}, \Psi|_{s_\gamma=0}$. Therefore the three-body operator will be defined in this section using these boundary values.

The unperturbed operator $\mathbf{A}$ is defined in the Hilbert space

$$\mathbf{H} = L_2(\Lambda) \oplus \left(\oplus \sum_{\gamma=1}^3 L_2(\mathbf{R}_{t_\gamma}) \otimes h'_\gamma \right) \tag{7.28}$$

as follows

$$\mathbf{A} = -\Delta \oplus \left(\sum_{\gamma=1}^3 -\frac{\partial^2}{\partial t^2_\gamma} \otimes a'_\gamma \right). \tag{7.29}$$

In this formula we have used operators introduced in the previous section. To define the perturbed operator we consider first the restriction of the operator

$$\begin{aligned} -\Delta &= -\left(\frac{\partial^2}{\partial t^2_1} + \frac{\partial^2}{\partial s^2_1} \right) \\ &= -\left(\frac{\partial^2}{\partial t^2_2} + \frac{\partial^2}{\partial s^2_2} \right) \\ &= -\left(\frac{\partial^2}{\partial t^2_3} + \frac{\partial^2}{\partial s^2_3} \right) \end{aligned}$$

acting in the space $L_2(\Lambda)$ to the set $\mathrm{Dom}\,(-\Delta^0)$ of functions from $W^2_2(\Lambda)$ vanishing on the lines l_γ :

$$\mathrm{Dom}\,(-\Delta^0) = \{\Psi \in W^2_2(\Lambda) : \Psi(t_\gamma, s_\gamma)|_{s_\gamma=0} = 0, \gamma = 1, 2, 3\}. \tag{7.30}$$

The adjoint operator $(-\Delta)^{0*}$ is defined on the set $\mathrm{Dom}\,(-\Delta^{0*})$ of continuous functions from $W^2_2(\Lambda \setminus \{l_\gamma\}^3_{\gamma=1})$. Let us introduce the subspace

$$\mathrm{Dom}^\infty_0(\mathbf{A}^B) \subset W^2_2(\Lambda \setminus \{l_\gamma\}^3_{\gamma=1}) \oplus \left(\oplus \sum_{\gamma=1}^3 C^\infty_0 C^\infty_0(\mathbf{R}_{t_\gamma}) \otimes h'_\gamma \right)$$

of all four component functions $\Psi = (\Psi, \Psi'_1, \Psi'_2, \Psi'_3)$ satisfying the following conditions

1.   The restrictions of the component Ψ to each sector $\Lambda_{\alpha,\beta,\gamma}$ of the Λ plane coincides with the restriction of a certain $C_0^\infty(\Lambda)$ function to the same sector. (This condition implies that every function Ψ is infinitely many times differentiable inside each sector $\Lambda_{\alpha,\beta,\gamma}$ and all its derivatives are uniformly bounded. It follows that the boundary values $\Psi|_{s_\gamma=0}$, $[\partial\Psi/\partial s_\gamma]|_{s_\gamma=0}$ exist and are $C_0^\infty(\mathbf{R}_{t_\gamma})$ functions.)

2.   Each component $\Psi'_\gamma, \gamma = 1, 2, 3$, belongs to the corresponding space $C_0^\infty(\mathbf{R}_{t_\gamma} \setminus \{0\}) \otimes h'_\gamma \cap W_2^2(\mathbf{R}_{t_\gamma}) \otimes h'_\gamma$. (This condition implies that every internal component is infinitely many times differentiable outside the origin, is continuous at the origin and has continuous first derivative there.)

3.   The external and internal components satisfy the boundary conditions

$$\begin{cases} \Psi(t_\gamma, s_\gamma)|_{s_\gamma=+0} = \Psi(t_\gamma, s_\gamma)|_{s_\gamma=-0}, \\ \langle \varphi', \Psi'(t_\gamma) \rangle_{h'_\gamma} = \beta_{21}\Psi(t_\gamma, s_\gamma)|_{s_\gamma=0} + \beta_{22}[\partial\Psi(t_\gamma, s_\gamma)/\partial s_\gamma]|_{s_\gamma=0}, \end{cases}$$

$$\gamma = 1, 2, 3.$$

$$(7.31)$$

Then the three-body operator $\mathbf{A}^B$ with the generalized two-body interactions is defined on the domain $\mathrm{Dom}_0^\infty(\mathbf{A}^B)$ by the formula

$$\mathbf{A}^B \begin{pmatrix} \Psi \\ \Psi'_1 \\ \Psi'_2 \\ \Psi'_3 \end{pmatrix} = \begin{pmatrix} (-\Delta)^{0*}\Psi \\ a'_1\Psi'_1 + (\beta_{11}\Psi|_{s_1=0} + \beta_{12}[\partial\Psi/\partial s_1]|_{s_1=0}) \varphi' \\ a'_2\Psi'_2 + (\beta_{11}\Psi|_{s_2=0} + \beta_{12}[\partial\Psi/\partial s_2]|_{s_2=0}) \varphi' \\ a'_3\Psi'_3 + (\beta_{11}\Psi|_{s_3=0} + \beta_{12}[\partial\Psi/\partial s_1]|_{s_3=0}) \varphi' \end{pmatrix}. \qquad (7.32)$$

The operator $\mathbf{A}^B$ is essentially self–adjoint on the domain $\mathrm{Dom}_0^\infty(\mathbf{A}^B)$. One can easily check by direct calculations that the operator $\mathbf{A}^B$ is symmetric on this domain. Moreover $\mathbf{A}^B$ is bounded from below on this domain. Therefore the corresponding self–adjoint operator can be defined as the Friedrichs extension of the symmetric operator $\mathbf{A}^B$ restricted to $\mathrm{Dom}_0^\infty(\mathbf{A}^B)$ (without using the fact that the restricted operator is actually essentially self–adjoint). We are going to prove in the following section that the symmetric operator is bounded from below for the case of indistinguishable particles, i.e. the system of particles possessing boson or fermion symmetry. Semiboundedness of the operator describing arbitrary particles can be proven similarly.

Thus we have explicitly defined a self–adjoint operator describing three one dimensional particles with generalized delta interactions.

In contrast to the case of Section 7.1.3, the eigenfunctions of the self-adjoint three–body operator with such an interaction cannot be constructed using the Bethe Ansatz even in the case of identical particles. In fact, the two–body scattering matrix for this interaction is given by

$$T_\gamma(k) = \frac{2ik}{D_\gamma(k^2) + 2ik},$$
$$R_\gamma(k) = \frac{-D_\gamma(k^2)}{D_\gamma(k^2) + 2ik}, \tag{7.33}$$

where $D_\gamma(\lambda) = \frac{\beta_{11}^\gamma F_\gamma'(\lambda) + \beta_{21}^\gamma}{\beta_{12}^\gamma F_\gamma'(\lambda) + \beta_{22}^\gamma}$, $F_\gamma'(\lambda) = \langle \varphi_\gamma', \frac{1}{a_\gamma' - \lambda} \varphi_\gamma' \rangle_{h_\gamma'}$, $\gamma = 1, 2, 3$. If all the parameters defining the two–body interactions are equal, i.e

$$h_1' = h_2' = h_3' \equiv h'$$

$$a_1' = a_2' = a_3' \equiv a',$$

$$\varphi_1' = \varphi_2' = \varphi_3' \equiv \varphi' \tag{7.34}$$

$$\beta_{ik}^1 = \beta_{ik}^2 = \beta_{ik}^3 \equiv \beta_{ik},$$

then the two–body scattering matrix satisfies conditions (7.14), but does not satisfy conditions (7.15) and (7.16). Therefore the set of plane waves can be discontinuous inside certain sectors. One can calculate the eigenfunctions of the constructed operator by adding certain Fresnel integrals. We are going to concentrate our attention on the case where the Bethe Anstaz gives a continuous set of plane waves. Therefore we are going to restrict our consideration to the systems of particles having boson or fermion symmetries. Let us study first the symmetry group for the operator constructed here.

If the dimension of the internal space h' is finite, then the two-body reflection and transition coefficients are rational functions of the spectral parameter k. One can easily prove that the two-body scattering matrix has no more than $2\dim h' + 1$ singularities. We suppose in the following that all these singularities are situated on the imaginary axis (on physical and nonphysical half planes). (In general one can prove that at least $2\dim h' - 1$ of the singularities are situated on this line.) We need this assumption only to simplify our calculations.

7.2.3 Symmetry group

We consider here a system of identical particles, i.e. we suppose that conditions (7.34) hold. The component Ψ of the total wave function $\mathbf{\Psi} =$

$(\Psi, \Psi'_1, \Psi'_2, \Psi'_3)$ can be considered in three different coordinate systems related to the three cluster decompositions of the three particles. We denote the corresponding functions by the index $1, 2$ or 3 in such a way that

$$\Psi^1(t_1, s_1) = \Psi^2(t_2, s_2) = \Psi^3(t_3, s_3).$$

The symmetries of the system of three identical particles interacting via an even potential are described by the dihedral group D_{12} [451] generated by two elements s and t such that

$$s^6 = 1, \quad t^2 = 1, \quad tst = s^{-1}.$$

The model symmetric operator constructed here has the same symmetry group. The element s of order 6 corresponds to the rotation of the plane Λ by the angle $\pi/3$

$$s\mathbf{U} = \mathbf{V} \quad \Rightarrow \quad \begin{pmatrix} V^3(t_3, s_3) \\ V'_1(t_1) \\ V'_2(t_2) \\ V'_3(t_3) \end{pmatrix} = \begin{pmatrix} U^2(-t_3, -s_3) \\ U'_3(-t_1) \\ U'_1(-t_2) \\ U'_2(-t_3) \end{pmatrix}.$$

The element t can be chosen equal to the operator Z_γ of the transposition of the particles α and β

$$t\mathbf{U} = Z_\gamma \mathbf{U} = \mathbf{V} \Rightarrow \begin{pmatrix} V^\gamma(t_\gamma, s_\gamma) \\ V'_1(t_1) \\ V'_2(t_2) \\ V'_3(t_3) \end{pmatrix} = \begin{pmatrix} U^\gamma(t_\gamma, -s_\gamma) \\ U'_1(t_1) \\ U'_2(t_2) \\ U'_3(t_3) \end{pmatrix}.$$

The transpositions Z_γ generate the subgroup of permutations $\mathcal{P}_3$, which consists of six elements [451].

The element s generates an important cyclic subgroup, namely the group of central rotations on the plane Λ by the angles $n\pi/3$. If the operator $\mathbf{A}^B$ commutes with the rotations s^n: $\mathbf{A}^B s^n = s^n \mathbf{A}^B$ the Hilbert space is decomposable into the orthogonal sum of Hilbert spaces of functions, which are quasi invariant with respect to the rotations s:

(3.13) $$s\mathbf{U} = e^{-im\pi/3}\mathbf{U}, \quad m = 0, 1, 2, 3, 4, 5.$$

Let us denote by P_m the projector on the subspaces of quasi–invariant elements. Every element from such a subspace is defined by the values of the first component Ψ in the sectors Λ_{123} on the plane Λ and the values of the component Ψ'_1 on the positive half axis. Let us untroduce the following restriction operator

$$P\mathbf{U} \rightarrow (U, U') \in L_2(\Lambda_{123}) \oplus L_2(\mathbf{R}_+, h')$$

$$PU = (U|_{\Lambda_{123}}, U'_1|_{\mathbf{R}_+})$$

where P_m denotes the projector on the set of quasi–invariant elements. Then the transformation

$$T_m = P \, P_m$$

is invertible on quasi–invariant functions, i.e. the total quasi–invariant function can be reconstructed from its restriction. The operator $6T_m$ is norm preserving. The Hilbert space $\mathbf{H}$ and the operator $\mathbf{A}^B$ can be decomposed as follows

$$\mathbf{H}^B = \oplus \sum_{m=0}^{5} T_m^{-1} \mathbf{H}_m, \ \ \mathbf{H}_m = L_2(\Lambda_{123}) \oplus L_2(\mathbf{R}_+, h');$$

$$\mathbf{A}^B = \oplus \sum_{m=0}^{5} T_m^{-1} \mathbf{A}_m^B T_m.$$

To define the operators $\mathbf{A}_m^B$ we are going to use the polar coordinates introduced in the previous section: $\Lambda_{123} = \{(r, \varphi) : 1 \leq r < \infty, 0 \leq \varphi \leq \pi/3\}$. Starting from the domain $\mathrm{Dom}_0^\infty(\mathbf{A}^B)$ we get the domains $\mathrm{Dom}_0^\infty(\mathbf{A}_m^B)$ on which the operators $\mathbf{A}_m^B$ are essentially selfajoint. Let us define such essentially self–adjoint operators.

Lemma 7.2.1 *Let $\mathbf{A}^B$ be an essentially self–adjoint operator defined on the domain $\mathrm{Dom}_0^\infty(\mathbf{A}^B)$. Then the operator $\mathbf{A}_m^B = T_m \mathbf{A}^B T_m^{-1}$ is defined by the formula*

$$\mathbf{A}_m^B \begin{pmatrix} U \\ U' \end{pmatrix} = \begin{pmatrix} -\left(\dfrac{\partial^2}{\partial r^2} + \dfrac{1}{r}\dfrac{\partial}{\partial r} + \dfrac{1}{r^2}\dfrac{\partial^2}{\partial \varphi^2}\right) U \\ \left(-\dfrac{\partial^2}{\partial r^2} + a'\right) U' + \ell(U)\varphi' \end{pmatrix}, \quad (7.35)$$

where

$$\ell(U) = \frac{\beta_{11}}{2}\left(U\,|_{\varphi=0} + e^{-im\pi/3}U\,|_{\varphi=\pi/3}\right) + \frac{\beta_{12}}{r}\left(\frac{\partial U}{\partial \varphi}\,|_{\varphi=0} - e^{-im\pi/3}\frac{\partial U}{\partial \varphi}\,|_{\varphi=\pi/3}\right)$$

on the domain $\mathrm{Dom}_0^\infty(\mathbf{A}_m^B) = T_m\mathrm{Dom}_0^\infty(\mathbf{A}^B)$ of functions from

$$P_{\Lambda_{123}}C_0^\infty(\Lambda) \oplus P_{\mathbf{R}_+}C_0^\infty(\mathbf{R}, h')$$

satisfying the boundary conditions

$$\langle \varphi', U'(r)\rangle_{h'} = \frac{\beta_{21}}{2}\left(U(r,0) + e^{-im\pi/3}U(r,\pi/3)\right)$$

$$+ \frac{\beta_{22}}{r}\left(\frac{\partial U}{\partial \varphi}(r,0) - e^{-im\pi/3}\frac{\partial U}{\partial \varphi}(r,\pi/3)\right), \quad (7.36)$$

$$U(r,0) = e^{-im\pi/3}U(r,\pi/3) \qquad\qquad (7.37)$$

and

$$U'(0) = 0 \quad \text{for} \quad m = 1, 3, 5;$$
$$\qquad\qquad (7.38)$$
$$\frac{\partial U'}{\partial r}(0) = 0 \quad \text{for} \quad m = 0, 2, 4.$$

Proof The domain of the operator A_m^B coincides with the set $T_m \text{Dom}_0^\infty(A^B)$ consisting of all quasi–invariant functions from the domain of the operator A^B. Every such function satisfies the boundary conditions (7.36) and (7.37), as can easily be derived from the fact that the component U of every quasi–invariant function $\mathbf{U}$ multiplies by the factor $e^{im\pi/3}$ after a rotation by the angle $\pi/3$ and boundary conditions (7.31) are satisfied. Each component U' of the function from the domain $\text{Dom}_0^\infty(A^B)$ belongs to the space $W_2^2(\mathbf{R}, h')$ and therefore is continuous at the origin and its first derivative is continuous also. Then the boundary condition (7.38) follows from the following property of the component U' for every quasi–invariant function

$$U'(-r) = e^{im\pi}U'(r).$$

Similarly formula (7.35) for the operator A_m^B follows from (7.32). The lemma is proven.

$$\square$$

The following lemma proves that the corresponding self–adjoint operator can be defined using the Friedrichs extension.

Lemma 7.2.2 *The operator A_m^B defined on the domain $\text{Dom}_0^\infty(A_m^B)$ is symmetric and bounded from below.*

Proof Integrating by parts one gets the following formula for the boundary form of the operator A_m^B for arbitrary $\mathbf{U}, \mathbf{V} \in \text{Dom}_0^\infty(A_m^B)$

$$\langle U, A_m^B V \rangle_H - \langle A_m^B U, V \rangle_H$$

$$
\begin{aligned}
= & \int_0^\infty \overline{U(r,0)} \frac{1}{r} \frac{\partial V}{\partial \varphi}(r,0) dr - \int_0^\infty \overline{U(r,\pi/3)} \frac{1}{r} \frac{\partial V}{\partial \varphi}(r,\pi/3) dr \\
& + \langle U'(0), \frac{\partial V'}{\partial r}(0) \rangle_{h'} + \int_0^\infty \langle U', \varphi' \rangle_{h'} \ell(V) dr \\
& - \int_0^\infty \frac{1}{r} \overline{\frac{\partial U}{\partial \varphi}(r,0)} V(r,0) dr + \int_0^\infty \frac{1}{r} \overline{\frac{\partial U}{\partial \varphi}(r,\pi/3)} V(r,\pi/3) dr \\
& - \langle \frac{\partial U'}{\partial r}(0), V'(0) \rangle_{h'} - \int_0^\infty \overline{\ell(U)} \langle \varphi', V' \rangle_{h'} dr.
\end{aligned}
\tag{7.39}
$$

This boundary form vanishes on all functions satisfying boundary conditions (7.36), (7.37) and (7.38). Thus we have proven that the operator is symmetric.

We are going to estimate the quadratic form of the operator in the case where $\beta_{12} = \beta_{21} = 0$:

$$\langle U, A_m^B U \rangle_H$$

$$
\begin{aligned}
= & \langle U, -\Delta U \rangle_{L_2(\Lambda_{123})} + \left\langle U', -\frac{\partial^2}{\partial r^2} U' \right\rangle_{L_2(\mathbf{R}_+, h')} \\
& + \langle U', a' U' \rangle_{L_2(\mathbf{R}_+, h')} + \langle U', \ell(U) \rangle_{L_2(\mathbf{R}_+, h')}
\end{aligned}
$$

$$
\begin{aligned}
= & \| \nabla U \|_{L_2(\Lambda_{123})}^2 + \int_0^\infty \overline{U(r,0)} \frac{1}{r} \frac{\partial U}{\partial \varphi}(r,0) dr \\
& - \int_0^\infty \overline{U(r,\pi/3)} \frac{1}{r} \frac{\partial U}{\partial \varphi}(r,\pi/3) dr + \left\| \frac{\partial U'}{\partial r} \right\|_{L_2(\mathbf{R}_+, h')}^2 \\
& + \langle U'(0), \frac{\partial U'}{\partial r}(0) \rangle_{h'} + \langle U', a' U' \rangle_{L_2(\mathbf{R}_+, h')} \\
& + \int_0^\infty \langle U'(r), \varphi' \rangle_{h'} \frac{\beta_{11}}{2} (U(r,0) + e^{-im\pi/3} U(r,\pi/3)) dr
\end{aligned}
$$

$$= \ \parallel \nabla U \parallel^2_{L_2(\Lambda_{123})} + \left\parallel \frac{\partial U'}{\partial r}\right\parallel^2_{L_2(\mathbf{R_+}, h')} \tag{7.40}$$

$$+ \langle U', a'U'\rangle_{L_2(\mathbf{R_+}, h')} + 2\beta_{11} \int_0^\infty \Re \left(\overline{U(r,0)}\langle \varphi', U'(r)\rangle_{h'}\right) dr.$$

To derive the equality we have used the fact that the function $\mathbf{U}$ satisfies boundary conditions (7.36), (7.37), (7.38). The latter integral in this formula can be estimated using the inequality

$$|f(0)|^2 \leq \frac{1}{\epsilon} \int_0^\infty |f(x)|^2 dx + \epsilon \int_0^\infty \left|\frac{df}{dx}\right|^2 dx \tag{7.41}$$

valid for any $f \in C_0^\infty(\mathbf{R})$ and $\epsilon > 0$ as follows

$$\left| 2\beta_{11} \int_0^\infty \Re \left(\overline{U(r,0)}\langle \varphi', U'(r)\rangle\right) dr \right|$$

$$\leq |\beta_{11}| \left(\parallel U', U' \parallel^2_{L_2(\mathbf{R_+}, h')} + \parallel U(r,0) \parallel^2_{L_2(\mathbf{R})}\right) \tag{7.42}$$

$$\leq |\beta_{11}| \left(\parallel U' \parallel^2_{L_2(\mathbf{R_+}, h')} + \frac{1}{\epsilon} \parallel U \parallel^2_{L_2(\Lambda_{123})} + \epsilon \parallel \nabla U \parallel^2_{L_2(\Lambda_{123})}\right).$$

This estimate is obtained by integrating (7.41) over the lines parallel to the line $\varphi = \pi/3$. Then the quadratic form of the operator can be estimated as follows

$$\langle \mathbf{U}, \mathbf{A}_m^B \mathbf{U}\rangle_{\mathbf{H}} \ \geq \ \langle U', a'U'\rangle_{L_2(\mathbf{R_+}, h')} - |\beta_{11}| \parallel U' \parallel^2_{L_2(\mathbf{R_+}, h')}$$

$$- \frac{|\beta_{11}|}{\epsilon} \parallel U \parallel^2_{L_2(\Lambda_{123})}, \tag{7.43}$$

provided $\epsilon|\beta_{11}| < 1$. The operator $\mathbf{A}_m^B$ is bounded from below, since every element in the latter formula can be estimated by the norm of the element $\mathbf{U}$. (We remark that the operator a' is a finite dimensional matrix.) The lemma is proven.

□

In the general situation the sets of plane waves constructed using the Bethe Ansatz are continuous only for the operators $\mathbf{A}_0^B$ and $\mathbf{A}_3^B$. These operators describe systems of particles possessing boson and fermion symmetries.

In what follows we are going to study in more detail the operator decribing a system of three boson particles in one dimension. The operator describing

three fermions with point interactions is not interesting, since every continuous antisymmetric function and the jump of its derivative are equal to zero
at the origin. Therefore the restriction of the operator $\mathbf{A}^B$ to the set of all
functions with fermion symmetry gives an orthogonal sum of operators, and
no interaction between the internal and external channels occurs. The restriction of the operator to the original physical Hilbert space coincides with
the restriction of the Laplace operator on Λ to the set of functions possessing fermion symmetry. Therefore we restrict our consideration to the case
of bosons. To define the self-adjoint operator describing such a system we
have to carry out a reduction similar to that carried out for quasi–invariant
subspaces.

Suppose that the wave function Ψ is invariant under all transpositions
Z_γ. Then it can be reconstructed from the values of the components Ψ in
the sector Λ_{123}, Ψ'_1 on the positive half axis $[0, \infty)$ and Ψ'_3 on the negative
half axis $(-\infty, 0]$. Actually all three internal components of boson elements
are equal: $U'_1 = U'_2 = U'_3$, and in principle one can consider only one internal
component defined on the entire real axis. But it will be more convenient
in what follows to consider two internal components defined on the positive
half axis

$$\begin{cases} U'(r) & = & U'_1(r)/\sqrt{2} \\[2mm] U''(r) & = & U'(-r)/\sqrt{2}. \end{cases} \tag{7.44}$$

Let us denote by P_b the projector to the set of functions invariant under the
group of transpositions and by $\mathbf{P}$ the restriction operator

$$\mathbf{P} : \mathbf{H} \to \mathbf{H}_b = L_2(\Lambda_{123}) \oplus L_2(\mathbf{R}_+, h') \oplus L_2(\mathbf{R}_-, h')$$

$$\mathbf{P}U = (U|_{\Lambda_{123}}, U', U'').$$

Then the transformation

$$T_b = \mathbf{P}\, P_b$$

is invertible on functions having boson symmetry. Then the boson operator
is defined as follows

$$\mathbf{A}_b^B = T_b \mathbf{A}^B T_b^{-1} \tag{7.45}$$

in the Hilbert space $\mathbf{H}_b$. The following lemma describes the domain where
the operator $\mathbf{A}_b^B$ is essentially self–adjoint.

Lemma 7.2.3 *Let $\mathbf{A}^B$ be an essentially self–adjoint operator defined on the
domain $\mathrm{Dom}_0^\infty(\mathbf{A}^B)$. Then the operator $\mathbf{A}_b^B = T_b \mathbf{A}^B T_b^{-1}$ is defined by the*

formula

$$
\mathbf{A}_b^B \begin{pmatrix} U \\ U' \\ U'' \end{pmatrix} = \begin{pmatrix} -\left(\dfrac{\partial^2}{\partial r^2} + \dfrac{1}{r}\dfrac{\partial}{\partial r} + \dfrac{1}{r^2}\dfrac{\partial^2}{\partial \varphi^2} \right) U \\[2mm] \left(-\dfrac{\partial^2}{\partial r^2} + a' \right) U' + \dfrac{\beta_{11}}{\sqrt{2}} U|_{\varphi=0}\varphi' \\[2mm] \left(-\dfrac{\partial^2}{\partial r^2} + a' \right) U'' + \dfrac{\beta_{11}}{\sqrt{2}} U|_{\varphi=\pi/3}\varphi' \end{pmatrix} \tag{7.46}
$$

on the domain $\mathrm{Dom}_0^\infty(\mathbf{A}_b^B) = T_b\mathrm{Dom}_0^\infty(\mathbf{A}^B)$ *of functions from*

$$
P_{\Lambda_{123}} C_0^\infty(\Lambda) \oplus P_{\mathbf{R}_+} C_0^\infty(\mathbf{R}, h') \oplus P_{\mathbf{R}_+} C_0^\infty(\mathbf{R}, h')
$$

satisfying the boundary conditions

$$
\langle \varphi', U'(r) \rangle_{h'} = \sqrt{2}\frac{\beta_{22}}{r}\frac{\partial U}{\partial \varphi}(r,0); \tag{7.47}
$$

$$
\langle \varphi', U''(r) \rangle_{h'} = -\sqrt{2}\frac{\beta_{22}}{r}\frac{\partial U}{\partial \varphi}(r,\pi/3); \tag{7.48}
$$

$$
\begin{cases} U'(0) &= U''(0), \\[2mm] \dfrac{\partial U'}{\partial r}(0) &= -\dfrac{\partial U''}{\partial r}(0). \end{cases} \tag{7.49}
$$

Proof The proof of this lemma follows the same lines as the proof of Lemma 7.2.1. One has to take into account that the internal components U', U'' have been defined using an extra factor $1/2$ (see formula (7.44)).

□

Similarly one can prove that the operator $\mathbf{A}_b^B$ is symmetric and bounded from below.

Lemma 7.2.4 *The operator* $\mathbf{A}_b^B$ *defined on the domain* $\mathrm{Dom}_0^\infty(\mathbf{A}_b^B)$ *is symmetric and bounded from below.*

Proof This just coincides with the proof of Lemma 7.2.2.

□

Thus we can define the self–adjoint operator describing the system of three one dimensional bosons with the generalized delta interaction as the Friedrichs extension of the operator $\mathbf{A}_b^B$ defined on the domain $\mathrm{Dom}_0^\infty(\mathbf{A}_b^B)$. We are going to keep the same notation $\mathbf{A}_b^B$ for the self–adjoint operator. Its domain will be denoted by $\mathrm{Dom}\,(\mathbf{A}_b^B)$. The set of plane waves constructed using the Bethe Anstaz is continuous inside the sector. One can calculate the corresponding internal components satisfying the boundary conditions (7.47) and (7.48). But the internal components defined in this way do not satisfy the boundary conditions (7.49) at the origin. Therefore the total eigenfunction is equal to the sum of plane waves and an outgoing spherical wave. In the following section we are going to discuss how to calculate this outgoing spherical wave.

7.2.4   Outgoing wave and Sommerfeld–Maluzhinetz transformation

This section is devoted to the construction of the outgoing spherical wave $\mathbf{G}$ corresponding to the energy $k^2 = E$. This outgoing wave is a solution of the differential equation

$$
\begin{pmatrix}
-\left(\dfrac{\partial^2}{\partial r^2} + \dfrac{1}{r}\dfrac{\partial}{\partial r} + \dfrac{1}{r^2}\dfrac{\partial^2}{\partial \varphi^2}\right)G \\[2mm]
\left(-\dfrac{\partial^2}{\partial r^2} + a'\right)G' + \dfrac{\beta_{11}}{\sqrt{2}}G|_{\varphi=0}\varphi' \\[2mm]
\left(-\dfrac{\partial^2}{\partial r^2} + a'\right)G'' + \dfrac{\beta_{11}}{\sqrt{2}}G|_{\varphi=\pi/3}\varphi'
\end{pmatrix}
= k^2
\begin{pmatrix}
G \\ G' \\ G''
\end{pmatrix}
\tag{7.50}
$$

from the Hilbert space $\mathbf{H}_b$ satisfying the boundary conditions (7.47) and (7.48). The internal components of the spherical wave do not necessarily satisfy boundary conditions (7.49). We are going to calculate the function which contains only outgoing waves in the asymptotics.

The Sommerfeld–Maluzhinetz integral representation [739, 681, 680] will be used to solve the system of equations (7.50). To calculate the outgoing wave we are going to consider the limit of the solution of the corresponding equation where k is a complex parameter with positive imaginary part when the imaginary part is vanishing. This will help us to separate the outgoing spherical wave from the incoming one.

We suppose that the components of the function $\mathbf{G}$ can be represented by the following integrals over the plane waves

$$
G(k) = \frac{1}{2\pi i}\int_\Gamma e^{ikr\cos\alpha}\left\{\tilde{G}_+(\alpha+\varphi) + \tilde{G}_-(\alpha+\pi/3-\varphi)\right\}d\alpha,
\tag{7.51}
$$

$$
G'(k) = \frac{1}{2\pi i}\int_\Gamma e^{ikr\cos\alpha}\tilde{G}'(\alpha)d\alpha,
\tag{7.52}
$$

$$G''(k) = \frac{1}{2\pi i} \int_\Gamma e^{ikr\cos\alpha} \tilde{G}'''(\alpha) d\alpha, \qquad (7.53)$$

where Γ is a contour in the complex plane α. The contour goes to infinity for real positive k in the upper half plane in the strips

$$(2m-1)\pi < \Re\alpha < 2\pi m, \quad m = 0, 1.$$

The contour Γ is not closed and has two infinite branches. We first choose the contour Γ in such a way that no singularity of the density function is situated inside the contour. This is possible because the singularities are situated at a finite distance from the real axis. This assumption will be justified later when the integral density is calculated. The integral densities are supposed to be analytic functions in α in the region of deformation of the contour Γ. Moreover we are going to carry out integration by parts during these calculations. We assume that the contribution of the boundary terms is equal to zero. These calculations will be justified later for the calculated solution only. The contour Γ will be chosen in a special way later, but for the moment we fix some contour Γ_1, which is situated in the upper half plane $\Im\alpha > Q, Q > 0$. The positive real number Q will be determined after the calculation of the solution.

The densities $\tilde{G}_+(\alpha)$ and $\tilde{G}_-(\alpha)$ are analytic functions, and $\tilde{G}'(\alpha)$ and $\tilde{G}''(\alpha)$ are vector analytic functions with values in the finite dimensional Hilbert space h'.

The component G satisfies the first equation (7.50) due to the special dependence of the integral density on the angle φ. The proof of this fact can be carried out by integration by parts which is possible due to our assumption. The second and third equations (7.50) for the components G', G'' give the following equations for the integral densities $\tilde{G}', \tilde{G}''$:

$$(a' - k^2 \sin^2\alpha)\tilde{G}'(\alpha) = -\frac{\beta_{11}}{\sqrt{2}} \left[\tilde{G}_+(\alpha) + \tilde{G}_-(\alpha + \pi/3) \right] \varphi'; \qquad (7.54)$$

$$(a' - k^2 \sin^2\alpha)\tilde{G}''(\alpha) = -\frac{\beta_{11}}{\sqrt{2}} \left[\tilde{G}_+(\alpha + \pi/3) + \tilde{G}_-(\alpha) \right] \varphi'. \qquad (7.55)$$

We do not get all solutions of the differential equation by solving these matrix equations. The general solution contains an additional decreasing exponent which belongs to the kernel of the Sommerfeld–Maluzhinetz transformation

$$\exp\left(i\sqrt{\lambda - a'}r\right)\theta. \qquad (7.56)$$

Here θ is an arbitrary vector from h'. The matrix exponential function can be represented by the Sommerfeld integral with the following integral density (see [681, 680])

$$-\frac{k\sin\alpha}{k\cos\alpha - \sqrt{\lambda - a'}}\theta. \tag{7.57}$$

Therefore the general solution of the second and third equations (7.50) in the Sommerfeld representation is given by

$$\tilde{G}'(\alpha) = -\frac{k\sin\alpha}{k\cos\alpha - \sqrt{k^2 - a'}}\theta'$$
$$-\frac{\beta_{11}}{\sqrt{2}}\left(\tilde{G}_+(\alpha) + \tilde{G}_-(\alpha + \pi/3)\right)\frac{1}{a' - k^2\sin^2\alpha}\varphi'; \tag{7.58}$$

$$\tilde{G}''(\alpha) = -\frac{k\sin\alpha}{k\cos\alpha - \sqrt{k^2 - a'}}\theta''$$
$$-\frac{\beta_{11}}{\sqrt{2}}\left(\tilde{G}_+(\alpha + \pi/3) + \tilde{G}_-(\alpha)\right)\frac{1}{a' - k^2\sin^2\alpha}\varphi'; \tag{7.59}$$

where θ' and θ'' are two arbitrary vectors from h' parametrizing the solutions. Consider the projections of the two latter solutions on the element φ' :

$$\langle\varphi', \tilde{G}'(\alpha)\rangle_{h'} = -\left\langle\varphi', \frac{k\sin\alpha}{k\cos\alpha - \sqrt{k^2 - a'}}\theta'\right\rangle_{h'}$$
$$-\frac{\beta_{11}}{\sqrt{2}}\left(\tilde{G}_+(\alpha) + \tilde{G}_-(\alpha + \pi/3)\right)\left\langle\varphi', \frac{1}{a' - k^2\sin^2\alpha}\varphi'\right\rangle_{h'}; \tag{7.60}$$

$$\langle\varphi', \tilde{G}''(\alpha)\rangle_{h'} = -\left\langle\varphi', \frac{k\sin\alpha}{k\cos\alpha - \sqrt{k^2 - a'}}\theta''\right\rangle_{h'}$$
$$-\frac{\beta_{11}}{\sqrt{2}}\left(\tilde{G}_+(\alpha + \pi/3) + \tilde{G}_-(\alpha)\right)\left\langle\varphi', \frac{1}{a' - k^2\sin^2\alpha}\varphi'\right\rangle_{h'}. \tag{7.61}$$

Using integration by parts boundary conditions (7.47) and (7.48) can be written as follows in terms of the densities:

$$\langle\varphi', \tilde{G}'(\alpha)\rangle_{h'} = \sqrt{2}\beta_{22}ik\sin\alpha\left(\tilde{G}_+(\alpha) - \tilde{G}_-(\alpha + \pi/3)\right);$$
$$\langle\varphi', \tilde{G}''(\alpha)\rangle_{h'} = -\sqrt{2}\beta_{22}ik\sin\alpha\left(\tilde{G}_+(\alpha + \pi/3) - \tilde{G}_-(\alpha)\right). \tag{7.62}$$

Substituting equalities (7.60),(7.61) into boundary conditions (7.62) we can exclude the internal component and obtain the following system of functional equations on the densities $\tilde{G}_\pm(\alpha)$

$$\tilde{G}_-(\alpha + \pi/3) = \Pi(\alpha)\tilde{G}_+(\alpha) + f'(\alpha); \tag{7.63}$$

$$\tilde{G}_+(\alpha + \pi/3) = \Pi(\alpha)\tilde{G}_-(\alpha) + f''(\alpha), \tag{7.64}$$

where the following notation is used

$$\Pi(\alpha) = \frac{2\beta_{22}ik\sin\alpha + \beta_{11}F'(k^2\sin^2\alpha)}{2\beta_{22}ik\sin\alpha - \beta_{11}F'(k^2\sin^2\alpha)}$$

$$= (T(k\sin\alpha) + R(k\sin\alpha))^{-1}, \tag{7.65}$$

$$f'(\alpha) = \frac{\sqrt{2}\langle\varphi', \frac{k\sin\alpha}{k\cos\alpha - \sqrt{k^2-a'}}\theta'\rangle_{h'}}{2\beta_{22}ik\sin\alpha - \beta_{11}F'(k^2\sin^2\alpha)}. \tag{7.66}$$

$$f''(\alpha) = \frac{\sqrt{2}\langle\varphi', \frac{k\sin\alpha}{k\cos\alpha - \sqrt{k^2-a'}}\theta''\rangle_{h'}}{2\beta_{22}ik\sin\alpha - \beta_{11}F'(k^2\sin^2\alpha)}. \tag{7.67}$$

The two-body transition and reflection coefficients appear in formula (7.65).

The system of difference equations (7.63) and (7.64) can be transformed into a system of two independent difference equations by considering the following functions

$$\tilde{G}_s(\alpha) = \frac{\tilde{G}_+(\alpha) + \tilde{G}_-(\alpha)}{2},$$

$$\tilde{G}_a(\alpha) = \frac{\tilde{G}_+(\alpha) - \tilde{G}_-(\alpha)}{2}. \tag{7.68}$$

Then the functions $\tilde{G}_s, \tilde{G}_a$ are solutions to the difference equations

$$\tilde{G}_s(\alpha + \pi/3) = \Pi(\alpha)\tilde{G}_s(\alpha) + f_s(\alpha),$$

$$\tilde{G}_a(\alpha + \pi/3) = -\Pi(\alpha)\tilde{G}_a(\alpha) - f_a(\alpha), \tag{7.69}$$

where

$$f_s(\alpha) = \frac{f'(\alpha) + f''(\alpha)}{2} = \sqrt{2}\frac{\langle\varphi', \frac{k\sin\alpha}{k\cos\alpha - \sqrt{k^2-a'}}\theta'_s\rangle_{h'}}{2\beta_{22}ik\sin\alpha - \beta_{11}F'(k^2\sin^2\alpha)},$$

$$f_a(\alpha) = \frac{f'(\alpha) - f''(\alpha)}{2} = \sqrt{2}\frac{\langle\varphi', \frac{k\sin\alpha}{k\cos\alpha - \sqrt{k^2-a'}}\theta'_a\rangle_{h'}}{2\beta_{22}ik\sin\alpha - \beta_{11}F'(k^2\sin^2\alpha)}, \tag{7.70}$$

where we have used the notation

$$\theta'_s = \frac{\theta' + \theta''}{2},$$

$$\theta'_a = \frac{\theta' - \theta''}{2}.$$

(7.71)

The solutions of these functional equations will be studied in the following section.

Consider the following Sommerfeld integrals with densities given by the functions $\tilde{G}_s$ and $\tilde{G}_a$:

$$G_s(k, r, \varphi) = \frac{1}{2\pi i} \int_\Gamma e^{ikr \cos \alpha} \left\{ \tilde{G}_s(\alpha + \varphi) + \tilde{G}_s(\alpha + \pi/3 - \varphi) \right\} d\alpha,$$

$$G_a(k, r, \varphi) = \frac{1}{2\pi i} \int_\Gamma e^{ikr \cos \alpha} \left\{ \tilde{G}_a(\alpha + \varphi) - \tilde{G}_a(\alpha + \pi/3 - \varphi) \right\} d\alpha.$$

(7.72)

These function are symmetric or antisymmetric with respect to the diagonal of the sector Λ_{123}

$$G_s(k, r, \pi/3 - \varphi) = G_s(k, r, \varphi),$$

$$G_a(k, r, \pi/3 - \varphi) = -G_a(k, r, \varphi).$$

(7.73)

In particular these functions satisfy Neumann (G_s) or Dirichlet (G_a) boundary condition on the line $\varphi = \pi/6$. If $\tilde{G}_s$ and $\tilde{G}_a$ are solutions for the functional equations then it is possible to reconstruct internal components $G_s{}', G_a{}'$ in such a way that the functions $\mathbf{G}_s = (G_s, G_s{}', G_s{}')$ and $\mathbf{G}_a = (G_a, G_a{}', -G_a{}')$ are solutions to the system of differential equations (7.50) satisfying the boundary conditions (7.47) and (7.48). In fact these functions can be represented by Sommerfeld integrals with the densities

$$\tilde{G}_s{}' = \frac{1}{2}(\tilde{G}' + \tilde{G}''),$$

$$\tilde{G}_a{}' = \frac{1}{2}(\tilde{G}' - \tilde{G}'').$$

(7.74)

The densities $\tilde{G}'_s, \tilde{G}'_a$ will be calculated in the following section.

We managed to separate the system of functional equations (7.63), (7.64) because the original problem is invariant under the symmetry transformation I with respect to the diagonal of the sector

$$IG = F \Rightarrow \left\{ \begin{array}{ccc} F(r, \varphi) & = & G(r, \pi/3 - \varphi) \\ F'(r) & = & G''(r) \\ F''(r) & = & G'(r). \end{array} \right.$$

The the elements G_s and G_a have the following properties

$$IG_s = G_s;$$

$$IG_a = -G_a.$$

7.2.5   Solution of the functional equations

The difference functional equations (7.69) are investigated in the present section. These two equations have a similar structure and can be solved using the same method. The solution of the first equation will be discussed in detail. The final formula for the solution of the second equation will be presented.

Method of iterations for the functional equation

Consider the functional equation

$$\tilde{G}(\alpha + \pi/3) = \Pi(\alpha)\tilde{G}(\alpha) + f_s(\alpha) \tag{7.75}$$

with coefficients possessing the properties

$$\Pi(\alpha + \pi) = \Pi(-\alpha) = \Pi^{-1}(\alpha), \tag{7.76}$$

$$\Pi(\alpha + 2\pi) = \Pi(\alpha), \tag{7.77}$$

$$f_s(\alpha + 2\pi) = f_s(\alpha). \tag{7.78}$$

Suppose that the singularities and zeroes of the coefficients in th equation are situated at a finite distance from the real axis. All these assumptions hold for the coefficients of the functional difference equations (7.75). In fact the function

$$\Pi(\alpha) = (T(k \sin \alpha) + R(k \sin \alpha))^{-1}$$

is a unimodular rational function of $k \sin \alpha$. The singularities of the function

$$T(k) + R(k) = \frac{2ik - D(k^2)}{2ik + D(k^2)}$$ coincide with the two-body bound states and

the resonances. Under the assumptions made in Section 7.2.2 all two-body resonances are pure imaginary. The zeroes and singularities of $\Pi(\alpha)$ are symmetric with respect to each other in accordance with the property (7.76). Then the singularities and zeroes of $\Pi(\alpha)$ are situated on the lines $\Re\alpha = \pi s, s \in \mathbf{Z}$, in the complex plane α. The zeroes of the function $\Pi(\alpha)$ on the line $\Re\alpha = 0$ will be denoted by $i\gamma_m, m = -\dim h', ..., -2, -1, 0, 1, 2, ..., \dim h'$ in such a way that positive values of m corresponds to the two–body bound states. Then the zeroes of the function $\Pi(\alpha)$ are situated at the points

$$(-1)^{s+1} i\gamma_m + s\pi, \quad s \in \mathbf{Z}, \quad m = -\dim h', ..., -2, -1, 0, 1, ..., \dim h'.$$

The function $\Pi(\alpha)$ possesses a remarkable Blaschke representation. Let us denote by $i\chi_n$ the two-body resonances and bound states on the k-plane in such a way, that positive values of m correspond to the bound states. Then the following representation holds:

$$\Pi(\alpha) = \prod_{n=-\dim h'}^{\dim h'} \frac{ik\sin\alpha + \chi_n}{ik\sin\alpha - \chi_n}. \tag{7.79}$$

We are going to use this representation in the following section to calculate the residues of the functions $\Pi(\alpha)$ at the singular points.

We are interested in solutions of the equation which are analytic functions in a neighborhood of infinity, i.e. in the region

$$\Im\alpha > Q = \max_{-\dim h' \le m \le \dim h'} |\gamma_m|,$$

where we have fixed the parameter Q introduced in the previous section. The solution of the functional equation can be expressed in terms of elementary functions using the properties of the coefficients outlined above. The solution has to be a meromorphic function on the plane α.

The general solution of the functional equation is formed as a sum of a particular solution of the inhomogeneous equation and the general solution of the homogeneous equation. The general solution of the homogeneous equation

$$y(\alpha + \pi/3) = \Pi(\alpha)y(\alpha) \tag{7.80}$$

is represented by the product of one particular solution and an arbitrary $\pi/3$ periodic function.

To derive the particular solution of the inhomogeneous equation we iterate this equation five times. All solutions of the inhomogeneous equation satisfy this new equation

$$y(\alpha + 2\pi) = y(\alpha) + \sigma(\alpha), \tag{7.81}$$

where

$$\sigma(\alpha) \; = \; f_s(\alpha + 5\pi/3)$$

$$+\Pi(\alpha + 5\pi/3)f_s(\alpha + 4\pi/3)$$

$$+\Pi(\alpha + 5\pi/3)\Pi(\alpha + 4\pi/3)f_s(\alpha + \pi)$$

$$+\Pi(\alpha + 5\pi/3)\Pi(\alpha + 4\pi/3)\Pi(\alpha + \pi)f_s(\alpha + 2\pi/3)$$

$$+\Pi(\alpha + 4\pi/3)\Pi(\alpha + \pi)f_s(\alpha + \pi/3)$$

$$+\Pi(\alpha + \pi)f_s(\alpha).$$

The function $\sigma(\alpha)$ is a 2π periodic function. Consequently one of the solutions $y^*(\alpha)$ of equation (7.81) is equal to

$$y^*(\alpha) = \frac{\alpha}{2\pi}\sigma(\alpha). \tag{7.82}$$

The general solution of the homogeneous equation (7.81) is a 2π periodic function. Hence we arrive at the following Ansatz for the solution of the functional equation (7.75)

$$\tilde{g}_0^s(\alpha) = y^*(\alpha) + y_0(\alpha), \tag{7.83}$$

where $y_0(\alpha)$ is a 2π periodic function. Substitution of this Ansatz (7.83) into equation (7.75) gives the following equation for the periodic function $y_0(\alpha)$

$$y_0(\alpha + \pi/3) = \Pi(\alpha)y_0(\alpha) + f(\alpha) - \frac{1}{6}\sigma(\alpha + \pi/3). \tag{7.84}$$

Here we have used the fact that the function $\sigma(\alpha)$ satisfies the homogeneous equation

$$\sigma(\alpha + \pi/3) = \Pi(\alpha)\sigma(\alpha). \tag{7.85}$$

The solution of (7.84) can be calculated using the Ansatz

$$y_0(\alpha) \;=\; a_1 f_s(\alpha + 5\pi/3)$$

$$+ a_2 \Pi(\alpha + 5\pi/3) f_s(\alpha + 4\pi/3)$$

$$+ a_3 \Pi(\alpha + 5\pi/3)\Pi(\alpha + 4\pi/3) f_s(\alpha + \pi)$$

$$+ a_4 \Pi(\alpha + 5\pi/3)\Pi(\alpha + 4\pi/3)\Pi(\alpha + \pi) f_s(\alpha + 2\pi/3)$$

$$+ a_5 \Pi(\alpha + 4\pi/3)\Pi(\alpha + \pi) f_s(\alpha + \pi/3)$$

$$+ a_6 \Pi(\alpha + \pi) f_s(\alpha).$$

Substitution of this representation into (7.84) gives the following relations for the constants $\{a_j\}_{j=1}^{6}$

$$a_1 = a_6 + 5/6,$$

$$a_j = a_{j+1} + 1/6.$$

So we have derived a one–parameter set of solutions of the functional equation. Thus the following theorem has been proven.

Theorem 7.2.1 *The function*

$$\tilde{G}_s(\alpha) \;=\; (\tfrac{\alpha}{2\pi} + t) f_s(\alpha + 5\pi/3)$$

$$+ (\tfrac{\alpha}{2\pi} + t - 1/6)\Pi(\alpha + 5\pi/3) f_s(\alpha + 4\pi/3)$$

$$+ (\tfrac{\alpha}{2\pi} + t - 2/6)\Pi(\alpha + 5\pi/3)\Pi(\alpha + 4\pi/3) f_s(\alpha + \pi)$$

$$+ (\tfrac{\alpha}{2\pi} + t - 3/6)\Pi(\alpha + 5\pi/3)\Pi(\alpha + 4\pi/3)\Pi(\alpha + \pi) f_s(\alpha + 2\pi/3)$$

$$+ (\tfrac{\alpha}{2\pi} + t - 4/6)\Pi(\alpha + 4\pi/3)\Pi(\alpha + \pi) f_s(\alpha + \pi/3)$$

$$+ (\tfrac{\alpha}{2\pi} + t - 5/6)\Pi(\alpha + \pi) f_s(\alpha)$$

$$(7.86)$$

for every value of the parameter t is a solution of the difference functional equation (7.75), which is analytic in the region $\Im \alpha > Q$.

The set of functions $\tilde{G}_s(\alpha)$ does not coincide with the set of all solutions of the functional equation (7.75). One can easily write down the complete set of meromorphic solutions of this equation but we are not going to do

that here. The one–parameter family we derived contains the solution we
are searching for. The solution of the second functional equation for the
function $\tilde{G}_a(\alpha)$ can be calculated using a similar method:

$$\tilde{G}_a(\alpha) = -(\tfrac{\alpha}{2\pi} + t)f_a(\alpha + 5\pi/3)$$

$$+(\tfrac{\alpha}{2\pi} + t - 1/6)\Pi(\alpha + 5\pi/3)f_a(\alpha + 4\pi/3)$$

$$-(\tfrac{\alpha}{2\pi} + t - 2/6)\Pi(\alpha + 5\pi/3)\Pi(\alpha + 4\pi/3)f_a(\alpha + \pi)$$

$$+(\tfrac{\alpha}{2\pi} + t - 3/6)\Pi(\alpha + 5\pi/3)\Pi(\alpha + 4\pi/3)\Pi(\alpha + \pi)f_a(\alpha + 2\pi/3)$$

$$-(\tfrac{\alpha}{2\pi} + t - 4/6)\Pi(\alpha + 4\pi/3)\Pi(\alpha + \pi)f_a(\alpha + \pi/3)$$

$$+(\tfrac{\alpha}{2\pi} + t - 5/6)\Pi(\alpha + \pi)f_a(\alpha).$$

$$(7.87)$$

To calculate the total outgoing wave the internal components have to be
reconstructed. It is enough to calculate the densities $\tilde{G}'_s(\alpha), \tilde{G}'_a$.

Lemma 7.2.5 *Let $\mathbf{G}_s = (G_s, G'_s, G'_s)$ and $\mathbf{G}_a = (G_a, G'_a, -G'_a)$ be solutions
of the differential equation (7.50) satisfying the boundary conditions (7.47)
and (7.48). Let $\mathbf{G}_s$ and $\mathbf{G}_a$ be represented by Sommerfeld integrals with density $\tilde{G}_s$ and $\tilde{G}_a$ given by (7.86) and (7.87) respectively. Then the internal
components G'_s and G'_a of the outgoing waves are equal to Sommerfeld integrals with densities given by*

$$\tilde{G}'_s(\alpha) = -\frac{k\sin\alpha}{k\cos\alpha - \sqrt{k^2 - a'}}\theta'_s$$

$$-\frac{\beta_{11}}{\sqrt{2}}\left\{\tilde{G}_s(\alpha)\frac{4ik\sin\alpha}{2ik\sin\alpha - D(k^2\sin^2\alpha)} + f_s(\alpha)\right\}\frac{1}{a' - k^2\sin^2\alpha}\varphi';$$

$$(7.88)$$

$$\tilde{G}'_a(\alpha) = -\frac{k\sin\alpha}{k\cos\alpha - \sqrt{k^2 - a'}}\theta'_a$$

$$-\frac{\beta_{11}}{\sqrt{2}}\left\{\tilde{G}_a(\alpha)\frac{4ik\sin\alpha}{2ik\sin\alpha - D(k^2\sin^2\alpha)} + f_a(\alpha)\right\}\frac{1}{a' - k^2\sin^2\alpha}\varphi'.$$

$$(7.89)$$

Proof The integral density corresponding to the wave $\mathbf{G}_s$ can be reconstructed using equations (7.58) and (7.74) as follows

$$
\tilde{G}'_s(\alpha) = -\frac{k\sin\alpha}{k\cos\alpha - \sqrt{\lambda - A_{12}}}\theta'_s
$$

$$
-\frac{1}{2}\frac{\beta_{11}}{\sqrt{2}}\left\{\tilde{G}_+(\alpha) + \tilde{G}_-(\alpha + \pi/3) + \tilde{G}_+(\alpha + \pi/3) + \tilde{G}_-(\alpha)\right\}
$$

$$
\times \frac{1}{a' - k^2\sin^2\alpha}\varphi'
$$

$$
= -\frac{k\sin\alpha}{k\cos\alpha - \sqrt{\lambda - A_{12}}}\theta'_s
$$

$$
-\frac{\beta_{11}}{\sqrt{2}}\left\{\tilde{G}_s(\alpha) + \tilde{G}_s(\alpha + \pi/3)\right\}\frac{1}{a' - k^2\sin^2\alpha}\varphi'.
$$

The latter formula can be simplified using the fact that $\tilde{G}_s(\alpha)$ is a solution of (7.75). In this way we obtain formula (7.88). The density $\tilde{G}'_a(\alpha)$ can be calculated using a similar method. The lemma is proven.

$\square$

Analytical properties of the solution

The analytical properties of the derived solution are described by the following:

Theorem 7.2.2 *The integral densities* $\tilde{G}_s(\alpha+\varphi)+\tilde{G}_s(\alpha+\pi/3-\varphi)$, $\tilde{G}_a(\alpha+\varphi)-\tilde{G}_a(\alpha+\pi/3-\varphi)$, $\tilde{G}'_s(\alpha)$ *and* $\tilde{G}'_a(\alpha)$ *are meromorphic on the whole complex* α*-plane. The singularities of the functions* $\tilde{G}_s(\alpha+\varphi)+\tilde{G}_s(\alpha+\pi/3-\varphi)$ *and* $\tilde{G}_a(\alpha+\varphi)-\tilde{G}_a(\alpha+\pi/3-\varphi)$ *are poles of finite multiplicity on the lattice of points*

$$
-\varphi + (-1)^{s+1}i\gamma_m - n\pi/3 + s\pi, \quad s = 0, \pm 1, \pm 2, ...; \quad n = 0, 1, 2,
$$

$$
-\pi/3 + \varphi + (-1)^{s+1}i\gamma_m - n\pi/3 + s\pi, \quad s = 0, \pm 1, \pm 2, ...; \quad n = 0, 1, 2.
$$

The same is true for the densities $\tilde{G}'_s$ *and* $\tilde{G}'_a$ *with the lattice of points*

$$
(-1)^{s+1}i\gamma_m - n\pi/3 + s\pi, \quad s = 0, \pm 1, \pm 2, ...; \quad n = 0, 1, 2,
$$

$$\pi/3 + (-1)^{s+1} i\gamma_m - n\pi/3 + s\pi, \quad s = 0, \pm 1, \pm 2, ...; \quad n = 0, 1, 2.$$

Proof The singularities of the function $\tilde{G}_s(\alpha)$ are caused by the singularities of the functions $\Pi(\alpha)$ and $f_s(\alpha)$. The singularities of the function $f_s(\alpha)$ are situated at the points where the denominator is equal to zero:

$$2\beta_{22} ik \sin\alpha - \beta_1 1 F'(k^2 \sin^2\alpha) = 0 \Rightarrow 2ik\sin\alpha = D(\lambda\sin^2\alpha).$$

These singularities coincide with the singularities of the function $\Pi(\alpha)$. Some additional singularities can be caused by the singularities of the numerator $\left\langle \varphi', \dfrac{k\sin\alpha}{k\cos\alpha - \sqrt{k^2 - a'}} \theta'_s \right\rangle_{h'}$ but these singularities cancel with the singularities of the denominator. As a result the function $f_s(\alpha)$ is analytic in a neighborhood of these points. Hence the function $\tilde{G}_s(\alpha)$ has singularities at the points

$$..., i\gamma_m - 2\pi/3, i\gamma_m - \pi/3, i\gamma_m, -i\gamma_m + \pi/3, -i\gamma_m + 2\pi/3, ...$$

The integral density $\tilde{G}_s(\alpha + \varphi) + \tilde{G}_s(\alpha + \pi/3 - \varphi)$ has singularities at the points

$$i\gamma_m - 2\pi/3 - \varphi, i\gamma_m - \pi/3 - \varphi, i\gamma_m - \varphi, -i\gamma_m + \pi/3 - \varphi, -i\gamma_m + 2\pi/3 - \varphi, ...$$

$$i\gamma_m - \pi + \varphi, i\gamma_m - 2\pi/3 + \varphi, i\gamma_m - \pi/3 + \varphi, -i\gamma_m + \varphi, -i\gamma_m + \pi/3 + \varphi, ...$$

The singularities of the function $\tilde{G}'_s(\alpha)$ are caused by the singularities of the functions $\tilde{G}_s(\alpha), \tilde{G}_s(\alpha + \pi/3)$. Additional singularities can appear at the points corresponding to the eigenvalues a'_j of the operator a', i.e. at the points $k^2 \sin^2\alpha_j = a'_j$. The function $D(k^2 \sin^2\alpha)$ has poles at these points.

The first term in (7.88) has the following singularity near the point α_j

$$-\frac{k\sin\alpha}{k\cos\alpha - \sqrt{k^2 - a'}}\theta'_s$$

$$\underset{\alpha \to \alpha_j}{=} -\frac{k\sin\alpha}{k\cos\alpha - \sqrt{k^2 - a'_j}}\langle e_j, \theta'_s \rangle_{h'} e_j + O(1),$$

where e_j is the normed eigenvector of a' corresponding to the eigenvalue a'_j. The first term in the braces in (7.88) has a second order zero at the points $\alpha = \alpha_j$. Thus the second term of (7.88) possesses the following representation

$$\frac{\beta_{11}}{\sqrt{2}}\left\{ \tilde{G}_s(\alpha)\frac{4ik\sin\alpha}{2ik\sin\alpha - D(\lambda\sin^2\alpha)} + f_s(\alpha) \right\} \frac{1}{a' - k^2 \sin^2\alpha}\varphi'$$

$$\sim_{\alpha \to \alpha_j} -\beta_{11} \dfrac{\langle \varphi', e_j \rangle_{h'} \dfrac{k \sin \alpha}{k \cos \alpha - \sqrt{k^2 - a'_j}} \langle e_j, \theta'_s \rangle_{h'}}{\beta_{11} \langle \varphi', e_j \rangle_{h'} \dfrac{1}{a'_j - k^2 \sin^2 \alpha} \langle \varphi', e_j \rangle_{h'}}$$

$$\dfrac{1}{a'_j - k^2 \sin^2 \alpha} \langle \varphi', e_j \rangle_{h'} e_j + O(1)$$

$$\sim_{\alpha \to \alpha_j} \dfrac{k \sin \alpha}{k \cos \alpha - \sqrt{k^2 - a'_j}} \langle e_j, \theta'_s \rangle_{h'} e_j + O(1).$$

Thus the function $\tilde{G}'_s(\alpha)$ is bounded in a neighborhood of the point α_j.

Th proof for the antisymmetric wave is similar. The lemma is proven.

$\square$

The singularities of all integral densities are situated on a finite distance from the real axis. The integral densities are analytic everywhere in the region $\Im \alpha > Q = \max_{-\dim h' \le m \le \dim h'} |\gamma_m|$, thus the corresponding integrals are solutions of the system of equations for outgoing wave. However the functions are exponentially increasing for large r and consequently do not belong to the Hilbert space even for $\Im k^2 > 0$. The asymptotic behavior of the integrals for $r \to \infty$ will be discussed in the following section. It will be shown that the integration contour can be changed in a special way to obtain functions decreasing at infinity.

7.2.6   Properties of the outgoing wave

We discuss here the properties of the functions $\mathbf{G}_s$ and $\mathbf{G}_a$ derived in the previous section. It will be shown that the Sommerfeld integrals with the calculated densities and contour Γ_1 are functions which are not decreasing at infinity. Another contour of integration will be chosen. It will be proven that for a special choice of the parameter t the asymptotics of the external components G_s and G_a contain only an outgoing spherical wave for $\varphi \ne 0, \pi/3$. In other words the corresponding integrals will be elements of the Hilbert space for $k^2 = \lambda : \Im \lambda > 0$. The asymptotic behavior of the outgoing wave for large $r \to \infty$ and for small $r \to 0$ will be investigated.

Asymptotics for large r

Consider the symmetric wave $\mathbf{G}_s$. The asymptotic behavior at infinity will be studied with the help of the steepest descent method. The saddle points for the Sommerfeld integral are $\alpha = 0$ and $\alpha = \pi$. These two critical points define outgoing and incoming spherical waves in the asymptotics of the component G_s :

$$\alpha = 0 : \quad \frac{1}{2\pi i}\sqrt{\frac{2\pi}{kr}}e^{-i\pi/4}e^{ikr}\left(\tilde{G}_s(\varphi) + \tilde{G}_s(\pi/3 - \varphi)\right) ;$$

$$\alpha = \pi : \quad \frac{1}{2\pi i}\sqrt{\frac{2\pi}{kr}}e^{i\pi/4}e^{-ikr}\left(\tilde{G}_s(\pi + \varphi) + \tilde{G}_s(4\pi/3 - \varphi)\right) . \tag{7.90}$$

The second point defines an exponentially increasing function for k with positive imaginary part. The solution of the difference equation contains the free parameter t. This parameter can be chosen in a special way to make the amplitude of the incoming spherical wave equal to zero. Let the parameter t be equal to $-1/6$. Then the amplitude of the incoming spherical wave vanishes (see Lemma 7.2.6).

Consider now the antisymmetric wave $\mathbf{G}_a$. The amplitudes of the outgoing and incoming spherical waves can again be calculated using the steepest descent method:

$$\alpha = 0 : \quad \frac{1}{2\pi i}\sqrt{\frac{2\pi}{kr}}e^{-i\pi/4}e^{ikr}\left(\tilde{G}_a(\varphi) - \tilde{G}_a(\pi/3 - \varphi)\right) ;$$

$$\alpha = \pi : \quad \frac{1}{2\pi i}\sqrt{\frac{2\pi}{kr}}e^{i\pi/4}e^{-ikr}\left(\tilde{G}_a(\pi + \varphi) - \tilde{G}_a(4\pi/3 - \varphi)\right) . \tag{7.91}$$

Similarly the amplitude of the incoming wave vanishes if the parameter t is equal to $-1/6$.

Lemma 7.2.6 *If* $t = -1/6$ *then the solutions* $\tilde{G}_s$ *and* $\tilde{G}_a$ *of the difference equations (7.69) satisfy the following equations*

$$\tilde{G}_s(\pi + \varphi) + \tilde{G}_s(4\pi/3 - \varphi) = 0,$$

$$\tilde{G}_a(\pi + \varphi) - \tilde{G}_a(4\pi/3 - \varphi) = 0, \tag{7.92}$$

for every φ.

Proof The lemma will be proven for a symmetric element only, since the proof for antisymmetric functions is similar. The coefficients of the difference equation possess the following properties

$$f_s(-\alpha) = \Pi(\alpha + \pi)f_s(\alpha),$$
$$\Pi(-\alpha) = \Pi(\alpha + \pi).$$

Then the following calculations can be performed

$$
\begin{aligned}
\tilde{G}_s(\pi + \varphi) = {}& (\tfrac{\varphi}{2\pi} + 2/6)f(\varphi + 2\pi/3) \\
& + (\tfrac{\varphi}{2\pi} + 1/6)\Pi(\varphi + 2\pi/3)f(\varphi + \pi/3) \\
& + (\tfrac{\varphi}{2\pi})\Pi(\varphi + 2\pi/3)\Pi(\varphi + \pi/3)f(\varphi) \\
& + (\tfrac{\varphi}{2\pi} - 1/6)\Pi(\varphi + 2\pi/3)\Pi(\varphi + \pi/3)\Pi(\varphi)f(\varphi - \pi/3) \\
& + (\tfrac{\varphi}{2\pi} - 2/6)\Pi(\varphi + \pi/3)\Pi(\varphi)f(\varphi - 2\pi/3) \\
& + (\tfrac{\varphi}{2\pi} - 3/6)\Pi(\varphi)f(\varphi - \pi) \\[2mm]
= {}& (\tfrac{\varphi}{2\pi} + 2/6)\Pi(-\varphi + \pi/3)f(-\varphi - 2\pi/3) \\
& + (\tfrac{\varphi}{2\pi} + 1/6)\Pi(-\varphi + 2\pi/3)\Pi(-\varphi + \pi/3)f(-\varphi - \pi/3) \\
& + (\tfrac{\varphi}{2\pi})\Pi(-\varphi + \pi)\Pi(-\varphi + 2\pi/3)\Pi(-\varphi + \pi/3)f(-\varphi) \\
& + (\tfrac{\varphi}{2\pi} - 1/6)\Pi(-\varphi + \pi)\Pi(-\varphi + 2\pi/3)f(-\varphi + \pi/3) \\
& + (\tfrac{\varphi}{2\pi} - 2/6)\Pi(-\varphi + \pi)f(-\varphi + 2\pi/3) \\
& + (\tfrac{\varphi}{2\pi} - 3/6)f(-\varphi + \pi) = \\[2mm]
= {}& -\tilde{G}_s(-\varphi + 4\pi/3).
\end{aligned}
$$

The proof of the lemma is accomplished.

$\square$

The amplitude of the incoming spherical wave for $t = -1/6$ is equal to zero. However the contour Γ_1 cannot be deformed to the steepest descent contour in the region of analyticity of the solution $\tilde{G}_s(\alpha + \varphi) + \tilde{G}_s(\alpha + \pi/3 - \varphi)$ and $\tilde{G}_a(\alpha + \varphi) - \tilde{G}_a(\alpha + \pi/3 - \varphi)$. The residues at the poles of the integral density would add exponentially increasing terms into the asymptotics. This means that the Sommerfeld integral over the contour Γ_1 is not an element of the Hilbert space. Hence a new contour of integration must be chosen.

The new contour Γ_2 goes to infinity in the same strips as the contour Γ_1 and passes the saddle points $\alpha = 0$ and $\alpha = \pi$. It surrounds all singularities corresponding to the resonances in the region $\Im\alpha > 0, \pi \geq \Re\alpha \geq 0$, and all singularities corresponding to the bound states in the region $\Im\alpha > 0, 2\pi \geq \Re\alpha \geq \pi$. No other singularities are situated inside the contour.

Lemma 7.2.7 *The asymptotics of the integral*

$$G_s(r,\varphi) = \frac{1}{2\pi i} \int_{\Gamma_2} e^{ikr\cos\alpha}(\tilde{G}_s(\alpha+\varphi) + \tilde{G}_s(\alpha+\pi/3-\varphi))d\alpha$$

are given by

$$G_s(r,\varphi) \sim_{r\to\infty} \frac{1}{2\pi i}\sqrt{\frac{2\pi}{kr}}e^{-i\pi/4}e^{ikr}\left(\tilde{G}_s(\varphi) + \tilde{G}_s(\pi/3-\varphi)\right)$$

$$+ \sum_{m=1}^{\dim h'} A_s^m(\theta_s')\left\{e^{ikr\cos(i\gamma_m-\varphi)} + e^{ikr\cos(i\gamma_m+\varphi-\pi/3)}\right\},$$

(7.93)

where

$$
\begin{aligned}
A_s^m(\theta_s') &= (f_s(\alpha) + \Pi(\alpha+4\pi/3)f_s(\alpha+\pi/3) \\
&\quad + \Pi(\alpha+5\pi/3)\Pi(\alpha+4\pi/3)f_s(\alpha+2\pi/3) \\
&\quad + \Pi(\alpha)\Pi(\alpha+4\pi/3)\Pi(\alpha+5\pi/3)f_s(\alpha+\pi))\,|_{\alpha=i\gamma_m} \\
&\quad\quad 2\tan i\gamma_m \Pi^m,
\end{aligned}
$$

(7.94)

and $\Pi^m = \prod_{n\neq m} \dfrac{\chi_m + \chi_n}{\chi_m - \chi_n}$. *The asymptotics of the integral*

$$G_a(r,\varphi) = \frac{1}{2\pi i} \int_{\Gamma_2} e^{ikr\cos\alpha}(\tilde{G}_a(\alpha+\varphi) - \tilde{G}_a(\alpha+\pi/3-\varphi))d\alpha$$

are given by

$$G_a(r,\varphi) \sim_{r\to\infty} \frac{1}{2\pi i}\sqrt{\frac{2\pi}{kr}}e^{-i\pi/4}e^{ikr}\left(\tilde{G}_a(\varphi) - \tilde{G}_a(\pi/3-\varphi)\right)$$

$$+ \sum_{m=1}^{\dim h'} A_a^m(\theta_a')\left\{e^{ikr\cos(i\gamma_m-\varphi)} - e^{ikr\cos(i\gamma_m+\varphi-\pi/3)}\right\},$$

(7.95)

where

$$
\begin{aligned}
A_a^m(\theta_a') &= (f_a(\alpha) - \Pi(\alpha+4\pi/3)f_a(\alpha+\pi/3) \\
&\quad + \Pi(\alpha+5\pi/3)\Pi(\alpha+4\pi/3)f_a(\alpha+2\pi/3) \\
&\quad - \Pi(\alpha)\Pi(\alpha+4\pi/3)\Pi(\alpha+5\pi/3)f_a(\alpha+\pi))\,|_{\alpha=i\gamma_m} \\
&\quad\quad 2\tan i\gamma_m \Pi^m.
\end{aligned}
$$

(7.96)

Comment The amplitudes A_s^m and A_a^m depend on the parameters θ_s' and θ_a', since the functions $f_s(\alpha)$ and $f_a(\alpha)$ depend on these parameters also.

Proof If $\varphi \neq 0, \pi/3$ then the asymptotics of the integral are given by the steepest descent method in accordance with Lemma 7.2.6 and formulas (7.90).

If $\varphi = 0$ or $\varphi = \pi/3$ then the contour Γ_2 cannot be transformed to the steepest descent contour without passing through the singularities of the integral density. The asymptotics of the integral for $\varphi = 0$ contain, in addition to the spherical outgoing wave (7.90), the outgoing surface waves which are determined by the residues at the points $i\gamma_m, i\gamma_m + 2\pi, m = 1, 2, ..., \dim h'$ for $\varphi = 0$,

$$\sum_{m=1}^{\dim h'} \left(\mathrm{Res} \left(\tilde{G}_s(\alpha + \varphi) + \tilde{G}_s(\alpha + \pi/3 - \varphi) \right) \Big|_{\alpha = i\gamma_m + 2\pi - \varphi} \right.$$

$$-\mathrm{Res}\left(\tilde{G}_s(\alpha + \varphi) + \tilde{G}_s(\alpha + \pi/3 - \varphi) \right) \Big|_{\alpha = i\gamma_m - \varphi} \right) e^{ikr \cos(i\gamma_m - \varphi)}$$

$$= \sum_{m=1}^{\dim h'} \left(\mathrm{Res}\left(\tilde{G}_s(\alpha) \right) \Big|_{\alpha = i\gamma_m + 2\pi} - \mathrm{Res}\left(\tilde{G}_s(\alpha) \right) \Big|_{\alpha = i\gamma_m} \right) e^{ikr \cos(i\gamma_m - \varphi)}$$

$$= \sum_{m=1}^{\dim h'} \mathrm{Res}\left(\tilde{G}_s(\alpha + 2\pi) - \tilde{G}_s(\alpha) \right) \Big|_{\alpha = i\gamma_m} e^{ikr \cos(i\gamma_m - \varphi)}.$$

$$\tag{7.97}$$

The residue in the latter formula can be calculated by taking into account that the function $\tilde{G}_s(\alpha)$ satisfies the difference equation (7.81). It follows that

$$\mathrm{Res}\left(\tilde{G}_s(\alpha + 2\pi) - \tilde{G}_s(\alpha) \right) \Big|_{\alpha = i\gamma_m}$$

$$= \mathrm{Res}\left(\sigma(\alpha) \right) \Big|_{\alpha = i\gamma_m}$$

$$= (f_s(\alpha) + \Pi(\alpha + 4\pi/3) f_s(\alpha + \pi/3)$$

$$+ \Pi(\alpha + 5\pi/3)\Pi(\alpha + 4\pi/3) f_s(\alpha + 2\pi/3)$$

$$+ \Pi(\alpha)\Pi(\alpha + 4\pi/3)\Pi(\alpha + 5\pi/3) f_s(\alpha + \pi)) \Big|_{\alpha = i\gamma_m}$$

$$\mathrm{Res}\left(\Pi(\alpha + \pi) \right) \Big|_{\alpha = i\gamma_m} .$$

Here we also have used the fact that the function $\Pi(\alpha)f_s(\alpha + \pi)$ is regular in a neighborhood of $i\gamma_m$ and we have used the equality $f_s(\alpha + \pi) = \Pi(\alpha + \pi)\Pi(\alpha)f_s(\alpha + \pi)$. The residue of the function $\Pi(\alpha + \pi)$ can be calculated using the Blaschke representation (7.79):

$$\mathrm{Res}\,\Pi(\alpha + \pi)\,|_{\alpha=i\gamma_m}$$

$$= \prod_{n \neq m} \frac{ik\sin\alpha - \chi_n}{ik\sin\alpha + \chi_n}\,|_{\alpha=i\gamma_m}\,\mathrm{Res}\left(\frac{ik\sin\alpha - \chi_m}{ik\sin\alpha + \chi_m}\right)|_{\alpha=i\gamma_m}$$

$$= 2\tan i\gamma_m \prod_{n \neq m} \frac{\chi_m + \chi_n}{\chi_m - \chi_n}$$

$$= 2\tan i\gamma_m \Pi^m.$$

The calculated residues determine the surface waves in the asymptotics. These functions decrease exponentially inside the sector. In fact

$$|\,e^{ikr\cos(i\gamma_m - \varphi)}\,| = e^{-kr\sin\varphi(\exp(\gamma_m) - \exp(-\gamma_m))/2}$$

is an exponentially decreasing function for $\gamma_m > 0, \pi/3 > \varphi > 0$. But this function does not decrease exponentially for $\varphi = 0$ and real λ. The residues at the points $i\gamma_m - \pi/3, i\gamma_m + 5\pi/3$ can be analysed in the same way. Thus the asymptotics of the integral are given by (7.93).

$\square$

A similar method can be applied to investigate the properties of the elements $G'_s(r)$ and $G'_a(r)$. The difference is that the saddle points do not give the main contribution to the asymptotics in this case.

Lemma 7.2.8 *The asymptotics of the integrals*

$$G'_s(r) = \frac{1}{2\pi i}\int_{\Gamma_2} \tilde{G}'_s(\alpha)e^{ikr\cos\alpha}d\alpha,$$

$$G'_a(r) = \frac{1}{2\pi i}\int_{\Gamma_2} \tilde{G}'_a(\alpha)e^{ikr\cos\alpha}d\alpha$$

are given by

$$G'_s(r) \sim_{r\to\infty} -\frac{\beta_{11}}{2}\sum_{m=1}^{\dim h'} A_s^m(\theta'_s)e^{ikr\cos(i\gamma_m)}\frac{1}{a' + \chi_m^2}\varphi';$$

(7.98)

$$G'_a(r) \sim_{r\to\infty} -\frac{\beta_{11}}{2}\sum_{m=1}^{\dim h'} A_a^m(\theta'_a)e^{ikr\cos(i\gamma_m)}\frac{1}{a' + \chi_m^2}\varphi'.$$

Proof The proof of this lemma is quite similar to the proof of Lemma 7.2.7.

$\square$

Theorem 7.2.3 *The functions* $\mathbf{G}_s = (G_s, G'_s, G'_s)$ *and* $\mathbf{G}_a = (G_a, G'_a, -G'_a)$ *are symmetric and antisymmetric solutions of equation (7.50) and satisfy boundary conditions (7.47) and (7.48). The asymptotics of these elements contain only outgoing spherical and surface waves with the amplitudes given by Lemmas 7.2.7 and 7.2.8.*

Proof Again we are going to prove the theorem for a symmetric element only. The proof for antisymmetric elements is similar. The integrals over the contour Γ_1 form a solution of the differential equations and satisfy the boundary conditions (7.47) and (7.48). This is true because the integration by parts (which we have supposed is possible) gives no boundary terms, since the function $e^{ik\cos\alpha}$ decreases exponentially in the strips where the contour Γ_1 tends to ∞. The integral over the contour Γ_2 differs from the integral over the original contour Γ_1 by the residues at the points

$$i\gamma_m + 2\pi - \varphi, i\gamma_m + 5\pi/3 - \varphi, i\gamma_m + 4\pi/3 - \varphi, \qquad 1 \le m \le \dim h';$$
$$i\gamma_m + 5\pi/3 + \varphi, i\gamma_m + 4\pi/3 + \varphi, i\gamma_m + \pi + \varphi,$$

$$-i\gamma_m + \pi - \varphi, -i\gamma_m + 2\pi/3 - \varphi, -i\gamma_m + \pi/3 - \varphi, \qquad -\dim h' \le m \le 0.$$
$$-i\gamma_m + 2\pi/3 + \varphi, -i\gamma_m + \pi/3 + \varphi, -i\gamma_m + \varphi,$$

The residues at the points corresponding to $m > 0$ give the set of surface waves coming along the boundary of the sector from infinity and going away after two reflections. This set of functions is similar to the set of surface waves which will be obtained in the following section. The residues for $m \le 0$ correspond to the analogous set of resonance functions. Both sets of functions satisfy the differential equations and boundary conditions. It follows that the integrals over the contour Γ_2 form a solution of the differential equations satisfying the boundary conditions.

$\square$

The calculated outgoing waves depend on the parameters $\theta'_a, \theta'_s \in h'$. In general these elementa do not satisfy the boundary conditions (7.49) at the origin. We are going to calculate the boundary values of these elements at the origin.

Boundary values of the outgoing wave

We are going to study the behavior of the calculated functions in a neighborhood of the point zero. All components of the outgoing waves are bounded

functions. Therefore we are going to calculate only the boundary values
of the internal component G' at the origin. We start with the symmetric
outgoing wave.

Lemma 7.2.9 *The boundary values of the component G'_s at the origin are
given by the formulas*

$$
G'_s(0) = \theta'_s - \frac{\beta_{11}}{\sqrt{2}} \sum_{m=1}^{\dim h'} \sum_{n=0}^{3} \frac{1}{a' - k^2 \sin^2(i\gamma_m + \pi + n\pi/3)} \varphi'
$$

$$
\times \mathrm{Res}\left(\tilde{G}_s(\alpha) + \tilde{G}_s(\alpha + \pi/3)\right)|_{\alpha = i\gamma_m + \pi + n\pi/3}
$$

$$
- \frac{\beta_{11}}{\sqrt{2}} \sum_{m=-\dim h'}^{0} \sum_{n=0}^{3} \frac{1}{a' - k^2 \sin^2(-i\gamma_m + n\pi/3)} \varphi'
$$

$$
\times \mathrm{Res}\left(\tilde{G}_s(\alpha) + \tilde{G}_s(\alpha + \pi/3)\right)|_{\alpha = -i\gamma_m + n\pi/3};
$$

(7.99)

$$
\frac{\partial G'_s}{\partial r}(0) = i\sqrt{k^2 - a'}\theta'_s - \frac{\beta_{11}}{\sqrt{2}} \sum_{m=1}^{\dim h'} \sum_{n=0}^{3} ik\cos(i\gamma_m + \pi + n\pi/3)
$$

$$
\times \frac{1}{a' - k^2 \sin^2(i\gamma_m + \pi + n\pi/3)} \varphi'
$$

$$
\times \mathrm{Res}\left(\tilde{G}_s(\alpha) + \tilde{G}_s(\alpha + \pi/3)\right)|_{\alpha = i\gamma_m + \pi + n\pi/3}
$$

$$
- \frac{\beta_{11}}{\sqrt{2}} \sum_{m=-\dim h'}^{0} \sum_{n=0}^{3} ik\cos(-i\gamma_m + n\pi/3)
$$

(7.100)

$$
\times \frac{1}{a' - k^2 \sin^2(-i\gamma_m + n\pi/3)} \varphi'
$$

$$
\times \mathrm{Res}\left(\tilde{G}_s(\alpha) + \tilde{G}_s(\alpha + \pi/3)\right)|_{\alpha = -i\gamma_m + n\pi/3}.
$$

Comment The residues

$$
\mathrm{Res}\left(\tilde{G}_s(\alpha) + \tilde{G}_s(\alpha + \pi/3)\right)|_{\alpha = i\gamma_m + \pi + n\pi/3}
$$

and

$$
\mathrm{Res}\left(\tilde{G}_s(\alpha) + \tilde{G}_s(\alpha + \pi/3)\right)|_{\alpha = -i\gamma_m + n\pi/3}
$$

can easily be calculated analytically. For example the residue

$$\text{Res}\left(\tilde{G}_s(\alpha) + \tilde{G}_s(\alpha + \pi/3)\right)|_{\alpha=i\gamma_m+\pi}$$

is equal to

$$
\begin{aligned}
&\text{Res}\left(\tilde{G}_s(\alpha) + \tilde{G}_s(\alpha + \pi/3)\right)|_{\alpha=i\gamma_m+\pi} \\
&= \left\{ \tfrac{\alpha}{2\pi}\Pi(\alpha+\pi)f_s(\alpha) + (\tfrac{\alpha}{2\pi} - 1/6)f_s(\alpha+5\pi/3) \right. \\
&\quad (\tfrac{\alpha}{2\pi} - 2/6)\pi(\alpha+5\pi/3)f_s(\alpha+4\pi/3) \\
&\quad \left. (\tfrac{\alpha}{2\pi} - 3/6)\Pi(\alpha+5\pi/3)\Pi(\alpha+4\pi/3)f_s(\alpha+3\pi/3) \right\}|_{\alpha=i\gamma_m+\pi} 2i\tan\gamma_m\Pi^m.
\end{aligned}
$$

Proof   The value $U(0)$ of the Sommerfeld integral

$$U(r) = \frac{1}{2\pi i}\int_{\Gamma} e^{ikr\cos\alpha}\tilde{U}(\alpha)d\alpha$$

is related to the behavior of the integral density $\tilde{U}(\alpha)$ at infinity. In particular, the following equation is valid for every even function analytic inside the contour of integration [681, 680]

$$U(0) = \frac{1}{i}\lim_{\alpha\to i\infty}\tilde{U}(\alpha).$$

Thus the integral over the initial contour Γ_1 with density $\tilde{G}_s(\alpha)$ defined by (7.88) is equal to θ'_s at the origin. The first derivative of this integral at the origin is equal to $i\sqrt{k^2 - a'}\theta'_s$. To calculate the boundary values of the integral over the contour Γ_2, the residues at the singular points situated between the contour Γ_1 and Γ_2 have to be taken into account. We get formulas (7.99) and (7.100). The lemma is proven.

$\square$

The boundary values of the first component are linear functions of the vector θ'_s. Therefore the matrices B_s and $\hat{B}_s$ can be introduced as follows

$$
\begin{cases}
G'_s(0) &= (1 + B_s(k))\theta'_s \\[2mm]
\dfrac{\partial G'_s}{\partial r}(0) &= (i\sqrt{k^2 - a'} + \hat{B}_s(k))\theta'_s.
\end{cases}
\tag{7.101}
$$

We suppose that the matrix $i\sqrt{k^2 - a'} + \hat{B}_s(k)$ is invertible, i.e. no resonance occurs for positive k. This assumption holds for small values of the parameter β_{11}, since the matrix a' has only negative eigenvalues. If this condition is not satisfied then there exists a nontrivial solution of the differential equation (7.50) satisfying the boundary conditions and containing only outgoing waves in the asymptotics.

Consider now the antisymmetric wave. Similarly the following lemma can be proven.

Lemma 7.2.10 *The boundary values of the component G'_a at the origin are given by the following formulas*

$$
G'_a(0) = \theta'_a - \frac{\beta_{11}}{\sqrt{2}} \sum_{m=1}^{\dim h'} \sum_{n=0}^{3} \frac{1}{a' - k^2 \sin^2(i\gamma_m + \pi + n\pi/3)} \varphi'
$$

$$
\times \operatorname{Res} \left(\tilde{G}_a(\alpha) - \tilde{G}_a(\alpha + \pi/3) \right) |_{\alpha = i\gamma_m + \pi + n\pi/3}
$$

$$
- \frac{\beta_{11}}{\sqrt{2}} \sum_{m=-\dim h'}^{0} \sum_{n=0}^{3} \frac{1}{a' - k^2 \sin^2(-i\gamma_m + n\pi/3)} \varphi'
$$

$$
\times \operatorname{Res} \left(\tilde{G}_a(\alpha) - \tilde{G}_a(\alpha + \pi/3) \right) |_{\alpha = -i\gamma_m + n\pi/3};
$$

(7.102)

$$
\frac{\partial G'_a}{\partial r}(0) = i\sqrt{k^2 - a'}\,\theta'_a - \frac{\beta_{11}}{\sqrt{2}} \sum_{m=1}^{\dim h'} \sum_{n=0}^{3} ik\cos(i\gamma_m + \pi + n\pi/3)
$$

$$
\times \frac{1}{a' - k^2 \sin^2(i\gamma_m + \pi + n\pi/3)} \varphi'
$$

$$
\times \operatorname{Res} \left(\tilde{G}_a(\alpha) - \tilde{G}_a(\alpha + \pi/3) \right) |_{\alpha = i\gamma_m + \pi + n\pi/3}
$$

(7.103)

$$
- \frac{\beta_{11}}{\sqrt{2}} \sum_{m=-\dim h'}^{0} \sum_{n=0}^{3} ik\cos(-i\gamma_m + n\pi/3)
$$

$$
\times \frac{1}{a' - k^2 \sin^2(-i\gamma_m + n\pi/3)} \varphi'
$$

$$
\times \operatorname{Res} \left(\tilde{G}_a(\alpha) - \tilde{G}_a(\alpha + \pi/3) \right) |_{\alpha = -i\gamma_m + n\pi/3}.
$$

Proof The proof of this lemma follows the same lines as the proof of Lemma 7.2.9. $\qquad\square$

All the residues appearing in the formulas of Lemma 7.2.10 can be calculated analytically. We introduce, similarly, the matrices

$$
\left\{
\begin{aligned}
G'_a(0) &= (1 + B_a(k))\theta'_a \\[2ex]
\frac{\partial G'_a}{\partial r}(0) &= (i\sqrt{k^2 - a'} + \hat{B}_a(k))\theta'_a.
\end{aligned}
\right.
\tag{7.104}
$$

We suppose that no resonance occurs, i.e. that the matrix $1 + B_a(k)$ is invertible. This condition is again satisfied for small values of the parameter β_{11}.

7.2.7   Spectrum and scattering matrix

The spectrum of the operator $\mathbf{A}_b^B$ consists of the branch of the continuous spectrum $[0, \infty)$ corresponding to the processes with three free particles, branches of the continuous spectrum $[-\chi_m^2, \infty)$, $m = 1, 2, ..., N_{12}$, corresponding to the two-body bound states, and probably some eigenvalues. The scattering matrix will be calculated from the asymptotics of the continuous spectrum eigenfunctions of the operator $\mathbf{A}_b^B$.

Definition of the scattering matrix

We are going to calculate the continuous spectrum eigenfunctions $\mathbf{U} = (U, U', U'')$ of the operator $\mathbf{A}_b^B$. These functions are generalized solutions of equation (7.50) satisfying the boundary conditions (7.47), (7.48) and (7.49). In contrast to the outgoing wave the asymptotics of the eigenfunction contain not only outgoing sperical and plane waves but also incoming waves. But the eigenfunction satisfies boundary conditions (7.49) at the origin. (The outgoing wave does not necessarily satisfy these conditions. We have already supposed that no resonance occurs, i.e. these conditions are not satisfied by the outgoing wave.)

The continuous spectrum eigenfunctions can be separated into two classes in accordance with the type of the incoming channel. The eigenfunctions of the first type correspond to the three-body incoming channel. Such eigenfunctions contain only one incoming plane wave in the asymptotics of the component U. The eigenfunctions of the second type contain only an incoming surface wave in the asymptotics of the external component U. In addition to these incoming waves the asymptotics of the external component contains a set of outgoing waves: plane, spherical and surface waves. This set of eigenfunctions will be called *incoming*.

The second complete set of eigenfunctions, the so–called *outgoing* set, is determined by different outgoing waves. The eigenfunctions contain in their asymptotics only one outgoing wave and a set of incoming waves. The scattering matrix can be defined as an operator connecting the spectral representations with respect to these two sets of eigenfunctions. We are going to define the scattering matrix from the asymptotics of the incoming set of eigenfunctions.

The scattering matrix is a matrix integral operator of the form

$$S(\lambda) = \left\{ \begin{matrix} S_{33} & S_{32} \\ S_{23} & S_{22} \end{matrix} \right\},$$ (7.105)

acting in the space $L_2(0, \pi/6) \oplus N, \dim N = 2 \dim h'$. The dimension of the space N' is equal to twice the number of two–body eigenvalues. The finite dimensional space N is equal to the orthogonal sum of two finite dimensional spaces $N = N_1 \oplus N_2$, $\dim N_i = \dim h', i = 1, 2$. The spaces N_1 and N_2 are associated with the surface two-body waves concentrated near the boundaries of the sector Λ_{123} given by the equations $\varphi = 0$ and $\varphi = \pi/3$ respectively.

The operator $S_{33} : L_2(0, \pi/3) \to L_2(0, \pi/3)$ is an integral operator with the kernel $s_{33}(\lambda, \varphi, \varphi_0)$:

$$\left(S_{33}(\lambda) f \right)(\varphi) = \int_0^{\pi/3} s_{33}(\lambda, \varphi, \varphi_0) f(\varphi_0) d\varphi_0.$$ (7.106)

The operator $S_{23} = S_{23}^1 \oplus S_{23}^2 : L_2(0, \pi/3) \to N^1 \oplus N^2$ is equal to the orthogonal sum of the integral operators with the kernels $s_{23}^1(\lambda, m, \varphi_0)$, $m = 1, 2, ..., \dim h'$ and $s_{23}^2(\lambda, m, \varphi_0)$, $m = 1, 2, ..., \dim h'$:

$$\left(S_{23}^i f \right)_m = \int_0^{\pi/3} s_{23}^i(\lambda, m, \varphi_0) f(\varphi_0) d\varphi_0, \ i = 1, 2.$$ (7.107)

The operator $S_{32} : N = N^1 \oplus N^2 \to L_2(0, \pi/3)$ is the operator of multiplication by the vector $s_{32}(\lambda, \varphi) = \begin{pmatrix} s_{32}^1(\lambda, \varphi) \\ s_{32}^2(\lambda, \varphi) \end{pmatrix}$, where $s_{32}^i(\lambda, \varphi) = \{s_{32}^i(\lambda, \varphi, m)\}_{m=1}^{\dim h'}$:

$$\left(S_{32} \begin{pmatrix} f^1 \\ f^2 \end{pmatrix} \right)(\varphi) = \sum_{m=1}^{\dim h'} \left(s_{32}^1(\lambda, \varphi, m) f_m^1 + s_{32}^2(\lambda, \varphi, m) f_m^2 \right).$$ (7.108)

The operator S_{22} coincides with the operator of multiplication by a matrix of dimension $2 \dim h' \times 2 \dim h'$ and possesses the natural representation

$$S_{22}(\lambda) = \begin{pmatrix} S_{22}^{11}(\lambda) & S_{22}^{12}(\lambda) \\ S_{22}^{21}(\lambda) & S_{22}^{22}(\lambda) \end{pmatrix},$$ (7.109)

where $S_{22}^{ij}(\lambda)$ are $\dim h' \times \dim h'$ matrices: $S_{22}^{ij}(\lambda) = \{s_{22}^{ij}(\lambda, n, m)\}_{n,m=1}^{\dim h'}$.

Consider first the incoming eigenfunction determined by the incoming plane wave

(7.4) $\Psi_{in}(\lambda, \varphi_0, r, \varphi) = \dfrac{\exp(ikr\cos(\varphi - \varphi_0))}{2\pi\sqrt{2k}}, \quad k^2 = \lambda, \quad \pi \le \varphi_0 \le 4\pi/3.$

Then the asymptotics at infinity $r \to \infty$ of the external component of this function contain the following outgoing waves

$$R_{33}(\lambda, \varphi_0)\dfrac{\exp(ikr\cos(\varphi + \varphi_0 + 2\pi/3))}{2\pi\sqrt{2k}}$$

$$+ a_{33}(\lambda, \varphi, \varphi_0)\dfrac{1}{2k}\dfrac{e^{-i\pi/4}}{\sqrt{\pi r}}e^{ikr}$$

$$+ \sum_{m=1}^{\dim h'} s_{23}^1(\lambda, m, \varphi_0)\dfrac{c_m}{\sqrt{2\pi}\sqrt[4]{\lambda + \chi_m^2}}e^{ikr\cos(i\gamma_m - \varphi)}$$

$$+ \sum_{m=1}^{\dim h'} s_{23}^2(\lambda, m, \varphi_0)\dfrac{c_m}{\sqrt{2\pi}\sqrt[4]{\lambda + \chi_m^2}}e^{ikr\cos(i\gamma_m + \varphi - \pi/3)}.$$

(7.110)

The normalizing constant c_m for the surface wave was determined in Chapter 2 by (2.97); $-\chi_m^2$ are the energies of the two-body bound state. The real number γ_m is given by the formula $i\gamma_m = \arctan\dfrac{i\chi_m}{\sqrt{\lambda + \chi_m^2}}$. The scattering amplitude $a_{33}(\lambda, \varphi, \varphi_0)$ and the three-body reflection coefficient $R_{33}(\lambda, \varphi_0)$ form the kernel $s_{33}(\lambda, \varphi, \varphi_0)$ of the scattering matrix

(7.6) $s_{33}(\lambda, \varphi, \varphi_0) = R_{33}(\lambda, \varphi_0)\delta(\varphi - \varphi_0) + a_{33}(\lambda, \varphi_0, \varphi).$

The scattering amplitudes $s_{23}^1(\lambda, m, \varphi_0)$ and $s_{23}^2(\lambda, m, \varphi_0)$ are just the kernels of the operators S_{23}^1 and S_{23}^2 respectively.

The eigenfunction determined by the incoming surface wave

$$\Psi_{in}^1(\lambda, n, r, \varphi) = \dfrac{c_n}{\sqrt{2\pi}\sqrt[4]{\lambda + \chi_n^2}}e^{-ikr\cos(i\gamma_n - \varphi)},$$

(7.111)

$$n = 1, 2, ..., \dim h', \quad \lambda > -\chi_n^2$$

concentrated near the boundary $\varphi = 0$, has asymptotics at infinity, which contain the following outgoing waves

$$s_{32}^1(\lambda, \varphi, n)\frac{1}{2k}\frac{e^{-i\pi/4}}{\sqrt{\pi r}}e^{ikr}$$

$$+ \sum_{m=1}^{\dim h'} s_{22}^{11}(\lambda, m, n)\frac{c_m}{\sqrt{2\pi}\sqrt[4]{\lambda + \chi_m^2}}e^{ikr\cos(i\gamma_m - \varphi)} \qquad (7.112)$$

$$+ \sum_{m=1}^{\dim h'} s_{22}^{21}(\lambda, m, n)\frac{c_m}{\sqrt{2\pi}\sqrt[4]{\lambda + \chi_m^2}}e^{ikr\cos(i\gamma_m + \varphi - \pi/3)}.$$

Similarly the eigenfunction induced by an incoming suface wave along the other boundary

$$\Psi_{in}^2(\lambda, n, r, \varphi) = \frac{c_n}{\sqrt{2\pi}\sqrt[4]{\lambda + \chi_n^2}}e^{-ikr\cos(i\gamma_n + \varphi - \pi/3)},$$
$$(7.113)$$

$$n = 1, 2, ..., \dim h', \ \lambda > -\chi_n^2$$

has the following outgoing waves in the asymptotics of the external component at infinity

$$s_{32}^2(\lambda, \varphi, n)\frac{1}{2k}\frac{e^{-i\pi/4}}{\sqrt{\pi r}}e^{ikr}$$

$$+ \sum_{m=1}^{\dim h'} s_{22}^{12}(\lambda, m, n)\frac{c_m}{\sqrt{2\pi}\sqrt[4]{\lambda + \chi_m^2}}e^{ikr\cos(i\gamma_m - \varphi)} \qquad (7.114)$$

$$+ \sum_{m=1}^{\dim h'} s_{22}^{22}(\lambda, m, n)\frac{c_m}{\sqrt{2\pi}\sqrt[4]{\lambda + \chi_m^2}}e^{ikr\cos(i\gamma_m + \varphi - \pi/3)}.$$

The scattering amplitudes $s_{32}^i(\lambda, \varphi, n)$, $s_{22}^{ij}(\lambda, n, m)$ are just the coefficients of the matrix operators S_{32} and S_{22}. In the following section we are going to calculate the eigenfunctions induced by incoming plane and surface waves. The scattering matrix will be calculated from the asymptotics of the eigenfunctions.

7.2.8   Calculation of the scattering matrix

The scattering matrix defined in the previous section will be calculated using continuous spectrum eigenfunctions of the operator A_b^B. Every such eigenfunction $\Psi(\lambda, \varphi_0)$ determined by a certain incoming plane wave $\Psi_{in}(\lambda, \varphi_0, r, \varphi)$

is equal to the sum of the set of plane waves determined by the incoming wave and two outgoing waves calculated in Section 7.2.5:

$$\Psi(\lambda, \varphi_0) = \Psi^p(\lambda, \varphi_0) + G_s(\theta'_s) + G_a(\theta'_a). \qquad (7.115)$$

The function Ψ^p coincides with the set of plane waves constructed using the laws of geometrical optics. The two outgoing waves $G_s(\theta'_s)$ and $G_a(\theta'_a)$ are determined by two vector parameters $\theta'_s, \theta'_s \in h'$ depending on the incoming plane wave $\Psi(\lambda, \varphi_0)$. These parameters will be chosen in such a way that the total wave function $\Psi = \Psi^p + G_s + G_a$ satisfies the boundary conditions (7.49) at the origin.

The set of plane waves Ψ^p is the result of multiple reflections of the incoming plane wave $\Psi_{in}(\lambda, \varphi_0)$ from the boundaries of the sector in accordance with the laws of geometrical optics. The total number of the reflected waves is equal to 5. The reflection coefficient from the boundaries is determined by the two-body scattering matrix only: $P(k_\perp) = T(k_\perp) + R(k_\perp)$. Here $k_\perp$ denotes the perpendicular component of the wave vector. We note that the reflection coefficient $P(k)$ is related to the coefficient $\Pi(\alpha)$ in the difference functional equation as follows: $\Pi(\alpha) = P(k \sin \alpha)^{-1}$.

The incoming eigenfunctions determined by the three-body incoming plane wave can be parameterized by the energy $\lambda = k^2, 0 \leq \lambda < \infty$, and the angle $\varphi_0, \pi \leq \varphi_0 \leq 4\pi/3$ of the incoming wave $\Psi_{in}(\lambda, \varphi_0)$. The set of induced plane waves consists of the six waves with the following external component:

$$\psi^p(\lambda, \varphi_0, r, \varphi)$$

$$= \frac{1}{2\pi\sqrt{2k}} \Big\{ e^{ikr\cos(\varphi-\varphi_0)} + P(k\sin(\varphi_0 - \pi))e^{ikr\cos(\varphi+\varphi_0)}$$

$$+ P(k\sin(\varphi_0 - \pi))P(k\sin(\varphi_0 - 2\pi/3))e^{ikr\cos(\varphi-\varphi_0-2\pi/3)}$$

$$+ P(k\sin(\varphi_0 - \pi/3))e^{ikr\cos(\varphi+\varphi_0-2\pi/3)} \qquad (7.116)$$

$$+ P(k\sin(\varphi_0 - 2\pi/3))P(k\sin(\varphi_0 - \pi/3))e^{ikr\cos(\varphi-\varphi_0+2\pi/3)}$$

$$+ P(k\sin(\varphi_0 - \pi))P(k\sin(\varphi_0 - 2\pi/3))P(k\sin(\varphi_0 - \pi/3))$$

$$e^{ikr\cos(\varphi+\varphi_0+2\pi/3)} \Big\}.$$

The internal components for the set of plane waves are given by

$$\Psi^{p'}(\lambda, \varphi_0, r)$$

$$= -\frac{\beta_{11}}{4\pi\sqrt{k}} \Bigg\{ (1 + P(k\sin(\varphi_0 - \pi))) \frac{e^{ikr\cos\varphi_0}}{a' - k^2\sin^2\varphi_0}\varphi'$$

$$+ P(k\sin(\varphi_0 - \pi))P(k\sin(\varphi_0 - 2\pi/3))$$

$$(1 + P(k\sin(\varphi_0 - \pi/3))) \frac{e^{ikr\cos(\varphi_0 + 2\pi/3)}}{a' - k^2\sin^2(\varphi_0 + 2\pi/3)}\varphi' \qquad (7.117)$$

$$+ P(k\sin(\varphi_0 - \pi/3)) (1 + P(k\sin(\varphi_0 - 2\pi/3)))$$

$$\frac{e^{ikr\cos(\varphi_0 - 2\pi/3)}}{a' - k^2\sin^2(\varphi_0 - 2\pi/3)}\varphi' \Bigg\},$$

$$\Psi^{p''}(\lambda, \varphi_0, r)$$

$$= -\frac{\beta_{11}}{4\pi\sqrt{k}} \Bigg\{ (1 + P(k\sin(\varphi_0 - \pi/3))) \frac{e^{ikr\cos(\varphi_0 - \pi/3)}}{a' - k^2\sin^2(\varphi_0 - \pi/3)}\varphi'$$

$$+ P(k\sin(\varphi_0 - 2\pi/3))P(k\sin(\varphi_0 - \pi/3))$$

$$(1 + P(k\sin(\varphi_0 - \pi))) \frac{e^{ikr\cos(\varphi_0 - \pi)}}{a' - k^2\sin^2(\varphi_0 - \pi)}\varphi'$$

$$+ P(k\sin(\varphi_0 - \pi)) (1 + P(k\sin(\varphi_0 - 2\pi/3)))$$

$$\frac{e^{ikr\cos(\varphi_0 + \pi/3)}}{a' - k^2\sin^2(\varphi_0 + \pi/3)}\varphi' \Bigg\}.$$

$$(7.118)$$

Consider the boundary values of the internal components at the origin:

$\Psi^{p'}(\lambda, \varphi_0, 0)$

$$
\begin{aligned}
= \ & -\frac{\beta_{11}}{4\pi\sqrt{k}} \left\{ (1 + P(k\sin(\varphi_0 - \pi))) \frac{1}{a' - k^2\sin^2\varphi_0} \varphi' \right. \\[2mm]
& + P(k\sin(\varphi_0 - \pi))P(k\sin(\varphi_0 - 2\pi/3)) \\[2mm]
& (1 + P(k\sin(\varphi_0 - \pi/3))) \frac{1}{a' - k^2\sin^2(\varphi_0 + 2\pi/3)} \varphi' \\[2mm]
& + P(k\sin(\varphi_0 - \pi/3))\,(1 + P(k\sin(\varphi_0 - 2\pi/3))) \\[2mm]
& \left. \frac{1}{a' - k^2\sin^2(\varphi_0 - 2\pi/3)} \varphi' \right\},
\end{aligned}
$$

(7.119)

$\Psi^{p''}(\lambda, \varphi_0, 0)$

$$
\begin{aligned}
= \ & -\frac{\beta_{11}}{4\pi\sqrt{k}} \left\{ (1 + P(k\sin(\varphi_0 - \pi/3))) \frac{1}{a' - k^2\sin^2(\varphi_0 - \pi/3)} \varphi' \right. \\[2mm]
& + P(k\sin(\varphi_0 - 2\pi/3))P(k\sin(\varphi_0 - \pi/3)) \\[2mm]
& (1 + P(k\sin(\varphi_0 - \pi))) \frac{1}{a' - k^2\sin^2(\varphi_0 - \pi)} \varphi' \\[2mm]
& + P(k\sin(\varphi_0 - \pi))\,(1 + P(k\sin(\varphi_0 - 2\pi/3))) \\[2mm]
& \left. \frac{1}{a' - k^2\sin^2(\varphi_0 + \pi/3)} \varphi' \right\}.
\end{aligned}
$$

(7.120)

$$\frac{\partial \Psi^{p\prime}(\lambda, \varphi_0, r)}{\partial r}\Big|_{r=0}$$

$$= -\frac{\beta_{11}}{4\pi\sqrt{k}}\left\{(1 + P(k\sin(\varphi_0 - \pi)))\frac{ik\cos\varphi_0}{a' - k^2\sin^2\varphi_0}\varphi'\right.$$

$$+P(k\sin(\varphi_0 - \pi))P(k\sin(\varphi_0 - 2\pi/3))$$

$$(1 + P(k\sin(\varphi_0 - \pi/3)))\frac{ik\cos(\varphi_0 + 2\pi/3)}{a' - k^2\sin^2(\varphi_0 + 2\pi/3)}\varphi'$$

$$+P(k\sin(\varphi_0 - \pi/3))\,(1 + P(k\sin(\varphi_0 - 2\pi/3)))$$

$$\left.\frac{ik\cos(\varphi_0 - 2\pi/3)}{a' - k^2\sin^2(\varphi_0 - 2\pi/3)}\varphi'\right\},$$

$$\frac{\partial \Psi^{p\prime\prime}(\lambda, \varphi_0, r)}{\partial r}\Big|_{r=0}$$

$$= -\frac{\beta_{11}}{4\pi\sqrt{k}}\left\{(1 + P(k\sin(\varphi_0 - \pi/3)))\frac{ik\cos(\varphi_0 - \pi/3)}{a' - k^2\sin^2(\varphi_0 - \pi/3)}\varphi'\right.$$

$$+P(k\sin(\varphi_0 - 2\pi/3))P(k\sin(\varphi_0 - \pi/3))$$

$$(1 + P(k\sin(\varphi_0 - \pi)))\frac{ik\cos(\varphi_0 - \pi)}{a' - k^2\sin^2(\varphi_0 - \pi)}\varphi'$$

$$+P(k\sin(\varphi_0 - \pi))\,(1 + P(k\sin(\varphi_0 - 2\pi/3)))$$

$$\left.\frac{ik\cos(\varphi_0 + \pi/3)}{a' - k^2\sin^2(\varphi_0 + \pi/3)}\varphi'\right\}.$$

(7.121)

(7.122)

In general the boundary values of the set of plane waves do not satisfy the boundary conditions (7.49). Therefore we have to add a certain outgoing wave $\mathbf{G}_s(\theta'_s(\lambda, \varphi_0)) + \mathbf{G}_a(\theta'_a(\lambda, \varphi_0))$. The wave $\Psi^p + \mathbf{G}_s + \mathbf{G}_a$ satisfies these boundary conditions at the origin only if the following equalities hold

$$G'_a(0) = -\frac{\Psi^{p\prime}(0) + \Psi^{p\prime\prime}(0)}{2};$$

$$\frac{\partial G'_s}{\partial r}(0) = -\frac{\frac{\partial \Psi^{p\prime}}{\partial r}\Big|_{r=0} + \frac{\partial \Psi^{p\prime\prime}}{\partial r}\Big|_{r=0}}{2}.$$

(7.123)

Now the vectors θ'_s and θ'_a can be calculated using (7.101) and (7.104):

$$\theta'_a(\lambda, \varphi_0) = -\frac{1}{2}(1 + B_a)^{-1}\left(\Psi^{p'}(0) + \Psi^{p''}(0)\right),$$

$$\theta'_s(\lambda, \varphi_0) = -\frac{1}{2}\left(i\sqrt{k^2 - a'} + \hat{B}_s\right)^{-1}\left(\frac{\partial\Psi^{p'}}{\partial r}\Big|_{r=0} + \frac{\partial\Psi^{p''}}{\partial r}\Big|_{r=0}\right). \tag{7.124}$$

The first two components of the scattering matrix can now be calculated from the asymptotics of the solution

$$s_{33}(\lambda, \varphi, \varphi_0)$$

$$= \delta(\varphi + \varphi_0 + 2\pi/3)P(k\sin(\varphi_0 - \pi))P(k\sin(\varphi_0 - 2\pi/3))P(k\sin(\varphi_0 - \pi/3))$$

$$-(i\sqrt{2k})\left(\tilde{G}_s(\lambda, \theta'_s, \varphi)) + \tilde{G}_s(\lambda, \theta'_s, \pi/3 - \varphi)\right)$$

$$-(i\sqrt{2k})\left(\tilde{G}_a(\lambda, \theta'_a, \varphi) - \tilde{G}_a(\lambda, \theta'_a, \pi/3 - \varphi)\right); \tag{7.125}$$

$$s^1_{23}(\lambda, m, \varphi_0) = \frac{\sqrt{2\pi}\sqrt[4]{\lambda + \chi^2_m}}{c_m}\left(A^m_s(\theta'_s) + A^m_a(\theta'_a)\right); \tag{7.126}$$

$$s^2_{23}(\lambda, m, \varphi_0) = \frac{\sqrt{2\pi}\sqrt[4]{\lambda + \chi^2_m}}{c_m}\left(A^m_s(\theta'_s) - A^m_a(\theta'_a)\right).$$

The values of θ'_s, θ'_a appearing in the latter formula are given by (7.124).

The eigenfunctions induced by the surface waves can be considered in the same way. We consider here the waves induced by the surface waves coming along the boundary $\varphi = 0$. These eigenfunctions can be represented by the sum

$$\Psi_1(\lambda, m) = \Psi^{\mathrm{surf}}_1(\lambda, m) + \mathbf{G}_s(\theta'_{1s}) + \mathbf{G}_a(\theta'_{1a}). \tag{7.127}$$

The eigenfunctions are parameterized by the energy of the incoming surface wave $\lambda \in [-\chi^2_m, \infty)$ and the number m of the two-body bound state. This surface wave can be represented as a plane wave with complex wave vector

$$\frac{c_m}{\sqrt{2\pi}\sqrt[4]{\lambda + \chi^2_m}}e^{-ikr\cos(\varphi - i\gamma_m)}.$$

The set of surface waves can be constructed using the laws of geometrical optics and the two-body scattering data:

$$\Psi_1^{\text{surf}}(\lambda, m, r, \varphi)$$

$$= \frac{c_m}{\sqrt{2\pi}\sqrt[4]{\lambda + \chi_m^2}}\Big\{ e^{-ikr\cos(\varphi - i\gamma_m)}$$

$$+ P(k\sin(i\gamma_m + 2\pi/3))e^{ikr\cos(\varphi + i\gamma_m + \pi/3)} \tag{7.128}$$

$$+ P(k\sin(i\gamma_m + \pi/3))P(k\sin(i\gamma_m + 2\pi/3))e^{ikr\cos(\varphi - i\gamma_m - \pi/3)}\Big\}.$$

The first component of the set of plane waves can also be calculated:

$$\Psi_1^{\text{surf}'}(\lambda, m, r) = -\frac{c_m \beta_{11}}{2\sqrt{\pi}\sqrt[4]{\lambda + \chi_n^2}}\Bigg\{ \frac{e^{-ikr\cos i\gamma_m}}{a' - k^2\sin^2 i\gamma_m}\varphi'$$

$$+ P(k\sin(i\gamma_m + 2\pi/3))\,(1 + P(k\sin(i\gamma_m + \pi/3)))$$

$$\frac{e^{ikr\cos(i\gamma_m + \pi/3)}}{a' - k^2\sin^2(i\gamma_m + \pi/3)}\varphi'\Bigg\}; \tag{7.129}$$

$$\Psi_1^{\text{surf}''}(\lambda, m, r)$$

$$= -\frac{c_m \beta_{11}}{2\sqrt{\pi}\sqrt[4]{\lambda + \chi_m^2}}\times$$

$$\Bigg\{ (1 + P(k\sin(i\gamma_m + 2\pi/3)))\frac{e^{-ikr\cos(\varphi - i\gamma_m)}}{a' - k^2\sin^2(i\gamma_m + 2\pi/3)}\varphi'$$

$$+ P(k\sin(i\gamma_m + \pi/3))P(k\sin(i\gamma_m + 2\pi/3))\frac{e^{ikr\cos i\gamma_m}}{a' - k^2\sin^2 i\gamma_m}\varphi'\Bigg\}.$$
$$\tag{7.130}$$

The boundary values at the origin for the set of surface waves, i.e. the vectors

$$\Psi_1^{\text{surf}'}(\lambda, m, 0); \qquad \Psi_1^{\text{surf}''}(\lambda, m, 0);$$

$$\frac{\partial\Psi_1^{\text{surf}'}(\lambda, m, r)}{\partial r}\Big|_{r=0}; \quad \frac{\partial\Psi_1^{\text{surf}''}(\lambda, m, r)}{\partial r}\Big|_{r=0}; \tag{7.131}$$

can easily be calculated analytically. Therefore the vectors θ_{1s}' and θ_{1a}' can be reconstructed:

$$\theta'_{1a}(\lambda, m) = -\frac{1}{2}(1 + B_a)^{-1}\left(\Psi_1^{\text{surf}'}(0) + \Psi_1^{\text{surf}''}(0)\right),$$

$$\theta'_{1s}(\lambda, m) = -\frac{1}{2}\left(i\sqrt{k^2 - a'} + \hat{B}_s\right)^{-1}\left(\frac{\partial\Psi_1^{\text{surf}'}}{\partial r}\bigg|_{r=0} + \frac{\partial\Psi_1^{\text{surf}''}}{\partial r}\bigg|_{r=0}\right).$$

$$(7.132)$$

The components S_{22}, S_{32} of the scattering matrix can be calculated from the asymptotics of the constructed eigenfunction

$$s_{22}^{21}(\lambda, n, m) = \delta_{nm}\Pi^{-1}(2\pi/3 - i\gamma_m)\Pi^{-1}(\pi/3 - i\gamma_m)$$
$$+ \frac{\sqrt{2\pi}\sqrt[4]{\lambda + \chi_n^2}}{c_n}\left(A_s^n(\theta'_{1s}) + A_a^n(\theta'_{1a})\right); \qquad (7.133)$$

$$s_{32}^1(\lambda, \varphi, m) = -i\sqrt{2k}\left\{\tilde{G}_s(\lambda, \theta'_{1s}, \varphi) + \tilde{G}_s(\lambda, \theta'_{1s}, \pi/3 - \varphi)\right.$$
$$\qquad (7.134)$$
$$\left. + \tilde{G}_a(\lambda, \theta'_{1a}, \varphi) - \tilde{G}_a(\lambda, \theta'_{1a}, \pi/3 - \varphi)\right\}\Theta(\lambda).$$

Here $\Theta(\lambda)$ is the Heaviside function. The values of θ'_{1s} and θ'_{1a} appearing in the latter formula are given by (7.132). The component s_{22}^{11} of the scattering matrix is equal to zero.

Similarly one can calculate the kernels s_{22}^{22}, s_{22}^{12} and s_{32}^2 of the scattering matrix by considering the eigenfunctions induced by the surface waves coming along the boundary $\varphi = \pi/3$.

We succeeded in giving a precise analytical calculation of the whole three-body scattering matrix for the case of identical particles.

The energies of the three-body bound states can be calculated as follows. The wave function of the three-body bound state is equal to the outgoing wave that is constructed. Therefore substituting the boundary values of the outgoing wave into the boundary conditions (7.49) we get two dispersion equation for the energies of the three particle bound states

$$\det(i\sqrt{\lambda - a'} + \hat{B}_s(k)) = 0 \qquad (7.135)$$

and

$$\det(1 + B_a(k)) = 0. \qquad (7.136)$$

Each solution of one of this equations determines the vector θ'_s or θ'_a such that the corresponding outgoing wave $\mathbf{G}_s(\theta'_s)$ or $\mathbf{G}_a(\theta'_a)$ satisfies the boundary conditions at the origin.

It should be emphasized that additional singularities in the scattering amplitudes are produced by the two-body bound states and resonances. A richer structure of the three-body bound states can be obtained by adding the three-body space of interaction. The bound state eigenfunctions are orthogonal to the continuous spectrum eigenfunctions.

Thus the investigation of the three-body model scattering problem has been accomplished. All eigenfunctions were presented by Sommerfeld integrals; however the scattering matrix was calculated in terms of elementary functions. This was possible due to the simple geometry of the problem. But the scattering matrix for nonidentical particles should contain some special functions. The components s_{23}, s_{32} of the scattering matrix are continuous functions of the angles for all φ. The component s_{33} contains a singularity corresponding to back scattering. The main difference with the standard three–particle problem with the delta interaction is that the two-body and three-body eigenfunctions are no longer orthogonal. Thus the components S_{23} and S_{32} of the scattering matrix are not trivial and the model describes break–up and capture processes.

Appendix A

Historical remarks.

A.1 Extension theory for symmetric operators

A.1.1 Extension theory and Nevanlinna R-functions

Probably the most interesting example of a singular rank one perturbation is the perturbation of the boundary condition for a second order ordinary differential operator. This is one of the first mathematical problems where the extension theory plays an indispensable role. In 1909–1910 H. Weyl investigated in his famous papers [954, 955] the behavior of the solutions to the second order differential equation under variation of the boundary condition. He was also the first to ask the question: How does the spectrum change under such a perturbation? He proved that the absolutely continuous spectrum is invariant under such a perturbation. The question by H.Weyl concerning other types of spectrum has been investigated by F. Wolf [964], N. Aronszajn [97], and N.Aronszajn and W.F.Donoghue [99]. See also the paper by V.A. Javrian [493]. So it was H.Weyl who was the first to understand the importance of this class of perturbations from the mathematical point of view. The first mathematically rigorous investigation of singular perturbations of partial differential operator was carried out by F.A.Berezin and L.D.Faddeev [135]. These authors have shown that such perturbations can be described using the extension theory of symmetric operators. This paper was extremely important because it clarified the relation between partial differential operators with point interactions and Krein's formula describing the resolvents of all self–adjoint extensions of a given symmetric operator. Considering the ordinary differential operator on an interval the boundary conditions at the limit points are necessary and it is not surprising that these conditions appeared in the very first papers on ordinary differential operators. Considering an arbitrary partial differential operator in a certain domain similar boundary conditions appear on the domain's boundary. But the point interactions for the partial differential operator are described by additional boundary conditions at the points situated inside the domain.

The abstract theory of self-adjoint extensions of symmetric operator due to J. von Neumann [199, 945] became an inalienable part of all standard textbooks on operator theory [154, 305, 842]. In this theory it has been shown, in particular, that a given symmetric operator can be extended to a self-adjoint operator if and only if the deficiency indices of the original operator are equal (perhaps infinite). In particular, if the deficiency indices are finite and equal, then all self-adjoint extensions can be described explicitly. We would like to point out that the extension theory had already been developed by H.Weyl but for the special case of the second order differential operator on the half axis [954, 955]. In 1943 M.G.Krein and M.A.Naimark (M.A.Neumark) described independently the set of all spectral functions of a closed symmetric operator with deficiency indices $(1,1)$ [565, 566, 567, 762]. It appeared that these extensions can be naturally described by Nevanlinna R-functions. Krein's formula relating the resolvents of two self-adjoint extensions of one symmetric operator appeared for the first time in these papers. Later on it was generalized for the case of arbitrary finite, respectively infinite, deficiency indices by M.G.Krein [569, 572] and Š.N.Saakjan [851], respectively.

Another chapter in the theory of self-adjoint extensions consists in the study of self-adjoint extensions of lower semibounded symmetric operators. This class of operators is of special importance. In fact each semibounded symmetric operator has equal deficiency indices and therefore can be extended to a self-adjoint operator. Moreover semibounded operators are interesting from the physical point of view. The existence of self-adjoint extensions for an arbitrary semibounded symmetric operator was first proven by M.N. Stone [925] and K. Friedrichs [384]. The set of all extensions of semibounded operators has been investigated by M.G.Krein [568, 570, 571], M.Sh.Birman [140], and M. Vishik [943]. See also review papers on this theory [80, 575, 576] and the paper by T.Kato [516]. The approach developed by M.G.Krein results from his investigation of n-th order ordinary differential operators [563, 564]. The Glazman–Krein–Naimark theory is discussed in many modern books on operator theory [8, 9, 10, 11, 425, 759, 760, 761].

The extension theory for operators with gaps was studied more recently by J.F.Brasche, V.Koshmanenko, H.Neidhardt and M.Malamud [177, 178, 677, 679].

The case of non–densely defined symmetric operator was considered by M. A. Krasnosel'skiĭ [560, 561, 562]. See also [585, 676, 678]. The extension theory of non–densely defined symmetric operators is closely related to the extension theory of binary operator relations. This approach has been developed in [274, 454, 535, 536]. The generalized resolvents were first studied by M.A.Naimark, M.G.Krein and A.V.Strauss [569, 757, 758, 762, 930] and later on for nondensely defined and/or positive operators in [583, 584, 586]. Recent developments in this direction were initiated by B.S. Pavlov, when he suggested in 1984 the use of generalized extensions with a finite dimensional extension space to derive exactly solvable models of quantum mechanical, optical and other physical problems [781]. Pavlov's method became a powerful tool in modern mathematical physics. It can be considered as an extension of the method of point interactions and Krein's theory. Two review papers on this subject are [785, 786]. We are going to discuss numerous applications of this model in what follows (Section A.2.2).

The theory of Nevanlinna R–functions was developed during the first half of the twentieth century. The integral representation for Nevanlinna functions can be obtained from the integral representation established by F.Riesz [845] and G.Herglotz [463] for functions holomorphic in the unit disk and taking their values in the right half plane using a linear–fractional transformation. The functions from the Nevanlinna class have been studied by M.Riesz [846], whereas functions regular in the half plane have been studied by I.I.Privalov [839, 840] and V.I.Krylov [590]. The representation of Nevanlinna functions has been investigated by I.Kac [498, 499] and by N.Aronszajn and W.F.Donoghue [99, 100]. The Riesz–Herglotz representation is almost equivalent to the spectral theorem for self–adjoint operators obtained by D. Hilbert and J. von Neumann [945, 946]. Different classes of Nevanlinna functions were introduced in the paper by I.Kac and M.Krein published as a supplement to the book [102] (see [500] for an English translation). In ivestigating subclasses of Nevanlinna functions the notion of Friedrichs extension has been generalized by S.Hassi and H. de Snoo for the case of not necessarily semibounded operators [452, 456, 457]. For recent studies of matrix–valued Herglotz functions and their applications to finite dimensional perturbations of self–adjoint operators see [400, 405]. Operator–valued functions with positive imaginary part have been studied by S.Naboko [749, 750, 751]. These functions appear in studies of infinite dimensional perturbations of self–adjoint operators [396].

Another generalization of the notion of Nevanlinna functions is related to the extension theory in Pontryagin or Krein spaces. This research was initiated by M.Krein and H.Langer [491, 578, 579, 580, 581, 582] and developed recently by V.A.Derkach, M.M.Malamud, S.Hassi and H. de Snoo [267, 268, 269, 271, 272].

The extension theory for operators in Pontryagin spaces and corresponding generalizations of Krein's formula have been developed in [82, 262, 263, 264, 265, 266, 275, 276, 277, 278].

This theory has been used by Yu.G. Shondin, A. Tip and J.F. van Deijen to construct exactly solvable quantum mechanical operators [885, 886, 887, 888, 889, 890, 891, 892, 941]. See also Section A.5.5 where the papers by B.S. Pavlov, A. Kiselev and I.Yu. Popov are discussed. This approach was used by above authors to define singular interaction which are not from the class $\mathcal{H}_{-2}$ in Pontryagin spaces. Singular interactions from this class were defined recently in a certain extended Hilbert space in [644, 86].

A.1.2   Symplectic structure and von Neumann construction

Self–adjoint extensions of symmetric operators have been described by J. von Neumann in terms of unitary operators acting in the deficiency subspaces. It was probably M.Krein who first realized that self–adjoint extensions can be described in terms of self–adjoint operators (or rather self–adjoint operator relations). This description is closely related to the description of self–adjoint extensions in terms of boundary conditions.

The boundary conditions describing self–adjoint operators first appeared during the investigation of ordinary differential operators. In fact an ordinary differential operator defined on infinitely differentiable functions with finite support can have nontrivial deficiency indices. Then the self–adjoint extensions of this minimal symmetric operator can be described by the boundary conditions connecting the value of the function and its derivatives on the boundary. Similar conditions can be used to describe interior singularities, as was been done recently by P.B.Bailey, P.Carlson, W.N.Everitt, L.Markus, C.Shubin, G.Stolz, R.Threadgill and A.Zettl [112, 203, 318, 319]. See also the book by W.N. Everitt and L. Markus [320]. Similar problems appear for the description of all self–adjoint extensions of Sturm–Liouville operators defined on graphs. Such problems have been studied by N.I.Gerasimenko and B.S. Pavlov [390, 391]. Dirichlet, von Neumann and other types of boundary conditions have been used to define for example self–adjoint Laplacians in two and three dimensional domains. Similarly as for the case of ordinary differential operators the interior singularities for partial differential operators can also be described by boundary conditions as it was shown by F.A.Berezin and L.D.Faddeev [135]. The self–adjoint extensions of an arbitrary symmetric operator have been described using some abstract boundary conditions by J.W.Calkin [199].

To describe all self–adjoint extensions of a certain symmetric operator A^0

in terms of abstract boundary conditions one should consider the boundary
form of the adjoint operator

$$\langle U, A^{0*}V \rangle - \langle A^{0*}U, V \rangle$$

which defines a symplectic structure in the deficiency subspace (since this
form vanishes on the elements from the domain of the symmetric operator
A^0). Then the self–adjoint extensions of the symmetric operator can be
characterized by Lagrangian subspaces (maximal isotropic subspaces) of the
symplectic boundary form. This approach has been proposed independently
by different mathematicians (M.Krein, I.Gelfand, B.S. Pavlov, G.Segal). See
for example recent papers by S.P.Novikov [774, 775]. For explicitly solvable
models it is important to combine abstract operators with differential operators. The standard method of construction of self–adjoint extensions for
differential operators (see H.Weyl [954, 955]) is based on the analysis of the
symplectic boundary forms and construction of the corresponding Lagrangian
planes. To construct a joint self–adjoint extension of an orthogonal sum of
an abstract and a differential operator one has to combine H.Weyl's theory
with the von Neumann construction. It became apparent that M.Krein's
parameterization of self–adjoint extensions by Hermitian operators (not by
isometric operators as in the von Neumann approach) can be used to select
joint self–adjoint extensions. This approach was first used by M.D. Faddeev
and B.S. Pavlov in [371] when constructing a solvable model for a resonator
with a small opening. It was M.Semenov–Tyan–Shanskii who drew Pavlov's
attention to the symplectic structure of the construction he had achieved.
This approach was used later by B.S. Pavlov to define the zero–range potential with internal structure [781]. See also the paper by Yu.E. Karpeshina
and B.S. Pavlov devoted to zero–radius interactions for the biharmonic and
polyharmonic equations [509]. Later on this idea was intensively used by the
St.Petersburg school to define exactly solvable models of quantum mechanical, optical and other physical problems. Two review articles where this idea
has been formulated in an abstract setting are by B.S.Pavlov [785, 786]. The
same idea has been employed by different authors to define self–adjoint extensions (see for example the papers by J. Boman, P. Kurasov, V. Kostrykin,
S.P. Novikov and R. Schrader [558, 636, 774, 775, 776]).

A similar approach has been developed by V.I. Gorbachuk, M.L. Gorbachuk, and A.N. Kochubei who described self–adjoint extensions using the
space of boundary values and the boundary operators [427, 429, 430, 535]
without discussion of the corresponding symplectic structure. This approach
has been used by A.N. Kochubei and V.A. Mikhailets to define point interactions for different operators [539, 540, 541, 718, 720, 723]. (See also
[315, 316].) The relations with M.Krein's theory of self–adjoint extensions

have been described in [535, 536, 537, 538]. General boundary conditions for elliptic operators had been considered earlier by M.Vishik and M.S.Birman [142, 143, 145, 943].

Energy dependent boundary conditions have been used by physicists to obtain realistic models of atomic and nuclear processes. The main disadvantage of this approach is that these models are not described by operators, but by operator pencils. In the spectral theory of operator pencils one cannot use so strong a tool as the Hilbert identity. To overcome this problem one can can try to reduce the spectral problem for an operator pencil to a simpler problem for a suitable operator. In particular such a reduction permits one to prove completeness and orthogonality of the scattering waves for many solvable models. Krein's theory of generalized extensions gives the possibility of describing such problems by self-adjoint operators acting in certain extended Hilbert spaces, as discussed for example by Yu.A.Kuperin, K.A.Makarov, S.P.Merkuriev, A.K.Motovilov, B.S. Pavlov, and R.Mennicken [606, 670, 741, 742, 743, 745, 746, 747].

See also [186, 279] where differential operators with energy dependent boundary conditions are studied. The problem of defining self-adjoint realizations of such operators is related to the study of matrix-value self-adjoint operators and has been studied by V.Adamyan, H.Langer, R.Mennicken, J.Saurer, A.K.Motovilov, and A.A.Shkalikov [4, 713, 714, 883, 884].

A.2 Finite rank perturbations

A.2.1 Definition of singular finite rank perturbations

Investigations of finite rank bounded perturbations can be found for example in the papers by T.Kato and S.T.Kuroda [517, 648] where such perturbations have been used to study nontrivial scattering problems. At the same time the problem of perturbation of differential operators has been discussed by N.Aronszajn and W.F.Donoghue [97, 99, 100]. In particular the relations between the extension theory of symmetric operators and finite rank perturbations has been clearly stated by W.F. Donoghue in [295]. Again the problem of H.Weyl played an important role in this investigation. Discussions of arbitrary finite rank perturbations can be found in the well-known book by T.Kato [522] and in the review paper [98]. Different models involving non-bounded finite rank perturbations have been considered by many mathematicians. An important step was made by F.Berezin and L.Faddeev who studied the Schödinger operator with delta potential in three dimensions as an example of a rank one unbounded perturbation [135]. The relations between unbounded rank one perturbations and the extension theory for symmetric operators have been clarified in this work. Later on new interest in these problems was aroused by the papers by B.Simon and T.Wolff [909] and F.Gesztesy and B.Simon [403] who studied $\mathcal{H}_{-1}$-perturbations of self-adjoint operators. The paper [904] gives a complete review of such perturbations which are form bounded and therefore can be defined using the standard form perturbation technique [522, 842]. Concerning $\mathcal{H}_{-2}$-perturbations we would like to mention the paper by V.A. Javrian [493] where the Weyl problem has been investigated. The first attempt to define the $\mathcal{H}_{-2}$-perturbation in an abstract setting is due to A. Kiselev and B. Simon [533] who generalized the approach first developed by F.A. Berezin and L.D. Faddeev for the Laplacian in $\mathbf{R}^3$ [135]. Abstract quadratic form formulation of the $\mathcal{H}_{-2}$ finite rank perturbation has been developed for bounded from below operators by V.Koshmanenko in [546, 549, 550, 551, 552]. In the operator formulation it appeared in paper [62] by S.Albeverio, W.Karwowski and V.Koshmanenko. Futher studies by S.Albeverio and V.Koshmanenko based on Krein resolvent formula can be found in [63, 64, 550, 551, 546, 547]. The $\mathcal{H}_{-2}$ construction has been used by W.Karwowski, V.Koshmanenko and S.Ota for perturbation of the Laplacian in $\mathbf{R}^3$ by dynamics supported by set of lower dimension [515].

The $\mathcal{H}_{-2}$ rank one perturbations without renormalization of the coupling constant were first defined by S. Albeverio and P. Kurasov for semibounded [68], respectively not necessarily semibounded [69] self-adjoint operators. Approximations of the perturbed operator (without renormalization of the coupling constant) have been defined. This work originates from earlier studies of point interactions for the second derivative operator in one dimension by P. Kurasov [631]. See also the paper by J. Boman and P. Kurasov devoted to studies of point interactions for the operator of the n-th derivative in one dimension [636]. Arbitrary finite rank $\mathcal{H}_{-2}$ perturbations of non-semibounded operators were studied in [70]. S. Hassi and H. de Snoo studied rank one singular perturbations using methods of the theory of analytic functions [453, 455]. An interesting version of the $\mathcal{H}_{-2}$ perturbation techniques not requiring the boundedness from below of the perturbed operator was suggested by S.T.Kuroda and developed by N.Nagatani and K.Watanabe [752, 753, 951, 952].

Perturbations of the absolutely continuous part of the spectrum under finite rank and trace class perturbations were studied by T. Kato, S.T. Kuroda and M. Birman [144, 146, 147, 148, 149, 150, 151, 152, 153, 517, 518, 519, 520, 521, 522, 523, 524, 645, 646, 648, 649, 650].

If the difference of the resolvents of two operators belongs to the trace class then one can prove that the absolutely continuous parts of the spectra of the two operators coincide. This can be proven by constructing the wave operators. The scattering operator for two such self–adjoint operators has been computed by T.Kato and S.T.Kuroda [517, 518, 519, 520, 521, 522, 523, 524, 645, 646, 647, 648, 649, 650] and by M.G.Krein, M.Birman, S.B.Entina and D.Yafaev [144, 146, 147, 148, 149, 150, 151, 152, 153, 156, 158].

Our approach to the scattering problem is based on the paper by V.M. Adamyan and B.S. Pavlov [5], where the scattering operator has been computed for two operators which are different self–adjoint extensions of a given symmetric operator with finite deficiency indices. See also [944].

Perturbations of the other parts of the discrete and continuous spectra under finite rank and infinite rank perturbations have been studied in [171, 173, 180, 181, 182, 238, 239, 240, 403, 431, 557, 904, 907, 908].

A.2.2   Generalized singular perturbations

The generalized extensions of symmetric operators first studied by M.G.Krein, M.A.Naimark and A.V.Strauss [569, 757, 758, 762, 930] can be used to construct exactly solvable perturbations of self–adjoint operators. The paper of F.A.Berezin and L.D.Faddeev [135] opened the way to using all the machinery of modern extension theory for the construction of solvable models in quantum mechanics. The first models with one dimensional extension spaces were constructed by Yu. M. Schirokov, Yu.G. Shondin and their coworkers [432, 885, 890, 940]. The first general models based on Krein's formula were constructed by B.S. Pavlov in 1984 [781]. It is worth mentioning here that M.G. Krein anticipated using his formula with a Hermitian parameter in quantum mechanics as an alternative to models proposed by M.S. Livšic [665] (private communication by V.M.Adamyan). This application had not been discovered earlier probably because before 1984 the symplectic technique was used mainly for differential operators and von Neumann's technique with unitary parameterization was used for abstract operators. B.S. Pavlov just proposed using symplectic technique in both cases and combined abstract operators with differential ones to construct solvable models with nontrivial spectral structure. The idea was to generalize the delta potential interaction for the Laplace operator in $\mathbf{R}^3$. The operator

$$L_\alpha = -\Delta + \alpha\delta(x)$$

in $L_2(\mathbf{R}^3)$ was first studied by F.A.Berezin and L.D.Faddeev and it was shown that the self–adjoint operator corresponding to the latter differential expression is equal to one of the self–adjoint extensions of the symmetric operator

$$L^0 = -\Delta|_{C_0^\infty(\mathbf{R}^3\backslash\{0\})}.$$

The family of self–adjoint extensions of L^0 can be characterized by one real parameter. Any self–adjoint extension of L^0 is a rank one perturbation of $-\Delta$. The self–adjoint operator L_α can have no more than one eigenvalue and its continuous spectrum coincides with the branch of the absolutely continuous spectrum $[0, \infty)$ of the Laplacian. To generalize the point interaction B.S. Pavlov considered generalized extensions of the symmetric operator L^0 with a higher dimensional extension space. It was realized that in all physical applications it is enough to consider only finite dimensional extensions of the Hilbert space $L_2(\mathbf{R}^3)$. Therefore B.S. Pavlov suggested defining the generalized extension of L^0 as a perturbation of a certain diagonal self–adjoint operator acting in the orthogonal sum of the original Hilbert space $L_2(\mathbf{R}^3)$ and a certain "internal" finite dimensional Hilbert space H^{in}. The generalized extension of L^0 can be defined by first restricting the operator $\mathcal{A} = (-\Delta) \oplus A^{in}$ (where A^{in} is a certain self–adjoint (Hermitian) operator acting in H^{in}) to the operator $L^0 \oplus A^{in,0}$ and then extending the latter operator to a certain self–adjoint operator, which does not possess the orthogonal decomposition associated with the decomposition of the total Hilbert space into the original and internal spaces. Such extensions were described in terms of Lagrangian planes for the symplectic boundary form of the "adjoint" operator. In particular such extensions can be described by certain boundary conditions. Solvable models with generalized interactions in this sense can have a rich spectral structure. Each operator can have an arbitrary (finite) number of bound states; the corresponding scattering matrix has an arbitrary number of resonances. Nevertheless the corresponding operator remain exactly solvable in the sense that all its eigenfunctions can be calculated analytically. The scattering matrix $S(k)$ is rational and can have physical behavior at infinity $S(k) \to_{k\to\infty} 1$. (It is known that the scattering matrix corresponding to the standard delta interaction in $\mathbf{R}^3$ tends to -1 at infinity, while the scattering matrix for local bounded potentials with compact support always tends to 1 for large energies.) Instead of one parameter defining the self–adjoint extensions of L^0, the extensions in the model with internal structure described above are defined by the internal matrix A^{in} and the boundary conditions (2×2 unitary matrix). The number of eigenvalues of the operator so obtained is determined by the number of negative eigenvalues of A^{in}, while the positive eigenvalues of A^{in} produce resonances of the total operator.

This model has been intensively used by the St. Petersburg group to construct exactly solvable models of different physical problems. B.S. Pavlov and A.A. Shushkov studied the case of several point interactions with internal structures [802, 894, 895, 896]. Different models of solid state physics have been constructed by B.S. Pavlov, N.V. Smirnov, P. Kurasov [627, 641, 782, 803] considering an infinite set of generalized point interactions with a periodic location. In these works it has been shown that the eigenvalues of the internal operator produce additional bands of the continuous spectrum and/or forbidden zones for the total operator. Models of surfaces were considered by B.S. Pavlov and P. Kurasov [788, 791]. Scattering on the random potential was investigated by B.S. Pavlov and A.E. Ryzhkov [799, 800, 801, 850]. Yu.A. Kuperin, K.A. Makarov, Yu.A. Melnikov, and B.S. Pavlov studied the Coulomb two–body problem with internal structure, scattering by a breathing bag and other interesting physical problems [599, 598, 600, 610, 612]. See also the papers by V. Evstratov, Yu.A. Kuperin, Yu.B. Melnikov, S.P. Merkuriev, M.A. Pankratov, B.S. Pavlov, and E.A. Yarevsky [321, 322, 323, 616, 625, 783, 784, 787, 792, 805]. Approximations of the generalized interactions by smooth potentials have been constructed by K.A. Makarov and S.E. Cheremshantsev [217, 672] (see Section A.3.2).

Different models of few–body problems constructed using generalizations of Pavlov's model form the most interesting applications of this approach. It appears that the three–body operator with generalized two–body delta interactions in $\mathbf{R}^3$ can be bounded from below, while the same operator with standard delta interactions is not bounded from below (and therefore is somewhat nonphysical). We discuss this approach developed by Yu.A. Kuperin, P. Kurasov, K.A. Makarov, A.K. Motovilov, S.P. Merkuriev, and B.S. Pavlov in detail in Section A.4. A similar approach was used by Yu.G. Shondin to define exactly solvable models of few–body scattering problems [885, 890]. A very interesting generalization of Pavlov's model has been worked out by D.R. Yafaev [970], who constructed a model describing a quantum particle interacting with the vacuum in the Fock space of a quantum field. A general approach to models with generalized point interactions has been developed by B.S. Pavlov in the two review papers [785, 786]. His approach is based essentially on Krein's formula for generalized resolvents. It is well known that in general this formula can define not only self–adjoint operators but also self–adjoint operator relations. P. Kurasov and B.S. Pavlov proved that no operator relation appears in models with generalized interactions [640].

A.2.3 Singular perturbations defined by forms

Singular finite rank perturbations can be defined using the form perturbation technique. In fact, if the perturbation is of type $\mathcal{H}_{-1}$ then the perturbation is infinitesimally form bounded with the relative bound zero with respect to the original operator and one can use the KLMN theorem to define the perturbation. This approach is especially useful for semibounded operators, when the self–adjoint operator is uniquely defined by its quadratic form. This approach has been developed by S. Albeverio, W. Karwowski and V. Koshmanenko in a series of papers [63, 64, 511, 512, 513, 514, 546, 548, 549, 554, 556]. Some of these results are summarized in the book [552]. Different singular form perturbations of the Laplacian in $L^2(\mathbf{R}^3)$ have been considered by M. Birman, J.F. Brasche, W. Karwowski, V. Koshmanenko, A. Teta, and A. Tip [141, 165, 166, 510, 555, 547]. One can define singular perturbations of self-adjoint operators using Dirichlet forms. This approach was first used by S.Albeverio, R.Høegh-Krohn and L.Streit to define singular perturbations of the Laplacian in $\mathbf{R}^n$ [58]. The relations between Schrödinger and Dirichlet forms have been discussed by S. Albeverio, J. Brasche, F. Gesztesy, R. Høegh-Krohn, W. Karwowski, S. Kusuoka, M. Röckner, and L. Streit [18, 45, 50, 71, 167, 168, 169, 173]. Diffusion processes with singular Dirichlet forms are studied in [73].

A.3 Point interactions

A.3.1 Definition of the point interactions in $\mathbf{R}^1, \mathbf{R}^2$ and $\mathbf{R}^3$

Point interactions represent a special case of singular perturbations, where the original operator is a differential operator and the perturbation has support of zero measure. These operators have been used by theoretical physicists since from the 1930s to obtain exactly solvable models of different physical phenomena. In the introduction and historical remarks of the book [39] these developments were described in great detail, so we refer to this source, except for some particular points we want to stress in connection with the present work and with developments which took place after the publication of [39]. In 1931 R. de L. Kronig and W.G.Penney [589] used this approach to model the (quantum mechanical) motion of an electron in a solid state. A few years later (in 1935) a similar model for the motion of nuclear particles was suggested by H.Bethe and R.Peierls [136] and L.H.Thomas [937]. At the same time E.Fermi indirectly suggested the use of the pseudopotential

$$-\alpha^{-1}\delta(r)\frac{\partial}{\partial r}r\Big|_{r=+0}$$

to describe neutron scattering in substances containing hydrogen [374]. G. Breit gave a description of the Fermi pseudopotential using boundary conditions [185]. These works established the method of point interactions, or zero-range potentials as a useful tool in nuclear and atomic physics. The interaction in all these models is described not by a potential but by certain boundary conditions on a discrete set of points. This operator was defined rigorously by F.A. Berezin and L.D. Faddeev [135] in 1961.

The case of the Coulomb plus point interaction was first studied by S. Albeverio, F. Gesztesy, R. Høegh–Krohn and L. Streit [33, 43] and J.Zorbas [976]. The same problem in L_p spaces was investigated by W.Caspers [206]. The case of Coulomb plus generalized point interaction was investigated by Yu.A.Kuperin, K.A.Makarov and Yu.B.Melnikov [598].

Point interactions in one dimension appear naturally as boundary conditions for functions from the domain of an ordinary differential operator. We have already mentioned that such perturbations were used by H. Weyl in 1910 [954, 955]. Two approaches have been developed to study such interactions. The first is based on the extension theory of symmetric operators and the corresponding interactions have been named point interactions. Consider the second derivative operator on the line $L = -\frac{d^2}{dx^2}$, Dom$(L) = W_2^2(\mathbf{R})$. Then all point interactions can be obtained first by restricting the original self–adjoint operator L to the set D_0 of functions having compact support separated from the set of points supporting the interaction. If the number of points is finite then the corresponding symmetric operator $L_0 = L|_{D_0}$ has finite deficiency indices and all its self–adjoint extensions can be calculated using, e.g., Krein's formula. Each point can be considered separately. It appears that all such operators can be defined using certain boundary conditions connecting the values of the functions and their first derivatives on the left and right hand sides of the points supporting the interaction. It is obvious that it is enough to describe all point interactions supported at the origin. Then the boundary conditions connect the values $\psi(\pm 0), \psi'(\pm 0)$. One can find the description of all point interactions for the one–dimensional Laplacian, e.g., in [39, 40, 42, 398, 859, 860, 864].

Following the second approach one tries to write the perturbed operator as the sum of the original operator and a certain singular operator

$$L + V.$$

This method can be named the method of singular perturbations. If the kernel of the singular operator V contains all functions from the set D_0

$$\mathrm{Ker}(V) \supset D_0$$

then every self–adjoint operator corresponding to the formal linear operator $L+V$ coincides with one of the self–adjoint extensions of the operator L_0. The first family of singular perturbations to be studied for the one dimensional Laplacian is formed by the δ potential

$$-\frac{d^2}{dx^2} + \alpha\delta = -\frac{d^2}{dx^2} + \alpha\langle\delta, \cdot\rangle\delta.$$

This family of singular perturbations is a family of rank one $\mathcal{H}_{-1}$ perturbations and the corresponding set of point interactions is uniquely defined. Another family of point interactions was named δ'-interactions and has been defined using the following boundary conditions [39, 40, 393]

$$\begin{pmatrix} \psi(+0) \\ \psi'(+0) \end{pmatrix} = \begin{pmatrix} 1 & \beta \\ 0 & 1 \end{pmatrix} \begin{pmatrix} \psi(-0) \\ \psi'(-0) \end{pmatrix}.$$

The relations between this set of point interactions and the set of interactions defined by the δ'-potential has been discussed in the literature (see [44, 863] and references therein). This problem has been solved independently in [307, 436] using additional assumptions on the symmetry properties of the interaction. In fact this interaction cannot be defined without such an assumption, since the singular interaction in this case is not from $\mathcal{H}_{-1}$. Different assumptions can lead to different one dimensional families of point interactions corresponding to the formal expression $-\frac{d^2}{dx^2} + \beta\delta'$. An interesting approach based on I.Segal's approach to singular interactions [480, 486] has been developed by P.Chernoff and R.Hughes [218]. Different authors have tried to obtain the whole four-parameter family of point interactions using approximations [20, 214, 226, 779, 847]. A four-parameter family of singular interactions describing the total set of point interactions has been obtained by P.Kurasov using symmetry arguments [631]. Therefore all point interactions have been classified using singular operators. This approach has been generalized for the operator of the n-th derivative by J. Boman and P. Kurasov [636]. The spectral properties of one dimensional point interactions have been studied recently by D. Buschmann, A.N. Kochubei, M. Maioli, V. Mikhailets, L.P. Nizhnik, A. Sacchetti, C. Shubin, G. Stolz, and J. Weidmann [193, 194, 540, 541, 667, 718, 719, 720, 721, 722, 723, 724, 773, 893]. The symmetries of one dimensional point interactions were studied by S. Albeverio, P. Kurasov, L. Dąbrowski, and J. Boman [22, 70, 636]

Sturm–Liouville problems related to one dimensional point interactions have been studied by W.N. Everitt and C. Shubin and G. Stolz and A. Zettl [203, 319]. See also the recent book by W.N. Everitt and L. Markus [320].

Two–dimensional point interactions are described in detail in [39]. These interactions were first defined by A. Grossman, R. Høegh–Krohn, and M. Mebkhout [443] and studied later in [38, 445, 471]. Physical applications have been considered by P. Exner and P. Šeba [337, 341]. Two dimensional point interactions play an important role in studies of operators with magnetic fields [93, 211, 406] and Aharonov–Bohm Hamiltonians (see our comments in Section A.5.6). Nonlocal point interactions in two dimensions were introduced by L.Dąbrowski and H.Grosse [228]. Surface interactions in two dimension have been used in [690] to model thin films. The path integral interpretation of such point interactions can be found in [439]. These interactions have been studied recently in [461, 778].

The first mathematically rigorous definition of the point interaction in dimension three was given by F.A. Berezin and L.D. Faddeev [135]. Hamiltonians in $L_2(\mathbf{R}^3)$ with a finite number of such interactions have been studied by S. Albeverio, H. English, A. Grossman, R. Høegh–Krohn, M. Mebkhout, P. Šeba, L.E. Thomas, and T.T. Wu, [49, 443, 445, 862, 870, 934, 935]. We have already discussed Pavlov's model of point interactions with internal structure in Section A.2.2.

Applications to diffraction models and quantum mechanics are discussed in [349, 341]. Models of non–rotationally symmetric point interactions were suggested by Yu.G. Shondin [886, 887, 888]. Nonlocal point interactions have been studied in [228]. A rigorous path integral approach has been developed in [234] (see also [439] for previous physical work).

Spectral problems involving singular prseudodifferential operators in $\mathbf{R}^n$ in bounded (irregular) domains including in particular extensions of Weyl's type results on the distribution of eigenvalues have been discussed extensively in important work by H.Triebel [939]. In terms of perturbations of Laplace operators the techniques in [939] are different from those used in our work, in particular $\mathcal{H}_{-2}$-perturbations are not covered.

A.3.2   Approximations of point interactions

Approximations of singularly perturbed operators by operators with bounded and/or infinitesimally form bounded perturbations are important in order to compare the spectral properties of these two classes of operators. The approximations of singular perturbations by short range interactions have been discussed in many particular situations and the book [39] contains numerous examples of such approximations. For example approximations of the delta interaction in three dimensions have been considered by F.A. Berezin and L.D. Faddeev [135] and by S. Albeverio and R. Høegh–Krohn [49, 54]. S.E. Cheremshantsev and K.A. Makarov used a similar technique to obtain approximations of the generalized point interactions [217, 672]. In particular it was proven that if the spectrum of the generalized operator contains no negative eigenvalues then it can be approximated in the weak resolvent limit by Schrödinger operators with smooth potentials.

The approximation procedure for finite rank perturbations described in the book has been developed by S. Albeverio and P. Kurasov [68, 69, 70].

Recently J.F. Brasche, R. Figari, and A. Teta considered approximations of perturbations defined by measures [175]. Different regularizations of singular interactions were used by H. Neidhardt and V.A. Zagrebnov to define singularly perturbed operators [764, 765, 766, 767, 768].

The approximations of the delta potentials using path integrals have been discussed recently by M. de Faria and L. Streit [234] (see also [114, 437, 438, 439, 440, 441] for physical work).

A.3.3 Point interactions and inverse problems

Solutions of the inverse spectral and scattering problems for the one dimensional Schrödinger operator on the half axis were first obtained by I.M. Gelfand, B.M. Levitan, V.A. Marchenko, and L.D. Faddeev [7, 358, 359, 363, 389, 688]. The uniqueness result for the inverse spectral problem was first established by G. Borg [160, 161]. The inverse scattering problem on the whole axis was first solved by L.D. Faddeev [365, 367, 368, 369]. Faddeev's approach has been further developed by V.A. Marchenko [689], P. Deift and E. Trubowitz [235, 237] and A. Melin [707]. Developed originally for locally integrable potentials, the inverse scattering method has been generalized to include δ–like potentials (see for example [658] and references therein). The relations between the inverse spectral problem and self–adjoint extensions have been studied recently by S. Albeverio, J.F. Brasche, and H. Neidhardt [17, 169, 179]. Continuous potentials defining the same scattering matrix as generalized delta interactions in one dimension have been calculated by P. Kurasov [629]. It appears that these potentials do not belong to Faddeev's class but rather have long–range tails like Wigner–von Neumann potentials having positive eigenvalues. The Faddeev–Marchenko approach to the inverse scattering problem has been generalized to include such potentials [630, 634]. These potentials coincide with so–called positon solutions of the Korteweg–de Vries equation obtained independently by V.B. Matveev [692, 693, 694]. The relations between generalized point interactions, inverse scattering problem and positon solutions have been investigated recently by P. Kurasov and K. Packalén [629, 632, 639].

A.4 Few–body problems

A.4.1 Three–body problems in $\mathbf{R}^3$ and $\mathbf{R}^2$

The method of zero–range potentials was first developed by physicists in order to describe atomic scattering. Therefore it was natural to try to apply this method to construct exactly solvable models for few–body scattering problems. The first realistic model of this sort was suggested by L.D. Landau and investigated by G.V. Skorniakov and K.A. Ter–Martirosian [910, 911] in 1957. A system of three particles in three dimensional space was described by the Laplacian in $L_2(\mathbf{R}^6)$ defined on the functions satisfying certain boundary conditions on the planes where the coordinates of any two of the three particles coincide. It was shown by G.S. Danilov that the integral equations which appear in the latter paper do not have a unique solution [232]. Therefore the operator defined is not self–adjoint but only symmetric as was shown by L.D. Faddeev and R.A. Minlos [730, 731]. It has been proven that the operator has deficiency indices $(1,1)$ and therefore possesses self–adjoint extensions. All such extensions of the Skorniakov–Ter–Martirosian operator have been described and it was proven that all these operators are not bounded from below. It thus appeared that the Skorniakov–Ter–Martirosian Hamiltonian is unphysical, since only lower semibounded operators can be used in quantum theory to describe atomic processes. We would like to mention that the papers [730, 731] also served as a starting point for the investigation of the quantum mechanical few–body problem by L.D. Faddeev [360, 361, 362, 364, 366, 370]. Studies of the problem were continued by R.A. Minlos and his school. Systems of particles with different statistics have been considered by A.M. Melnikov, R.A. Minlos and M.H. Shermatov [708, 709, 728, 729, 732, 876] (The papers by L.D. Faddeev and R.A. Minlos deal essentially with a system of three bosons.) The attractors which appear in the point interaction model were studied by S. Albeverio and K. Makarov [75, 76, 77]. In particular the relations with the Efimov effect have been illuminated [60].

But it was still a challenging task to construct a few–body operator bounded from below and including pointwise interactions. The first such Hamiltonian was defined in 1977 by S. Albeverio, R. Høegh–Krohn and L. Streit using Dirichlet forms [58]. But it appeared that the operator obtained by S. Albeverio, R. Høegh–Krohn and L. Streit differs from the Skorniakov–Ter–Martirosian operator by a certain three–body interaction. Another interesting model with a nontrivial point–like two–body interaction was suggested by L.E. Thomas [936]. It was obtained by connecting together by certain boundary conditions the four free Hamiltonians describing the system of three free particles and the three systems where two particles are bounded and the third particle is free. The two–body interactions of a special type allowed to construct a three–body lower bounded operator.

The corresponding problem in two dimensions did not attract the attention of many scientists. A system of several particles in $\mathbf{R}^2$ interacting through local and translationally invariant zero–range interactions was investigated for the first time by G.F. Dell'Antonio, R. Figari, and A. Teta [241, 242, 246]. The few–body interactions are not infinitesimally form bounded in this case, but nevertheless the interaction is strongly separable and the three–body operator can be defined. Studies of the two dimensional problems allowed the authors to look at the three dimensional problem from a new and very interesting perspective [241].

In 1982 Yu.G. Shondin used Shirokov's technique of singular potentials in the momentum representation to define a three body operator bounded from below in $\mathbf{R}^3$. The operator was defined using the theory of generalized extensions with a nontrivial extension space, corresponding to each two–body interaction. It has been shown that the total three body operator is bounded from below if the corresponding two–body scattering matrices have the correct physical behaviour at infinity [885, 887, 886].

A similar idea was used by the St.Petersburg school when Pavlov's model of generalized potentials was applied to construct models of three–body problems. We should mention that essential difficulties already appeared on the stage in the definition of the operator describing several particles. In a series of papers different few–body models were considered by Yu.A. Kuperin, K.A. Makarov, S.P. Merkuriev, A.K. Motovilov, and B.S. Pavlov [591, 601, 602, 603, 604, 607, 609]. See also the review article by Yu.A. Kuperin and S.P. Merkuriev [621]. The method of Faddeev equations has been developed for singular potentials (see also [715]). The main result of these investigations was the construction of a three–body operator bounded from below with pointwise two–body interactions. This operator was first suggested and investigated by B.S. Pavlov [786]. He discovered that the semiboundedness of the corresponding three–body operator is related to the physical behaviour of the two–body scattering matrices at infinity. The two–body scattering matrix determined by a standard point interaction in $\mathbf{R}^3$ tends to -1 for large energies, whereas the scattering matrix for perturbations determined by regular potentials always has limit 1. Pavlov's models constructed using generalized point interactions have scattering matrices with limits ± 1 at infinity. Models with two–body scattering matrices tending to 1 were named "physical". It has been proven that if all three two–body scattering matrices have physical behaviour at infinity, then the three–body operator is bounded from below. Unfortunately the "two–body interaction" used in the latter paper did not have a true two–body nature. K.A. Makarov suggested another model [669], but again the two–body interactions were constructed in such a way that the rearrangement processes were forbidden. Correct models with nontrivial two–body interactions were constructed by K.A. Makarov in 1992 [671] and investigated in collaboration with A.K. Motovilov and V.V. Melezhik [673, 674]. K.A. Makarov and V.V. Melezhik proved that the three–body operator is bounded from below if only two of the three two–body scattering matrices have physical behaviour at infinity. Finally using Pavlov's method of generalized interactions an operator bounded from below, describing a system of three particles in $\mathbf{R}^3$ with pointwise potentials, has been constructed. A system of N arbitrary particles interacting via two–body generalized point interactions has been studied in [600, 616, 621, 651, 717]. A recent paper by P. Kurasov and B.S. Pavlov proves the semiboundedness of the N–body operator with generalized point interactions [643]. The authors show that the N–body Schrödinger operator with generalized two–body delta interactions in $\mathbf{R}^3$ is semibounded from below even in the case where not all of the two–body scattering matrices have physical behaviour at infinity. Using similar methods a model of few–body systems in a lattice has been constructed in [622].

A.4.2   Few–body problems in $\mathbf{R}^1$

The first few–body problem with delta interactions in one dimension was investigated by J.B. McGuire in 1964 [701]. It appeared that this quantum mechanical problem is similar to the diffraction problem on several screens intersecting at one point. The problem has been solved in the case where one of the particles has infinite mass, the strength of the interaction is infinite or when all particles have equal masses. The three–body problem in the case where one of the particles has infinite mass and there is a break–up process has been analyzed by S. Albeverio [12]. Relations with the diffraction problem allows one to use a technique going back to the Sommerfeld integral to obtain exact solutions for these problems. This method has been developed by G.D. Malyuzhinets [680, 681, 682, 683] and the inversion formula for the Sommerfeld integral has been proven. (See also the book [190] where the Malyuzhinets results are presented.) One can also use the Kantorovich–Lebedev transformation [652] to solve the problem. One needs to use these integral representations to solve few–body problems in the case where the particles have different masses. But if the particles are equal then the solution can be presented as a sum of elementary functions (exponentials). Exactly this case is interesting in applications to statistical physics. The first two papers in this direction were written by J.B. McGuire [702, 703], where a system of one dimensional fermions was investigated. The exact solution of the one dimensional few–body problem has been used by C.N. Yang and C.P. Yang [971, 972, 973] to calculate the scattering matrix. It appears that this solution can be obtained using the Yang–Baxter equation or the Bethe Ansatz. One can find numerous applications of this model in the book by M. Gaudin [386]. Properties of the scattering matrix for three one–dimensional particles have been studied by V.S. Buslaev, S.P. Merkuriev, and S.P. Salikov [197, 198]. In particular the following question has been studied: When does the wave function corresponding to the three–body scattering process contain no out-going Fresnel waves, but only plane waves? All two–body scattering matrices defining such three–body problems have been characterized. This approach is based on the paper [195] by V.S. Buslaev and has been generalized to the case of N particles by V.S. Buslaev and N.A. Kaliteevski [196].

The investigation of a system of three arbitrary one dimensional particles interacting via pairwise interactions given by delta functions was continued by K. Lipszyc. He obtained solutions to some problems of this type (two identical particles interacting with a third particle, two particles interacting with the wall) using the Sommerfeld–Malyuzhinets transformation [659, 660, 663]. This author has also shown how to use the Faddeev equations to solve these problems [661, 662]. A solution to the scattering problem for three impenetrable particles was obtained by J.B. McGuire and C.A. Hurst [705]. The same authors also studied the problem of three arbitrary particles and showed that its solution is equivalent to a certain matrix Riemann–Hilbert functional equation, but the solution to the latter problem appears to be extremely complicated [706]. See also recent papers by J.B. McGuire [700, 704]. Several other problems related to one dimensional particles have been considered recently [426, 446, 638, 655, 748].

A system of three one dimensional particles with generalized interactions was studies by the St.Petersburg group. In the first paper the pairwise in-teractions were modeled by Neumann boundary conditions on the screen and a three–body generalized interaction was been introduced by Yu.A. Ku-perin, K.A. Makarov, and B.S. Pavlov [608]. The problem of three identical particles with generalized two–body interactions was studied by P. Kurasov [628, 635] and its solution was obtained using the Sommerfeld–Malyuzhinets transformation. This approach is discussed in detail in Chapter 7. It has been extended by L.A.Dmitrieva, Yu.A.Kuperin, and G.E.Rudin [287].

We would like to point out that the models of few–body problems con-structed with the help of point interactions are closely related to models with hard–core interactions. Such models were used for example by T.D. Lee, K. Huang and C.N. Yang in order to carry out exact statistical calculations for many–body systems [478, 479, 653]. The interaction in these models is introduced using standard boundary conditions on low–dimensional mani-folds. Many problems which appear in point interaction models are trivial in hard–core models. For example the corresponding three–body operator in $\mathbf{R}^3$ is always bounded from below. Point interaction models can be ob-tained as limits of hard–core models when the radius of the core tends to zero. Hard–core models were used to construct physically realistic models with point interactions (see e.g. [716, 740]). They have been studied recently by A. Boutet de Monvel, V. Georgescu, and A. Soffer [162].

On the other hand, models with point interactions sometimes possess exact solutions and can therefore also serve as a starting point for the investigation few–body Schrödinger operators with other potentials. For example Faddeev's theory of few–body operators has been developed using knowledge of the three–body problem with delta interactions in $\mathbf{R}^3$ [360, 361, 362, 364, 366, 370]. Similarly the point interaction model helped Yu.A. Kuperin, P. Kurasov, Yu.B. Melnikov, S.P. Merkuriev, and B.S. Pavlov to develop an adiabatic approach to few–body problems [592, 614, 617, 620, 638].

A.5   Recent developments

A.5.1   Sphere interactions

One of the disadvantages of point interaction models in $\mathbf{R}^3$ is that the interaction is spherically symmetric and gives rise to a scattering matrix which is nontrivial in the s–channel only. We have already mentioned that by extending the Hilbert space to a certain Pontryagin space exactly solvable nonspherically symmetric point interaction models can be constructed. Another approach to construct such interactions is related to so–called sphere interactions, where the interaction is supported by a small ball of finite radius. The perturbed operator is defined by the boundary conditions on the sphere imposed on the function from the domain of the operator. The first model of this type was suggested by J.-P. Antoine, F. Gesztesy, and J. Shabani [89]. These and other types of sphere interactions were studied by L. Dąbrowski and J. Shabani [229, 873, 874]. The case of several concentric spheres supporting delta interactions was first described by J. Shabani [872] and later investigated by M.N. Hounkonnou, M. Hounkpe, and J. Shabani [476]. The case of Coulomb plus a sphere interaction is studied in [477]. J. Dittrich, F. Domínguez–Adame, P. Exner, and P. Šeba studied the Dirac equation with a surface delta interaction [282, 283, 288]. Dirac Hamiltonians with Coulomb potential and a sphere interaction were investigated in [284, 285].

A.5.2   Interactions on low dimensional manifolds

The method of point interactions has been used to model potentials with support concentrated near certain manifolds in $\mathbf{R}^3$ having dimensions 1 and 2. A special important class of such models is formed by lattice models, where the interaction is defined by an infinite number of point interactions situated on a plane or on a line. Such models were first used by Yu.N. Demkov and R. Subramanian [255, 258] to study the motion of a quantum particle in a linear periodic array of small radius potential wells. The same model was used later to study scattering processes [137, 231, 257]. A mathematically rigorous theory for this type of interactions including interactions forming a two dimensional plane was developed by A. Grossmann, R. Høegh–Krohn, M. Mebkhout [444] and Yu.E. Karpeshina [502, 503, 507], see also [39] and the paper by B.S. Pavlov and I.Yu. Popov [797]. The case of generalized interactions supported by a superlattice has been investigated by P. Kurasov and B.S. Pavlov [627, 641, 788].

The case of an interaction supported by a low dimensional manifold has been studied by many mathematicians and physicists. For example such interactions are described in the review article by B.S. Pavlov [786], in the paper by W. Karwowski [510] and in papers by S. Albeverio, J.E. Fenstad, R. Høegh–Krohn, W. Karwowski, and T. Lindstrøm [28, 29, 32] (see also [168]).

The model of generalized interactions supported by a two dimensional plane was considered by P. Kurasov and B.S. Pavlov [791]. J.F. Brasche and A. Teta studied interactions supported by a regular curve [184]. W.Karwowski, V.Koshmanenko and S.Ota used the $\mathcal{H}_{-2}$ construction to define interactions of the Laplace operator in $\mathbf{R}^3$ with a perturbation concentrated on sets of lower dimensions. As an example they described perturbations by dynamics living on a one–dimensional segment [515]. H. Nagatani, Sh. Shimada and S.T. Kuroda used Kuroda's version of the $\mathcal{H}_{-2}$ construction to define surface and line interactions in $\mathbf{R}^3$ [752, 753, 882]. See also [508, 492, 690], where a similar approach has been used to model surface phenomena, and [28, 29, 32, 216], where the case of a perturbation supported by a Brownian path is investigated. The surface states for a crystal have been studied by M. Steslicka and her coworkers [915, 917, 919, 920, 921, 923, 924].

A.5.3 Relativistic point interactions

The Dirac operator with point interactions has been investigated, for example, in [106, 917]. P. Šeba described and investigated two one–parameter subfamilies of the four–parameter family of point interactions for the Dirac operator [865]. The Dirac operator with the sphere interaction and Coulomb plus delta potentials were investigated by J. Dittrich, P. Exner, and P. Šeba [282, 283, 284, 285]. F. Domínguez–Adame and his group carried out an intensive study of the whole family of point interactions including surface interactions and Kronig–Penney models [288, 289, 290, 291, 292, 293]. S. Benvegnù and L. Dąbrowski studied the four parameter family of point interactions for the Dirac operator and proved that in the nonrelativistic limit one gets the four–parameter family of point interactions for the Schrödinger equation [125, 126]. R.J. Hughes considered the renormalization procedure for the relativistic delta potential and its approximations by smooth potentials [481, 483]. Relativistic Kronig–Penney Hamiltonians have been investigated by J. Avron, A. Grossmann, F. Domínguez-Adame, B. Mendez and R.J. Hughes [106, 293, 485, 917]. Point interactions for general relativistic Hamiltonians have also been described by S. Albeverio and P. Kurasov [67].

A.5.4   Self–adjoint operators on graphs

Any graph with several infinite components can be considered as an open system. For the general theory of quantum mechanical open systems see the book by M.S. Livšic [664, 665]. The first mathematically rigorous study of the scattering problem on a graph was carried out by N.I. Gerasimenko and B.S. Pavlov in 1988 [390, 391]. To construct the model the authors used the extension theory for symmetric operators. Compact graphs and graphs having several infinite branches have been considered. The scattering problem for graphs with infinite branches has been formulated and studied in detail. In particular the inverse scattering problem has been investigated. Similar methods were used by P. Exner and P. Šeba to study other problems of quantum mechanics on graphs [329, 330, 331, 343, 344, 345, 346, 348, 351, 352, 353]. In particular infinite graph superlattices have been investigated [328, 332].

The mathematical study of the scattering problem on graphs has been continued by V.M. Adamyan who calculated the wave operators and the scattering matrix for the microscheme represented by several semi–infinite straight lines attached to a "black box". Recently V. Kostrykin and R. Schrader generalized the Gerasimenko–Pavlov model of the graphs by considering the most general boundary conditions joining the functions at the graph branching points [558]. Again the scattering problem has been investigated. It was emphasized that the self–adjoint operators corresponding to the Laplacian on a branching graph can be characterized using Lagrangian planes of the Laplacian's symplectic boundary form. Yu.B. Melnikov and B.S. Pavlov constructed and investigated the two–body scattering problem on the graph consisting of three half lines joined together at one point [710, 712]. These authors have shown that this problem is equivalent to a three–body scattering problem, since the center of mass motion of two particles moving on a nontrivial graph cannot be separated. J. Avron and L. Sadun used an adiabatic approach to study similar problems [109]. Several models of transport in nanosystems have been considered recently by S. Albeverio, V.A. Geyler, O.G. Kostrov, I.Yu. Popov and S.L. Popova [46, 418, 421]. Differential operators on graphs have been investigated by R. Carlson [200, 201, 202], who paid special attention to Sturm–Liouville operators on graphs with unbounded potentials. The self–adjoint extensions of the minimal differential operator on the graph have been characterized in terms of its adjoint. Similar problems for discrete Schrödinger operators on graphs have been investigated by S.P. Novikov and I.A. Dynnikov [774, 776, 775].

A.5.5 Acoustic problems

Different acoustic models were constructed using point interaction models. We have already mentioned in Section A.4.2 the relations between the one dimensional three–body problem and diffraction problems in two dimensions. The point interaction method was used to construct a model of small opening in the resonator suggested by B.S. Pavlov and I.Yu. Popov [794, 795, 796, 798]. It appeared that such models can be constructed for non–Neumann boundary conditions only by using spaces with indefinite metrics. This approach has been developed by I.Yu. Popov and his coworkers [421, 447, 810, 811, 812, 813, 814, 816, 817, 818, 819, 820, 821, 822, 823, 825, 828, 830, 834, 975]. This method was also used by B.S. Pavlov, A. Kiselev, and I.Yu. Popov to construct a model of two or more connected resonators [528, 529, 531, 532].

A.5.6 Magnetic field, Aharonov–Bohm effect and anyons

The method of point interactions has been applied by P. Šťovíček to construct a model of the Aharonov–Bohm effect [926, 927]. The Sommerfeld integral approach has been used by Yu.A. Kuperin, R.V. Romanov, and H.E. Rudin to calculate the scattering matrix for the charged quantum particle on the hyperbolic plane in the Abelian Aharonov–Bohm field [624]. D.K. Park used a Green's function approach to study the spin 1/2 Aharonov–Bohm problem [778]. Recently the whole family of Hamilton operators corresponding to the model of the Aharonov–Bohm effect has been described independently by L. Dąbrowski, and P. Šťovíček [230] and by R. Adami and A. Teta [1]. A periodic system of Aharonov–Bohm vortices was investigated by V.A. Geyler and A.V. Popov [417]. The scattering theory for anyons was studied by C. Korff, G. Lang and R. Schrader [545].

For magnetic Hamiltonians with point interactions see the papers by V.A. Geyler and his coworkers [93, 380, 406, 410, 411, 412, 414, 416, 420].

A similar problem in connection with the quantum Hall effect was studied by F. Gesztesy, H. Holden and P. Šeba [395]. Applications to nanosystems were considered by S. Albeverio, V.A. Geyler, and O.G. Kostrov [46]. R.M.

Cavalcanti and C. de Carvalho studied the two dimensional electron gas in a uniform magnetic field [211].

C. Manuel, P. Roy, and R. Tarrach studied point interactions for anyons [685, 686, 849]. The system of N anyons with point interactions has been described by P. Šťovíček [929].

A.5.7   Time dependent interactions

The Schrödinger equation with a time dependent point interaction was investigated first by M.R. Sayapova and D.R. Yafaev [853, 854, 967, 968]. The propagators for the time dependent Schrödinger operator in dimensions one, respectively three, were calculated by B. Gaveau, L.S. Schulman and E.B. Manoukian [387, 684, 858] respectively by S. Scarlatti and A. Teta [856]. The propagators for the Schrödinger and heat equations in all dimensions less than or equal to 3 have been investigated in detail by S. Albeverio, Z. Brzeźniak, and L. Dąbrowski [19, 20]. The case of several point sources of time dependent strengths has been analyzed by G.F. Dell'Antonio, R. Figari, and A. Teta [244, 245]. Moving point interactions were studied by R.J. Hughes in the case of constant speed [482] using I. Segal's approach to singular interactions. The case of N point interactions moving on preassigned non–intersecting paths has been analyzed by J.F. Dell'Antonio, R. Figari, and A. Teta [247] (in $\mathbf{R}^3$). The corresponding one–dimensional case was analyzed by V.A. Geyler and I.V. Chudaev [407]. The relations between self–adjoint extensions and evolution equations were studied in the abstract situation by H. Neidhardt [763]. See also the paper by Th. Kovar and Ph.A. Martin who studied periodically kicked interactions [559] and the paper by Yu.N. Demkov, P. Kurasov, and V.N. Ostrovsky devoted to one exactly solvable model involving an infinite number of equidistant potential curves [253].

A.5.8   Solid state

Among the most important applications of methods of singular interactions are in the construction of models of solid state physics. These models, starting from R. de L. Kronig and W.G. Penney [589], helped us to better understand solid state Hamiltonians. The reader can find an extended review on this subject in [39]. The mathematical theory of Hamiltonians of this type has been developed by S.Albeverio, J. Avron, F. Gesztesy, A. Grossman, H. Holden, S. Johannesen, R. Høegh–Krohn, Yu.E. Karpeshina, W. Kirsch, C. Macedo, M. Mebkhout, L. Streit, and T. Wentzel–Larsen [38, 41, 49, 105, 394, 399, 444, 466, 474, 502, 503, 504, 505, 506, 507]. Spectral properties and scattering theory were analyzed in addition in [103, 104, 107, 108, 110,

122, 123, 309, 310, 311, 325, 326, 388, 422, 423, 485, 642]. Explicitly solvable Hamiltonians involving point interactions were used for the construction of models for disordered (random) systems in [34, 56, 120] (see also references in [39]). Models involving generalized interactions have been constructed and investigated by V.V. Evstratov, Yu.A. Kuperin, P. Kurasov, K. Makarov, and B.S. Pavlov [321, 322, 323, 611, 627, 641, 782, 783, 784, 803, 805]. It has been shown, in particular, that the resonances of single atoms can produce gaps in the spectrum of the periodic problem. Stark–Wannier Hamiltonians are considered in [101, 116, 118, 120, 121, 124, 286, 327, 434, 668].

Impurities in solids have been investigated particularly by S. Albeverio, F. Bentosela, L.A. Dmitrieva, R. Høegh-Krohn, H. Holden, Yu.A. Kuperin, F. Martinelli, M. Mebkhout, Yu.B. Melnikov, B.S. Pavlov, S. Sengupta, M. Steslicka, A.V. Strepetov, and L.E. Thomas, see e.g. [57, 115, 286, 467, 805, 923, 933].

The methods discussed were successfully applied to study random and stochastic problems. Exact solvability of the model Hamiltonians can help to carry out statistical calculations. These problems are discussed in [39].

A.5.9 Further developments

Point interactions for pseudodifferential operators and differential operators on manifolds were studied by Y. Colin de Verdière [221, 222], S. Albeverio and P. Kurasov [67], A. Kiselev [527] and V.N. Kapshai and T.A. Alferova [501]. Perturbations of the polyharmonic operator by delta–like potentials were investigated by A.K. Fragela [383, 382, 381] and by Yu.E. Karpeshina and B.S. Pavlov [509]. Point interactions for Maxwell equations have been introduced by H.J.S. Dorren and A. Tip [297]. Nonlinear models involving point interactions were constructed for example by R. Adami and A. Teta [2] and by W. Caspers and Ph. Clément [205, 207].

The sum rules for a quantum particle in the presence of point interactions have been considered by Yu.N.Demkov [256].

The relations between nonstandard analysis and point interactions have been described in the books [29, 39] following the paper [26]. L. Dąbrowski and H.Grosse studied nonlocal point interactions [228]. The same question in connection with Dirichlet forms was investigated by J.F. Brasche [167] (see also [494]). Point interactions in L_p spaces were considered by Ph. Clément and W. Caspers [205, 206, 207, 208] and by S. Albeverio, Z. Brzeźniak, and L. Dąbrowski [21]. Applications of methods of point interactions to problems of quantum field theory have been discussed in [27, 29, 32, 51, 433, 687, 914, 931].

For the discussion of physical applications we refer to the book by Yu.N.

Demkov and V.N. Ostrovsky [255] which contains numerous examples. We have added some most recent references, but the main physical references can be found in the book [255], especially references to the papers, where the method described has been applied to atomic and nuclear physics. We would like to recall the classical papers [185, 248, 249, 250, 251, 252, 254, 937] and the recent paper [87]. The method of generalized interactions has been used particularly by Yu.A. Kuperin, S.B. Levin, K. Makarov, S.P. Merkuriev, Yu.B. Melnikov, A.K. Motovilov, B.S. Pavlov and E.A. Yarevsky [593, 594, 595, 596, 597, 605, 606, 609, 615, 618, 619, 626]. The operator method to exclude forbidden states has been suggested in [599].

The method of point interactions and Krein's formula have been used to study from a physical point of view scattering problems on multiple wells [215, 224, 225, 280, 469, 475, 487, 488, 489, 490, 879, 880, 881]. Applications to acoustic and diffraction problems are studied in [12, 350, 790, 794, 797, 796, 810, 811, 812, 813, 816, 817, 818, 819, 821, 823, 825, 828, 830, 832, 833, 834, 835, 837, 975]. In particular small openings in resonators have been investigated [448, 528, 529, 531, 532, 795, 798, 811, 816, 817, 818, 820, 822, 823, 835]. Applications to fluid mechanics can be found in [448, 449, 798, 807, 814, 826, 827, 829, 832].

Models of different nanoelectronic devices have been considered in [336, 340, 343, 356, 418, 421, 712, 815, 834, 836, 838]. Quantum mechanical models of composite devices formed by elements having different dimensions have been considered by P. Exner and P. Šeba [324, 337, 338, 339, 340, 341].

Applications of the methods described in the present book to other physical problems can be found in [219] (optics), [317] (superconductivity), [610, 612] (high energy physics), [78, 299, 333, 355, 357, 497, 675, 866, 867, 869, 868, 871, 878] (quantum chaotic systems), [16] (Lippmann–Schwinger equation), and [48] (mesoscopic physics).

Models for disordered systems involving random point interactions have been discussed, e.g., in [34, 41, 163] (crystals with random defects), [56, 120] (crystals with random electromagnetic interactions), [176] (energy distribution in random systems), [308, 312, 313, 314, 435] (systems with random ergodic potentials), [424, 787] (random matrices and random media), [176, 314] (systems with random boundary conditions), [799, 800, 801, 850] (systems with random magnetic field), and [792, 793] (electron transport in random media). See also the book by L. Pastur and A. Figotin [780] and references therein.

Bibliography

[1] R. Adami and A. Teta. On the Aharonov–Bohm Hamiltonian. *Lett. Math. Phys.*, 43(1):43–53, 1998.

[2] R. Adami and A. Teta. A simple model of concentrated nonlinearity. In *Mathematical Results in Quantum Mechanics, Prague, June 22–26, 1998*. Eds. *J.Dittrich, P.Exner, and M.Tater*, pages 183–189. Birkhäuser, 1999.

[3] V. Adamyan. Scattering matrices for microschemes. In *Operator Theory and Complex Analysis (Sapporo, 1991)*, volume 59 of *Oper. Theory Adv. Appl.*, pages 1–10. Birkhäuser, Basel, 1992.

[4] V. Adamyan, H. Langer, R. Mennicken, and J. Saurer. Spectral components of selfadjoint block operator matrices with unbounded entries. *Math. Nachr.*, 178:43–80, 1996.

[5] V. Adamyan and B. Pavlov. Zero–radius potentials and M. G. Krein's formula for generalized resolvents. *Zap. Nauchn. Sem. Leningrad. Otdel. Mat. Inst. Steklov. (LOMI)*, 149(Issled. Linein. Teor. Funktsii. XV):7–23, 186, 1986.

[6] V. Adamyan and N. Pušek. A Schrödinger operator with singular attracting potential. *Dokl. Akad. Nauk SSSR*, 249(1):81–85, 1979.

[7] Z. S. Agranovich and V. A. Marchenko. *The Inverse Problem of Scattering Theory*. Gordon and Breach Science Publishers, New York, 1963. Translated from the Russian by B. D. Seckler.

[8] N. I. Akhiezer and I. M. Glazman. *Theory of Linear Operators in Hilbert Space. Vol I.* Vishcha Shkola, Kharkov, 1977. Third edition, corrected and augmented. (In Russian.)

[9] N. I. Akhiezer and I. M. Glazman. *Theory of Linear Operators in Hilbert Space. Vol II.* Vishcha Shkola, Kharkov, 1978. Third edition, corrected and augmented. (In Russian.)

[10] N. I. Akhiezer and I. M. Glazman. *Theory of Linear Operators in Hilbert Space. Vol. I*, volume 9 of *Monographs and Studies in Mathematics*. Pitman

(Advanced Publishing Program), Boston, Mass., 1981. Translated from the third Russian edition by E. R. Dawson, translation edited by W. N. Everitt.

[11] N. I. Akhiezer and I. M. Glazman. *Theory of Linear Operators in Hilbert Space. Vol. II*, volume 10 of *Monographs and Studies in Mathematics*. Pitman (Advanced Publishing Program), Boston, Mass., 1981. Translated from the third Russian edition by E. R. Dawson, translation edited by W. N. Everitt.

[12] S. Albeverio. Analytische Lösung eines idealisierten Stripping- oder Beugungs-problems. *Helv. Phys. Acta*, 40:135–184, 1967. (In German).

[13] S. Albeverio. On bound states in the continuum of N-body systems and the virial theorem. *Ann. Physics*, 71:167–276, 1971.

[14] S. Albeverio and T. Arede. The relation between quantum mechanics and classical mechanics: A survey of some mathematical aspects. In *Chaotic Behavior in Quantum System. Theory and Applications*. Ed. by G. Casati, pages 37–76. Plenum Press, New York, 1985.

[15] S. Albeverio, D. Bollé, F. Gesztesy, R. Høegh-Krohn, and L. Streit. Low–energy parameters in nonrelativistic scattering theory. *Ann. Physics (NY)*, 148:308–326, 1983.

[16] S. Albeverio, J. F. Brasche, and V. Koshmanenko. Lippmann–Schwinger equation for singularly perturbed operators. *Methods Fun. Anal. Topol.*, 3:1–27, 1997.

[17] S. Albeverio, J. F. Brasche, and H. Neidhardt. On inverse spectral theory for self–adjoint extensions: mixed types of spectra. *J. Fun. Anal.*, 154:130–173, 1998.

[18] S. Albeverio, J. F. Brasche, and M. Röckner. Dirichlet forms and generalized Schrödinger operators. In *Schrödinger Operators (Sønderborg, 1988)*, volume 345 of *Lecture Notes in Phys.*, pages 1–42. Springer, Berlin, 1989.

[19] S. Albeverio, Z. Brzeźniak, and L. Dąbrowski. Time–dependent propagator with point interaction. *J. Phys. A*, 27(14):4933–4943, 1994.

[20] S. Albeverio, Z. Brzeźniak, and L. Dąbrowski. Fundamental solution of the heat and Schrödinger equations with point interaction. *J. Funct. Anal.*, 130(1):220–254, 1995.

[21] S. Albeverio, Z. Brzeźniak, and L. Dąbrowski. The heat equation with point interaction in L^p spaces. *Int. Eq. Oper. Theory*, 21(2):127–138, 1995.

[22] S. Albeverio, L. Dąbrowski, and P. Kurasov. Symmetries of Schrödinger operators with point interactions. *Lett. Math. Phys.*, 45(1):33–47, 1998.

[23] S. Albeverio and R.-Z. Fan. Representation of martingale additive functionals and absolute continuity of infinite–dimensional symmetric diffusions. In *Dirichlet Forms and Stochastic Processes (Beijing, 1993)*, pages 25–45. de Gruyter, Berlin, 1995.

[24] S. Albeverio, S.M. Fei, and L. Dąbrowski. One dimensional many–body problems with point interactions. Technical report, SISSA 139/FM/98, Trieste, Italy, 1998.

[25] S. Albeverio, S.M. Fei, and P. Kurasov. Gauge fields, point interactions and few–body problems in one dimension. Technical report, Bonn Univ., Germany, 1999.

[26] S. Albeverio, J. E. Fenstad, and R. Høegh-Krohn. Singular perturbations and nonstandard analysis. *Trans. Amer. Math. Soc.*, 252:275–295, 1979.

[27] S. Albeverio, J. E. Fenstad, R. Høegh-Krohn, W. Karwowski, and T. Lindstrøm. Perturbations of the Laplacian supported by null sets, with applications to polymer measures and quantum fields. *Phys. Lett. A*, 104(8):396–400, 1984.

[28] S. Albeverio, J. E. Fenstad, R. Høegh-Krohn, W. Karwowski, and T. Lindstrøm. Schrödinger operators with potentials supported by null sets. In *Ideas and Methods in Quantum and Statistical Physics (Oslo, 1988)*, pages 63–95. Cambridge Univ. Press, Cambridge, 1992.

[29] S. Albeverio, J. E. Fenstad, R. Høegh-Krohn, and T. Lindstrøm. *Nonstandard Methods in Stochastic Analysis and Mathematical Physics*, volume 122 of *Pure and Applied Mathematics*. Academic Press, Orlando, Fla., 1986.

[30] S. Albeverio, J. E. Fenstad, H. Holden, and T. Lindstrøm, editors. *Ideas and Methods in Mathematical Analysis, Stochastics, and Applications.* Cambridge University Press, Cambridge, 1992. In memory of Raphael Høegh–Krohn (1938–1988). Vol. 1, Papers from the Symposium on Ideas and Methods in Mathematics and Physics held at the University of Oslo, Oslo, September 1988.

[31] S. Albeverio, J. E. Fenstad, H. Holden, and T. Lindstrøm, editors. *Ideas and Methods in Quantum and Statistical Physics.* Cambridge University Press, Cambridge, 1992. In memory of Raphael Høegh–Krohn (1938–1988). Vol. 2, Papers from the Symposium on Ideas and Methods in Mathematics and Physics held at the University of Oslo, Oslo, September 1988.

[32] S. Albeverio, J.E. Fenstad, R. Høegh-Krohn, and T. Lindstrøm. *Нестандартные методы в стохастическом анализе и математическои физике.* Mir, Moscow, 1990. Translated from the English by A. K. Zvonkin,

translation edited by M. A. Shubin, With a preface by Zvonkin and Shubin. (In Russian.)

[33] S. Albeverio, L. S. Ferreira, F. Gesztesy, R. Høegh-Krohn, and L. Streit. Model dependence of Coulomb–corrected scattering lengths. *Phys. Rev. C,* 29:680–683, 1984.

[34] S. Albeverio, R. Figari, F. Gesztesy, R. Høegh-Krohn, H. Holden, and W. Kirsch. Point interaction Hamiltonians for crystals with random defects. In *Applications of Selfadjoint Extensions in Quantum Physics (Dubna, 1987),* volume 324 of *Lecture Notes in Phys.,* pages 87–99. Springer, Berlin, 1989.

[35] S. Albeverio, R. Figari, E. Orlandi, and A. Teta, editors. *Advances in Dynamical Systems and Quantum Physics,* River Edge, NJ, 1995. World Scientific.

[36] S. Albeverio, F. Gesztesy, and R. Høegh-Krohn. On the universal low-energy limit in nonrelativistic scattering theory. *Acta Phys. Austriaca Suppl.,* 23:577–585, 1981.

[37] S. Albeverio, F. Gesztesy, and R. Høegh-Krohn. The low–energy expansion in nonrelativistic scattering theory. *Ann. Inst. H. Poincaré, Sect. A,* 37:1–28, 1982.

[38] S. Albeverio, F. Gesztesy, R. Høegh-Krohn, and H. Holden. Point interactions in two dimensions: basic properties, approximations and applications to solid state physics. *J. Reine Angew. Math.,* 380:87–107, 1987.

[39] S. Albeverio, F. Gesztesy, R. Høegh-Krohn, and H. Holden. *Solvable Models in Quantum Mechanics.* Texts and Monographs in Physics. Springer, New York, 1988.

[40] S. Albeverio, F. Gesztesy, R. Høegh-Krohn, and H. Holden. *Решаемые модели в квантовои механике.* Mir, Moscow, 1991. Translated from the 1988 English original by V. A. Geyler, Yu. A. Kuperin and K. Makarov, with an introduction by B. S. Pavlov and an appendix by K. Makarov. (In Russian.)

[41] S. Albeverio, F. Gesztesy, R. Høegh-Krohn, H. Holden, and W. Kirsch. The Schrödinger operator for a particle in a solid with deterministic and stochastic point interactions. In *Schrödinger Operators, Aarhus 1985,* volume 1218 of *Lecture Notes in Math.,* pages 1–38. Springer, Berlin, 1986.

[42] S. Albeverio, F. Gesztesy, R. Høegh-Krohn, and W. Kirsch. On point interactions in one dimension. *J. Oper. Theory,* 12(1):101–126, 1984.

[43] S. Albeverio, F. Gesztesy, R. Høegh-Krohn, and L. Streit. Charged particles with short range interactions. *Ann. Inst. H. Poincaré Sect. A (N.S.)*, 38(3):263–293, 1983.

[44] S. Albeverio, F. Gesztesy, and H. Holden. Comments on a recent note on the Schrödinger equation with a δ'-interaction by B. H. Zhao: "Comments on the Schrödinger equation with δ'-interaction in one dimension" [J. Phys. A 25 (1992), no. 10, 1617–1618]. *J. Phys. A*, 26(15):3903–3904, 1993.

[45] S. Albeverio, F. Gesztesy, W. Karwowski, and L. Streit. On the connection between Schrödinger and Dirichlet forms. *J. Math. Phys.*, 26(10):2546–2553, 1985.

[46] S. Albeverio, V. A. Geyler, and O. G. Kostrov. Quasi–one–dimensional nanosystems in a uniform magnetic field: explicitly solvable model. *Reports Math. Phys.*, 1999. (To appear.)

[47] S. Albeverio and V.A. Geyler. The band structure of the general periodic Schrödinger operator with point interactions. *Comm. Math. Phys.*, 1999.

[48] S. Albeverio, F. Haake, P. Kurasov, M. Kuś, and P. Šeba. *S*-matrix, resonances, and wave functions for transport through billiards with leads. *J. Math. Phys.*, 37(10):4888–4903, 1996.

[49] S. Albeverio and R. Høegh-Krohn. Point interactions as limits of short range interactions. *J. Oper. Theory*, 6(2):313–339, 1981.

[50] S. Albeverio and R. Høegh-Krohn. Some remarks on Dirichlet forms and their applications to quantum mechanics and statistical mechanics. In *Functional Analysis and Markov Processes (Katata/Kyoto, 1981)*, volume 923 of *Lecture Notes in Math.*, pages 120–132. Springer, Berlin, 1982.

[51] S. Albeverio and R. Høegh-Krohn. Diffusion fields, quantum fields, and fields with values in Lie groups. In *Stochastic Analysis and Applications*. Ed. M. A. Pinsky, pages 1–98. Marcel Dekker, New York, 1984.

[52] S. Albeverio and R. Høegh-Krohn. Perturbation of resonances in quantum mechanics. *J. Math. Anal. Appl.*, 101(2):491–513, 1984.

[53] S. Albeverio and R. Høegh-Krohn. The resonance expansion for the Green's function of the Schrödinger and wave equations. In *Resonances–Models and Phenomena (Bielefeld, 1984)*, volume 211 of *Lecture Notes in Phys.*, pages 105–127. Springer, Berlin, 1984.

[54] S. Albeverio and R. Høegh-Krohn. Schrödinger operators with point interactions and short range expansions. *Phys. A*, 124(1–3):11–27, 1984. Mathematical Physics, VII (Boulder, Colo., 1983).

[55] S. Albeverio, R. Høegh-Krohn, F. Gesztesy, and H. Holden. Some exactly solvable models in quantum mechanics and the low energy expansions. In *Proceedings of the Second International Conference on Operator Algebras, Ideals, and their Applications in Theoretical Physics (Leipzig, 1983)*, volume 67 of *Teubner-Texte Math.*, pages 12–28, Leipzig, 1984. Teubner.

[56] S. Albeverio, R. Høegh-Krohn, W. Kirsch, and F. Martinelli. The spectrum of the three–dimensional Kronig–Penney model with random point defects. *Adv. in Appl. Math.*, 3(4):435–440, 1982.

[57] S. Albeverio, R. Høegh-Krohn, and M. Mebkhout. Scattering by impurities in a solvable model of a three–dimensional crystal. *J. Math. Phys.*, 25(5):1327–1334, 1984.

[58] S. Albeverio, R. Høegh-Krohn, and L. Streit. Energy forms, Hamiltonians, and distorted Brownian paths. *J. Math. Phys.*, 18:907–917, 1977.

[59] S. Albeverio, R. Høegh-Krohn, and L. Streit. Regularization of Hamiltonians and processes. *J. Math. Phys.*, 21:1636–1642, 1980.

[60] S. Albeverio, R. Høegh-Krohn, and T. T. Wu. A class of exactly solvable three–body quantum mechanical problems and universal low energy behavior. *Phys. Lett. A*, 83(3):105–109, 1981.

[61] S. Albeverio, W. Hunziker, W. Schneider, and R. Schrader. A note on L. D. Faddeev's three–particle theory. *Helv. Phys. Acta*, 40:745–748, 1967.

[62] S. Albeverio, W. Karwowski, and V. Koshmanenko. Square powers of singularly perturbed operators. *Math. Nachr.*, 173:5–24, 1995.

[63] S. Albeverio and V. Koshmanenko. Form–sum approximations of singular perturbations of selfadjoint operators. Technical report, BIBOS Preprint N771/4/97, 1997.

[64] S. Albeverio and V. Koshmanenko. On the problem of the right Hamiltonian under singular form–sum perturbations. Technical report, SFB 237 Preprint N375, Ruhr–Univ., Bochum, 1997. (To appear in Rev. Math. Phys.)

[65] S. Albeverio and V. Koshmanenko. Singular rank one perturbations of selfadjoint operators and Krein theory of selfadjoint extensions. Technical report, Ruhr–Univ., Bochum, 1998. (To appear in Potential Analysis.)

[66] S. Albeverio, V. Koshmanenko, and K. Makarov. Eigenfunction expansions under singular perturbations. 1999. (In preparation.)

[67] S. Albeverio and P. Kurasov. Pseudo–differential operators with point interactions. *Lett. Math. Phys.*, 41(1):79–92, 1997.

[68] S. Albeverio and P. Kurasov. Rank one perturbations, approximations, and selfadjoint extensions. *J. Funct. Anal.*, 148(1):152–169, 1997.

[69] S. Albeverio and P. Kurasov. Rank one perturbations of not semibounded operators. *Int. Eq. Oper. Theory*, 27(4):379–400, 1997.

[70] S. Albeverio and P. Kurasov. Finite rank perturbations and distribution theory. *Proc. Amer. Math. Soc.*, 127(4):1151–1161, 1999.

[71] S. Albeverio, S. Kusuoka, and L. Streit. Convergence of Dirichlet forms and associated Schrödinger operators. *J. Funct. Anal.*, 68:130–148, 1986.

[72] S. Albeverio, S. Lakaev, and K. Makarov. The Efimov effect and extended Szegö-Kac limit theorem. *Lett. Math. Phys.*, 43:73–85, 1998.

[73] S. Albeverio and Z. M. Ma. Diffusion processes with singular Dirichlet forms. In *Stochastic Analysis and Applications (Lisbon, 1989)*, volume 26 of *Progr. Probab.*, pages 11–28. Birkhäuser Boston, 1991.

[74] S. Albeverio and Z. M. Ma. Additive functionals, nowhere Radon and Kato class smooth measures associated with Dirichlet forms. *Osaka J. Math.*, 29(2):247–265, 1992.

[75] S. Albeverio and K. Makarov. The three–body quantum problem with short-range interactions from the point of view of harmonic analysis. In *Proceedings of the Workshop on Singular Schroedinger Operators, Trieste, 29 September–1 October 1994*. Eds. G.F.Dell'Antonio, R.Figari and A.Teta. SISSA, Trieste, 1995.

[76] S. Albeverio and K. Makarov. Attractors in a model related to the three body quantum problem. *C. R. Acad. Sci. Paris Sér. I Math.*, 323(6):693–698, 1996.

[77] S. Albeverio and K. Makarov. Nontrivial attractors in a model related to the three–body quantum problem. *Acta Appl. Math.*, 48(2):113–184, 1997.

[78] S. Albeverio and P. Šeba. Wave chaos in quantum systems with point interaction. *J. Statist. Phys.*, 64(1–2):369–383, 1991.

[79] A. Alonso. Shrinking potentials in the Schrödinger equation. PhD thesis, Princeton University, Princeton, NJ, 1978.

[80] A. Alonso and B. Simon. The Birman–Kreĭn–Vishik theory of selfadjoint extensions of semibounded operators. *J. Oper. Theory*, 4(2):251–270, 1980.

[81] A. Alonso and B. Simon. Addenda to: "The Birman– Kreĭn– Vishik theory of selfadjoint extensions of semibounded operators" [J. Oper. Theory 4 (1980), no. 2, 251–270]. *J. Oper. Theory*, 6(2):407, 1981.

[82] D. Alpay, A. Dijksma, J. Rovnyak, and H. de Snoo. *Schur Functions, Operator Colligations, and Reproducing Kernel Pontryagin spaces*, volume 96 of *Operator Theory: Advances and Applications*. Birkhäuser, Basel, 1997.

[83] I. V. Andronov. Application of zero-range models to the problem of diffraction by a small hole in elastic plate. *J. Math. Phys.*, 34(6):2226–2241, 1993.

[84] I. V. Andronov. Application of null-range potentials in problems of diffraction by small inhomogeneities in elastic plates. *Prikl. Mat. Mekh.*, 59(3):451–463, 1995.

[85] I. V. Andronov. Zero-range potentials in boundary-contact problems of acoustics. *Appl. Anal.*, 68(1-2):3–29, 1998.

[86] I.V. Andronov. Zero–range potential model of a protruding stiffener. *J. Phys. A*, 32:L231–L238, 1999.

[87] J-P. Antoine, P. Exner, and P. Šeba. A mathematical model of heavy-quarkonia mesonic decays. *Ann. Phys. (NY)*, 233:1–16, 1994.

[88] J.-P. Antoine, P. Exner, P. Šeba, and J. Shabani. A Fermi–type rule for contact embedded–eigenvalue perturbations. In *Mathematical Results in Quantum Mechanics (Blossin, 1993)*, volume 70 of *Oper. Theory Adv. Appl.*, pages 79–87. Birkhäuser, Basel, 1994.

[89] J.-P. Antoine, F. Gesztesy, and J. Shabani. Exactly solvable models of sphere interactions in quantum mechanics. *J. Phys. A*, 20(12):3687–3712, 1987.

[90] J.-P. Antoine and A. Grossmann. Partial inner product spaces. I. General properties. *J. Funct. Anal.*, 23(4):369–378, 1976.

[91] J.-P. Antoine and A. Grossmann. Partial inner product spaces. II. Operators. *J. Funct. Anal.*, 23(4):379–391, 1976.

[92] M. A. Antonets. The transmission problem for differential operators and contact interactions of quantum particles. 1998. (In preparation.)

[93] M. A. Antonets and V. A. Geyler. A quasi–two–dimensional charged particle in a tilted magnetic field: asymptotic properties of the spectrum. *Russian J. Math. Phys.*, 3(4):413–422, 1995.

[94] I. Antoniou, I. Prigogine, and Yu. A. Kuperin, editors. *Computational Tools of Complex Systems. I.* Pergamon Press, Exeter, 1997. Dynamical Systems, Chaos and Control, Comput. Math. Appl. **34** (1997), no. 2–4.

[95] I. Antoniou, I. Prigogine, and Yu. A. Kuperin, editors. *Computational Tools of Complex Systems. II.* Pergamon Press, Exeter, 1997. Quantum Systems and Operators, Comput. Math. Appl. **34** (1997), no. 5–6.

[96] Jonathan Arazy and Leonid Zelenko. Finite-dimensional perturbations of self-adjoint operators. *Integral Equations Operator Theory*, 34(2):127–164, 1999.

[97] N. Aronszajn. On a problem of Weyl in the theory of singular Sturm–Liouville equations. *Amer. J. Math.*, 79:597–610, 1957.

[98] N. Aronszajn and R. D. Brown. Finite–dimensional perturbations of spectral problems and variational approximation methods for eigenvalue problems. I: Finite–dimensional perturbations. *Studia Math.*, 36:1–76, 1970.

[99] N. Aronszajn and W. F. Donoghue. On exponential representations of analytic functions in the upper half–plane with positive imaginary part. *J. Anal. Math.*, 5:321–388, 1956.

[100] N. Aronszajn and W. F. Donoghue. A supplement to the paper on exponential representations of analytic functions in the upper half–plane with positive imaginary part. *J. Anal. Math.*, 12:113–127, 1964.

[101] J. Asch, P. Duclos, and P. Exner. Stark–Wannier Hamiltonians with pure point spectrum. In *Proceedings of the Conference on Differential Equations, Asymptotic Analysis, and Mathematical Physics (Potsdam 1996)*, pages 10–25. Akademie, Berlin, 1997.

[102] F. Atkinson. *Distrete and Continuous Boundary Problems*. Izdat. Mir, Moscow, 1968. Translated from the English by I. S. Iohvidov and G. A. Karal'nik. Edited and supplemented by I. S. Kac and M. G. Krein. (In Russian.)

[103] J. Avron. Bragg scattering from point interactions: An explicit formula for the reflection coefficient. In *Mathematical Problems in Theoretical Physics, Proceedings, Berlin (West), 1981*. Ed. by R. Schrader, R. Seiler, D. A. Uhlenbrock. *Lecture Notes in Physics, Vol. 153.*, pages 126–128. Springer, Berlin, 1982.

[104] J. Avron, P. Exner, and Y. Last. Periodic Schrödinger operators with large gaps and Wannier–Stark ladders. *Phys. Rev. Lett.*, 72:896–899, 1994.

[105] J. Avron, A. Grossman, and R. Høegh-Krohn. The reflection from a semi–infinite crystal of point scatterers. *Phys. Lett. A*, 94(1):42–44, 1983.

[106] J. Avron and A. Grossmann. The relativistic Kronig–Penney Hamiltonian. *Phys. Lett. A*, 56:55–57, 1976.

[107] J. Avron, A. Grossmann, and R. Rodriguez. Hamiltonians in one–electron theory of solids. I. *Rep. Math. Phys.*, 5(1):113–120, 1974.

[108] J. Avron, A. Grossmann, and R. Rodriguez. Spectral properties of reduced Bloch Hamiltonians. *Ann. Physics*, 103(1):47–63, 1977.

[109] J. Avron and L. Sadun. Adiabatic quantum transport in networks with macroscopic components. *Ann. Physics*, 206(2, part 1):440–493, 1991.

[110] J. Avron and B. Simon. Analytic properties of band functions. *Ann. Physics*, 110(1):85–101, 1978.

[111] J. Avron and B. Simon. Almost periodic Schrödinger operators. II. The integrated density of states. *Duke Math. J.*, 50(1):369–391, 1983.

[112] P. B. Bailey, W. N. Everitt, and A. Zettl. Regular and singular Sturm–Liouville problems with coupled boundary conditions. *Proc. Roy. Soc. Edinburgh Sect. A*, 126(3):505–514, 1996.

[113] G. Barton and D. Waxman. Wave equations with point–support potentials having dimensionless strength parameters. Technical report, Univ. of Sussex, England, 1993.

[114] D. Bauch. The path integral for a particle moving in a δ–function potential. *Nouvo Cimento*, 85(86B), 1985.

[115] F. Bentosela. Scattering from impurities in a crystal. *Comm. Math. Phys.*, 46(2):153–166, 1976.

[116] F. Bentosela. Electrons of a solid in an external electric field. In *Stochastic Aspects of Classical and Quantum Systems, Proceedings, Marseille 1983*. Ed. by S. Albeverio, P. Combe, M. Sirugue–Collin. *Lecture Notes in Mathematics, Vol. 1109*, pages 32–38. Springer, Berlin, 1985.

[117] F. Bentosela. Resonance states in disordered systems. In *Lyapunov Exponents (Bremen, 1984)*, volume 1186 of *Lecture Notes in Math.*, pages 246–251. Springer, Berlin, 1986.

[118] F. Bentosela. Stark–Wannier resonant states. In *Recent Developments in Quantum Mechanics (Poiana Braşov, 1989)*, volume 12 of *Math. Phys. Stud.*, pages 85–95. Kluwer, Dordrecht, 1991.

[119] F. Bentosela. Propagation in irregular optic fibres. In *Mathematical Results in Quantum Mechanics (Blossin, 1993)*, volume 70 of *Oper. Theory Adv. Appl.*, pages 293–298. Birkhäuser, Basel, 1994.

[120] F. Bentosela, R. Carmona, P. Duclos, B. Simon, B. Souillard, and R. Weder. Schrödinger operators with an electric field and random or deterministic potentials. *Comm. Math. Phys.*, 88(3):387–397, 1983.

[121] F. Bentosela and V. Grecchi. Stark Wannier ladders. *Comm. Math. Phys.*, 142(1):169–192, 1991.

[122] F. Bentosela, V. Grecchi, and F. Zironi. Approximate ladder for resonances in a semi–infinite crystal. *J. Phys. C*, 15:7119–7131, 1982.

[123] F. Bentosela, V. Grecchi, and F. Zironi. Oscillations of Wannier resonances. *Phys. Rev. Lett.*, 50:84–86, 1983.

[124] F. Bentosela, V. Grecchi, and F. Zironi. Stark–Wannier states in disordered systems. *Phys. Rev. B*, 31:6909–6912, 1985.

[125] S. Benvegnù and L. Dąbrowski. Relativistic point interaction. *Lett. Math. Phys.*, 30(2):159–167, 1994.

[126] S. Benvegnù and L. Dąbrowski. Relativistic point interaction in one dimension. In *Proceedings of the Workshop on singular Schroedinger operators, Trieste, 29 September–1 October 1994*. Eds. G.F.Dell'Antonio, R.Figari and A.Teta. SISSA, Trieste, 1995.

[127] Yu. M. Berezanski. On certain normed rings constructed from orthogonal polynomials. *Ukrain. Mat. Žurnal*, 3:412–432, 1951.

[128] Yu. M. Berezanski. On expansion according to eigenfunctions of general self–adjoint differential operators. *Dokl. Akad. Nauk SSSR (N.S.)*, 108:379–382, 1956.

[129] Yu. M. Berezanski. Eigenfunction expansions of self–adjoint operators. *Mat. Sb. N.S.*, 43(85):75–126, 1957.

[130] Yu. M. Berezanski. On an eigenfunction expansion for self–adjoint operators. *Ukrain. Mat. Ž.*, 11:16–24, 1959.

[131] Yu. M. Berezanski. *Разложение по собственным функциям самосопряженных операторов*. Наукова Думка, Kiev, 1965. Академия Наук Украинской ССР. Институт Математики. (In Russian.)

[132] Yu. M. Berezanski. *Expansions in Eigenfunctions of Selfadjoint Operators*. American Mathematical Society, Providence, R.I., 1968. Translated from the Russian by R. Bolstein, J. M. Danskin, J. Rovnyak and L. Shulman. Translations of Mathematical Monographs, Vol. 17.

[133] Yu. M. Berezanski. A remark concerning the essential selfadjointness of the powers of an operator. *Ukrain. Mat. Ž.*, 26:790–793, 862, 1974.

[134] Yu. M. Berezanski and G. F. Us. Eigenfunction expansions of operators admitting separation of an infinite number of variables. *Rep. Math. Phys.*, 7(1):103–126, 1975.

[135] F. A. Berezin and L. D. Faddeev. Remark on the Schrödinger equation with singular potential. *Dokl. Akad. Nauk SSSR*, 137:1011–1014, 1961.

[136] H. Bethe and R. Peierls. Quantum theory of diplon. *Proc. Roy. Soc. London*, 148A:146–156, 1935.

[137] K.V. Bhagwat and R. Subramanian. On the use of zero–range potentials in some energy band problems. *Phys. Stat. Sol. (b)*, 47:317–323, 1971.

[138] K.V. Bhagwat and R. Subramanian. Relativistic effects on impurity states. *Physica*, 62:614–622, 1972.

[139] M. Š. Birman. On the theory of self–adjoint extensions of positive definite operators. *Doklady Akad. Nauk SSSR (N.S.)*, 91:189–191, 1953. (In Russian.)

[140] M. Š. Birman. On the theory of self–adjoint extensions of positive definite operators. *Mat. Sb. N.S.*, 38(80):431–450, 1956. (In Russian.)

[141] M. Š. Birman. Perturbations of quadratic forms and the spectrum of singular boundary value problems. *Dokl. Akad. Nauk SSSR*, 125:471–474, 1959. (In Russian.)

[142] M. Š. Birman. On the spectrum of singular boundary–value problems. *Mat. Sb. (N.S.)*, 55 (97):125–174, 1961. (In Russian.)

[143] M. Š. Birman. Perturbation of the spectrum of a singular elliptic operator under variation of the boundary and boundary conditions. *Soviet Math. Dokl.*, 2:326–328, 1961.

[144] M. Š. Birman. Conditions for the existence of wave operators. *Sov. Math. Dokl.*, 3:408–411, 1962.

[145] M. Š. Birman. Perturbations of the continuous spectrum of a singular elliptic operator by varying the boundary and the boundary conditions. *Vestnik Leningrad. Univ.*, 17(1):22–55, 1962. (In Russian.)

[146] M. Š. Birman. A test for the existence of wave operators. *Soviet Math. Dokl.*, 3:1747–11748, 1962.

[147] M. Š. Birman. Existence conditions for wave operators. *Izv. Akad. Nauk SSSR Ser. Mat.*, 27:883–906, 1963. (In Russian.)

[148] M. Š. Birman. A local criterion for the existence of wave operators. *Soviet Math. Dokl.*, 5:1505–1509, 1964.

[149] M. Š. Birman. A local test for the existence of wave operators. *Izv. Akad. Nauk SSSR Ser. Mat.*, 32:914–942, 1968. (In Russian.)

[150] M. Š. Birman. Some applications of a local criterion for the existence of wave operators. *Soviet Math. Dokl.*, 10:393–397, 1969.

[151] M. Š. Birman and S. B. Èntina. A stationary approach in the abstract theory of scattering. *Soviet Math. Dokl.*, 5:432–435, 1965.

[152] M. Š. Birman and S. B. Èntina. Stationary approach in abstract scattering theory. *Izv. Akad. Nauk SSSR Ser. Mat.*, 31:401–430, 1967. (In Russian.)

[153] M. Š. Birman and M. G. Krein. On the theory of wave operators and scattering operators. *Soviet Math. Dokl.*, 3:740–744, 1962. (In Russian.)

[154] M. Š. Birman and M. Z. Solomjak. *Spectral Theory of Selfadjoint Operators in Hilbert Space*. Mathematics and its Applications (Soviet Series). Reidel, Dordrecht, 1987. Translated by S. Khrushchev and V. Peller from the Russian edition: Спектральная теория самосопряженных операторов в Гильбертовом пространстве, изд-во Ленинградского университета, Ленинград, 1980.

[155] M. Š. Birman and M. Z. Solomyak. Tensor product of a finite number of spectral measures is always a spectral measure. *Integ. Eq. Oper. Theory*, 24(2):179–187, 1996.

[156] M. Š. Birman and D. R. Yafaev. Spectral properties of the scattering matrix. *Algebra i Analiz*, 4(6):1–27, 1992. (In Russian.)

[157] M. Š. Birman and D. R. Yafaev. The spectral shift function. The papers of M. G. Krein and their further development. *Algebra i Analiz*, 4(5):1–44, 1992. (In Russian.)

[158] M. Š. Birman and D. R. Yafaev. A general scheme in stationary scattering theory. In *Wave Propagation. Scattering Theory*, volume 157 of *Amer. Math. Soc. Transl. Ser. 2*, pages 87–112. Amer. Math. Soc., Providence, RI, 1993.

[159] J. Blank, P. Exner, and M. Havlíček. *Hilbert space operators in quantum physics*. AIP Series in Computational and Applied Mathematical Physics. American Institute of Physics, New York, 1994.

[160] G. Borg. Eine Umkehrung der Sturm–Liouvilleschen Eigenwertaufgabe. Bestimmung der Differentialgleichung durch die Eigenwerte. *Acta Math.*, 78:1–96, 1946.

[161] G. Borg. Uniqueness theorems in the spectral theory of $y'' + (\lambda - q(x))y = 0$. In *Den 11te Skandinaviske Matematikerkongress, Trondheim, 1949*, pages 276–287. Johan Grundt Tanums Forlag, Oslo, 1952.

[162] A. Boutet de Monvel, V. Georgescu, and A. Soffer. N–body Hamiltonians with hard–core interactions. *Rev. Math. Phys.*, 6(4):515–596, 1994.

[163] A. Boutet de Monvel and V. Grinshpun. Exponential localization for multi-dimensional Schrödinger operator with random point potential. *Rev. Math. Phys.*, 9(4):425–451, 1997.

[164] L. J. Boya, H. C. Rosu, A. J. Segui-Santonja, and J. Socorro. Darboux strictly isospectral attractive delta potentials. *quant-ph/9709019*, 1997.

[165] J. F. Brasche. Störungen von Schrödingeroperatoren durch Maße. Master's thesis, University of Bielefeld, Germany, 1983.

[166] J. F. Brasche. Perturbation of Schrödinger Hamiltonians by measures-selfadjointness and lower semiboundedness. *J. Math. Phys.*, 26(4):621–626, 1985.

[167] J. F. Brasche. Non-local point interactions and associated Dirichlet forms. Technical report, Preprint, University of Bielefeld, Germany, 1986.

[168] J. F. Brasche. Dirichlet forms and nonstandard Schrödinger operators. In *Schrödinger Operators, Standard and Nonstandard (Dubna, 1988)*, pages 42–57. World Scientific, Teaneck, NJ, 1989.

[169] J. F. Brasche. Generalized Schrödinger operators, an inverse problem in spectral analysis and the Efimov effect. In *Stochastic Processes, Physics and Geometry (Ascona and Locarno, 1988)*, pages 207–244. World Scientific, Teaneck, NJ, 1990.

[170] J. F. Brasche. On extension theory in L^2-spaces. In *Rigorous Results in Quantum Dynamics (Liblice, 1990)*, pages 298–301. World Scientific, River Edge, NJ, 1991.

[171] J. F. Brasche. On the spectral properties of generalized Schrödinger operators. In *Mathematical Results in Quantum Mechanics (Blossin, 1993)*, volume 70 of *Oper. Theory Adv. Appl.*, pages 73–78. Birkhäuser, Basel, 1994.

[172] J. F. Brasche. On extension theory in L^2-spaces. *Potential Anal.*, 4(3):297–307, 1995.

[173] J. F. Brasche. On the spectral properties of singular perturbed operators. In *Dirichlet forms and stochastic processes (Beijing, 1993)*, pages 65–72. de Gruyter, Berlin, 1995.

[174] J. F. Brasche, P. Exner, Yu. A. Kuperin, and P. Šeba. Schrödinger operators with singular interactions. *J. Math. Anal. Appl.*, 184(1):112–139, 1994.

[175] J. F. Brasche, R. Figari, and Teta A. Singular Schrödinger operators as limits of point interactions. *Potential Anal.*, 8:163–178, 1998.

[176] J. F. Brasche and W. Karwowski. On boundary theory for Schrödinger operators and stochastic processes. In *Order, Disorder and Chaos in Quantum Systems (Dubna, 1989)*, volume 46 of *Oper. Theory Adv. Appl.*, pages 199–208. Birkhäuser, Basel, 1990.

[177] J. F. Brasche, V. Koshmanenko, and H. Neidhardt. New aspects of Krein's extension theory. *Ukraïn. Mat. Ž.*, 46(1–2):37–54, 1994.

[178] J. F. Brasche and H. Neidhardt. Some remarks on Krein's extension theory. *Math. Nachr.*, 165:159–181, 1994.

[179] J. F. Brasche and H. Neidhardt. On inverse spectral theory and point interactions. In *Proceedings of the Workshop on Singular Schroedinger Operators, Trieste, 29 September–1 October 1994*. Eds. G.F.Dell'Antonio, R.Figari and A.Teta. SISSA, Trieste, 1995.

[180] J. F. Brasche and H. Neidhardt. On the absolutely continuous spectrum of self-adjoint extensions. *J. Funct. Anal.*, 131(2):364–385, 1995.

[181] J. F. Brasche and H. Neidhardt. On the singular continuous spectrum of self-adjoint extensions. *Math. Z.*, 222(4):533–542, 1996.

[182] J. F. Brasche, H. Neidhardt, and J. Weidmann. On the point spectrum of selfadjoint extensions. *Math. Z.*, 214(2):343–355, 1993.

[183] J. F. Brasche, H. Neidhardt, and J. Weidmann. On the spectra of selfadjoint extensions. In *Operator Extensions, Interpolation of Functions and Related Topics (Timişoara, 1992)*, volume 61 of *Oper. Theory Adv. Appl.*, pages 29–45. Birkhäuser, Basel, 1993.

[184] J. F. Brasche and A. Teta. Spectral analysis and scattering theory for Schrödinger operators with an interaction supported by a regular curve. In *Ideas and Methods in Quantum and Statistical Physics (Oslo, 1988)*, pages 197–211. Cambridge Univ. Press, Cambridge, 1992.

[185] G. Breit. The scattering of slow neutrons by bound protons I. Methods of calculation. *Phys. Rev.*, 71:215–231, 1947.

[186] V. M. Bruk. A certain class of boundary value problems with a spectral parameter in the boundary condition. *Mat. Sb. (N.S.)*, 100(142)(2):210–216, 1976.

[187] E. Brüning, M. Demuth, and F. Gesztesy. Invariance of the essential spectra for perturbations with unbounded hard cores. *Lett. Math. Phys.*, 13(1):69–77, 1987.

[188] E. Brüning and F. Gesztesy. On essential spectra of hard–core type Schrödinger operators. *J. Phys. A*, 18(1):L7–L11, 1985.

[189] J. Brüning and V.A. Geyler. On the spectrum of gauge periodic point perturbations on Lobachevsky plane. Technical report, SFB 288, Preprint 341, Berlin, 1998.

[190] B. Budaev. *Diffraction by Wedges.* Longman, Harlow, 1995.

[191] W. Bulla, P. Falkensteiner, and H. Grosse. On the calculation of the Berry phase in a solvable model. *Phys. Lett. B*, 215(2):359–363, 1988.

[192] W. Bulla and F. Gesztesy. Deficiency indices and singular boundary conditions in quantum mechanics. *J. Math. Phys.*, 26(10):2520–2528, 1985.

[193] D. Buschmann. Eindimnsionale Schrödingeroperatoren mit lokalen Punktwechselwirkungen. Master's thesis, Dept. of Math., Univ. Frankfurt am Main, 1994. (in German)

[194] D. Buschmann, G. Stolz, and J. Weidmann. One–dimensional Schrödinger operators with local point interactions. *J. Reine Angew. Math.*, 467:169–186, 1995.

[195] V. S. Buslaev. Trace formula and singularities in the scattering matrix for a system of three one–dimensional particles. The third group integral. *Teoret. Mat. Fiz.*, 16(2):247–259, 1973.

[196] V. S. Buslaev and N. A. Kaliteevski. Principal singularities of the scattering matrix for a system of one–dimensional particles. *Teoret. Mat. Fiz.*, 70(2):266–277, 1987.

[197] V. S. Buslaev, S. P. Merkuriev, and S. P. Salikov. Description of pair potentials for which the scattering in the quantum system of three one–dimensional particles is free of diffractional effects. *Zap. Nauchn. Sem. Leningrad. Otdel. Mat. Inst. Steklov. (LOMI)*, 84:16–22, 310, 316, 1979. Boundary Value Problems of Mathematical Physics and Related Questions in the Theory of Functions, 11.

[198] V. S. Buslaev, S. P. Merkuriev, and S. P. Salikov. Diffraction characteristics of scattering in a quantum system of three one–dimensional particles. In *Scattering Theory. Theory of Oscillations (In Russian.)*, volume 9 of *Probl. Mat. Fiz.*, pages 14–30, 183. Leningrad. Univ., Leningrad, 1979.

[199] J. W. Calkin. Abstract symmetric boundary conditions. *Trans. Amer. Math. Soc.*, 45:369–442, 1939.

[200] R. Carlson. Inverse eigenvalue problems on directed graphs. *Trans. Amer. Math. Soc.* (To appear.)

[201] R. Carlson. Hearing point masses in a string. *SIAM J. Math. Anal.*, 26(3):583–600, 1995.

[202] R. Carlson. Adjoint and self–adjoint differential operators on graphs. *Electron. J. Differential Equations*, 1998:No. 6, 10 pp. (electronic), 1998.

[203] R. Carlson, R. Threadgill, and C. Shubin. Sturm–Liouville eigenvalue problems with finitely many singularities. *J. Math. Anal. Appl.*, 204(1):74–101, 1996.

[204] C. Carvalho. On the relation between two different methods to treat Hamiltonians with point interactions. *Int. Eq. Oper. Theory*, 14:342–358, 1991.

[205] W. Caspers. On point interactions. PhD thesis, Delft Tech. Univ., The Netherlands, 1992.

[206] W. Caspers. Perturbation of the Laplacian by the Coulomb potential and a point interaction in $L^p(\mathbf{R}^3)$. *Potential Anal.*, 1(4):401–409, 1992.

[207] W. Caspers and Ph. Clément. A bifurcation problem for point interactions in $L^2(\mathbf{R}^3)$. In *Semigroups of Linear and Nonlinear Operations and Applications (Curaçao, 1992)*, pages 99–108. Kluwer, Dordrecht, 1993.

[208] W. Caspers and Ph. Clément. Point interactions in L^p. *Semigroup Forum*, 46(2):253–265, 1993.

[209] W. Caspers and Ph. Clément. A different approach to singular solutions. *Diff. Integr. Eq.*, 7(5):1227–1240, 1994.

[210] W. Caspers and G. Sweers. Point interactions on bounded domains. *Proc. Roy. Soc. Edinburgh Sect. A*, 124(5):917–926, 1994.

[211] R. M. Cavalcanti and C. de Carvalho. Two–dimensional electron gas in a uniform magnetic field in the presence of a δ–impurity. *J. Phys. A*, 31:2391–2399, 1998.

[212] T. Cheon and T. Shigehara. Geometric phase in quantum billiards with a pointlike scatterer. *Phys. Rev. Lett.*, 76(11):1770–1773, 1996.

[213] T. Cheon and T. Shigehara. Flux string in quantum billiards with two particles. *Phys. Lett. A*, 233(1–2):11–16, 1997.

[214] T. Cheon and T. Shigehara. Realizing discontinuous wave function with renormalized short–range potentials. *quant-ph/9709035*, 1997.

[215] T. Cheon and T. Shigehara. Resonance tunneling in double–well billiards with a point–like scatterer. *Phys. Lett. A*, 228(3):151–158, 1997.

[216] S. E. Cheremshantsev. Hamiltonians with zero–range interactions supported by a Brownian path. *Ann. Inst. Henri Poincaré*, 56(1):1–25, 1992.

[217] S. E. Cheremshantsev and K. Makarov. Point interactions with an internal structure as limits of nonlocal separable potentials. In *Order, Disorder and Chaos in Quantum Systems (Dubna, 1989)*, volume 46 of *Oper. Theory Adv. Appl.*, pages 179–182. Birkhäuser, Basel, 1990.

[218] P. R. Chernoff and R. J. Hughes. A new class of point interactions in one dimension. *J. Funct. Anal.*, 111(1):97–117, 1993.

[219] Ch. W. Clark. Frequency–dependent polarizability of an electron bound by a zero–range potential. *J.Opt.Soc.Am. B*, 7(4):488–, 1990.

[220] E. Coddington and H. de Snoo. Positive selfadjoint extensions of positive symmetric subspaces. *Math. Z.*, 159:203–214, 1978.

[221] Y. Colin de Verdière. Pseudo–laplaciens. I. *Ann. Inst. Fourier (Grenoble)*, 32(3):xiii, 275–286, 1982.

[222] Y. Colin de Verdière. Pseudo–laplaciens. II. *Ann. Inst. Fourier (Grenoble)*, 33(2):87–113, 1983.

[223] J.-M. Combes, P. Duclos, M. Klein, and R. Seiler. The shape resonance. *Comm. Math. Phys.*, 110(2):215–236, 1987.

[224] J.-M. Combes, P. Duclos, and R. Seiler. Krein's formula and one–dimensional multiple–well. *J. Funct. Anal.*, 52(2):257–301, 1983.

[225] J.-M. Combes and I. Sigal. Dynamical behaviour for some multiple well Schrödinger operators. In *Rigorous Results in Quantum Dynamics (Liblice, 1990)*, pages 41–48. World Scientific, River Edge, NJ, 1991.

[226] F. A. B. Coutinho, Y. Nogami, and J. F. Perez. Generalized point interactions in one–dimensional quantum mechanics. *J. Phys. A: Math. Gen.*, 30:3937–3945, 1997.

[227] H. L. Cycon, R. G. Froese, W. Kirsch, and B. Simon. *Schrödinger Operators with Application to Quantum Mechanics and Global Geometry*. Texts and Monographs in Physics. Springer, Berlin, 1987.

[228] L. Dąbrowski and H. Grosse. On nonlocal point interactions in one, two, and three dimensions. *J. Math. Phys.*, 26(11):2777–2780, 1985.

[229] L. Dąbrowski and J. Shabani. Finitely many sphere interactions in quantum mechanics: nonseparated boundary conditions. *J. Math. Phys.*, 29(10):2241–2244, 1988.

[230] L. Dąbrowski and P. Šťovíček. Aharonov–Bohm effect with δ–type interaction. *J. Math. Phys.*, 39(1):47–62, 1998.

[231] F.I. Dalidchik and G.K. Ivanov. An electron in a periodic chain with impurities. *Theoret. Exper. Chem.*, 7:123–127, 1971.

[232] G. S. Danilov. On the three–body problem with short–range forces. *Soviet Physics JETP*, 13:349–355, 1961.

[233] E. B. Davies and B. Simon. Spectral properties of Neumann Laplacian on horns. *Geom. Funct. Anal.*, 2(1):105–117, 1992.

[234] M. de Faria and L. Streit. Some recent advances in white noise analysis. In *Stochastic Analysis on Infinite-dimensional Spaces (Baton Rouge, LA, 1994)*, pages 52–59. Longman, Harlow, 1994.

[235] P. Deift. Inverse scattering on the line–an overview. In *Differential Equations and Mathematical Physics (Birmingham, AL, 1990)*, volume 186 of *Math. Sci. Engrg.*, pages 45–62. Academic Press, Boston, MA, 1992.

[236] P. Deift and B. Simon. Almost periodic Schrödinger operators. III. The absolutely continuous spectrum in one dimension. *Comm. Math. Phys.*, 90(3):389–411, 1983.

[237] P. Deift and E. Trubowitz. Inverse scattering on the line. *Comm. Pure Appl. Math.*, 32(2):121–251, 1979.

[238] R. del Rio, S. Jitomirskaya, Y. Last, and B. Simon. Operators with singular continuous spectrum. IV. Hausdorff dimensions, rank one perturbations, and localization. *J. Anal. Math.*, 69:153–200, 1996.

[239] R. del Rio, N. Makarov, and B. Simon. Operators with singular continuous spectrum. II. Rank one operators. *Comm. Math. Phys.*, 165(1):59–67, 1994.

[240] R. del Rio and B. Simon. Point spectrum and mixed spectral types for rank one perturbations. *Proc. Amer. Math. Soc.*, 125(12):3593–3599, 1997.

[241] G. F. Dell'Antonio, R. Figari, and A. Teta. Hamiltonians for systems of N particles interacting through point interactions. *Ann. Inst. H. Poincaré Phys. Théor.*, 60(3):253–290, 1994.

[242] G. F. Dell'Antonio, R. Figari, and A. Teta. N–particle systems with zero–range interactions. In *Stochastic Processes, Physics and Geometry II, (Locarno 1991)*. World Scientific, 1995.

[243] G. F. Dell'Antonio, R. Figari, and A. Teta, editors. *Proceedings of the Workshop on Singular Schroedinger Operators*. SISSA, Trieste, 1995.

[244] G. F. Dell'Antonio, R. Figari, and A. Teta. Time dependent point interactions. In *Proceedings of the Workshop on Singular Schroedinger Operators, Trieste, 29 September–1 October 1994*. Eds. G.F.Dell'Antonio, R.Figari and A.Teta. SISSA, Trieste, 1995.

[245] G. F. Dell'Antonio, R. Figari, and A. Teta. A limit evolution problem for time–dependent point interactions. *J. Funct. Anal.*, 142(1):249–275, 1996.

[246] G. F. Dell'Antonio, R. Figari, and A. Teta. Statistics in space dimension two. *Lett. Math. Phys.*, 40(3):235–256, 1997.

[247] G. F. Dell'Antonio, R. Figari, and A. Teta. Diffusion of a particle in the presence of N moving point sources. *Ann. Inst. H. Poincaré Phys. Théor.*, 69(4):413–424, 1998.

[248] Yu. N. Demkov. Detachment of electrons in slow collisions between negative ions and atoms. *Soviet Phys. JETP*, 19:762–768, 1964.

[249] Yu. N. Demkov. Electron detachment in slow collisions between negative ions and atoms. II. Account of finite size of the system. *Soviet Phys. JETP*, 22:615–621, 1966.

[250] Yu. N. Demkov and G.F. Drukarev. Decay and polarizability of negative ions in electric field. *Soviet Phys. JETP*, 20:614–618, 1965.

[251] Yu. N. Demkov and G.F. Drukarev. Particle of low binding energy in a magnetic field. *Soviet Phys. JETP*, 22:182–186, 1965.

[252] Yu. N. Demkov, G.F. Drukarev, and V.V. Kuchinskii. Negative ion decay in the short–range potential approximation. *Soviet Phys. JETP*, 31:509–512, 1970.

[253] Yu. N. Demkov, P. Kurasov, and V. N. Ostrovsky. Doubly periodical in time and energy exactly soluble system with two interacting systems of states. *J. Phys. A*, 28(15):4361–4380, 1995.

[254] Yu. N. Demkov and V.N. Ostrovsky. Zero–radius potential approximation and inelastic scattering of electrons by a molecule. The $e + H_2$ scattering. *Soviet Phys. JETP*, 32:959–963, 1970.

[255] Yu. N. Demkov and V.N. Ostrovsky. *Zero–range Potentials and their Applications in Atomic Physics*. Plenum, New York, 1988. Translation from the 1975 Russian original: Демков, Ю.Н., Островский, В.Н., Метод потенциалов нулевого радиуса в атомной физике, Изд–во Ленинградского Университета, Ленинград, 1975.

[256] Yu.N Demkov. Sum rules for the problem of quantum particle in the presence of zero range potentials. In *Topics in the Quantum Theory of Atoms, Molecules and Solid State*, pages 143–149. St.Petersburg Univ. Press, St.Petersburg, 1996.

[257] Yu.N. Demkov, V.N. Ostrovsky, and E.A. Soloviev. Scattering of electrons by long linear molecules. The effect of approximate translational symmetry. *Soviet. Phys. JETP*, 39:239–242, 1974.

[258] Yu.N. Demkov and R. Subramanian. Motion of a particle in a linear periodic array of small–radius potential wells in quantum mechanics. *Soviet. Phys. JETP*, 30:381–383, 1970.

[259] M. Demuth. Scattering by potentials infinite on unbounded regions in R^n. *Math. Nachr.*, 107:315–325, 1982.

[260] M. Demuth, P. Exner, H. Neidhardt, and V. Zagrebnov, editors. *Mathematical Results in Quantum Mechanics*, volume 70 of *Operator Theory: Advances and Applications*. Birkhäuser, Basel, 1994.

[261] M. Demuth, F. Jeske, and W. Kirsch. Rate of convergence for large coupling limits by Brownian motion. *Ann. Inst. H. Poincaré Phys. Théor.*, 59(3):327–355, 1993.

[262] V. A. Derkach. Extensions of a Hermitian operator in a Krein space. *Dokl. Akad. Nauk Ukrain. SSR Ser. A*, 5:5–9, 86, 1988.

[263] V. A. Derkach. Extensions of a Hermitian operator that is not densely defined in a Krein space. *Dokl. Akad. Nauk Ukrain. SSR Ser. A*, 10:15–19, 84, 1990.

[264] V. A. Derkach. Generalized resolvents of a class of Hermitian operators in a Krein space. *Dokl. Akad. Nauk SSSR*, 317(4):807–812, 1991.

[265] V. A. Derkach. On Weyl function and generalized resolvents of a Hermitian operator in a Krein space. *Integ. Eq. Oper. Theory*, 23(4):387–415, 1995.

[266] V. A. Derkach. On extensions of Laguerre operator in spaces with indefinite metric. *Mat. Zametky*, 63, 1998. (In Russian.) (To appear.)

[267] V. A. Derkach, S. Hassi, and H. de Snoo. Operator models associated with Kac subclasses of generalized Nevanlinna functions. In *Methods of Functional Analysis and Topology*. (To appear.)

[268] V. A. Derkach, S. Hassi, and H. de Snoo. Rank one perturbations in a Pontryagin space with one negative square. 1998. (In preparation.)

[269] V. A. Derkach and M. M. Malamud. Some classes of analytic operator–valued functions with a nonnegative imaginary part. *Dokl. Akad. Nauk Ukrain. SSR Ser. A*, 3:13–17, 87, 1989. (In Russian.)

[270] V. A. Derkach and M. M. Malamud. Generalized resolvents and the boundary value problems for Hermitian operators with gaps. *J. Funct. Anal.*, 95(1):1–95, 1991.

[271] V. A. Derkach and M. M. Malamud. On a generalization of the Krein–Stieltjes class of functions. *Izv. Akad. Nauk Armenii Mat.*, 26(2):115–137, 181 (1992), 1991. (In Russian.)

[272] V. A. Derkach and M. M. Malamud. Characteristic functions of linear operators. *Dokl. Akad. Nauk*, 323(5):816–822, 1992.

[273] V. A. Derkach and M. M. Malamud. Inverse problems for Weyl functions and preresolvent and resolvent matrices of Hermitian operators. *Dokl. Akad. Nauk*, 326(1):12–18, 1992.

[274] A. Dijksma and H. de Snoo. Self-adjoint extensions of symmetric subspaces. *Pacific J. Math.*, 54:71–100, 1974.

[275] A. Dijksma and H. de Snoo. Symmetric and selfadjoint relations in Krein spaces. I. In *Operators in Indefinite Metric Spaces, Scattering Theory and Other Topics (Bucharest, 1985)*, volume 24 of *Oper. Theory: Adv. Appl.*, pages 145–166. Birkhäuser, Basel, 1987.

[276] A. Dijksma and H. de Snoo. Symmetric and selfadjoint relations in Krein spaces. II. *Ann. Acad. Sci. Fenn. Ser. A I Math.*, 12(2):199–216, 1987.

[277] A. Dijksma, H. Langer, and H. de Snoo. Addendum: "Selfadjoint π_κ-extensions of symmetric subspaces: an abstract approach to boundary problems with spectral parameter in the boundary conditions". *Integ. Eq. Oper. Theory*, 7(6):905, 1984.

[278] A. Dijksma, H. Langer, and H. de Snoo. Selfadjoint π_κ-extensions of symmetric subspaces: an abstract approach to boundary problems with spectral parameter in the boundary conditions. *Integ. Eq. Oper. Theory*, 7(4):459–515, 1984.

[279] A. Dijksma, H. Langer, and H. de Snoo. Symmetric Sturm–Liouville operators with eigenvalue depending on boundary conditions. In *Oscillations, Bifurcation and Chaos (Toronto, Ont., 1986)*, volume 8 of *CMS Conf. Proc.*, pages 87–116. Amer. Math. Soc., Providence, RI, 1987.

[280] J. Dittrich and P. Exner. Tunneling through a singular potential barrier. *J. Math. Phys.*, 26(8):2000–2008, 1985.

[281] J. Dittrich and P. Exner, editors. *Rigorous Results in Quantum Dynamics*. World Scientific, River Edge, NJ, 1991. Papers from the conference held in Liblice, June 10–15, 1990.

[282] J. Dittrich, P. Exner, and P. Šeba. Dirac Hamiltonian with contact interaction on a sphere. In *Schrödinger Operators, Standard and Nonstandard (Dubna, 1988)*, pages 190–204. World Scientific, Teaneck, NJ, 1989.

[283] J. Dittrich, P. Exner, and P. Šeba. Dirac operators with a spherically symmetric δ-shell interaction. *J. Math. Phys.*, 30(12):2875–2882, 1989.

[284] J. Dittrich, P. Exner, and P. Šeba. Dirac Hamiltonian with Coulomb potential and contact interaction on a sphere. In *Order, Disorder and Chaos in Quantum Systems (Dubna, 1989)*, volume 46 of *Oper. Theory Adv. Appl.*, pages 209–219. Birkhäuser, Basel, 1990.

[285] J. Dittrich, P. Exner, and P. Šeba. Dirac Hamiltonian with Coulomb potential and spherically symmetric shell contact interaction. *J. Math. Phys.*, 33(6):2207–2214, 1992.

[286] L. A. Dmitrieva, Yu. A. Kuperin, and Yu. Melnikov. One–dimensional discrete Stark Hamiltonian and resonance scattering by an impurity. *J. Phys. A*, 30(9):3087–3099, 1997.

[287] L. A. Dmitrieva, Yu. A. Kuperin, and G. E. Rudin. Extended class of Dubrovin's equations related to the one–dimensional quantum three–body problem. *Comput. Math. Appl.*, 34(5–6):571–585, 1997. Computational Tools of Complex Systems, II.

[288] F. Domínguez-Adame. Exact solutions of the Dirac equation with surface delta interactions. *J. Phys. A*, 23(11):1993–1999, 1990.

[289] F. Domínguez-Adame. Spectroscopy of a perturbed Dirac oscillator. *Europhys. Lett.*, 15(6):569–574, 1991.

[290] F. Domínguez-Adame. Localized solutions of one–dimensional nonlinear Dirac equations with point interaction potentials. *J. Phys. A*, 26(15):3863–3868, 1993.

[291] F. Domínguez-Adame and E. Maciá. Bound states and confining properties of relativistic point interaction potentials. *J. Phys. A*, 22(10):L419–L423, 1989.

[292] F. Domínguez-Adame and B. Méndez. A solvable two–body Dirac equation in one space dimension. *Canad. J. Phys.*, 69(7):780–785, 1991.

[293] F. Domínguez-Adame and A. Sánchez. Relativistic effects in Kronig–Penney models on quasiperiodic lattices. *Phys. Lett. A*, 159(3):153–157, 1991.

[294] Sh. J. Dong and C. N. Yang. Bound states between two particles in a two- or three–dimensional infinite lattice with attractive Kronecker δ–function interaction. *Rev. Math. Phys.*, 1(1):139–146, 1989.

[295] W. F. Jr. Donoghue. On the perturbation of spectra. *Comm. Pure Appl. Math.*, 18:559–579, 1965.

[296] T. C. Dorlas, N. Macris, and J. V. Pulé. The nature of the spectrum for a Landau Hamiltonian with delta impurities. *J. Statist. Phys.*, 87(3-4):847–875, 1997.

[297] H. J. S. Dorren and A. Tip. Maxwell's equations for nonsmooth media; fractal–shaped and pointlike objects. *J. Math. Phys.*, 32(11):3060–3070, 1991.

[298] G.F. Drukarev and Yu.N Demkov. Potential curves for the system AB^- in the zero-range potential approximation for the p state. In *Topics in the Theory of Atomic Collisions (vyp. 3)*, pages 184–195. St.Petersburg Univ. Press, St.Petersburg, 1986.

[299] P. Duclos and P. Šťovíček. Quantum Fermi accelerators with pure–point quasi–spectrum. In *Partial Differential Operators and Mathematical Physics (Holzhau, 1994)*, volume 78 of *Oper. Theory Adv. Appl.*, pages 109–118. Birkhäuser, Basel, 1995.

[300] P. Duclos and P. Šťovíček. Floquet Hamiltonians with pure point spectrum. *Comm. Math. Phys.*, 177(2):327–347, 1996.

[301] M. Dudek, S. Giller, and P. Milczarski. Rectangular well as perturbation. *J. Math. Phys.*, 40:1163–1179, 1999.

[302] N. Dudkin and V. Koshmanenko. Commutativity properties of singularly perturbed operators. *Teoret. Mat. Fiz.*, 102(2):183–197, 1995.

[303] C. B. Duke and E. P. Wigner. Approximation method in collision theory based on R–matrix theory. *Rev. Modern Phys.*, 36:584–589, 1964.

[304] N. Dunford and J. T. Schwartz. *Linear Operators. Part I.* Wiley Classics Library. Wiley, New York, 1988. General theory, With the assistance of William G. Bade and Robert G. Bartle. Reprint of the 1958 original.

[305] N. Dunford and J. T. Schwartz. *Linear Operators. Part II.* Wiley Classics Library. Wiley, New York, 1988. Spectral theory. Selfadjoint operators in Hilbert space, With the assistance of William G. Bade and Robert G. Bartle. Reprint of the 1963 original.

[306] N. Dunford and J. T. Schwartz. *Linear Operators. Part III.* Wiley Classics Library. Wiley, New York, 1988. Spectral operators, With the assistance of William G. Bade and Robert G. Bartle. Reprint of the 1971 original.

[307] N. Elander and P. Kurasov. On the δ'–interaction in one dimension. Technical report, MSI, Stockholm, 1993.

[308] H. Englisch. One–dimensional Schrödinger operators with ergodic potential. *Z. Anal. Anwendungen*, 2(5):411–426, 1983.

[309] H. Englisch. There is no energy interval lying in the spectra of all one-dimensional periodic alloys. *Phys. Stat. Soll (b)*, 118:K17–K19, 1983.

[310] H. Englisch. Instabilities and δ-distributions in solid state physics. *Phys. Stat. Soll (b)*, 125:393–400, 1984.

[311] H. Englisch. Lyapunov exponents and one–dimensional alloys. In *Lyapunov Exponents (Bremen, 1984)*, volume 1186 of *Lecture Notes in Math.*, pages 242–245. Springer, Berlin, 1986.

[312] H. Englisch, W. Kirsch, M. Schröder, and B. Simon. Random Hamiltonians ergodic in all but one direction. *Comm. Math. Phys.*, 128(3):613–625, 1990.

[313] H. Englisch and K.-D. Kürsten. Infinite representability of Schrödinger operators with ergodic potential. *Z. Anal. Anwendungen*, 3(4):357–366, 1984.

[314] H. Englisch and M. Schröder. Schrödinger operators with random boundary conditions. In *Localization in Disordered Systems (Bad Schandau, 1986)*, volume 16 of *Teubner-Texte Phys.*, pages 223–230. Teubner, Leipzig, 1988.

[315] H. Englisch, M. Schröder, and P. Šeba. The free Laplacian with attractive boundary conditions. *Ann. Inst. H. Poincaré Phys. Théor.*, 46(4):373–382, 1987.

[316] H. Englisch and P. Šeba. The stability of the Dirichlet and Neumann boundary conditions. *Rep. Math. Phys.*, 23(3):341–348, 1986.

[317] H. Englisch and P. Šeba. Cooper pairs with point interaction. *Rep. Math. Phys.*, 30(1):9–13 (1992), 1991.

[318] W. N. Everitt and L. Markus. The Glazman–Krein–Naimark theorem for ordinary differential operators. In *New Results in Operator Theory and its Applications*, pages 118–130. Birkhäuser, Basel, 1997.

[319] W. N. Everitt, C. Shubin, G. Stolz, and A. Zettl. Sturm–Liouville problems with an infinite number of interior singularities. In *Spectral Theory and Computational Methods of Sturm–Liouville Problems (Knoxville, TN, 1996)*, volume 191 of *Lecture Notes in Pure and Appl. Math.*, pages 211–249. Dekker, New York, 1997.

[320] W.N. Everitt and L. Markus. *Boundary Value Problems and Symplectic Algebra for Ordinary Differential and Quasi–differential Operators*. American Mathematical Society, Boston, 1998.

[321] V. V. Evstratov and B. S. Pavlov. Electron–phonon scattering, polaron and bipolaron: a solvable model. In *Schrödinger Operators, Standard and Nonstandard (Dubna, 1988)*, pages 214–240. World Scientific, Teaneck, NJ, 1989.

[322] V. V. Evstratov and B. S. Pavlov. Electron–phonon scattering, and an explicitly solvable model of a polaron and a bipolaron. In *Differential Equations. Spectral Theory. Wave Propagation (*In Russian.*)*, volume 13 of *Probl. Mat. Fiz.*, pages 265–304, 309. Leningrad. Univ., Leningrad, 1991.

[323] V. V. Evstratov and B. S. Pavlov. Lattice models of solids. In *Ideas and Methods in Quantum and Statistical Physics (Oslo, 1988)*, pages 212–226. Cambridge Univ. Press, Cambridge, 1992.

[324] P. Exner. A solvable model of two–channel scattering. *Helv. Phys. Acta*, 64(5):592–609, 1991.

[325] P. Exner. The absence of the absolutely continuous spectrum for δ' Wannier–Stark ladders. *J. Math. Phys.*, 36(9):4561–4570, 1995.

[326] P. Exner. Lattice Kronig–Penney models. *Phys. Rev. Lett.*, 74:3503–3506, 1995.

[327] P. Exner. Wannier–Stark systems with a singular interaction. In *Proceedings of the Workshop on Singular Schroedinger Operators, Trieste, 29 September – 1 October 1994*. Eds. G.F.Dell'Antonio, R.Figari and A.Teta. SISSA, Trieste, 1995.

[328] P. Exner. Contact interactions on graph superlattices. *J. Phys. A*, 29(1):87–102, 1996.

[329] P. Exner. Weakly coupled states on branching graphs. *Lett. Math. Phys.*, 38(3):313–320, 1996.

[330] P. Exner. A duality between Schrödinger operators on graphs and certain Jacobi matrices. *Ann. Inst. H. Poincaré Phys. Théor.*, 66(4):359–371, 1997.

[331] P. Exner. Magnetoresonances on a lasso graph. *Found. Phys.*, 27(2):171–190, 1997.

[332] P. Exner and R. Gawlista. Band spectra of rectangular graph superlattices. *Phys.Rev.B*, 253:7275–7286, 1996.

[333] P. Exner, R. Gawlista, P. Šeba, and M. Tater. Point interactions in a strip. *Ann. Physics*, 252(1):133–179, 1996.

[334] P. Exner and D. Krejcirik. Quantum waveguides with a lateral semitransparent barrier: spectral and scattering properties. *J. Phys. A*, 32:4475–4494, 1999.

[335] P. Exner and H. Neidhardt, editors. *Order, Disorder and Chaos in Quantum Systems*, volume 46 of *Operator Theory: Advances and Applications*, Basel, 1990. Birkhäuser.

[336] P. Exner, A. F. Sadreev, P. Šeba, P. Feher, and P. Středa. Topologically induced vortex magnetance of a quantum device. *Phys. Rev. Lett.* (To appear.)

[337] P. Exner and P. Šeba. Quantum motion on two planes connected at one point. *Lett. Math. Phys.*, 12(3):193–198, 1986.

[338] P. Exner and P. Šeba. Quantum motion on a half–line connected to a plane. *J. Math. Phys.*, 28(2):386–391, 1987.

[339] P. Exner and P. Šeba. Mathematical models for quantum point–contact spectroscopy. *Czechoslovak J. Phys. B*, 38(1):1–11, 1988.

[340] P. Exner and P. Šeba. Quantum–mechanical splitters: how should one understand them? *Phys. Lett. A*, 128(9):493–496, 1988.

[341] P. Exner and P. Šeba. A simple model of thin–film point contact in two and three dimensions. *Czechoslovak J. Phys. B*, 38(10):1095–1110, 1988.

[342] P. Exner and P. Šeba, editors. *Applications of Selfadjoint Extensions in Quantum Physics*, volume 324 of *Lecture Notes in Physics*, Berlin, 1989. Springer.

[343] P. Exner and P. Šeba. Electrons in semiconductor microstructures: a challenge to operator theorists. In *Schrödinger Operators, Standard and Nonstandard (Dubna, 1988)*, pages 78–100. World Scientific, Teaneck, NJ, 1989.

[344] P. Exner and P. Šeba. Free quantum motion on a branching graph. *Rep. Math. Phys.*, 28(1):7–26, 1989.

[345] P. Exner and P. Šeba. Quantum junctions and selfadjoint extensions theory. In *Applications of Selfadjoint Extensions in Quantum Physics (Dubna, 1987)*, volume 324 of *Lecture Notes in Phys.*, pages 203–217. Springer, Berlin, 1989.

[346] P. Exner and P. Šeba. Quantum waveguides modelled by graphs. In *Stochastic Methods in Mathematics and Physics (Karpacz, 1988)*, pages 375–384. World Scientific, Teaneck, NJ, 1989.

[347] P. Exner and P. Šeba, editors. *Schrödinger Operators, Standard and Nonstandard*. World Scientific, Teaneck, NJ, 1989. Papers from the conference held in Dubna, September 6–10, 1988.

[348] P. Exner and P. Šeba. Schroedinger operators on unusual manifolds. In *Ideas and Methods in Quantum and Statistical Physics (Oslo, 1988)*, pages 227–253. Cambridge Univ. Press, Cambridge, 1992.

[349] P. Exner and P. Šeba. Point interactions in two and three dimensions as models of small scatterers. *Phys. Lett. A*, 222(1-2):1-4, 1996.

[350] P. Exner and P. Šeba. Resonance statistics in a microwave cavity with a thin antenna. *Phys. Lett. A*, 228(3):146-150, 1997.

[351] P. Exner, P. Šeba, and P. Štovíček. Quantum interference on graphs controlled by an external electric field. *J. Phys. A*, 21(21):4009-4019, 1988.

[352] P. Exner and E. Šerešová. Appendix resonances on a simple graph. *J. Phys. A*, 27(24):8269-8278, 1994.

[353] P. Exner and M. Tater. A one-band model for a weakly coupled quantum-wire resonator. *Phys. Rev. B*, 50:18350-18354, 1994.

[354] P. Exner and M. Tater. Evanescent modes in a multiple scattering factorization. *Czech. J. Phys.*, 47, 1998. (To appear.)

[355] P. Exner and A. Truman. Models of K-capture decay: stochastic vs. quantum mechanics. In *Proceedings of the Conference on Stochastic and Quantum Mechanics (Swansea 1990)*, pages 130-150. World Scientific, Singapore, 1992.

[356] P. Exner and P. Šeba. A new type of quantum interference transistor. *Phys. Lett. A*, 129:477-480, 1988.

[357] P. Exner and P. Šeba. Probability current tornado loops in three-dimensional scattering. *Phys. Lett.*, A245:35-39, 1998.

[358] L. D. Faddeev. On the relation between S-matrix and potential for the one-dimensional Schrödinger operator. *Dokl. Akad. Nauk SSSR*, 121:63-66, 1958.

[359] L. D. Faddeev. The inverse problem in the quantum theory of scattering. *Uspehi Mat. Nauk*, 14(4 (88)):57-119, 1959.

[360] L. D. Faddeev. Scattering theory for a three-particle system. *Soviet Physics. JETP*, 12:1014-1019, 1960.

[361] L. D. Faddeev. The resolvent of the Schroedinger operator for a system of three particles interacting in pairs. *Soviet Physics Dokl.*, 6:384-386, 1961.

[362] L. D. Faddeev. The construction of the resolvent of the Schrödinger operator for a three-particle system, and the scattering problem. *Soviet Physics Dokl.*, 7:600-602, 1963.

[363] L. D. Faddeev. The inverse problem in the quantum theory of scattering. *J. Math. Phys.*, 4:72-104, 1963.

[364] L. D. Faddeev. Mathematical questions in the quantum theory of scattering for a system of three particles. *Trudy Mat. Inst. Steklov.*, 69:122, 1963.

[365] L. D. Faddeev. Properties of the S-matrix of the one–dimensional Schrödinger equation. *Trudy Mat. Inst. Steklov.*, 73:314–336, 1964.

[366] L. D. Faddeev. *Mathematical Aspects of the Three–body Problem in Quantum Scattering Theory*. Israel Program for Scientific Translations Jerusalem, 1965. Translated from the Russian by Ch. Gutfreund. Translation edited by L. Meroz.

[367] L. D. Faddeev. Properties of the S-matrix of the one–dimensional Schrödinger equation. *Amer. Math. Soc. Transl.*, 65:139–166, 1967.

[368] L. D. Faddeev. The inverse problem in the quantum theory of scattering. II. In *Current Problems in Mathematics, Vol. 3* (In Russian.), pages 93–180, 259. Akad. Nauk SSSR Vsesojuz. Inst. Naučn. i Tehn. Informacii, Moscow, 1974.

[369] L. D. Faddeev. Lectures on Quantum Inverse Scattering Method. In *Integrable Systems (Tianjin, 1987)*, Nankai Lectures Math. Phys., pages 23–70. World Scientific, Teaneck, NJ, 1990.

[370] L. D. Faddeev and S. P. Merkuriev. *Quantum Scattering Theory for Several Particle Systems*, volume 11 of *Mathematical Physics and Applied Mathematics*. Kluwer, Dordrecht, 1993. Translated from the 1985 Russian original.

[371] M.D. Faddeev and B.S. Pavlov. Scattering by a resonator with a small opening. *J. Sov. Math.*, 27:2527–2533, 1984.

[372] S. Fassari. An estimate regarding one dimensional point interactions. *Helv. Phys. Acta*, 68:121–125, 1995.

[373] S. Fassari and G. Inglese. Spectroscopy of a three dimensional isotropic harmonic oscillator with a δ-type perturbation. *Helv. Phys. Acta*, 68:130–140, 1995.

[374] E. Fermi. Sul moto dei neutroni nelle sostanze idrogenate. *Ricerca Scientifica*, 7:13–52, 1936. (In Italian.), English translation in E.Fermi, Collected papers, vol. I, Italy 1921–1938, Univ. of Chicago Press, Chicago, 1962, pp. 980–1016.

[375] C. J. Fewster. Generalized point interactions for the radial Schrödinger equation via unitary dilations. *J. Phys. A*, 28(4):1107–1127, 1995.

[376] R. Figari, H. Holden, and A. Teta. A law of large numbers and a central limit theorem for the Schrödinger operator with zero–range potentials. *J. Statist. Phys.*, 51(1–2):205–214, 1988.

[377] R. Figari, E. Orlandi, and Teta A. The Laplacian in regions with many small obstacles: fluctuations around the limit operator. *J. Statist. Phys.*, 41:465–487, 1985.

[378] R. Figari and A. Teta. Effective potential and fluctuations for a boundary value problem on a randomly perforated domain. *Lett. Math. Phys.*, 26(4):295–305, 1992.

[379] R. Figari and A. Teta. A boundary value problem of mixed type on perforated domains. *Asymptotic Anal.*, 6(3):271–284, 1993.

[380] L. I. Filina, V. A. Geyler, V. A. Margulis, and O. B. Tomilin. Magnetic moment of a three–dimensional well in a quantizing magnetic field. *Phys. Letters*, A244:295–302, 1998.

[381] A. K. Fragela. Perturbation of a polyharmonic operator by potentials with small supports. *Dokl. Akad. Nauk SSSR*, 245(1):34–36, 1979.

[382] A. K. Fragela. Perturbation of a polyharmonic operator by potentials of delta–function type. *Funktsional. Anal. i Prilozhen.*, 15(1):86–87, 1981.

[383] A. K. Fragela. Perturbation of a polyharmonic operator with delta–like potentials. *Mat. Sb. (N.S.)*, 130(172)(3):386–393, 431–432, 1986.

[384] K. Friedrichs. Spektraltheorie halbbeschränkter Operatoren und Anwendung auf die Spektralzergung von Differentialoperatoren. *Math. Ann.*, 109:465–487, 1934.

[385] G. V. Galunov and V. L. Oleinik. Analysis of the dispersion equation for a negative Dirac "comb". *St.Petersburg Math.J.*, 4(4):707–, 1993.

[386] M. Gaudin. *La Fonction d'onde de Bethe*. Masson, 1983.

[387] B. Gaveau and L. S. Schulman. Explicit time–dependent Schrödinger propagators. *J. Phys. A*, 19(10):1833–1846, 1986.

[388] M. M. Gekhtman and I. V. Stankevich. The generalized Kronig–Penney problem. *Funk. Anal. Pril.*, 11:61–62, 1977.

[389] I. M. Gelfand and B. M. Levitan. On the determination of a differential equation from its spectral function. *Amer. Math. Soc. Transl. Ser. 2*, 1:253–304, 1955. Tranlation of И.М.Гельфанд и Б.М.Левитан, Об определении дифференциального уравнения по его спектральной функции, Известия АН СССР, сер. матем. **15** (1951), 309–360.

[390] N. I. Gerasimenko. The inverse scattering problem on a noncompact graph. *Teoret. Mat. Fiz.*, 75(2):187–200, 1988.

[391] N. I. Gerasimenko and B. S. Pavlov. A scattering problem on noncompact graphs. *Teoret. Mat. Fiz.*, 74(3):345–359, 1988.

[392] F. Gesztesy. Scattering theory for one–dimensional systems with nontrivial spatial asymptotics. In *Schrödinger Operators, Aarhus 1985*, volume 1218 of *Lecture Notes in Math.*, pages 93–122. Springer, Berlin, 1986.

[393] F. Gesztesy and H. Holden. A new class of solvable models in quantum mechanics describing point interactions on the line. *J. Phys. A*, 20(15):5157–5177, 1987.

[394] F. Gesztesy, H. Holden, and W. Kirsch. On energy gaps in a new type of analytically solvable model in quantum mechanics. *J. Math. Anal. Appl.*, 134(1):9–29, 1988.

[395] F. Gesztesy, H. Holden, and P. Šeba. On point interactions in magnetic field systems. In *Schrödinger Operators, Standard and Nonstandard (Dubna, 1988)*, pages 146–164. World Scientific, Teaneck, NJ, 1989.

[396] F. Gesztesy, N.J. Kalton, K. Makarov, and E. Tsekanovskii. Some applications of operator–valued Herglotz functions. *Operator Theory: Advances and Applications*, pages 1–45. (To appear.)

[397] F. Gesztesy and G. Karner. On three–body scattering near thresholds. *SIAM J. Math. Anal.*, 18(4):1064–1086, 1987.

[398] F. Gesztesy and W. Kirsch. One–dimensional Schrödinger operators with interactions singular on a discrete set. *J. Reine Angew. Math.*, 362:28–50, 1985.

[399] F. Gesztesy, C. Macedo, and L. Streit. An exactly solvable periodic Schrödinger operator. *J. Phys. A*, 18(9):L503–L507, 1985.

[400] F. Gesztesy, K. Makarov, and E. Tsekanovskii. An addendum to Krein's formula. *J. Math. Anal. Appl.*, 222:594–606, 1998.

[401] F. Gesztesy and L. Pittner. Two–body scattering for Schrödinger operators involving zero–range interactions. *Rep. Math. Phys.*, 19(2):143–154, 1984.

[402] F. Gesztesy and P. Šeba. New analytically solvable models of relativistic point interactions. *Lett. Math. Phys.*, 13(4):345–358, 1987.

[403] F. Gesztesy and B. Simon. Rank–one perturbations at infinite coupling. *J. Funct. Anal.*, 128(1):245–252, 1995.

[404] F. Gesztesy, B. Simon, and G. Teschl. Spectral deformations of one–dimensional Schrödinger operators. *J. Anal. Math.*, 70:267–324, 1996.

[405] F. Gesztesy and E. Tsekanovskii. On matrix–valued Herglotz functions. *Math. Nachr.*, pages 1–81. (To appear.)

[406] V. A. Geyler. The two–dimensional Schrödinger operator with a homogeneous magnetic field and its perturbations by periodic zero–range potentials. *Algebra i Analiz*, 3(3):1–48, 1991.

[407] V. A. Geyler and I. V. Chudaev. Schrödinger operators with moving point interactions and related solvable models of quantum mechanical systems. *Z. Anal.Anw.*, 17:37–55, 1998.

[408] V. A. Geyler and V. V. Demidov. The spectrum of the three–dimensional Landau operator perturbed by a periodic point source. *Teoret. Mat. Fiz.*, 103(2):283–294, 1995.

[409] V. A. Geyler and V. V. Demidov. On the Green function of the Landau operator and its properties related to point interactions. *Z. Anal. Anwendungen*, 15(4):851–863, 1996.

[410] V. A. Geyler and V. A. Margulis. Spectrum of the Bloch electron in a magnetic field in a two–dimensional lattice. *Teoret. Mat. Fiz.*, 58(3):461–472, 1984.

[411] V. A. Geyler and V. A. Margulis. The structure of the spectrum of the Bloch electron in a magnetic field and in a two–dimensional lattice. *Teoret. Mat. Fiz.*, 61(1):140–149, 1984.

[412] V. A. Geyler and V. A. Margulis. Anderson localization in the nondiscrete Maryland model. *Teoret. Mat. Fiz.*, 70(2):192–201, 1987.

[413] V. A. Geyler, V. A. Margulis, and I. I. Chuchaev. Zero–range potentials and Carleman operators. *Sibirsk. Mat. Zh.*, 36(4):828–841, ii, 1995.

[414] V. A. Geyler, V. A. Margulis, and L. I. Filina. Conductance of a quantum wire in a longitudinal magnetic field. *J.Exp.Theor.Phys.*, 86:751–762, 1998.

[415] V. A. Geyler, B. S. Pavlov, and I. Yu. Popov. Spectral properties of a charged particle in antidot array: a limiting case of quantum billiard. *J. Math. Phys.*, 37(10):5171–5194, 1996.

[416] V. A. Geyler, B.S. Pavlov, and I. Yu. Popov. One–particle spectral problem for superlattice with a constant magnetic field. *Atti. Sem. Mat. Fis. Univ. Modena*, XLVI:79–124, 1998.

[417] V. A. Geyler and A. V. Popov. An explicitly solvable model with a periodic system of the Aharonov–Bohm vortices. In *Fourth Intern. Conf. on Difference Equations and Applications, August 27–31, 1998, Poznan*. Zaklad Graficzny Politechnichki Poznanskiej, 1998.

[418] V. A. Geyler and I. Yu. Popov. Ballistic transport in nanostructures: explicitly solvable models. *Teoret. Mat. Fiz.*, 107(1):12–20, 1996.

[419] V. A. Geyler and I. Yu. Popov. Eigenvalues imbedded in the band spectrum for a periodic array of quantum dots. *Rep. Math. Phys.*, 39(2):275–281, 1997.

[420] V. A. Geyler and I. Yu. Popov. Solvable model of a double quantum electron layer in a magnetic field. *Proc. R. Soc. Lond. A*, 1998.

[421] V. A. Geyler, I. Yu. Popov, and S. L. Popova. Transmission coefficient for ballistic transport through quantum resonator. *Rep. Math. Phys.*, 40(3):531–538, 1997.

[422] V. A. Geyler and M. M. Senatorov. Periodic potentials for which all the gaps are nonzero. *Funktsional. Anal. i Prilozhen.*, 31(1):67–70, 1997.

[423] P. K. Ghosh. The Kronig–Penney model on a generalized Fibonacci lattice. *Physics Letters A*, 161:153–157, 1991.

[424] V. L. Girko and W. Kirsch. Limit theorems for band random matrices whose entries have bounded variances. *Random Oper. Stochastic Equations*, 1(2):181–191, 1993.

[425] I. M. Glazman. *Direct Methods of Qualitative Spectral Analysis of Singular Differential Operators*. Israel Program for Scientific Translations, Jerusalem, 1965, 1966. Translated from the Russian by the IPST staff.

[426] F. Göhman, A. R. Its, and V. E. Korepin. Correlations in the impenetrable electron gas. *cond–mat/9809076*, 1998.

[427] M. L. Gorbachuk and V. I. Gorbachuk. The theory of selfadjoint extensions of symmetric operators; entire operators and boundary value problems. *Ukraïn. Mat. Zh.*, 46(1–2):55–62, 1994.

[428] M. L. Gorbachuk and V. I. Gorbachuk. *M. G. Krein's Lectures on Entire Operators*, volume 97 of *Operator Theory: Advances and Applications*. Birkhäuser, Basel, 1997.

[429] V. I. Gorbachuk and M. L. Gorbachuk. *Boundary Value Problems for Operator Differential Equations*. Kluwer, Dordrecht, 1991. Translated and revised from the 1984 Russian original.

[430] V. I. Gorbachuk, M. L. Gorbachuk, and A. N. Kochubei. The theory of extensions of symmetric operators, and boundary value problems for differential equations. *Ukraïn. Mat. Zh.*, 41(10):1299–1313, 1436, 1989.

[431] A. Ya. Gordon. Instability of dense point spectrum under finite rank perturbations. *Commun. Math. Phys.*, 187:583–595, 1997.

[432] O. G. Gorjaga and Ju. M. Širokov. Energy levels of an oscillator with a singular concentrated potential. *Teoret. Mat. Fiz.*, 46(3):321–324, 1981.

[433] P. Gosdzinsky and R. Tarrach. Learning quantum field theory from elementary quantum mechanics. *Amer. J. of Phys.*, 59:70–74, 1991.

[434] V. Grecchi, M. Maioli, and A. Sacchetti. Stark ladders and perturbation theory. In *Mathematical Results in Quantum Mechanics (Blossin, 1993)*, volume 70 of *Oper. Theory Adv. Appl.*, pages 33–36. Birkhäuser, Basel, 1994.

[435] S. A. Gredeskul, L. Pastur, and P. Šeba. Transmission properties of random point scatterers for waves with two–band dispersion law. *J. Statist. Phys.*, 58(5–6):795–816, 1990.

[436] D. J. Griffiths. Boundary conditions at the derivative of a delta function. *J. Phys. A*, 26(9):2265–2267, 1993.

[437] Ch. Grosche. Path integrals for potential problems with δ–function perturbation. *J. Phys. A*, 23(22):5205–5234, 1990.

[438] Ch. Grosche. δ–function perturbations and boundary problems by path integration. *Ann. Physik*, 2:557–589, 1993.

[439] Ch. Grosche. Path integrals for two– and three–dimensional δ–function perturbations. *Ann. Physik (8)*, 3(4):283–312, 1994.

[440] Ch. Grosche. Boundary conditions in path integrals. In *Proceedings of the Workshop on Singular Schroedinger Operators, Trieste, 29 September–1 October 1994*. Eds. G.F.Dell'Antonio, R.Figari and A.Teta. SISSA, Trieste, 1995.

[441] Ch. Grosche. δ'–function perturbations and Neumann boundary conditions by path integration. *J. Phys. A*, 28(3):L99–L105, 1995.

[442] A. Grossman and T. T. Wu. A class of potentials with extremely narrow resonances. I. Case with discrete rotational symmetry. Technical report, Preprint CPT–81/PE. 1291, Centre de Physique Theorique, CNRS–Luminy, Marseille, 1981.

[443] A. Grossmann, R. Høegh-Krohn, and M. Mebkhout. A class of explicitly soluble, local, many–center Hamiltonians for one–particle quantum mechanics in two and three dimensions. I. *J. Math. Phys.*, 21(9):2376–2385, 1980.

[444] A. Grossmann, R. Høegh-Krohn, and M. Mebkhout. The one particle theory of periodic point interactions. Polymers, monomolecular layers, and crystals. *Comm. Math. Phys.*, 77(1):87–110, 1980.

[445] A. Grossmann and T. T. Wu. Fermi pseudopotential in higher dimensions. *J. Math. Phys.*, 25(6):1742–1745, 1984.

[446] C. H. Gu and C. N. Yang. A one–dimensional N fermion problem with factorized S matrix. *Comm. Math. Phys.*, 122(1):105–116, 1989.

[447] Yu. V. Gugel and I. Yu. Popov. A model of the suction of a boundary layer through a small opening, and the shift of neutral curves. In *Differential equations. Spectral theory. Wave propagation (In Russian.)*, volume 13 of *Probl. Mat. Fiz.*, pages 117–125, 306–307. Leningrad. Univ., Leningrad, 1991.

[448] Yu. V. Gugel, I. Yu. Popov, and S. L. Popova. Hydroton: creep and slip. *Fluid Dynamics Res.*, 18:199–210, 1996.

[449] V. V. Gusarov and I. Yu. Popov. Flows in two–dimensional non–autonomous phases in polycrystalline systems. *Nuovo Cimento*, 18:799–805, 1996.

[450] G. A. Hagedorn. Analysis of a nontrivial, explicitly solvable multichannel scattering system. *Ann.Inst.Henri Poincaré*, 51(1):1–22, 1989.

[451] G. G. Hall. *Applied Group Theory*. Elsevier, Inc., New York, 1967.

[452] S. Hassi and H. de Snoo. On some subclasses of Nevanlinna functions. *Z. Anal. Anwendungen*, 15(1):45–55, 1996.

[453] S. Hassi and H. de Snoo. On rank one perturbations of selfadjoint operators. *Integ. Eq. Oper. Theory*, 29(3):288–300, 1997.

[454] S. Hassi and H. de Snoo. One–dimensional graph perturbations of selfadjoint relations. *Ann. Acad. Sci. Fenn. Math.*, 22(1):123–164, 1997.

[455] S. Hassi, H. de Snoo, and A. D. I. Willemsma. Smooth rank one perturbations of selfadjoint operators. *Proc. Amer. Math. Soc.*, 126(9):2663–2675, 1998.

[456] S. Hassi, M. Kaltenbäck, and H. de Snoo. A characterization of semibounded selfadjoint operators. *Proc. Amer. Math. Soc.*, 125(9):2681–2692, 1997.

[457] S. Hassi, M. Kaltenbäck, and H. de Snoo. Triplets of Hilbert spaces and Friedrichs extensions associated with the subclass n_1 of Nevanlinna functions. *J. Operator Theory*, 37(1):155–181, 1997.

[458] S. Hassi, M. Kaltenbäck, and H. de Snoo. Generalized Krein–von Neumann extensions and associated operator models. *Acta Sci. Math. (Szeged)*, 64:627–655, 1998.

[459] S. Hassi, H. Langer, and H. de Snoo. Selfadjoint extensions for a class of symmetric operators with defect numbers $(1,1)$. In *Topics in Operator Theory, Operator Algebras and Applications (Timişoara, 1994)*, pages 115–145. Rom. Acad., Bucharest, 1995.

[460] R. J. Henderson and S. G. Rajeev. Renormalized path integral in quantum mechanics. *J.Math.Phys.*, 38(5):2171, 1997.

[461] R. J. Henderson and S. G. Rajeev. Renormalized contact potential in two dimensions. *J. Math. Phys.*, 39:749–759, 1998.

[462] J. Herczyński. On Schrödinger operators with distributional potentials. *J. Operator Theory*, 21(2):273–295, 1989.

[463] G. Herglotz. Über Potenzreihen mit positiven reellen Teil im Einheitskreise. *Ber. Verh. Sächs. Acad. Wiss. Leipzig*, 63:501–511, 1911.

[464] M. Herrmann. A relativistic method for the nonrelativistic gap problem. *Lett. Math. Phys.*, 18(1):19–25, 1989.

[465] K. Hikami. Notes on the structure of the δ–function interacting gas. Intertwining operator in the degenerate affine Hecke algebra. *J. Phys. A: Math. Gen.*, 31:L85–L91, 1998.

[466] R. Høegh-Krohn, H. Holden, S. Johannesen, and T. Wentzel-Larsen. The Fermi surface for point interactions. *J. Math. Phys.*, 27(1):385–405, 1986.

[467] R. Høegh-Krohn, H. Holden, and F. Martinelli. The spectrum of defect periodic point interactions. *Lett. Math. Phys.*, 7(3):221–228, 1983.

[468] R. Høegh-Krohn and M. Mebkhout. The $1/r$ expansion for the critical multiple well problem. *Comm. Math. Phys.*, 91:65–73, 1983.

[469] R. Høegh-Krohn and M. Mebkhout. The multiple well problem–asymptotic behavior of the eigenvalues and resonances. In *Trends and Developments in the Eighties (Bielefeld, 1982/1983)*, pages 244–272. World Scientific, Singapore, 1985.

[470] H. Holden. Konvergens mot punkt–interaksjoner (in norwegian). cand. real. thesis. Master's thesis, University of OSIO, Norway, 1981.

[471] H. Holden. On coupling constant thresholds in two dimensions. *J. Operator Theory*, 14(2):263–276, 1985.

[472] H. Holden. Point interaction and the short–range expansion. A solvable model in quantum mechanics and its approximation. Dr. Philos. Dissertation. PhD thesis, University of Oslo, Norway, 1985.

[473] H. Holden, R. Høegh-Krohn, and S. Johannesen. The short range expansion. *Adv. in Appl. Math.*, 4(4):402–421, 1983.

[474] H. Holden, R. Høegh-Krohn, and S. Johannesen. The short–range expansion in solid state physics. *Ann. Inst. H. Poincaré Phys. Théor.*, 41(4):335–362, 1984.

[475] H. Holden, R. Høegh-Krohn, and M. Mebkhout. The short–range expansion for multiple well scattering theory. *J. Math. Phys.*, 26(1):145–151, 1985.

[476] M. N. Hounkonnou, M. Hounkpe, and J. Shabani. Scattering theory for finitely many sphere interactions supported by concentric spheres. *J. Math. Phys.*, 38(6):2832–2850, 1997.

[477] M. Hounkpe, P. C. Rutomera, and J. Shabani. Scattering theory for a Coulomb plus a sphere ineteraction. 1998. (In preparation.).

[478] K. Huang and C. N. Yang. Quantum–mechanical many–body problem with hard–sphere interaction. *Phys. Rev. (2)*, 105:767–775, 1957.

[479] K. Huang, C. N. Yang, and J. M. Luttinger. Imperfect Bose gas with hard–sphere interaction. *Phys. Rev. (2)*, 105:776–784, 1957.

[480] R. J. Hughes. Singular perturbations in the interaction representation. II. *J. Funct. Anal.*, 49(3):293–314, 1982.

[481] R. J. Hughes. Renormalization of the relativistic delta potential in one dimension. *Lett. Math. Phys.*, 34(4):395–406, 1995.

[482] R. J. Hughes. Unitary propagators for time–dependent Hamiltonians with singular potentials. *J. Math. Anal. Appl.*, 193(2):447–464, 1995.

[483] R. J. Hughes. Relativistic point interactions: approximation by smooth potentials. *Rep. Math. Phys.*, 39(3):425–432, 1997.

[484] R. J. Hughes. Generalized Kronig–Penney Hamiltonians. *J.Math.Anal.Appl.*, 222:151–166, 1998.

[485] R. J. Hughes. Relativistic Kronig–Penney–type Hamiltonians. *Integr. Equ. Oper. Theory*, 31:436–448, 1998.

[486] R. J. Hughes and I. E. Segal. Singular perturbations in the interaction representation. *J. Funct. Anal.*, 38(1):71–98, 1980.

[487] T. Ikebe. The Schrödinger operator with a penetrable wall interaction. *J. Anal. Math.*, 59:37–43, 1992. Festschrift on the Occasion of the 70th birthday of Shmuel Agmon.

[488] T. Ikebe and Sh. Shimada. Scattering with penetrable wall potentials. In *Differential Equations and Mathematical Physics (Birmingham, Ala., 1986)*, volume 1285 of *Lecture Notes in Math.*, pages 211–214. Springer, Berlin, 1987.

[489] T. Ikebe and Sh. Shimada. Spectral and scattering theory for the Schrödinger operators with penetrable wall potentials. *J. Math. Kyoto Univ.*, 31(1):219–258, 1991.

[490] T. Ikebe and Sh. Shimada. Resolvent convergence of the Schrödinger operator with a penetrable wall interaction. *Sūrikaisekikenkyūsho Kōkyūroku*, 873:133–139, 1994. Spectra, Scattering Theory and Related Topics (In Japanese.) (Kyoto, 1993).

[491] I. S. Iohvidov, M. G. Krein, and H. Langer. *Introduction to the Spectral Theory of Operators in Spaces with an Indefinite Metric*, volume 9 of *Mathematical Research*. Akademie, Berlin, 1982.

[492] V. Jakšić, S. Molchanov, and L. Pastur. On the propagation properties of surface waves. In *Wave Propagation in Complex Media (Minneapolis, MN, 1994)*, volume 96 of *IMA Vol. Math. Appl.*, pages 143–154. Springer, New York, 1998.

[493] V. A. Javrian. On an inverse problem for Sturm–Liuville operators. *Izv. Akad. Nauk Armjan. SSR Ser. Mat.*, 6:246–251, 1971. (In Russian, English summary.)

[494] A. Jensen. Some remarks on eigenfunction expansions for Schrödinger operators with non–local potentials. *Math. Scand.*, 41(2):347–357, 1977.

[495] A. Jensen. Resonances in an abstract analytic scattering theory. *Ann. Inst. H. Poincaré Sect. A (N.S.)*, 33(2):209–223, 1980.

[496] A. Jensen. Spectral properties of Schrödinger operators and time–decay of the wave functions results in $L^2(r^m)$, $m \geq 5$. *Duke Math. J.*, 47(1):57–80, 1980.

[497] W. John, B. Milek, H. Schanz, and P. Šeba. Statistical properties of resonances in quantum irregular scattering. *Phys. Rev. Lett.*, 67(15):1949–1952, 1991.

[498] I. S. Kac. On integral representations of analytic functions mapping the upper half–plane onto a part of itself. *Uspehi Mat. Nauk (N.S.)*, 11(3(69)):139–144, 1956.

[499] I. S. Kac. Integral and exponential representations of analytic functions mapping the upper half–plane into itself. In *Problems of Mathematical Physivs and Theory of Functions, II (In Russian.)*, pages 51–62. Naukova Dumka, Kiev, 1964.

[500] I.S. Kac and M.G Krein. R–functions–analytic functions mapping the upper halfplane into itself. *Amer. Math. Soc. Transl.*, Series 2, 103:1–102, 1974.

[501] V.N. Kapshai and T.A. Alferova. Relativistic two–particle one–dimensional scattering problem for superposition of δ–potentials. *J. Phys. A*, 32:5329–5342, 1999.

[502] Yu. E. Karpeshina. An eigenfunction expansion theorem for the problem of scattering on homogeneous periodic supports of chain type in three-dimensional space. In *Spectral Theory. Wave Processes*, volume 10 of *Probl. Mat. Fiz.*, pages 137–163, 299. Leningrad. Univ., Leningrad, 1982.

[503] Yu. E. Karpeshina. Eigenfunctions of the Schrödinger operator in a three-dimensional space with periodic point potential of two–dimensional lattice type. *Tohoku Math. J.*, 34:58–64, 126, 1982.

[504] Yu. E. Karpeshina. The spectrum and eigenfunctions of the Schrödinger operator in a three–dimensional space with point–like potential of the homogeneous two–dimensional lattice type. *Teoret. Mat. Fiz.*, 57(3):414–423, 1983.

[505] Yu. E. Karpeshina. Spectrum and eigenfunctions of the Schrödinger operator with point potential of uniform lattice type in a three–dimensional space. *Teoret. Mat. Fiz.*, 57(2):304–313, 1983.

[506] Yu. E. Karpeshina. A theorem for eigenfunction expansion of the Schrödinger operator with an exact potential of lattice type. In *Operator Theory and Function Theory, No. 1*, pages 100–111. Leningrad. Univ., Leningrad, 1983.

[507] Yu. E. Karpeshina. An eigenfunction expansion theorem for the Schrödinger operator with a homogeneous simple two–dimensional lattice of potentials of zero radius in a three–dimensional space. *Vestnik Leningrad. Univ. Mat. Mekh. Astronom.*, vyp. 1:11–17, 1984.

[508] Yu. E. Karpeshina. A model of a crack in a plate. In *Linear and Nonlinear Boundary Value Problems. Spectral Theory.*, volume 10 of *Probl. Mat. Anal.*, pages 139–153, 214. Leningrad. Univ., Leningrad, 1986. (In Russian.)

[509] Yu. E. Karpeshina and B. S. Pavlov. Interaction of the zero radius for the biharmonic and the polyharmonic equation. *Mat. Zametki*, 40(1):49–59, 140, 1986.

[510] W. Karwowski. Hamiltonians with additional kinetic energy terms on hypersurfaces. In *Applications of Selfadjoint Extensions in Quantum Physics (Dubna, 1987)*, volume 324 of *Lecture Notes in Phys.*, pages 203–217. Springer, Berlin, 1989.

[511] W. Karwowski and V. Koshmanenko. Additive regularization of singular bilinear forms. *Ukrain. Mat. Zh.*, 42(9):1199–1204, 1990.

[512] W. Karwowski and V. Koshmanenko. On the definition of singular bilinear forms and singular linear operators. *Ukrain. Mat. Zh.*, 45(8):1084–1089, 1993.

[513] W. Karwowski and V. Koshmanenko. Regular restrictions of singular bilinear forms. *Funktsional. Anal. i Prilozhen.*, 29(2):79–81, 1995.

[514] W. Karwowski and V. Koshmanenko. Singular quadratic forms: regularization by restriction. *J. Funct. Anal.*, 143(1):205–220, 1997.

[515] W. Karwowski, V. Koshmanenko, and S. Ota. Schrödinger operator perturbed by operators related to null–sets. *Positivity*, 2:77–99, 1998.

[516] T. Kato. Perturbation theory of semi–bounded operators. *Math. Ann.*, 125:435–447, 1953.

[517] T. Kato. On finite–dimensional perturbations of self–adjoint operators. *J. Math. Soc. Japan*, 9:239–249, 1957.

[518] T. Kato. Perturbation of continuous spectra by trace class operators. *Proc. Japan Acad.*, 33:260–264, 1957.

[519] T. Kato. Perturbation of a scattering operator and its continuous spectrum. *Sugaku*, 9:75–84, 1957/58.

[520] T. Kato. Wave operators and unitary equivalence. *Pacific J. Math.*, 15:171–180, 1965.

[521] T. Kato. Scattering theory and perturbation of continuous spectra. In *Actes du Congrès International des Mathématiciens (Nice, 1970), Tome 1*, pages 135–140. Gauthier–Villars, Paris, 1971.

[522] T. Kato. *Perturbation Theory for Linear Operators*. Springer–Verlag, Berlin, second edition, 1976. Grundlehren der Mathematischen Wissenschaften, Band 132.

[523] T. Kato and S. T. Kuroda. A remark on the unitarity property of the scattering operator. *Nuovo Cimento (10)*, 14:1102–1107, 1959.

[524] T. Kato and S. T. Kuroda. The abstract theory of scattering. *Rocky Mountain J. Math.*, 1(1):127–171, 1971.

[525] W. Kirsch. *Über Spektren stochastischer Schrödingeroperatoren (In German)*. *Ph.D. Dissertation*. PhD thesis, Ruhr–Univ. Bochum, Germany, 1981.

[526] W. Kirsch and F. Nitzschner. Lifshitz tails and non–Lifshitz tails for one–dimensional random point interactions. In *Order, Disorder and Chaos in Quantum Systems (Dubna, 1989)*, volume 46 of *Oper. Theory Adv. Appl.*, pages 179–182. Birkhäuser, Basel, 1990.

[527] A. A. Kiselev. Some examples in one–dimensional geometric scattering on manifolds. *J. Math. Anal. Appl.*, 212(1):263–280, 1997.

[528] A. A. Kiselev and B. S. Pavlov. Eigenfrequencies and eigenfunctions of the Laplace operator of the Neumann problem in a system of two connected resonators. *Teoret. Mat. Fiz.*, 100(3):354–366, 1994.

[529] A. A. Kiselev and B. S. Pavlov. The essential spectrum of the Laplace operator of the Neumann problem in a model domain of complex structure. *Teoret. Mat. Fiz.*, 99(1):3–19, 1994.

[530] A. A. Kiselev, B. S. Pavlov, N. N. Penkina, and M. G. Suturin. A technique of the theory of extensions using interaction symmetry. *Teoret. Mat. Fiz.*, 91(2):179–191, 1992.

[531] A. A. Kiselev and I. Yu. Popov. Higher moments in a zero–width aperture model. *Teoret. Mat. Fiz.*, 89(1):11–17, 1991.

[532] A. A. Kiselev and I. Yu. Popov. An indefinite metric and scattering by regions with a small aperture. *Mat. Zametki*, 58(6):837–850, 959, 1995.

[533] A. A. Kiselev and B. Simon. Rank one perturbations with infinitesimal coupling. *J. Funct. Anal.*, 130(2):345–356, 1995.

[534] F. Kleespies. Geometrische Aspekte der Spektraltheorie von Schrödinger Operatoren. Master's thesis, Frankfurt am Main Univ., 1998. (In German.)

[535] A. N. Kochubei. Extensions of symmetric operators and of symmetric binary relations. *Mat. Zametki*, 17:41–48, 1975.

[536] A. N. Kochubei. Extensions of a nondensely defined symmetric operator. *Sibirsk. Mat. Ž.*, 18(2):314–320, 478, 1977.

[537] A. N. Kochubei. Extensions and the characteristic function of a symmetric operator. In *Spectral theory of operators (Proceedings of the Second All-Union Summer Mathematical School, Baku, 1975)* (In Russian.), pages 114–121. "Èlm", Baku, 1979.

[538] A. N. Kochubei. Characteristic functions of symmetric operators and their extensions. *Izv. Akad. Nauk Armyan. SSR Ser. Mat.*, 15(3):219–232, 247, 1980.

[539] A. N. Kochubei. Elliptic operators with boundary conditions on a subset of measure zero. *Funktsional. Anal. i Prilozhen.*, 16(2):74–75, 1982.

[540] A. N. Kochubei. One–dimensional point interactions. *Ukrain. Mat. Zh.*, 41(10):1391–1395, 1439, 1989.

[541] A. N. Kochubei. Point interactions in one dimension. In *Schrödinger Operators, Standard and Nonstandard (Dubna, 1988)*, pages 78–100. World Scientific, Teaneck, NJ, 1989.

[542] A. N. Kochubei. Selfadjoint extensions of Schroedinger operators with singular potentials. In *Order, Disorder and Chaos in Quantum Systems (Dubna, 1989)*, pages 221–227. Birkhäuser, Basel, 1990.

[543] A. N. Kochubei. Selfadjoint extensions of the Schrödinger operator with a singular potential. *Sibirsk. Mat. Zh.*, 32(3):60–69, 1991.

[544] V. Korepin and N. Slavnov. Normal ordering in the theory of correlation functions of exactly solvable models. *hep-th/9709204*, 1997.

[545] C. Korff, G. Lang, and R. Schrader. Two–particle scattering theory for anyons. *J. Math. Phys.*, 40:1831–1869, 1999.

[546] V. Koshmanenko. Singular bilinear forms and selfadjoint extensions of symmetric operators. In *Spectral Analysis of Differential Operators*, pages 37–48, 133. Akad. Nauk Ukrain. SSR Inst. Mat., Kiev, 1980.

[547] V. Koshmanenko. δ–potential perturbations of an n–dimensional Laplace operator. In *The Spectral Theory of Operator-differential Equations (In Russian.)*, pages 70–79, v. Akad. Nauk Ukrain. SSR Inst. Mat., Kiev, 1986.

[548] V. Koshmanenko. Perturbations of selfadjoint operators by singular bilinear forms. *Ukrain. Mat. Zh.*, 41(1):3–19, 134, 1989.

[549] V. Koshmanenko. Singular perturbations defined by forms. In *Applications of Selfadjoint Extensions in Quantum Physics (Dubna, 1987)*, volume 324 of *Lecture Notes in Phys.*, pages 55–66. Springer, Berlin, 1989.

[550] V. Koshmanenko. One–dimensional singular perturbations of selfadjoint operators. In *Methods of Functional Analysis in Problems of Mathematical Physics (In Russian.)*, pages 110–122. Akad. Nauk Ukrain. SSR Inst. Mat., Kiev, 1990.

[551] V. Koshmanenko. Towards the rank–one singular perturbations of selfadjoint operators. *Ukrain. Mat. Zh.*, 43(11):1559–1566, 1991.

[552] V. Koshmanenko. *Singular Bilinear Forms in Perturbations of Selfadjoint Operators*. Naukova Dumka, Kiev, 1993. (English translation will be published by Kluwer).

[553] V. Koshmanenko. Singular Wick monomials in an orthogonally extended Fock space [translation of Boundary value problems for differential equations (In Russian), 78–82, iv, Akad. Nauk Ukrain. SSR, Inst. Mat., Kiev, 1988; MR 90j:81095]. *Selecta Math.*, 13(1):17–25, 1994. Selected translations.

[554] V. Koshmanenko. Singularly perturbed operators. In *Mathematical Results in Quantum Mechanics (Blossin, 1993)*, volume 70 of *Oper. Theory Adv. Appl.*, pages 347–351. Birkhäuser, Basel, 1994.

[555] V. Koshmanenko. Singularly perturbed operators of type $-\Delta + \lambda\delta$. In *Algebraic and Geometric Methods in Mathematical Physics (Kaciveli, 1993)*, volume 19 of *Math. Phys. Stud.*, pages 433–437. Kluwer, Dordrecht, 1996.

[556] V. Koshmanenko and Sh. Ota. On the characteristic properties of singular operators. *Ukraïn. Mat. Zh.*, 48(11):1484–1493, 1996.

[557] V. Koshmanenko and O. V. Samoĭlenko. Singular perturbations of finite rank. Point spectrum. *Ukraïn. Mat. Zh.*, 49(9):1186–1194, 1997.

[558] V. Kostrykin and R. Schrader. Scattering theory approach to random Schrödinger operators in one dimension. *Rev. Math. Phys.*, 11(2):187–242, 1999.

[559] Th. Kovar and Ph. A. Martin. Scattering with a periodically kicked interaction and cyclic states. *J. Phys. A: Math. Gen.*, 31:385–396, 1998.

[560] M. A. Krasnosel'skiĭ. On the deficiency numbers of closed operators. *Doklady Akad. Nauk SSSR (N. S.)*, 56:559–561, 1947.

[561] M. A. Krasnosel'skiĭ. On the extension of Hermitian operators with a non-dense domain of definition. *Doklady Akad. Nauk SSSR (N.S.)*, 59:13–16, 1948.

[562] M. A. Krasnosel'skiĭ. On self–adjoint extensions of Hermitian operators. *Ukraïn. Mat. Žurnal*, 1(1):21–38, 1949.

[563] M. G. Krein. Sur les développements des fonctions arbitraires en séries de fonctions fondamentales d'un problème aux limites quelconque. *Recueil Mathématique (Математический Сборник)*, 2 (44):923–933, 1937. (In French, Russian summary.)

[564] M. G. Krein. Sur les opérateurs différentiels autoadjoints et leurs fonctions de Green symétriques. *Recueil Mathématique (Математический Сборник)*, 2 (44):1023–1072, 1937. (In French, Russian summary.)

[565] M. G Krein. On a remarkable class of Hermitian operators. *Comptes Rendue (Doklady) Acad. Sci. URSS (N.S.)*, 44:175–179, 1944.

[566] M. G. Krein. On Hermitian operators whose deficiency indices are 1. *Comptes Rendus (Doklady) Acad. Sci. URSS (N.S.)*, 43:323–326, 1944.

[567] M. G. Krein. On Hermitian operators with deficiency indices equal to one. II. *Comptes Rendue (Doklady) Acad. Sci. URSS (N. S.)*, 44:131–134, 1944.

[568] M. G. Krein. On self–adjoint extensions of bounded and semi–bounded Hermitian transformations. *Comptes Rendue (Doklady) Acad. Sci. URSS (N. S.)*, 48:303–306, 1945.

[569] M. G. Krein. Concerning the resolvents of an Hermitian operator with the deficiency–index (m, m). *Comptes Rendue (Doklady) Acad. Sci. URSS (N.S.)*, 52:651–654, 1946.

[570] M. G. Krein. The theory of self–adjoint extensions of semi–bounded Hermitian transformations and its applications. I. *Rec. Math. [Mat. Sbornik] N.S.*, 20(62):431–495, 1947.

[571] M. G. Krein. The theory of self–adjoint extensions of semi–bounded Hermitian transformations and its applications. II. *Mat. Sbornik N.S.*, 21(63):365–404, 1947.

[572] M. G. Krein. The fundamental propositions of the theory of representations of Hermitian operators with deficiency index (m, m). *Ukrain. Mat. Žurnal*, 1(2):3–66, 1949.

[573] M. G. Krein. On determination of the potential of a particle from its S–function. *Dokl. Akad. Nauk SSSR (N.S.)*, 105:433–436, 1955.

[574] M. G. Krein. Integral representation of a continuous Hermitian–indefinite function with a finite number of negative squares. *Dokl. Akad. Nauk SSSR*, 125:31–34, 1959.

[575] M. G. Krein. The theory of self–adjoint extensions of semi–bounded Hermitian transformations and its applications. I. *Magyar Tud. Akad. Mat. Fiz. Oszt. Közl.*, 18:273–314, 1968.

[576] M. G. Krein. The theory of self–adjoint extensions of semi–bounded Hermitian transformations and its applications. II. *Magyar Tud. Akad. Mat. Fiz. Oszt. Közl.*, 19:131–188, 1970.

[577] M. G. Krein and M. A. Krasnosel'skiĭ. Stability of the index of an unbounded operator. *Mat. Sbornik N.S.*, 30(72):219–224, 1952.

[578] M. G. Krein and G. K. Langer. On the spectral function of a self–adjoint operator in a space with indefinite metric. *Dokl. Akad. Nauk SSSR*, 152:39–42, 1963.

[579] M. G. Krein and G. K. Langer. The defect subspaces and generalized resolvents of a Hermitian operator in the space π_κ. *Funkcional. Anal. i Priložen*, 5(3):54–69, 1971.

[580] M. G. Krein and G. K. Langer. The defect subspaces and generalized resolvents of a Hermitian operator in the space π_κ. *Funkcional. Anal. i Priložen*, 5(2):59–71, 1971.

[581] M. G. Krein and H. Langer. Über die verallgemeinerten Resolventen und die charakteristische Funktion eines isometrischen Operators im Raume π_κ. In *Hilbert Space Operators and Operator Algebras (Proceedings International Conference, Tihany, 1970)*, pages 353–399. Colloq. Math. Soc. János Bolyai, 5. North–Holland, Amsterdam, 1972.

[582] M. G. Krein and H. Langer. Über die Q–Funktion eines π–hermiteschen Operators im Raume π_κ. *Acta Sci. Math. (Szeged)*, 34:191–230, 1973.

[583] M. G. Krein and I. E. Ovčarenko. The generalized resolvents and resolvent matrices of positive Hermitian operators. *Dokl. Akad. Nauk SSSR*, 231(5):1063–1066, 1976.

[584] M. G. Krein and I. Ē. Ovčarenko. On the theory of generalized resolvents of nondensely defined Hermitian contractions. *Dokl. Akad. Nauk Ukrain. SSR Ser. A*, 10:881–885, 960, 1976.

[585] M. G. Krein and I. E. Ovčarenko. Q–functions and sc–resolvents of nondensely defined Hermitian contractions. *Sibirsk. Mat. Ž.*, 18(5):1032–1056, 1206, 1977.

[586] M. G. Krein and I. E. Ovčarenko. Inverse problems for Q–functions and resolvent matrices of positive Hermitian operators. *Dokl. Akad. Nauk SSSR*, 242(3):521–524, 1978.

[587] M. G. Krein and Š. N. Saakjan. Certain new results in the theory of resolvents of Hermitian operators. *Dokl. Akad. Nauk SSSR*, 169:1269–1272, 1966.

[588] M. G. Krein and Š. N. Saakjan. The resolvent matrix of a Hermitian operator and the characteristic functions connected with it. *Funkcional. Anal. i Priložen*, 4(3):103–104, 1970.

[589] R. de L. Kronig and W. G. Penney. Quantum mechanics of electrons in crystal lattices. *Proc. Soc. London*, 130:499–513, 1931.

[590] V. I. Krylov. Über Funktionen, die in der Halbebene regulär sind. *Recueil Mathématique Мат. Сборник*, 6(48):94–138, 1939. (In Russian, German summary.)

[591] Yu. A. Kuperin. Faddeev equations for three composite particles. In *Applications of Selfadjoint Extensions in Quantum Physics (Dubna, 1987)*, volume 324 of *Lecture Notes in Phys.*, pages 117–137. Springer, Berlin, 1989.

[592] Yu. A. Kuperin, P. Kurasov, Yu. Melnikov, and S. P. Merkuriev. Connections and effective S–matrix in triangle representation for quantum scattering. *Ann. Physics*, 205(2):330–361, 1991.

[593] Yu. A. Kuperin, S. B. Levin, and Yu. Melnikov. Generalized string–flip model for quantum cluster scattering. *J. Math. Phys.*, 35(1):71–95, 1994.

[594] Yu. A. Kuperin, S. B. Levin, and Yu. Melnikov. Quantum scattering with spin–spin interaction in a string–flip model. *Il Nuovo Cimento*, 107 A(12):2823–2835, 1994.

[595] Yu. A. Kuperin, S. B. Levin, Yu. Melnikov, and E. A. Yarevsky. $\bar{p}n$ scattering with annihilation channel in extended Hilbert space model. *Few–Body Systems Suppl.*, 8:462–467, 1995.

[596] Yu. A. Kuperin, S. B. Levin, Yu. Melnikov, and E. A. Yarevsky. Numerical analysis of the $\bar{p}pn$ system with continuous set of resonances in annihilation channel. *Phys. Atomic Nuclei*, 59(9):1648–1649, 1996.

[597] Yu. A. Kuperin, S. B. Levin, Yu. Melnikov, and E. A. Yarevsky. Application of extension theory to antiproton–nucleon systems with continuous set of resonances in annihilation channel. *Comput. Math. Appl.*, 34(5–6):559–570, 1997. Computational Tools of Complex Systems, II.

[598] Yu. A. Kuperin, K. Makarov, and Yu. Melnikov. Coulomb two–body problem with internal structure. *Theor. Math. Phys.*, 74:73–79, 1988.

[599] Yu. A. Kuperin, K. Makarov, and Yu. Melnikov. The operator method for excluding forbidden states. *Vestnik Leningrad. Univ. Fiz. Khim.*, vyp. 3:78–81, 124, 1989.

[600] Yu. A. Kuperin, K. Makarov, and Yu. Melnikov. A resonating–group model with extended channel spaces. In *Applications of Selfadjoint Extensions in Quantum Physics (Dubna, 1987)*, volume 324 of *Lecture Notes in Phys.*, pages 146–159. Springer, Berlin, 1989.

[601] Yu. A. Kuperin, K. Makarov, Yu. Melnikov, S. P. Merkurev, and E. A. Yarevsky. Faddeev's method for three–particle systems with additional degrees of freedom. *Ukrain. Fiz. Zh.*, 34(11):1613–1618, 1755, 1989.

[602] Yu. A. Kuperin, K. Makarov, S. P. Merkurev, and A. K. Motovilov. Algebraic theory of extensions in the quantum few–body problem for particles with internal structure. *Soviet J. Nuclear Phys.*, 48(2):224–231 (1989), 1988.

[603] Yu. A. Kuperin, K. Makarov, S. P. Merkurev, A. K. Motovilov, and B. S. Pavlov. The quantum problem of several particles with internal structure. I. The two–body problem. *Teoret. Mat. Fiz.*, 75(3):431–444, 1988.

[604] Yu. A. Kuperin, K. Makarov, S. P. Merkurev, A. K. Motovilov, and B. S. Pavlov. The quantum problem of several particles with internal structure. II. The three–body problem. *Teoret. Mat. Fiz.*, 76(2):242–260, 1988.

[605] Yu. A. Kuperin, K. Makarov, S. P. Merkuriev, A. K. Motovilov, and B. S. Pavlov. The model for the hadron resonance scattering at low and intermediate energies. In *Proceedings of Symposium on Nucleon–Nucleon and Hadron–Nucleon interactions at Intermediate Energies (Gatchina, April 21–23, 1986)*, pages 511–515. Leningrad: LINP, 1986. (in Russain).

[606] Yu. A. Kuperin, K. Makarov, S. P. Merkuriev, A. K. Motovilov, and B. S. Pavlov. Quantum theory of scattering by energy–dependent potentials. In *Properties of Few–Body and Quark–Hadronic systems, Part II (Proceedings of III All–Union School, Palanga, 30 September–9 October, 1986)*. Vilnuous, 1986. (In Russian.)

[607] Yu. A. Kuperin, K. Makarov, S. P. Merkuriev, A. K. Motovilov, and B. S. Pavlov. Extended Hilbert space approach to few–body problems. *J. Math. Phys.*, 31(7):1681–1690, 1990.

[608] Yu. A. Kuperin, K. Makarov, and B. S. Pavlov. One–dimensional model of three–particle resonances. *Teoret. Mat. Fiz.*, 63(1):78–87, 1985.

[609] Yu. A. Kuperin, K. Makarov, and B. S. Pavlov. Model of resonance scattering of compound particles. *Teoret. Mat. Fiz.*, 69(1):100–114, 1986.

[610] Yu. A. Kuperin, K. Makarov, and B. S. Pavlov. Scattering on a dynamic quark bag. *Vestnik Leningrad. Univ. Fiz. Khim.*, vyp. 4:60–62, 108, 1987.

[611] Yu. A. Kuperin, K. Makarov, and B. S. Pavlov. An exactly solvable model of a crystal with nonpoint atoms. In *Applications of Selfadjoint Extensions in Quantum Physics (Dubna, 1987)*, volume 324 of *Lecture Notes in Phys.*, pages 267–273. Springer, Berlin, 1989.

[612] Yu. A. Kuperin, K. Makarov, and B. S. Pavlov. An extensions theory setting for scattering by breathing bag. *J. Math. Phys.*, 31(1):199–201, 1990.

[613] Yu. A. Kuperin and Yu. Melnikov. Curved quantum waveguides: geometric setting for scattering problem. In *Topological Phases in Quantum Theory (Dubna, 1988)*, pages 146–172. World Scientific, Teaneck, NJ, 1989.

[614] Yu. A. Kuperin and Yu. Melnikov. Quantum scattering in gauge fields of adiabatic representations. *Mat. Sb.*, 182(2):236–282, 1991.

[615] Yu. A. Kuperin and Yu. Melnikov. Two–body resonance scattering and annihilation of composite charged particles. *J. Math. Phys.*, 33(8):2795–2807, 1992.

[616] Yu. A. Kuperin, Yu. Melnikov, and S. P. Merkurev. Quantum scattering for effective nonabelian interactions. *Trudy Mat. Inst. Steklov.*, 200:205–212, 1991. Number Theory, Algebra, Mathematical Analysis and their Applications (In Russian.)

[617] Yu. A. Kuperin, Yu. Melnikov, and S. P. Merkuriev. Total S–matrix and adi-
      abatic amplitudes for the three–body problem. *Lett. Math. Phys.*, 21(2):97–
      103, 1991.

[618] Yu. A. Kuperin, Yu. Melnikov, and A. K. Motovilov. Effects of the internal
      structure of alpha–particle in three–body problem of d(alpha)–scattering.
      Sov.J.Nucl.Phys., 52:433–446, 1990.

[619] Yu. A. Kuperin, Yu. Melnikov, and A. K. Motovilov. Three–body deuteron–
      nucleus scattering with extra resonance channels. *Il Nuovo Cimento*, 104
      A(3):299–324, 1991.

[620] Yu. A. Kuperin, Yu. Melnikov, and B. S. Pavlov. Quantum scattering prob-
      lem in triangle representation and induced gauge fields. In *Schrödinger Op-
      erators, Standard and Nonstandard (Dubna, 1988)*, pages 294–319. World
      Scientific, Teaneck, NJ, 1989.

[621] Yu. A. Kuperin and S. P. Merkurev. Selfadjoint extensions and scattering
      theory for several–body systems. In *Spectral Theory of Operators (Novgorod,
      1989)*, volume 150 of *Amer. Math. Soc. Transl. Ser. 2*, pages 141–176. Amer.
      Math. Soc., Providence, RI, 1992.

[622] Yu. A. Kuperin and B. S. Pavlov. Three particles in a lattice: a model
      of interaction and dynamical equations. In *Rigorous Results in Quantum
      Dynamics (Liblice, 1990)*, pages 152–163. World Scientific, River Edge, NJ,
      1991.

[623] Yu. A. Kuperin, B. S. Pavlov, G. E. Rudin, and S. I. Vinitsky. Spectral
      geometry: two exactly solvable models. *Phys. Lett. A*, 194(1–2):59–63, 1994.

[624] Yu. A. Kuperin, R. V. Romanov, and G. E. Rudin. Scattering on the hyper-
      bolic plane in the Aharonov–Bohm gauge field. *Lett. Math. Phys.*, 31(4):271–
      278, 1994.

[625] Yu. A. Kuperin and E. A. Yarevsky. Currents and the extensions theory. In
      Order, Disorder and Chaos in Quantum Systems (Dubna, 1989), volume 46
      of *Oper. Theory Adv. Appl.*, pages 229–233. Birkhäuser, Basel, 1990.

[626] Yu. A. Kuperin and E. A. Yarevsky. Internal degrees of freedom in low–
      energy nd–scattering. In *Proceedings of the International Workshop Mathe-
      matical Aspects of Scattering Theory and Applications, St.Petersburg*, 1991.

[627] P. Kurasov. A one–electron model of a long molecule. *Vestnik Leningrad.
      Univ. Fiz. Khim.*, vyp. 3:8–15, 122, 1989.

[628] P. Kurasov. Three one–dimensional bosons with an internal structure. In
      Schrödinger Operators, Standard and Nonstandard (Dubna, 1988), pages
      166–188. World Scientific, Teaneck, NJ, 1989.

[629] P. Kurasov. Zero–range potentials with internal structures and the inverse scattering problem. *Lett. Math. Phys.*, 25(4):287–297, 1992.

[630] P. Kurasov. On the inverse scattering problem for rational reflection coefficients. In *Inverse problems in Mathematical Physics (Saariselkä, 1992)*, volume 422 of *Lecture Notes in Phys.*, pages 126–133. Springer, Berlin, 1993.

[631] P. Kurasov. Distribution theory for discontinuous test functions and differential operators with generalized coefficients. *J. Math. Anal. Appl.*, 201(1):297–323, 1996.

[632] P. Kurasov. Inverse scattering problem on the half–line and positon solutions of the KdV equation. *J. Tech. Phys.*, 37(3–4):503–507, 1996. International Conference on Nonlinear Dynamics, Chaotic and Complex Systems (Zakopane, 1995).

[633] P. Kurasov. On the Coulomb potentials in one dimension. *J. Phys. A*, 29(8):1767–1771, 1996.

[634] P. Kurasov. Scattering matrices with finite phase shift and the inverse scattering problem. *Inverse Problems*, 12(3):295–307, 1996.

[635] P. Kurasov. Energy dependent boundary conditions and the few–body scattering problem. *Rev. Math. Phys.*, 9(7):853–906, 1997.

[636] P. Kurasov and J. Boman. Finite rank singular perturbations and distributions with discontinuous test functions. *Proc. Amer. Math. Soc.*, 126(6):1673–1683, 1998.

[637] P. Kurasov and S.T. Kuroda. Krein's formula in perturbation theory. Technical report, Gakushuin Univ., Japan, 1999.

[638] P. Kurasov and Yu. Melnikov. Three–body scattering in one dimension: a geometric version of an adiabatic approach. In *Topological Phases in Quantum Theory (Dubna, 1988)*, pages 204–217. World Scientific, Teaneck, NJ, 1989.

[639] P. Kurasov and K. Packalén. Inverse scattering transformation for positons. *J. Phys. A*, 32(7):1269–1278, 1999.

[640] P. Kurasov and B. Pavlov. Generalized perturbations and operator relations. Technical report, 1999.

[641] P. Kurasov and B. S. Pavlov. An electron in a homogeneous crystal of point–like atoms with internal structure. II. *Teoret. Mat. Fiz.*, 74(1):82–93, 1988.

[642] P. Kurasov and B. S. Pavlov. Localization effects in nonhomogeneous dielectrics. In *Order, Disorder and Chaos in Quantum Systems (Dubna, 1989)*, volume 46 of *Oper. Theory Adv. Appl.*, pages 307–313. Birkhäuser, Basel, 1990.

[643] P. Kurasov and B.S. Pavlov. Few–body Krein's formula. Technical report, 1998. (To be published in Proc. of Krein's conference.)

[644] P. Kurasov and K. Watanabe. On Rank One $\mathcal{H}_{-3}$-perturbations of Positive Self–adjoint Operators. Technical report, Stockholm Univ., Sweden, 1999.

[645] S. T. Kuroda. Perturbation of continuous spectra by unbounded operators. I. *J. Math. Soc. Japan*, 11:246–262, 1959.

[646] S. T. Kuroda. Perturbation of continuous spectra by unbounded operators. II. *J. Math. Soc. Japan*, 12:243–257, 1960.

[647] S. T. Kuroda. On a generalization of the Weinstein–Aronszajn formula and the infinite determinant. *Sci. Papers Coll. Gen. Ed. Univ. Tokyo*, 11:1–12, 1961.

[648] S. T. Kuroda. Finite–dimensional perturbation and a representaion of scattering operator. *Pacific J. Math.*, 13:1305–1318, 1963.

[649] S. T. Kuroda. On a stationary approach to scattering problem. *Bull. Amer. Math. Soc.*, 70:556–560, 1964.

[650] S. T. Kuroda. An abstract stationary approach to perturbation of continuous spectra and scattering theory. *J. Analyse Math.*, 20:57–117, 1967.

[651] A. A. Kvitsinsky, Yu. A. Kuperin, S. P. Merkuriev, A. K. Motovilov, and S. L. Yakovlev. The n – body quantum problem in configuration space. *Sov. J. Part. Nucl.*, 17(2):113–136, 1986.

[652] N. N. Lebedev. On the representation of arbitrary function by an integral involving Macdonald functions of complex order. *Doklady Akad. Nauk SSSR (N. S.)*, 58:1007–1010, 1947.

[653] T. D. Lee, K. Huang, and C. N. Yang. Eigenvalues and eigenfunctions of a Bose system of hard spheres and its low–temperature properties. *Phys. Rev. (2)*, 106:1135–1145, 1957.

[654] S. B. Levin and E. A. Yarevsky. $\bar{p}p$ and $\bar{p}n$ scattering with annihilation channel. *Hyperfine Interactions*, 101/102:511–515, 1996.

[655] B. Z. Li, S. Q. Lu, and F. C. Pu. Completeness of Bethe type eigenfunctions for the 1D N–body system with δ–function interactions. *Phys. Lett. A*, 110(2):65–67, 1985.

[656] E.H. Lieb. Some of the early history of exactly soluble models. In *Proceedings of the Conference on Exactly Soluble Models in Statistical Mechanics: Historical Perspectives and Current Status (Boston, MA, 1996)*, volume 11, pages 3–10, 1997.

[657] E.H. Lieb and W. Liniger. Simplified approach to the ground-state energy of an imperfect Bose gas. III. Application to the one-dimensional model. *Phys. Rev. (2)*, 134:A312–A315, 1964.

[658] J. C. Lin. Inverse scattering for the mixed spectrum of δ potentials. *J. Math. Phys.*, 29(10):2254–2255, 1988.

[659] K. Lipszyc. One–dimensional model of the rearrangement and dissociation processes. *Acta Phys. Polon. A*, 42:571–585, 1972.

[660] K. Lipszyc. One–dimensional model of the rearrangement and dissociation processes. Probability amplitudes and cross–sections. *Acta Phys. Polon. A*, 44:115–137, 1973.

[661] K. Lipszyc. One–dimensional model of the rearrangement process and the Faddeev equations. *J. Math. Phys.*, 15:133–138, 1974.

[662] K. Lipszyc. One–dimensional model of the rearrangement and dissociation processes and the Faddeev equations. II. *Phys. Rev. D (3)*, 11:1649–1661, 1975.

[663] K. Lipszyc. On the application of the Sommerfeld–Maluzhinetz transformation to some one–dimensional three–particle problems. *J. Math. Phys.*, 21(5):1092–1102, 1980.

[664] M. S. Livšic. *Операторы, колебания, волны. Открытые системы.* Nauka, Moscow, 1966. (In Russian.).

[665] M. S. Livšic. *Operators, Oscillations, Waves (Open Systems)*. American Mathematical Society, Providence, R.I., 1973. Translated from the Russian by Scripta Technica, Ltd. English translation edited by R. Herden. Translations of Mathematical Monographs, Vol. 34.

[666] V. È. Lyantse and O. G. Storozh. *Методы теории неограниченных операторов.* Naukova Dumka, Kiev, 1983. (In Russian.).

[667] M. Maioli and A. Sacchetti. One–dimensional many point interactions and stability of eigenvalues. *Boll. Un. Mat. Ital. A (7)*, 5(3):363–372, 1991.

[668] M. Maioli and A. Sacchetti. Absence of the absolutely continuous spectrum for Stark–Bloch operators with strongly singular periodic potentials. *J. Phys. A*, 28(4):1101–1106, 1995.

[669] K. Makarov. On delta–like interactions with internal structure and semi-bounded from below three–body Hamiltonian. *Preprint FUB/HEP*, 88–13:1–16, 1988.

[670] K. Makarov. Energy–dependent interactions and the extension theory. In *Applications of Selfadjoint Extensions in Quantum Physics (Dubna, 1987)*, volume 324 of *Lecture Notes in Phys.*, pages 28–39. Springer, Berlin, 1989.

[671] K. Makarov. Semiboundedness of the energy operator of a system of three particles with paired interactions of δ–function type. *Algebra i Analiz*, 4(5):155–171, 1992.

[672] K. Makarov and S. E. Cheremshantsev. Point interactions with internal structure as a limit of separable potentials. *Zap. Nauchn. Sem. Leningrad. Otdel. Mat. Inst. Steklov. (LOMI)*, 182(Kraev. Zadachi Mat. Fiz. i Smezh. Voprosy Teor. Funktsii. 21):113–122, 172, 1990.

[673] K. Makarov and V. V. Melezhik. Two sides of a coin: the Efimov effect and collapse in a three–body system with point interactions. I. *Teoret. Mat. Fiz.*, 107(3):415–432, 1996.

[674] K. Makarov, V. V. Melezhik, and A. K. Motovilov. Point interactions in the problem of three quantum particles with internal structure. *Teoret. Mat. Fiz.*, 102(2):258–282, 1995.

[675] K. Makarov and B. S. Pavlov. Quantum scattering on a Cantor bar. *J. Math. Phys.*, 35(4):1522–1531, 1994.

[676] M. M. Malamud. An approach to the theory of extensions of a nondensely defined Hermitian operator. *Dokl. Akad. Nauk Ukrain. SSR Ser. A*, 3:20–25, 87, 1990.

[677] M. M. Malamud. Boundary value problems for Hermitian operators with gaps. *Dokl. Akad. Nauk SSSR*, 313(6):1335–1340, 1990.

[678] M. M. Malamud. On a formula for the generalized resolvents of a non-densely defined Hermitian operator. *Ukraïn. Mat. Zh.*, 44(12):1658–1688, 1992.

[679] M. M. Malamud. Some classes of extensions of a Hermitian operator with lacunae. *Ukraïn. Mat. Zh.*, 44(2):215–233, 1992.

[680] G. D. Maljužinec. Connection between inverse formulas for the Sommerfeld integral and the formulas of Kantorovich–Lebedev. *Dokl. Acad. Nauk SSSR*, 119:49–51, 1958.

[681] G. D. Maljužinec. Inversion formula for Sommerfeld integral. *Dokl. Acad. Nauk SSSR*, 118:1099–1102, 1958.

[682] G. D. Maljužinec and A. A. Tuzilin. Diffraction of a plane acoustic wave on a thin semi-infinite elastic plate. *Ž. Vyčisl. Mat. i Mat. Fiz.*, 10:1210–1227, 1970.

[683] G. D. Malyuzhinets. Sound scattering by nonuniformities in a layer of discontinuity in the sea. *Soviet Physics. Acoust.*, 5(5):68–74 (70–76 Akust. Ž.), 1959.

[684] E. B. Manoukian. Explicit derivation of the propagator for a Dirac delta potential. *J. Phys. A*, 22(1):67–70, 1989.

[685] C. Manuel and R. Tarrach. Contact interactions of anyons. *Phys. Lett. B*, 268(2):222–226, 1991.

[686] C. Manuel and R. Tarrach. Contact interactions and Dirac anyons. *Phys. Lett. B*, 301:72–76, 1993.

[687] C. Manuel and R. Tarrach. Perturbative renormalization in quantum mechanics. *Phys. Lett. B*, 328:113–118, 1994.

[688] V. A. Marchenko. On reconstruction of the potential energy from phases of the scattered waves. *Dokl. Akad. Nauk SSSR (N.S.)*, 104:695–698, 1955.

[689] V.A. Marchenko. *Sturm–Liouville Operators and Applications*. Birkhäuser, Basel, 1986. Translated from the 1977 Russian original by A. Iacob.

[690] N. Markovska, J. Pop-Jordanov, and E. A. Soloviev. Canalized states in two-dimensional quantum model of thin films. *J.Phys.A:Math.Gen.*, 28:L201–L206, 1995.

[691] V. B. Matveev. Erratum: "Generalized Wronskian formula for solutions of the KdV equations: first applications". *Phys. Lett. A*, 168(5–6):463, 1992.

[692] V. B. Matveev. Generalized Wronskian formula for solutions of the KdV equations: first applications. *Phys. Lett. A*, 166(3–4):205–208, 1992.

[693] V. B. Matveev. Positon–positon and soliton–positon collisions: KdV case. *Phys. Lett. A*, 166(3–4):209–212, 1992.

[694] V.B. Matveev. Asymptotics of the multipositon–soliton τ function of the Korteweg–de Vries equation and supertransparency. *J. Math. Phys.*, 35(6):2955–2970, 1994.

[695] K. Maurin. *General Eigenfunction Expansions and Unitary Representations of Topological Groups*. PWN–Polish Scientific, Warsaw, 1968. Monografie Matematyczne, Tom 48.

[696] K. Maurin. A remark on the Berezanski version of the spectral theorem. *Studia Math.*, 34:165–167, 1970.

[697] K. Maurin. *Methods of Hilbert Spaces*. PWN–Polish Scientific, Warsaw, 1972. Second edition revised, Translated from the Polish by Andrzej Alexiewicz and Wacław Zawadowski, Monografie Matematyczne, Vol. 45. [Mathematical Monographs, Vol. 45].

[698] K. Maurin and L. Maurin. Spektrum und verallgemeinerte Eigenelemente separierbarer Operatoren. *Bull. Acad. Polon. Sci. Sér. Sci. Math. Astronom. Phys.*, 10:343–347, 1962.

[699] L. Maurin and K. Maurin. Spektraltheorie separierbarer Operatoren. *Studia Math.*, 23:1–29, 1963.

[700] J. B. McGuire. Distinguishable particles in delta interaction. In *Quantum Field Theory, Statistical Mechanics, Quantum Groups and Topology (Coral Gables, FL, 1991)*, pages 193–199. World Scientific, River Edge, NJ, 1992.

[701] J. B. McGuire. Study of exactly soluble one–dimensional N–body problems. *J. Math. Phys.*, 5:622–636, 1964.

[702] J. B. McGuire. Interacting fermions in one dimension. I. Repulsive potential. *J. Math. Phys.*, 6:432–439, 1965.

[703] J. B. McGuire. Interacting fermions in one dimension. II. Attractive potential. *J. Math. Phys.*, 7:123–132, 1966.

[704] J. B. McGuire. Delta interaction on a ring: the topological Bethe Ansatz. In *Proceedings of the Conference on Exactly Soluble Models in Statistical Mechanics: Historical Perspectives and Current Status (Boston, MA, 1996)*, volume 11, pages 161–168, 1997.

[705] J. B. McGuire and C. A. Hurst. The scattering of three impenetrable particles in one dimension. *J. Math. Phys.*, 13:1595–1607, 1972.

[706] J. B. McGuire and C. A. Hurst. Three interacting particles in one dimension: an algebraic approach. *J. Math. Phys.*, 29(1):155–168, 1988.

[707] A. Melin. Operator methods for inverse scattering on the real line. *Comm. Partial Differential Equations*, 10(7):677–766, 1985.

[708] A. M. Melnikov and R. A. Minlos. On the pointlike interaction of three different particles. In *Many–particle Hamiltonians: Spectra and Scattering*, volume 5 of *Adv. Soviet Math.*, pages 99–112. Amer. Math. Soc., Providence, RI, 1991.

[709] A. M. Melnikov and R. A. Minlos. Point interaction of three different particles. *Vestnik Moskov. Univ. Ser. I Mat. Mekh.*, 3:3–6, 110, 1991.

[710] Yu. Melnikov. Scattering on graphs as a quantum few–body problem. Technical report, IPRT, 09–93, St.Petersburg, 1993.

[711] Yu. Melnikov. On representation of functionals $\delta_z^{(m)}$. In *Generalized Functions, Operator Theory and Dynamical Systems* Eds. I.Antoniou and G.Lumer, Pitman Research Notes in Mathematics, Addison Wesley Longman, 1998.

[712] Yu. Melnikov and B. S. Pavlov. Two–body scattering on a graph and application to simple nanoelectronic devices. *J. Math. Phys.*, 36(6):2813–2825, 1995.

[713] R. Mennicken and A. K. Motovilov. Operator interpretation of resonances arising in spectral problems for 2×2 operator matrices. *Math. Nachr.*, 201:117–181, 1999.

[714] R. Mennicken and A.A. Shkalikov. Spectral decomposition of symmetric operator matrices. *Math. Nachr.*, 179:259–273, 1996.

[715] S. P. Merkuriev and A. K. Motovilov. Faddeev equations for simple–layer potential density. *Lett. Math. Phys.*, 7(6):497–503, 1983.

[716] S. P. Merkuriev and A. K. Motovilov. Scattering problem for three hard spheres and potential theory. In *Theory of Quantum Systems with Strong Interaction*, pages 95–116. Kalinin University Press, 1983.

[717] S. P. Merkuriev, A. K. Motovilov, and S. L. Yakovlev. The N–body problem in the boundary condition model, and quasipotentials. *Teoret. Mat. Fiz.*, 94(3):435–447, 1993.

[718] V. A. Mikhailets. Point interactions on the line. In *Proceedings of the XXV Symposium on Mathematical Physics (Toruń, 1992)*, volume 33, pages 131–135, 1993.

[719] V. A. Mikhailets. A criterion for the discreteness of the spectrum of the one–dimensional Schrödinger operator with δ–self–interactions. *Funktsional. Anal. i Prilozhen.*, 28(4):85–87, 1994.

[720] V. A. Mikhailets. A one–dimensional Schrödinger operator with point interactions. *Dokl. Akad. Nauk*, 335(4):421–423, 1994.

[721] V. A. Mikhailets. δ–interactions on the line. In *Spectral and Evolutional Problems, Vol. 4 (Sevastopol, 1993)*, pages 6–10. Simferopol. Gos. Univ., Simferopol, 1995.

[722] V. A. Mikhailets. Spectral properties of the one–dimensional Schrödinger operator with point intersections. In *Proceedings of the XXVII Symposium on Mathematical Physics (Toruń, 1994)*, volume 36, pages 495–500, 1995.

[723] V. A. Mikhailets. On the Schrödinger operator with point δ'-interactions. *Dokl. Akad. Nauk*, 348(6):727–730, 1996.

[724] V. A. Mikhailets. The structure of the continuous spectrum of the one-dimensional Schrödinger operator with point interactions. *Funktsional. Anal. i Prilozhen.*, 30(2):90–93, 1996.

[725] B. Milek and P. Šeba. Singular continuous quasi–energy spectrum in the kicked rotator with separable perturbation: onset of quantum chaos? In *Order, Disorder and Chaos in Quantum Systems (Dubna, 1989)*, volume 46 of *Oper. Theory Adv. Appl.*, pages 279–299. Birkhäuser, Basel, 1990.

[726] B. Milek and P. Šeba. Singular continuous quasi–energy spectrum in the kicked rotator with separable perturbation: possibility of the onset of quantum chaos. *Phys. Rev. A (3)*, 42(6):3213–3220, 1990.

[727] N. Minami. Level clustering in a one–dimensional finite system. *Progr. Theoret. Phys. Suppl.*, 116:359–368, 1994. Quantum and Chaos: How Incompatible? (Kyoto, 1993).

[728] R. A. Minlos. On the point interaction of three particles. In *Applications of Selfadjoint Extensions in Quantum Physics (Dubna, 1987)*, volume 324 of *Lecture Notes in Phys.*, pages 138–145. Springer, Berlin, 1989.

[729] R. A. Minlos. On pointlike interaction between n fermions and another particle. In *Proceedings of the Workshop on Singular Schroedinger Operators, Trieste, 29 September–1 October 1994*. Eds. G.F.Dell'Antonio, R.Figari and A.Teta. SISSA, Trieste, 1995.

[730] R. A. Minlos and L. D. Faddeev. Comment on the problem of three particles with point interactions. *Soviet Physics JETP*, 14:1315–1316, 1962.

[731] R. A. Minlos and L. D. Faddeev. On the point interaction for a three–particle system in quantum mechanics. *Soviet Physics Dokl.*, 6:1072–1074, 1962.

[732] R. A. Minlos and M. Kh. Shermatov. Point interaction of three particles. *Vestnik Moskov. Univ. Ser. I Mat. Mekh.*, 6:7–14, 97, 1989.

[733] T. Mizusaki, N. Yoshinaga, T. Shigehara, and T. Cheon. Chaos and symmetry in the interacting boson model. *Phys. Lett. B*, 269(1–2):6–12, 1991.

[734] I. M. Mladenov. On the Saxon–Hutner theorem. *C. R. Acad. Bulgare Sci.*, 38(8):993–996, 1985.

[735] I. M. Mladenov. Saxon–Hutner theorem for periodic multilayers. *C. R. Acad. Bulgare Sci.*, 40(10):35–38, 1987.

[736] I. M. Mladenov. An extension of the Saxon–Hutner theorem in the relativistic domain. *Phys. Lett. A*, 137(7–8):313–318, 1989.

[737] I. M. Mladenov. Geometry of the Saxon–Hutner theorem. *Phys. Lett. A*, 230(5–6):245–252, 1997.

[738] I. M. Mladenov. The matrizant and Saxon–Hutner theorem. *J. Phys. A*, 30(5):1689–1694, gg.

[739] Ph. M. Morse and H. Feshbach. *Methods of Theoretical Physics. 2 volumes.* McGraw-Hill, Inc., New York, 1953.

[740] A. K. Motovilov. Differential equations for the wave function components in three hard sphere problem. *Vestnik Leningrad. Univ. Fiz. Khim.*, 22:76–79, 1983. (In Russian.)

[741] A. K. Motovilov. The elimination of an energy from the energy–dependent potential. In *Physics of Elementary Interactions* Eds. Z.Ajduk, S.Pokorski, and A.K.Wròblewski, pages 494–499. World Scientific, 1990.

[742] A. K. Motovilov. Potentials appearing after the removal of an energy–dependence and scattering by them. In *Invited Talks of the International Workshop Mathematical Aspects of the Scattering Theory and Applications. St.Petersburg*, pages 101–108. St. Petersburg University Press, 1991.

[743] A. K. Motovilov. The removal of an energy dependence from the interaction in two-body systems. *J. Math. Phys.*, 32(12):3509–3518, 1991.

[744] A. K. Motovilov. The algebraic version of the theory of extensions for a quantum system with internal structure. *Teoret. Mat. Fiz.*, 97(2):163–181, 1993.

[745] A. K. Motovilov. Exclusion of energy from interactions that depend on energy as a resolvent. *Teoret. Mat. Fiz.*, 104(2):281–303, 1995.

[746] A. K. Motovilov. Removal of the resolvent–like energy dependence from interactions and invariant subspaces of a total Hamiltonian. *J. Math. Phys.*, 36(12):6647–6664, 1995.

[747] A. K. Motovilov. Removal of the resolvent–like dependence on the spectral parameter from interactions. *Zeitschrift für Angewandte Mathematik und Mechanik (ZAMM)*, Special Issue 2:229–232, 1996. (Proceedings of the International Congress on Industrial and Applied Mathematics, Hamburg, July 1995).

[748] J. G. Muga and R. F. Snider. Solvable three boson model with attractive delta function interactions. *quant-phys/9802068*, 1998.

[749] S. N. Naboko. Uniqueness theorems for operator–valued functions with positive imaginary part, and the singular spectrum in the selfadjoint Friedrichs model. *Ark. Mat.*, 25(1):115–140, 1987.

[750] S. N. Naboko. On the nontangential boundary values for operator–valued R–functions in the half–plane. *Algebra i Analiz*, 1(5):197–222, 1989.

[751] S. N. Naboko. On the structure of singularities of operator–valued functions with positive imaginary part. *Funktsional. Anal. i Prilozhen.*, 26(2):1–13, 1991.

[752] H. Nagatani. Scattering theory for solvable models – surface interaction and line interaction. Technical report, Dept. of Math., Gakushuin Univ., Tokyo, 1998. (In preparation.)

[753] H. Nagatani and S. T. Kuroda. $\mathcal{H}_{-2}$–construction and Hamiltonians with interactions of various type. *RIMS Kokyuroku*, 994:155–167, 1997. (In Japanese.)

[754] M. Naimark. Deficiency spaces of the direct product of symmetric operators. *C. R. (Doklady) Acad. Sci. URSS (N.S.)*, 28:209–210, 1940.

[755] M. Naimark. Self–adjoint extensions of the second kind of a symmetric operator. *Bull. Acad. Sci. URSS. Sér. Math. [Izvestià Akad. Nauk SSSR]*, 4:53–104, 1940.

[756] M. Naimark. Direkte Polynome von symmetrischen Operatoren und ihre selbst–adjungierte Fortsetzungen. I. *Rec. Math. [Mat. Sbornik] N.S.*, 9 (51):629–666, 1941.

[757] M. A. Naimark. Spectral functions of a symmetric operator. *Bull. Acad. Sci. URSS. Sér. Math. [Izvestia Akad. Nauk SSSR]*, 4:277–318, 1940.

[758] M. A. Naimark. On spectral functions of a symmetric operator. *Bull. Acad. Sci. URSS. Sér. Math. [Izvestia Akad. Nauk SSSR]*, 7:285–296, 1943.

[759] M. A. Naimark. *Linear Differential Operators*. Gosudarstv. Izdat. Tehn.–Teor. Lit., Moscow, 1954. (In Russian.)

[760] M. A. Naimark. *Linear Differential Operators. Part I: Elementary Theory of Linear Differential Operators*. Frederick Ungar, New York, 1967.

[761] M. A. Naimark. *Linear Differential Operators. Part II: Linear Differential Operators in Hilbert Space*. Frederick Ungar, New York, 1968. With additional material by the author, and a supplement by V. È. Ljance. Translated from the Russian by E. R. Dawson. English translation edited by W. N. Everitt.

[762] M.A. Naimark. On spectral functions of a symmetric operator. *Bull. Acad. Sciences URSS*, 7:285–296, 1943. (In Russian.)

[763] H. Neidhardt. Evolution equations and selfadjoint extensions. In *Applications of Selfadjoint Extensions in Quantum Physics (Dubna, 1987)*, volume 324 of *Lecture Notes in Phys.*, pages 12–27. Springer, Berlin, 1989.

[764] H. Neidhardt and V. Zagrebnov. About regularization and convergence for singular perturbations. In *Rigorous results in quantum dynamics (Liblice, 1990)*, pages 58–62. World Scientific, River Edge, NJ, 1991.

[765] H. Neidhardt and V. Zagrebnov. Regularization and convergence for singular perturbations. *Comm. Math. Phys.*, 149(3):573–586, 1992.

[766] H. Neidhardt and V. Zagrebnov. Singular perturbations, regularization and extension theory. In *Mathematical Results in Quantum Mechanics (Blossin, 1993)*, volume 70 of *Oper. Theory Adv. Appl.*, pages 299–305. Birkhäuser, Basel, 1994.

[767] H. Neidhardt and V. Zagrebnov. Towards the right Hamiltonian for singular perturbations via regularization and extension theory. *Rev. Math. Phys.*, 8(5):715–740, 1996.

[768] H. Neidhardt and V. Zagrebnov. On the right Hamiltonian for singular perturbations: general theory. *Rev. Math. Phys.*, 9(5):609–633, 1997.

[769] H. Neidhardt and V. Zagrebnov. Does each symmetric operator have a stability domain? Technical Report 6, 1998.

[770] H. Neidhardt and V. Zagrebnov. On semibounded restrictions of selfadjoint operators. *Integr. Equ. Oper. Theory*, 31:489–512, 1998.

[771] E. Nelson. Internal set theory: a new approach to nonstandard analysis. *Bull. Amer. Math. Soc.*, 83(6):1165–1198, 1977.

[772] E. Nelson. Zero range forces. In *Advances in Dynamical Systems and Quantum Physics (Capri, 1993)*, pages 201–208. World Scientific, River Edge, NJ, 1995.

[773] L.P. Nizhnik. On the point interaction in quantum mechanics. *Ukr. Mat. Zh.*, 49(11):1557–1560, 1997. (In Russian.)

[774] S. P. Novikov. The Schrödinger operator on graphs, and topology. *Uspekhi Mat. Nauk*, 52(6(318)):177–178, 1997.

[775] S. P. Novikov. Schrödinger operators on graphs and symplectic geometry. 1998. (In preparation.)

[776] S. P. Novikov and I. A. Dynnikov. Discrete spectral symmetries of small-dimensional differential operators and difference operators on regular lattices and two–dimensional manifolds. *Uspekhi Mat. Nauk*, 52(5(317)):175–234, 1997.

[777] V. L. Oleinik. Asymptotic behavior of energy band associated with a negative energy level. *J. Statist. Phys.*, 59(3–4):665–678, 1990.

[778] D. K. Park. Green's function approach to two- and three–dimensional delta-function potentials and application to the spin–1/2 Aharonov–Bohm problem. *J.Math.Phys.*, 36(10):5453–5464, 1995.

[779] D. K. Park. Proper incorporation of the self–adjoint extension method to the Green function formalism: one–dimensional δ'–function potential case. *J. Phys. A: Math. Gen.*, 29:6407–6411, 1996.

[780] L. Pastur and A. Figotin. *Spectra of Random and Almost–periodic Operators*. Springer, Berlin, 1992.

[781] B. S. Pavlov. A model of zero–radius potential with internal structure. *Teoret. Mat. Fiz.*, 59(3):345–353, 1984.

[782] B. S. Pavlov. An electron in a homogeneous crystal of point–like atoms with internal structure. I. *Teoret. Mat. Fiz.*, 72(3):403–415, 1987.

[783] B. S. Pavlov. An explicitly solvable one–dimensional model of electron-phonon scattering. *Vestnik Leningrad. Univ. Fiz. Khim.*, vyp. 2:60–66, 135, 1987.

[784] B. S. Pavlov. The spectral aspect of superconductivity–the pairing of electrons. *Vestnik Leningrad. Univ. Mat. Mekh. Astronom.*, vyp. 3:43–49, 127, 1987.

[785] B. S. Pavlov. The theory of extensions, and explicitly solvable models. *Uspekhi Mat. Nauk*, 42(6(258)):99–131, 247, 1987.

[786] B. S. Pavlov. Boundary conditions on thin manifolds and the semibounded-ness of the three–body Schrödinger operator with point potential. *Mat. Sb. (N.S.)*, 136(178)(2):163–177, 301, 1988.

[787] B. S. Pavlov. Coherent conductance in a random medium. In *Schrödinger Operators, Standard and Nonstandard (Dubna, 1988)*, pages 78–100. World Scientific, Teaneck, NJ, 1989.

[788] B. S. Pavlov. Thin lattices as waveguides. In *Applications of Selfadjoint Extensions in Quantum Physics (Dubna, 1987)*, volume 324 of *Lecture Notes in Phys.*, pages 241–256. Springer, Berlin, 1989.

[789] B. S. Pavlov. Zero–range interactions with an internal structure. In *Applications of Selfadjoint Extensions in Quantum Physics (Dubna, 1987)*, volume 324 of *Lecture Notes in Phys.*, pages 3–11. Springer, Berlin, 1989.

[790] B. S. Pavlov. Splitting of acoustic resonances in domains connected by a thin channel. *New Zealand J. Math.*, 25(2):199–216, 1996.

[791] B. S. Pavlov and P. Kurasov. Surfaces with an internal structure. In *Applications of Selfadjoint Extensions in Quantum Physics (Dubna, 1987)*, volume 324 of *Lecture Notes in Phys.*, pages 203–217. Springer, Berlin, 1989.

[792] B. S. Pavlov and M. A. Pankratov. Electron transport in a quasi–one-dimensional random structure. *J. Math. Phys.*, 33(8):2916–2922, 1992.

[793] B. S. Pavlov and A. A. Pokrovski. An explicitly solvable model of Mössbauer scattering. *Teoret. Mat. Fiz.*, 95(3):439–450, 1993.

[794] B. S. Pavlov and I. Yu. Popov. A model of diffraction by an infinitely narrow crack and the theory of extensions. *Vestnik Leningrad. Univ. Mat. Mekh. Astronom.*, vyp. 4:36–44, 1983.

[795] B. S. Pavlov and I. Yu. Popov. Scattering by resonators with small and point holes. *Vestnik Leningrad. Univ. Mat. Mekh. Astronom.*, vyp. 3:116–118, 1984.

[796] B. S. Pavlov and I. Yu. Popov. A traveling wave in a ring resonator. *Vestnik Leningrad. Univ. Fiz. Khim.*, vyp. 1:99–102, 124, 1985.

[797] B. S. Pavlov and I. Yu. Popov. Surface waves and extension theory. *Vestnik Leningrad. Univ. Mat. Mekh. Astronom.*, vyp. 4:105–107, 126, 1986.

[798] B. S. Pavlov and I. Yu. Popov. An acoustic model of zero–width slits and the hydrodynamic stability of a boundary layer. *Teoret. Mat. Fiz.*, 86(3):391–401, 1991.

[799] B. S. Pavlov and A. E. Ryzhkov. Neutron scattering by a point nucleus in a random magnetic field. I. In *Wave Propagation. Scattering Theory (In Russian.)*, volume 12 of *Probl. Mat. Fiz.*, pages 54–83, 257. Leningrad. Univ., Leningrad, 1987.

[800] B. S. Pavlov and A. E. Ryzhkov. Scattering on a random point potential. In *Applications of Selfadjoint Extensions in Quantum Physics (Dubna, 1987)*, volume 324 of *Lecture Notes in Phys.*, pages 100–114. Springer, Berlin, 1989.

[801] B. S. Pavlov and A. E. Ryzhkov. Neutron scattering by a point nucleus in a random magnetic field. I. In *Wave Propagation. Scattering Theory*, volume 157 of *Amer. Math. Soc. Transl. Ser. 2*, pages 51–77. Amer. Math. Soc., Providence, RI, 1993.

[802] B. S. Pavlov and A. A. Shushkov. The theory of extensions, and null–range potentials with internal structure. *Mat. Sb. (N.S.)*, 137(179)(2):147–183, 271, 1988.

[803] B. S. Pavlov and N. V. Smirnov. A crystal model consisting of potentials of zero radius with inner structure. In *Wave Propagation. Scattering Theory*, volume 157 of *Amer. Math. Soc. Transl. Ser. 2*, pages 151–160. Amer. Math. Soc., Providence, RI, 1993.

[804] B. S. Pavlov and A. V. Strepetov. Simultaneous completeness in the case of a continuous spectrum. *Funktsional. Anal. i Prilozhen.*, 20(1):33–36, 96, 1986.

[805] B. S. Pavlov and A. V. Strepetov. An explicitly solvable model of electron scattering by an inhomogeneity in a thin conductor. *Teoret. Mat. Fiz.*, 90(2):226–232, 1992.

[806] P. Perry, I. M. Sigal, and B. Simon. Spectral analysis of N–body Schrödinger operators. *Ann. of Math. (2)*, 114(3):519–567, 1981.

[807] I. Yu. Popov. A method of treatment of a three–dimensional laminar flow of a viscous fluid. *Godishnik Vissh. Uchebn. Zaved. Tekhn. Fiz.*, 19(1):87–94 (1983), 1982.

[808] I. Yu. Popov. On the serial structure of resonances for scattering by potentials of zero radius. In *Spectral Theory. Wave Processes*, volume 10 of *Probl. Mat. Fiz.*, pages 241–252, 301. Leningrad. Univ., Leningrad, 1982.

[809] I. Yu. Popov. Resonance states for the Schrödinger equation with potentials of zero radius. *Vestnik Leningrad. Univ. Mat. Mekh. Astronom.*, vyp. 1:35–39, 121, 1985.

[810] I. Yu. Popov. "Radiating" edges and scattering by domains with infinitely narrow cracks. In *Differential equations. Scattering theory (In Russian.)*, volume 11 of *Probl. Mat. Fiz.*, pages 222–232, 278. Leningrad. Univ., Leningrad, 1986.

[811] I. Yu. Popov. A slit of zero width and the Dirichlet condition. *Dokl. Akad. Nauk SSSR*, 294(2):330–334, 1987.

[812] I. Yu. Popov. The extension theory and diffraction problems. In *Applications of Selfadjoint Extensions in Quantum Physics (Dubna, 1987)*, volume 324 of *Lecture Notes in Phys.*, pages 218–229. Springer, Berlin, 1989.

[813] I. Yu. Popov. Justification of a model of cracks of zero width for the Dirichlet problem. *Sibirsk. Mat. Zh.*, 30(3):103–108, 218, 1989.

[814] I. Yu. Popov. The stability of boundary layer and the model of suction through a small opening. In *Schrödinger Operators, Standard and Nonstandard (Dubna, 1988)*, pages 78–100. World Scientific, Teaneck, NJ, 1989.

[815] I. Yu. Popov. Construction of an inelastic scatterer in nanoelectronics by the extension–theory methods. In *Order, Disorder and Chaos in Quantum Systems (Dubna, 1989)*, volume 46 of *Oper. Theory Adv. Appl.*, pages 179–182. Birkhäuser, Basel, 1990.

[816] I. Yu. Popov. Justification of a model of zero–width slits for the Neumann problem. *Dokl. Akad. Nauk SSSR*, 313(4):806–811, 1990.

[817] I. Yu. Popov. A model of zero–width slits and the real diffraction problem. In *Order, Disorder and Chaos in Quantum Systems (Dubna, 1989)*, volume 46 of *Oper. Theory Adv. Appl.*, pages 179–182. Birkhäuser, Basel, 1990.

[818] I. Yu. Popov. Integral equations in a model of apertures of zero width. *Leningrad Math. J.*, 2(5):1111–1119, 1991.

[819] I. Yu. Popov. Extension theory and localization of resonances for domains of trap type. *Math. USSR Sbornik*, 71(1):209–234, 1992.

[820] I. Yu. Popov. The extension theory and the opening in semitransparent surface. *J. Math. Phys.*, 33(5):1685–1689, 1992.

[821] I. Yu. Popov. The Helmholtz resonator and operator extension theory in a space with indefinite metric. *Mat. Sb.*, 183(3):3–27, 1992.

[822] I. Yu. Popov. A model of zero–width slits for an aperture in a semitransparent boundary. *Sibirsk. Mat. Zh.*, 33(5):121–126, 223, 1992.

[823] I. Yu. Popov. The resonator with narrow slit and the model based on the operator extensions theory. *J. Math. Phys.*, 33(11):3794–3801, 1992.

[824] I. Yu. Popov. Operator extensions theory and eddies in creeping flow. *Physica Scripta*, 47:682–686, 1993.

[825] I. Yu. Popov. Model of a quantum point as a cavity with semitransparent boundary. *Phys. Solid State*, 36:1046–1048, 1994.

[826] I. Yu. Popov. Hydrodynamic stability and perturbation of the Schrödinger operator. *Lett. Math. Phys.*, 35(2):155–161, 1995.

[827] I. Yu. Popov. Operator approach to three problems of fluid mechanics. *Arch. Mech. (Arch. Mech. Stos.)*, 47(6):1043–1056, 1995.

[828] I. Yu. Popov. The operator extension theory, semitransparent surface and short range potential. *Math. Proc. Cambridge Phil. Soc.*, 118(3):555–563, 1995.

[829] I. Yu. Popov. Stratified flow in electric field, Schrödinger equation and operator extension theory model. *Theor. Math. Phys.*, 103(2):535–542, 1995.

[830] I. Yu. Popov. The extension theory, domains with semitransparent surface and the model of quantum dot. *Proc. R. Soc. Lond. A*, 452:1505–1515, 1996.

[831] I. Yu. Popov. Operator extension theory models for periodic array of quantum dots and double quantum layer in a magnetic field. In *Proceedings of the XXVIII Symposium on Mathematical Physics (Toruń, 1995)*, volume 38, pages 349–356, 1996.

[832] I. Yu. Popov. Solvable model for the transmission of sound through a screen with narrow slit in the presence of a low–Mach–number bias flow. *Rep. Math. Phys.*, 37(3):419–426, 1996.

[833] I. Yu. Popov. Stokeslet and the operator extensions theory. *Rev. Mat. Univ. Complut. Madrid*, 9(1):235–258, 1996.

[834] I. Yu. Popov and S. L. Popova. Zero–width slit model and resonances in mesoscopic systems. *Europhys. Lett.*, 24:373–377, 1993.

[835] I. Yu. Popov and S. L. Popova. Model of zero–width gaps and resonance effects in a quantum waveguide. *Tech. Phys.*, 39:11–15, 1994.

[836] I. Yu. Popov and S. L. Popova. On the mesoscopic gate. *Acta Phys. Polon. A*, 88:1113–1117, 1995.

[837] I. Yu. Popov and S. L. Popova. Eigenvalues and bands imbedded in the continuous spectrum for a system of resonators and a waveguide: solvable model. *Phys. Lett. A*, 222(4):286–290, 1996.

[838] S. L. Popova. The possibility of making nanoelectronic devices using mesoscopic Fresnel zone plates. *Tech. Phys. Lett.*, 19:508–509, 1993.

[839] I. I. Privalov. *Boundary Properties of Analytical Functions (2nd edition)*. Gosudarstv. Izdat. Tehn.-Teor. Lit., Moscow–Leningrad, 1950. (In Russian.)

[840] I. I. Privalov. *Randeigenschaften analytischer Funktionen*. VEB Deutscher Verlag der Wissenschaften, Berlin, 1956. Zweite, unter Redaktion von A. I. Markuschewitsch überarbeitete und ergänzte Auflage. Hochschulbücher für Mathematik, Bd. 25.

[841] M. Reed and B. Simon. *Methods of Modern Mathematical Physics. I. Functional Analysis*. Academic Press, New York, 1972.

[842] M. Reed and B. Simon. *Methods of Modern Mathematical Physics. II. Fourier Analysis, Self-adjointness*. Academic Press, New York, 1975.

[843] M. Reed and B. Simon. *Methods of Modern Mathematical Physics. IV. Analysis of Operators.* Academic Press, New York, 1978.

[844] M. Reed and B. Simon. *Methods of Modern Mathematical Physics. III Scattering Theory.* Academic Press, New York, 1979.

[845] F. Riesz. Sur certaines systèmes singulières d'équations inteégrales. *Ann. Sci. École Norm. Sup.*, 28:33–62, 1911.

[846] M. Riesz. Sur certaines inégalités dans la théorie des fonctions. *Kungl. Fys. Sällsk. Lund Förh.*, 1:18–38, 1931.

[847] J. M. Román and R. Tarrach. The regulated four–parameter one-dimensional point interaction. *J. Phys. A*, 29(18):6073–6085, 1996.

[848] M. Rosenblum. Perturbation of the continuous spectrum and unitary equivalence. *Pacific J. Math.*, 7:997–1010, 1957.

[849] P. Roy and R. Tarrach. Supersymmetric anyon quantum mechanics. *Phys. Lett. B*, 274(1):59–64, 1992.

[850] A. E. Ryzhkov. Scattering of the acoustic waves on a Markovian point defect. In *Schrödinger Operators, Standard and Nonstandard (Dubna, 1988)*, pages 78–100. World Scientific, Teaneck, NJ, 1989.

[851] Š. N. Saakjan. Theory of resolvents of a symmetric operator with infinite defect numbers. *Akad. Nauk Armjan. SSR Dokl.*, 41:193–198, 1965.

[852] V. Zh. Sakbaev and P. E. Zhidkov. Schrödinger operators in spaces of multifunctions defined in multiply–connected domains. *J. Phys. A*, 28(21):L549–L555, 1995.

[853] M. R. Sayapova and D. R. Yafaev. Scattering theory for potentials of zero radius which are periodic with respect to time. In *Spectral Theory. Wave Processes*, volume 10 of *Probl. Mat. Fiz.*, pages 252–266, 301. Leningrad. Univ., Leningrad, 1982.

[854] M. R. Sayapova and D. R. Yafaev. The evolution operator for time-dependent potentials of zero radius. *Trudy Mat. Inst. Steklov.*, 159:167–174, 1983. Boundary Value Problems of Mathematical Physics, 12.

[855] M. Scandurra. The ground state energy of a massive scalar field in the backgorund of a semi–transparent spherical shall. *J. Phys. A*, 32:5679–5691, 1999.

[856] S. Scarlatti and A. Teta. Derivation of the time–dependent propagator for the three–dimensional Schrödinger equation with one–point interaction. *J. Phys. A*, 23(19):L1033–L1035, 1990.

[857] M. Schröder. Spectral properties of Laplacians with attractive boundary conditions. In *Applications of Selfadjoint Extensions in Quantum Physics (Dubna, 1987)*, volume 324 of *Lecture Notes in Phys.*, pages 203–217. Springer, Berlin, 1989.

[858] L. S. Schulman. Applications of the propagator for the delta function potential. In *Path Integrals from meV to MeV (Bielefeld, 1985)*, pages 302–311. World Scientific, Singapore, 1986.

[859] P. Šeba. Schrödinger particle on a half line. *Lett. Math. Phys.*, 10(1):21–27, 1985.

[860] P. Šeba. The generalized point interaction in one dimension. *Czechoslovak J. Phys. B*, 36(6):667–673, 1986.

[861] P. Šeba. Regularized potentials in nonrelativistic quantum mechanics. I. The one–dimensional case. *Czechoslovak J. Phys. B*, 36(4):455–461, 1986.

[862] P. Šeba. Regularized potentials in nonrelativistic quantum mechanics. II. The three–dimensional case. *Czechoslovak J. Phys. B*, 36(5):559–566, 1986.

[863] P. Šeba. Some remarks on the δ'–interaction in one dimension. *Rep. Math. Phys.*, 24(1):111–120, 1986.

[864] P. Šeba. A remark about the point interaction in one dimension. *Ann. Physik (7)*, 44(5):323–328, 1987.

[865] P. Šeba. Klein's paradox and the relativistic point interaction. *Lett. Math. Phys.*, 18(1):77–86, 1989.

[866] P. Šeba. Chaotic quantum billiards. In *Order, Disorder and Chaos in Quantum Systems (Dubna, 1989)*, volume 46 of *Oper. Theory Adv. Appl.*, pages 237–258. Birkhäuser, Basel, 1990.

[867] P. Šeba. Quantum chaos in the Fermi–accelerator model. *Phys. Rev. A (3)*, 41(5):2306–2310, 1990.

[868] P. Šeba. Wave chaos in singular quantum billiard. *Phys. Rev. Lett.*, 64(16):1855–1858, 1990.

[869] P. Šeba. Wave chaos in singular quantum systems. In *Quantum Chaos (Trieste, 1990)*, pages 169–178. World Scientific, River Edge, NJ, 1991.

[870] P. Šeba and H. Englisch. Perturbation theory for point interactions in three dimensions. *J. Phys. A*, 19(5):711–716, 1986.

[871] P. Šeba and K. Życzkowski. Wave chaos in quantized classically nonchaotic systems. *Phys. Rev. A (3)*, 44(6):3457–3465, 1991.

[872] J. Shabani. Finitely many δ interactions with supports on concentric spheres. *J. Math. Phys.*, 29(3):660–664, 1988.

[873] J. Shabani. Some properties of the Hamiltonian describing a finite number of δ'-interactions with support on concentric spheres. *Nuovo Cimento B (11)*, 101(4):429–437, 1988.

[874] J. Shabani. Finitely many sphere interactions in quantum mechanics: approximation by momentum cut–off Hamiltonians. *Ann. Soc. Sci. Bruxelles Sér. I*, 105(3):105–111 (1992), 1991.

[875] J. Shabani. Hamiltoniens quantiques perturbés par des potentiels de surface. In *Proceedings of the UMA Symposium (Porto–Novo, 1993)*, volume 2, pages 111–117, 1993.

[876] M. H. Shermatov. Point interaction of a three–particle system. In *Schrödinger Operators, Standard and Nonstandard (Dubna, 1988)*, pages 78–100. World Scientific, Teaneck, NJ, 1989.

[877] T. Shigehara, H. Mizoguchi, T. Mishima, and T. Cheon. Spectral properties of the two–dimensional Laplacian with a finite number of point interactions. Technical report, quant–ph/9710005, 1997.

[878] T. Shigehara, H. Mizoguchi, T. Mishima, and T. Cheon. Wave chaos in quantum pseudointegrable billiards. Technical report, quant–ph/9710006, 1997.

[879] Sh. Shimada. The approximation of the Schrödinger operators with penetrable wall potentials in terms of short range Hamiltonians. *J. Math. Kyoto Univ.*, 32(3):583–592, 1992.

[880] Sh. Shimada. The analytic continuation of the scattering kernel associated with the Schrödinger operator with a penetrable wall interaction. *J. Math. Kyoto Univ.*, 34(1):171–190, 1994.

[881] Sh. Shimada. Low energy scattering with a penetrable wall interaction. *J. Math. Kyoto Univ.*, 34(1):95–147, 1994.

[882] Sh. Shimada. A solvable model for line interaction. *RIMS Kokyuroku*, 994:168–183, 1997. (In Japanese.)

[883] A. A. Shkalikov. On the essential spectrum of matrix operators. *Mat. Zametki*, 58(6):945–949, 1995.

[884] A. A. Shkalikov and C. Tretter. Spectral analysis for linear pencils $N - \lambda P$ of ordinary differential operators. *Math. Nachr.*, 179:275–305, 1996.

[885] Yu. G. Shondin. On the three–particle problem with δ–potentials. *Teoret. Mat. Fiz.*, 51(2):181–191, 1982.

[886] Yu. G. Shondin. Generalized point interactions in $\mathbf{R}^3$ and related models with a rational S–matrix. I. $l = 0$. *Teoret. Mat. Fiz.*, 64(3):432–441, 1985.

[887] Yu. G. Shondin. Generalized point interactions in $\mathbf{R}^3$ and related models with a rational S–matrix. II. $l = 1$. *Teoret. Mat. Fiz.*, 65(1):24–34, 1985.

[888] Yu. G. Shondin. Quantum mechanical models in $\mathbf{R}^n$ connected with extensions of the energy operator in a Pontryagin space. *Teoret. Mat. Fiz.*, 74(3):331–344, 1988.

[889] Yu. G. Shondin. Perturbation of differential operators on high–codimension manifold and the extension theory for symmetric linear relations in an indefinite metric space. *Teoret. Mat. Fiz.*, 92(3):466–472, 1992.

[890] Yu. G. Shondin. On the semiboundness of δ–perturbations of the Laplacian supported by curves with angle points. *Theor. Math. Phys.*, 105(1):1189–1200, 1995.

[891] Yu. G. Shondin. Perturbation of elliptic operators on thin sets of high codimension, and extension theory in a space with an indefinite metric. *Zap. Nauchn. Sem. S.–Peterburg. Otdel. Mat. Inst. Steklov. (POMI)*, 222(Issled. po Linein. Oper. i Teor. Funktsii. 23):246–292, 310–311, 1995.

[892] Yu. G. Shondin. Semibounded local Hamiltonians for perturbations of the Laplacian on curves with angular points in $\mathbf{R}^4$. *Teoret. Mat. Fiz.*, 106(2):179–199, 1996.

[893] C. Shubin and G. Stolz. Spectral theory of one–dimensional Schrödinger operators with point interactions. *J. Math. Anal. Appl.*, 184(3):491–516, 1994.

[894] A. A. Shushkov. Structure of resonances for symmetric scatterers. *Teoret. Mat. Fiz.*, 64(3):442–449, 1985.

[895] A. A. Shushkov. Structure of resonances for the Schrödinger equation with symmetric potentials of zero range. *Vestnik Leningrad. Univ. Mat. Mekh. Astronom.*, vyp. 3:112–113, 134, 1985.

[896] A. A. Shushkov. Scattering on a two–center potential of zero radius equipped with an internal structure. *Vestnik Leningrad. Univ. Mat. Mekh. Astronom.*, vyp. 2:119, 135, 1987.

[897] B. Simon. *Quantum Mechanics for Hamiltonians Defined as Quadratic Forms*. Princeton University Press, Princeton, NJ, 1971.

[898] B. Simon. The bound state of weakly coupled Schrödinger operators in one and two dimensions. *Ann. Physics (NY)*, 97:279–288, 1976.

[899] B. Simon. Notes on infinite determinants of Hilbert space operators. *Adv. in Math.*, 24:244–273, 1977.

[900] B. Simon. *Functional Integration and Quantum Physics.* Academic Press, New York, 1979.

[901] B. Simon. *Trace Ideals and Their Applications.* Cambridge University Press, Cambridge, 1979.

[902] B. Simon. Almost periodic Schrödinger operators: a review. *Adv. in Appl. Math.*, 3(4):463–490, 1982.

[903] B. Simon. Singular spectrum: recent results and open questions. In *XIth International Congress of Mathematical Physics (Paris, 1994)*, pages 507–512. Internat. Press, Cambridge, Mass., 1995.

[904] B. Simon. Spectral analysis of rank one perturbations and applications. In *Mathematical Quantum Theory. II. Schrödinger Operators (Vancouver, BC, 1993)*, volume 8 of *CRM Proc. Lecture Notes*, pages 109–149. Amer. Math. Soc., Providence, RI, 1995.

[905] B. Simon. Bounded eigenfunctions and absolutely continuous spectra for one–dimensional Schrödinger operators. *Proc. Amer. Math. Soc.*, 124(11):3361–3369, 1996.

[906] B. Simon. Some Schrödinger operators with dense point spectrum. *Proc. Amer. Math. Soc.*, 125(1):203–208, 1997.

[907] B. Simon and T. Spencer. Trace class perturbations and the absence of absolutely continuous spectra. *Comm. Math. Phys.*, 125(1):113–125, 1989.

[908] B. Simon and G. Stolz. Operators with singular continuous spectrum. V. Sparse potentials. *Proc. Amer. Math. Soc.*, 124(7):2073–2080, 1996.

[909] B. Simon and T. Wolff. Singular continuous spectrum under rank one perturbations and localization for random Hamiltonians. *Comm. Pure Appl. Math.*, 39(1):75–90, 1986.

[910] G. V. Skorniakov. Three body problem for short range forces. II. *Soviet Phys. JEPT*, 4:910–917, 1957. Translation of ЖЭТФ **31** (1956), 1046–1054.

[911] G. V. Skorniakov and K. A. Ter-Martirosian. Three body problem for short range forces. I. Scattering of low energy neutrons by deuterons. *Soviet Phys. JEPT*, 4:648–661, 1957. Translation of ЖЭТФ **31** (1956), 775–790.

[912] S. L. Sobolev. *Избранные вопросы теории функционалных пространств и обобщенных функций.* Nauka, Moscow, 1989. With an English summary, Edited and with a preface by S. V. Uspenskiĭ.

[913] S. L. Sobolev. *Some Applications of Functional Analysis in Mathematical Physics*, volume 90 of *Translations of Mathematical Monographs*. American Mathematical Society, Providence, RI, 1991. Translated from the third Russian edition by Harold H. McFaden, with comments by V. P. Palamodov.

[914] S. N. Solodukhin. Exact solution for a quantum field with δ-like interaction. Technical report, hep–th/9801054v2, 1998.

[915] M. Steslicka. Kronig–Penney models for surface states. *Progr.Surface Science*, 5:157–259, 1974.

[916] M. Steslicka. Surface states in an external electric field. *Phys. Lett. A*, 57:255–256, 1976.

[917] M. Steslicka and S. G. Davison. Boundary conditions for the relativistic Kronig–Penney model. *Phys. Rev. B (3)*, 1:1858–1860, 1970.

[918] M. Steslicka, S. G. Davison, and U. Srinivasan. Electronic states of heavy diatomic crystals. *J. Phys. Chem. Solids*, 32:1917–1923, 1971.

[919] M. Steslicka and Z. Perkal. A possible role of surface states in field–ion microscopy of semiconductors. *Surface Science*, 62:406–414, 1977.

[920] M. Steslicka and Z. Perkal. A simple modelistic treatment of virtual surface states. *Solid State Comm.*, 35:349–352, 1980.

[921] M. Steslicka and M. Radny. Localized electronic states near the oxided crystal surface. *Physica B*, 124:239–246, 1984.

[922] M. Steslicka and M. Radny. Surface states and adsorption in an external electrlc field. *J. Physique Colloque C*, 9, 45:65–70, 1984.

[923] M. Steslicka and S. Sengupta. Kronig–Penney model for impurity states. *Physica*, 54:402–410, 1971.

[924] M. Steslicka and K. F. Wojciechowski. Surface states of a deformed one-dimensional crystal. *Physica*, 32:1274–1282, 1966.

[925] M. N. Stone. Linear transformations in Hilbert space. *Amer. Math. Soc. Colloquim Publications*, XV, 1932.

[926] P. Šťovíček. Green's function for the Aharonov–Bohm effect with a non-abelian gauge group. In *Order, Disorder and Chaos in Quantum Systems (Dubna, 1989)*, volume 46 of *Oper. Theory Adv. Appl.*, pages 183–193. Birkhäuser, Basel, 1990.

[927] P. Šťovíček. Krein's formula approach to the multisolenoid Aharonov–Bohm effect. *J. Math. Phys.*, 32(8):2114–2122, 1991.

[928] P. Šťovíček. Scattering on a finite chain of vortices. *Duke Math. J.*, 76(1):303–332, 1994.

[929] P. Šťovíček. Anyons defined by boundary conditions. In *Proceedings of the Workshop on Singular Schroedinger Operators, Trieste, 29 September–1 October 1994*. Eds. G.F.Dell'Antonio, R.Figari and A.Teta. SISSA, Trieste, 1995.

[930] A. V. Štraus. Generalized resolvents of symmetric operators. *Izvestiya Akad. Nauk SSSR. Ser. Mat.*, 18:51–86, 1954.

[931] R. Tarrach. Scalar fields and contact interactions. In *Proceedings of the Workshop on Singular Schroedinger Operators, Trieste, 29 September–1 October 1994*. Eds. G.F.Dell'Antonio, R.Figari and A.Teta. SISSA, Trieste, 1995.

[932] A. Teta. Quadratic forms for singular perturbations of the Laplacian. *Publ. Res. Inst. Math. Sci.*, 26(5):803–817, 1990.

[933] L. E. Thomas. Time dependent approach to scattering from impurities in a crystal. *Comm. Math. Phys.*, 33:335–343, 1973.

[934] L. E. Thomas. Birman–Schwinger bounds for the Laplacian with point interactions. *J. Math. Phys.*, 20(9):1848–1850, 1979.

[935] L. E. Thomas. Scattering from point interactions. In *Mathematical Methods and Applications of Scattering Theory, Washington, D.C., 1979*. Edited by J.A. De Santo, A.W. Saenz and W.W. Zachary. *Lecture Notes in Physics, Vol. 130*, pages 159–162. Springer, Berlin, 1980.

[936] L. E. Thomas. Multiparticle Schrödinger Hamiltonians with point interactions. *Phys. Rev. D (3)*, 30(6):1233–1237, 1984.

[937] L. H. Thomas. The interaction between a neutron and a proton and the structure of H^3. *Phys. Rev. D*, 30:903–909, 1935.

[938] A. Tip. Form perturbations of the Laplacian on $L^2(\mathbf{R})$ by a class of measures. *J. Math. Phys.*, 31(2):308–315, 1990.

[939] H. Triebel. *Fractals and Spectra*. Birkhäuser, Basel, 1997.

[940] I. S. Tsirova and Yu. M. Shirokov. A quantum delta–like potential acting in the P–state. *Teoret. Mat. Fiz.*, 46(3):310–315, 1981.

[941] J. F. van Diejen and A. Tip. Scattering from generalized point interactions using selfadjoint extensions in Pontryagin spaces. *J. Math. Phys.*, 32(3):630–641, 1991.

[942] A. M. Veselova and L. D. Faddeev. A singularity in Coulomb three–particle scattering on the threshold of ionization. *Studia Logica*, 40(3):42–46, 125, 1981.

[943] M. Vishik. On general boundary conditions for ellyptic differential equations. *Trudy Moskov. Mat. Obsc.*, 1:187–246, 1952. (In Russian.)

[944] M. Vollenberg, H. Neidhardt, and V. Koshmanenko. On the scattering problem in the theory of singular perturbations of selfadjoint operators. *Ukrain. Mat. Zh.*, 36(1):7–12, 1984.

[945] J. von Neumann. Algemeine Eigenwerttheorie Hermitescher Funktionaloperatoren. *Math. Ann.*, 102:49–131, 1929–1930.

[946] J. von Neumann. Eine Spektraltheorie für allgemeine Operatoren eines unitären Raumes. *Math. Nachr.*, 4:258–281, 1951.

[947] Y. Wang and F. C. Pu. A one–dimensional electron system in the Luther–Emery regime. *J. Phys. A*, 29(9):1979–1985, 1996.

[948] K. Watanabe. Smooth perturbations of the selfadjoint operator $|\delta|^{\alpha/2}$. *Tokyo J. Math.*, 14(1):239–250, 1991.

[949] K. Watanabe. Spectral concentration and resonances for unitary operators. *Proc. Japan Acad. Ser. A Math. Sci.*, 68(10):322–326, 1992.

[950] K. Watanabe. Spectral concentration and resonances for unitary operators: applications to self–adjoint problems. *Rev. Math. Phys.*, 7(6):979–1011, 1995.

[951] K. Watanabe. On the self–adjoint operators defined by the $\mathcal{H}^{-2}$–construction: finite rank and trace class perturbations. Technical report, Dept. of Mathematics, Gakushuin Univ., Tokyo, N12, 1998.

[952] K. Watanabe. Smooth perturbations of the selfadjoint operators defined by the $\mathcal{H}^{-2}$–construction. Technical report, Dept. of Math., Gakushuin Univ., Tokyo, 1998. (In preparation.)

[953] J. Weidmann. *Linear Operators in Hilbert spaces*, volume 68 of *Graduate Texts in Mathematics*. Springer, New York, 1980. Translated from the German by Joseph Szücs.

[954] H. Weyl. Über gewöhnliche lineare Differentialgleichungen mit singulären Stellen und ihre Eigenfunktionen. *Nachr. Akad. Wiss. Göttingen Math.-Phys.*, II:37–63, 1909.

[955] H. Weyl. Über gewöhnliche Differentialgleichungen mit Singularitäten und die zugehörigen Entwicklungen willkürlicher Funktionen. *Math. Ann.*, 68:220–269, 1910.

[956] E. P. Wigner. On the mass defect of helium. *Phys. Rev.*, 43:252–257, 1933.

[957] E. P. Wigner. On a class of analytic functions from the quantum theory of collisions. *Ann. of Math. (2)*, 53:36–67, 1951.

[958] E. P. Wigner. Derivative matrix and scattering matrix. *Rev. Mexicana Fis.*, 1:81–90, 1952. (In Spanish.)

[959] E. P. Wigner. Derivative matrix and scattering matrix. *Rev. Mexicana Fis.*, 1:91–101, 1952.

[960] E. P. Wigner. On the connection between the distribution of poles and residues for an R function and its invariant derivative. *Ann. of Math. (2)*, 55:7–18, 1952.

[961] E. P. Wigner. Causality, R–matrix, and collision matrix. In *Dispersion Relations and Their Connection with Causality (Proceedings International School of Physics "Enrico Fermi", Course XXIX, Varenna, 1963)*, pages 40–67. Academic Press, New York, 1964.

[962] E. P. Wigner and J. von Neumann. Significance of Loewner's theorem in the quantum theory of collisions. *Ann. of Math. (2)*, 59:418–433, 1954.

[963] N. K. Wilkin, J. M. F. Gunn, and R. A. Smith. Do attractive bosons condense? *Phys. Rev. Letters*, 80:2265–2268, 1998.

[964] F. Wolf. Perturbation by changes one–dimensional boundary conditions. *Indag. Math.*, 18:361–366, 1956.

[965] D. Würtz, M. P. Soerensen, and T. Schneider. Quasiperiodic Kronig–Penney model on a Fibonacci superlattice. *Helv. Phys. Acta*, 61(3):345–362, 1988.

[966] D. R. Yafaev. The virtual level of the Schrödinger equation. *J. Soviet Math.*, 11:501–510, 1979.

[967] D. R. Yafaev. "Eigenfunctions" of the nonstationary Schrodinger equation. *Theoret. Math. Phys.*, 43:428–436, 1980.

[968] D. R. Yafaev. Scattering theory for time–dependent zero–range potentials. *Ann. Inst. H. Poincaré Phys. Théor.*, 40(4):343–359, 1984.

[969] D. R. Yafaev. *Mathematical Scattering Theory*, volume 105 of *Translations of Mathematical Monographs*. American Mathematical Society, Providence, RI, 1992. General theory, translated from the Russian by J. R. Schulenberger.

[970] D. R. Yafaev. On a zero–range interaction of a quantum particle with the vacuum. *J. Phys. A*, 25(4):963–978, 1992.

[971] C. N. Yang. Some exact results for the many–body problem in one dimension with repulsive delta–function interaction. *Phys. Rev. Lett.*, 19:1312–1315, 1967.

[972] C. N. Yang. ß–matrix for the one–dimensional N–body problem with repulsive or attractive δ–function interaction. *Phys. Rev.*, 168:1920–1023, 1968.

[973] C. N. Yang and C. P. Yang. Thermodynamics of a one–dimensional system of bosons with repulsive delta–function interaction. *J. Mathematical Phys.*, 10:1115–1122, 1969.

[974] V. Zagrebnov. Singular potentials of interaction in quantum statistical mechanics. *Trudy Moskov. Mat. Obshch.*, 41:101–120, 1980.

[975] M. M. Zimnev and I. Yu. Popov. The choice of parameters in a model of cracks with zero width. *Zh. Vychisl. Mat. i Mat. Fiz.*, 27(3):466–470, 480, 1987.

[976] J. Zorbas. Perturbations of selfadjoint operators by Dirac distributions. *J.Math.Phys.*, 21:840–847, 1980.

Index

Underlined pages refer to main entries of the corresponding topics.

acoustic problems 349
admissible operators 121
Aharonov-Bohm Hamiltonian 339
anyons 349
approximations of rank one perturbations 41
   norm convergence 41
   in the sense of linear operators 42
   strong resolvent 42, $\underline{46}$
approximations of delta potential 53
atomic physics 337

Bethe Ansatz 277, 285, 290
biharmonic and polyharmonic equations 331
boson symmetry 285, 290, 291
boundary form 73
boundary map $\underline{196}$
boundary values 73,196
bounded perturbation 17
breathing bag 336

Cayley transform 66
cluster decompositions 227, 249, 257
connected extension 70
Coulomb interaction 346, $\underline{346}$

δ'-interaction 339
decomposable boundary operator 200
deficiency indices $(1, 1)$ 15, 21
densely defined restricted operator 70
derivative of delta function 16, 53
differential operators 49
diffraction problems 349
diffusion processes 337
Dirac operator 38, $\underline{60}$, 347
distribution theory 130
   for discontinuous testfunctions 134, $\underline{134}$

Efimov effect 342
extension theory 11, 21
extension space 69

few-body Hilbert space, structure of 227
few-body problems $\underline{227}$
   with $\mathcal{H}_{-1}$-interactions 232
   with $\mathcal{H}_{-2}$-interactions 238
   with delta interactions 247, 273
   with generalized interactions 248, $\underline{264}$
      infinitesimally separable interactions 266
   selfadjointness 244
   with hard-core interactions 345
finite rank perturbations
   form bounded 116
   form unbounded 120

generalized 125
form bounded perturbations 19
   infinitesimally 19
form unbounded perturbations 19
Friedrichs extension 284, 288, 293
functional equations 297, 298, 298, 301

Γ-modified operator 83
generalized perturbation 11
   of infinite rank 198
generalized point interactions 92
   spectrum 95
   in dimension one 99
generalized rank one perturbations 63, 64
generalized resolvent 64, 90
   orthogonal 64
   equivalence of 64
geometric optics 319
gauge field 142
graph norm 16, 17
graphs 348

$\mathcal{H}$–independent set of vectors 229
$\mathcal{H}_{-2}$ – perturbations 19
Herglotz function 11
homogeneous operators 35
hyperbolic plane 349

indefinite metrics 349
infinite coupling (constant) 11, 14
infinite deficiency indices 195
inner-cluster generalized interactions 250
interaction on low dimensional manifolds 345, 346
internal space 69
internal structure 69
invariance of absolutely continuous
spectrum 162
inverse scattering problems 341

Krein's formula 15, 23, 25, 40
Krein's Q-function 11, 27, 28

L_p spaces 338
Lagrangian subspace 75

magnetic fields 339
Maxwell equations 351

nanosystem 349
von Neumann construction 331
Nevanlinna function (class) 11, 13, 27, 29, 39
non densely defined operators 81
nonlocal point interactions 339
nonsingular component 233
nuclear physics 337

operator relation 14, 28, 41
operators with internal structure 90
ordinary differential operators of order n 140
outgoing wave 293, 293, 305, 306, 311, 312, 319

path integral 340
point interaction
   in dimension one 142
   in dimension three 49
perturbation of the first order derivative 58
physical sheet 96
Pontryagin space 346
positive eigenvalues 341
positon 341
Privalov's theorem 172

projective space 143, 143

R-function 11
rank one perturbations
    form bounded 15
    form unbounded 32, 35
rational transformation 12
regular elements 229
relativistic point interactions 347
renormalized coupling constant 49, 56
resolvent formulas 38
    for functionals 202
resonances 335
resonator 349

S-channel 98
scale of Hilbert spaces 17
scattering matrix 99
    singularities of 99
scattering matrix for selfadjoint extensions 184
    for rank two perturbations 190
    for generalized perturbations 191
scattering operator for selfadjoint extensions 175, 175
scattering theory 159
    for finite rank perturbations 188
second order differential operator in one dimension 142
selfadjoint (operator) relation 14
separable set of vectors 239
separated extension 69
simple spectrum 160
singular perturbation 17, 21, 32
singular rank one perturbations 30
singular set of vectors 229
Sobolev embedding theorem 16
solid state 337, 350, 350
Sommerfeld integral 280, 293

asymptotics 308
Sommerfeld-Maluzhinetz transformation 293, 294, 306
sphere interactions 346
standard extension 63
Stieltjes functions (class) 210
strongly separable set of vectors 239
superlattices 347, 348
symmetry group 285
symplectic structure 330

tensor decomposition 227, 250, 229, 249, 257, 266
three-body models in one dimension 273
time dependent interactions 350
    periodic 350
two body operator in dimension one 99, 205
    continuous spectrum 100, 106
two body problems
    with singular interaction 205
    with generalized interaction 219

wave operators 169

Printed in the United States
By Bookmasters